# AN INTRODUCTION TO RADIO ASTRONOMY

## *Fourth Edition*

Radio astronomy is an active and rapidly expanding field owing to advances in computing techniques, with several important new instruments on the horizon. This text provides a thorough introduction to radio astronomy and its contribution to our understanding of the Universe, bridging the gap between basic introductions and research-level treatments. It begins by covering the fundamental physics of radio techniques, before moving on to single-dish telescopes and aperture-synthesis arrays. Fully updated and extensively rewritten, this Fourth Edition places greater emphasis on techniques, with a detailed discussion of interferometry in particular and an introduction to digital techniques in the appendices. The science sections have been fully revised, with new author Peter Wilkinson bringing added expertise to the sections on quasars and active galaxies. Spanning the entirety of radio astronomy, this is an engaging introduction for students and researchers approaching radio astronomy for the first time.

BERNARD F. BURKE was William A. M. Burden Professor of Astrophysics, formerly of the Radio Astronomy Group of the MIT Research Laboratory of Electronics, and principal investigator at the MIT Kavli Institute for Astrophysics and Space Research. He was elected a member of the United States National Academy of Sciences in 1970 and served as AAS President from 1986 to 1988. He and Kenneth Franklin discovered that Jupiter is a source of radio waves while working at the Carnegie Institution for Science, and he was part of the six-member team credited with the discovery in 1988 of the first Einstein ring.

SIR FRANCIS GRAHAM-SMITH, FRS is Emeritus Professor at the University of Manchester. He was Astronomer Royal from 1982 to 1990 and Director of Jodrell Bank Observatory between 1981 and 1988. As Director of the Royal Greenwich Observatory between 1975 and 1981, Graham-Smith instituted the UK optical observatory on La Palma. In his student days at Cambridge he made the first accurate locations of cosmic radio sources, leading to their identification. At Jodrell Bank he discovered radio emission from cosmic ray showers and continues to work on pulsars, in which he discovered the polarization of the radio pulses. He is a co-author of *Pulsar Astronomy* (Cambridge University Press, Fourth Edition 2012).

PETER N. WILKINSON is Emeritus Professor of Radio Astronomy at the University of Manchester. He has been involved in the development of radio telescopes at Jodrell Bank Observatory since 1967, including five years spent jointly at the California Institute of Technology and the US National Radio Astronomy Observatory. In 1991 he wrote the first published paper describing the scientific rationale and outline structure of a radio interferometer array, which developed into the Square Kilometre Array (SKA) project. He is now working on a novel radio telescope to map the sky with a precision ten times better than that achieved by the discoverers of the cosmic microwave background. He is a leading member of the UK's Newton DARA Project, which teaches radio astronomy to students in Africa in preparation for hosting part of the Square Kilometre Array.

# AN INTRODUCTION TO RADIO ASTRONOMY

## Fourth Edition

**BERNARD F. BURKE**

*Massachusetts Institute of Technology*

**FRANCIS GRAHAM-SMITH**

*University of Manchester*

**PETER N. WILKINSON**

*University of Manchester*

CAMBRIDGE
UNIVERSITY PRESS

# CAMBRIDGE
## UNIVERSITY PRESS

University Printing House, Cambridge CB2 8BS, United Kingdom

One Liberty Plaza, 20th Floor, New York, NY 10006, USA

477 Williamstown Road, Port Melbourne, VIC 3207, Australia

314-321, 3rd Floor, Plot 3, Splendor Forum, Jasola District Centre, New Delhi - 110025, India

103 Penang Road, #05-06/07, Visioncrest Commercial, Singapore 238467

Cambridge University Press is part of the University of Cambridge.

It furthers the University's mission by disseminating knowledge in the pursuit of education, learning and research at the highest international levels of excellence.

www.cambridge.org
Information on this title: www.cambridge.org/9781107189416
DOI: 10.1017/9781316987506

First published 1996
Second edition 2002
Third edition 2010
Fourth edition 2019

*A catalogue record for this publication is available from the British Library*

*Library of Congress Cataloging in Publication data*
Names: Burke, Bernard F., 1928- author. | Graham-Smith, Francis, 1923-
author. | Wilkinson, Peter N., 1946- author.
Title: An introduction to radio astronomy / Bernard F. Burke (Massachusetts
Institute of Technology), Francis Graham-Smith (University of Manchester),
Peter N. Wilkinson (University of Manchester).
Description: Fourth edition. | Cambridge, United Kingdom ; New York, NY :
Cambridge University Press, 2019. | Includes bibliographical references
and index.
Identifiers: LCCN 2018057974 | ISBN 9781107189416 (hardback ; alk. paper) |
ISBN 1107189411 (hardback ; alk. paper)
Subjects: LCSH: Radio astronomy–Observations. | Radio astronomy–Methodology.
Classification: LCC QB476.5 .B87 2019 | DDC 522/.682–dc23
LC record available at https://lccn.loc.gov/2018057974

ISBN 978-1-107-18941-6 Hardback

Additional resources for this publication at www.cambridge.org/ira4

**In memoriam**

Our dear friend and colleague Bernard Burke died on 5 August 2018. He was co-author of the first edition of this *Introduction* in 1996. His own introduction to radio astronomy was in 1953, with FG-S at the Carnegie Institution of Washington. Bernie was known for his deep physical understanding, his good humour, and his love of history. He was widely consulted and gave wise advice on many projects in astronomy.

# Contents

*Preface*                                                                 *page* xvii

**Part I   The Emission, Propagation, and Detection of Radio Waves**

**1   The Role of Radio Observations in Astronomy**                        3
   1.1   The Discovery of Cosmic Radio Waves                 3
   1.2   The Origins of Radio Astronomy                      6
   1.3   Thermal and Non-Thermal Radiation Processes         8
   1.4   Radio Observations                                  10
   1.5   The Challenge of Manmade Radio Signals              12
   1.6   Further Reading                                     14
**2   Emission and General Properties of Radio Waves**                     15
   2.1   Electromagnetic Waves                               15
   2.2   Wave Polarization                                   17
      2.2.1   The Polarization Ellipse        18
   2.3   Blackbody Radiation                                 21
   2.4   Specific Intensity and Brightness                   24
   2.5   Radiative Transfer                                  26
   2.6   Free–Free Radiation                                 29
   2.7   Synchrotron Radiation                               32
      2.7.1   A Power-Law Energy Distribution 36
      2.7.2   Synchrotron Self-Absorption     38
   2.8   Inverse Compton Scattering                          39
   2.9   Further Reading                                     40
**3   Spectral Lines**                                                     41
   3.1   Radio Recombination Lines                           42
   3.2   Hyperfine Atomic Ground-State Transitions           44
   3.3   Rotational Lines                                    45
   3.4   Degeneracy Broken by Rotation                       47
   3.5   Detected Lines                                      49
   3.6   Linewidths                                          50
      3.6.1   Line Emission and Absorption    51

| | | |
|---|---|---|
| 3.7 | Masers | 53 |
| | 3.7.1 Common Masers | 56 |
| 3.8 | Further Reading | 57 |
| **4** | **Radio Wave Propagation** | **58** |
| 4.1 | Refractive Index | 58 |
| | 4.1.1 Dispersion and Group Velocity | 59 |
| 4.2 | Faraday Rotation | 60 |
| 4.3 | Scintillation | 62 |
| 4.4 | Propagation in the Earth's Atmosphere | 65 |
| 4.5 | Further Reading | 68 |
| **5** | **The Nature of the Received Radio Signal** | **69** |
| 5.1 | Gaussian Random Noise | 69 |
| 5.2 | Brightness Temperature and Flux Density | 71 |
| | 5.2.1 Brightness Temperatures of Astronomical Sources | 75 |
| 5.3 | Antenna Temperature | 76 |
| | 5.3.1 Adding Noise Powers | 78 |
| | 5.3.2 Sources of Antenna Noise | 79 |
| | 5.3.3 Measuring the Antenna Temperature | 80 |
| 5.4 | Further Reading | 81 |
| **6** | **Radiometers** | **82** |
| 6.1 | The Basic Radiometer | 83 |
| | 6.1.1 Impedance Matching and Power Transfer | 83 |
| | 6.1.2 Power Amplification | 84 |
| | 6.1.3 Bandwidth and Coherence | 84 |
| 6.2 | Detection and Integration | 85 |
| 6.3 | Post-Detection Signals | 87 |
| | 6.3.1 Time Series | 87 |
| | 6.3.2 Spectrum | 89 |
| | 6.3.3 Recognizing a Weak Source | 90 |
| 6.4 | System Noise Temperature | 91 |
| | 6.4.1 Receiver Temperature | 91 |
| | 6.4.2 Receivers for Millimetre and Sub-Millimetre Waves | 93 |
| | 6.4.3 System Equivalent Flux Density (SEFD) | 94 |
| 6.5 | Calibration of the System Noise | 95 |
| | 6.5.1 Receiver Noise Calibration | 95 |
| | 6.5.2 Secondary Methods | 96 |
| | 6.5.3 Relative and Absolute Calibration | 97 |
| 6.6 | Heterodyne Receivers | 97 |
| 6.7 | Tracing Noise Power through a Receiver | 100 |
| 6.8 | Gain Variations and Their Correction | 101 |
| | 6.8.1 Dicke Switched Radiometer | 103 |
| | 6.8.2 Correlation Radiometers | 105 |
| 6.9 | Digital Techniques | 107 |
| 6.10 | Further Reading | 107 |

**7    Spectrometers and Polarimeters**                                          108
   7.1    Spectrometers                                                          108
          7.1.1    Filter-Bank Spectrometers                                     109
   7.2    Autocorrelation Spectrometers                                          110
          7.2.1    Linewidth and the Window Function                             112
   7.3    Digital Autocorrelation Spectrometers                                  115
          7.3.1    Fast Fourier Transform Spectrometers                          116
   7.4    Polarimetry                                                            118
   7.5    Stokes Parameters                                                      118
          7.5.1    Choice of Orthogonal Polarizations                            122
   7.6    Polarized Waveguide Feeds                                              124
          7.6.1    Linear Feeds and Quadrature Hybrids                           125
   7.7    A Basic Polarimeter                                                    125
   7.8    Practical Considerations                                              126
   7.9    Further Reading                                                        127

**Part II    Radio Telescopes and Aperture Synthesis**

**8    Single-Aperture Radio Telescopes**                                        131
   8.1    Fundamentals: Dipoles and Horns                                        131
          8.1.1    Ground Planes                                                 135
          8.1.2    The Horn Antenna                                              136
          8.1.3    Wide-Band Antennas                                            138
   8.2    Phased Arrays of Elementary Antennas                                   139
   8.3    Antenna Beams                                                          145
          8.3.1    Aperture Distributions and Beam Patterns                      146
          8.3.2    Fraunhofer Diffraction at an Aperture                         147
          8.3.3    Effective Area                                                151
   8.4    Partially Steerable Telescopes                                         153
   8.5    Steerable Telescopes                                                   154
   8.6    Feed Systems                                                           156
          8.6.1    Twin-Beam Radiometry                                          158
   8.7    Focal Plane Arrays and Phased Array Feeds                              159
   8.8    Antenna Efficiency                                                     160
          8.8.1    Aperture Illumination                                         160
          8.8.2    Blockage of the Aperture                                      162
          8.8.3    Reflection at the Surface                                     163
          8.8.4    Summary                                                       165
   8.9    The Response to a Sky Brightness Distribution                          166
          8.9.1    Beam Smoothing and Convolution                               166
          8.9.2    Sampling in Angle                                             170
          8.9.3    Effects of Sidelobes                                          170
          8.9.4    Pointing Accuracy                                            171
          8.9.5    Source Confusion                                              172

|        | 8.9.6  | Source Positions                                    | 172 |
| 8.10   |        | State-of-the-Art Radio Telescopes                   | 173 |
|        | 8.10.1 | FAST and Arecibo                                    | 174 |
|        | 8.10.2 | Large Steerable Dishes with Active Surfaces         | 174 |
|        | 8.10.3 | Millimetre and Sub-Millimetre Wave Dishes           | 175 |
| 8.11   |        | Further Reading                                     | 175 |

**9  The Basics of Interferometry**                                    177

| 9.1    |       | The Basic Two-Element Interferometer                 | 178 |
|        | 9.1.1 | One-Dimensional Geometry                             | 180 |
|        | 9.1.2 | The Adding Interferometer                            | 181 |
|        | 9.1.3 | The Correlation Interferometer                       | 182 |
|        | 9.1.4 | Steps Towards Practicality                           | 185 |
|        | 9.1.5 | The Frequency Domain Approach                        | 186 |
| 9.2    |       | Finite Bandwidths and Fields of View                 | 187 |
| 9.3    |       | The Basis of Position Measurements                   | 189 |
| 9.4    |       | Dealing with Finite Source Sizes                     | 192 |
|        | 9.4.1 | The Situation in One Dimension                       | 192 |
|        | 9.4.2 | The Essence of Fourier Synthesis                     | 196 |
|        | 9.4.3 | Simple Sources and their Visibility Amplitudes       | 197 |
| 9.5    |       | Interferometry in Two Dimensions                     | 200 |
|        | 9.5.1 | Coordinate-Free Description                          | 201 |
|        | 9.5.2 | The $u, v$ Plane and the 2D Fourier Transform        | 202 |
| 9.6    |       | Coherence                                            | 207 |
| 9.7    |       | Propagation Effects                                  | 210 |
|        | 9.7.1 | Troposphere                                          | 211 |
|        | 9.7.2 | Ionosphere                                           | 212 |
| 9.8    |       | Practical Considerations                             | 214 |
|        | 9.8.1 | Point-Source Sensitivity                             | 214 |
|        | 9.8.2 | Amplitude and Phase Calibration: Basic Ideas         | 215 |
|        | 9.8.3 | Outline of a Practical Signal Path                   | 217 |
| 9.9    |       | Further Reading                                      | 219 |

**10  Aperture Synthesis**                                             220

| 10.1   |        | Interferometer Arrays                               | 220 |
| 10.2   |        | Recapitulation on the Visibility Function           | 224 |
| 10.3   |        | The Data from an Array                              | 224 |
| 10.4   |        | Conditions for a 2D Fourier Transform               | 225 |
| 10.5   |        | The Spatial Frequency Transfer Function, or $u, v$ Coverage | 227 |
| 10.6   |        | Filling the $u, v$ Plane                            | 229 |
|        | 10.6.1 | The $u, v$ Coverage and Earth Rotation              | 229 |
|        | 10.6.2 | The Effect of Incomplete $u, v$ Coverage            | 236 |
| 10.7   |        | Calibrating the Data                                | 238 |
|        | 10.7.1 | Flux and Bandpass Calibration                       | 239 |
|        | 10.7.2 | Amplitude Calibration                               | 239 |
|        | 10.7.3 | Phase Calibration                                   | 240 |

| | | |
|---|---|---|
| 10.8 | Producing the Initial Map | 242 |
| | 10.8.1 Gridding and Weighting the Data | 242 |
| | 10.8.2 Dimensions in the Image and the Visibility Planes | 244 |
| 10.9 | Non-Linear Deconvolution | 245 |
| | 10.9.1 The CLEAN Algorithm | 246 |
| 10.10 | Correcting the Visibility Data | 249 |
| | 10.10.1 Closure Quantities | 249 |
| | 10.10.2 Self-Calibration | 251 |
| 10.11 | Missing Short Spacings | 254 |
| 10.12 | Flux Density and Brightness Sensitivity | 254 |
| | 10.12.1 Source Confusion | 258 |
| 10.13 | Multifrequency Synthesis | 258 |
| 10.14 | Data Limitations | 259 |
| | 10.14.1 Flagging | 259 |
| | 10.14.2 Time Averaging | 260 |
| | 10.14.3 Frequency Averaging | 260 |
| | 10.14.4 Non-Closing Effects | 261 |
| 10.15 | Image Quality | 262 |
| | 10.15.1 Signal-to-Noise Limits | 262 |
| | 10.15.2 Dynamic Range | 263 |
| | 10.15.3 Fidelity | 264 |
| 10.16 | Further Reading | 265 |
| **11** | **Further Interferometric Techniques** | **266** |
| 11.1 | Spectral Line Imaging | 266 |
| | 11.1.1 Observational Choices | 267 |
| | 11.1.2 Data Analysis Issues | 268 |
| 11.2 | Polarization Imaging | 269 |
| | 11.2.1 Basic Formalism | 269 |
| | 11.2.2 Calibration for Linear Polarization | 270 |
| 11.3 | Aperture Synthesis at Millimetre Wavelengths | 273 |
| | 11.3.1 Phase and Amplitude Calibration | 274 |
| | 11.3.2 System Requirements and Current Arrays | 275 |
| 11.4 | Very Long Baseline Interferometry (VLBI) | 277 |
| | 11.4.1 VLBI versus Connected-Element Interferometry | 277 |
| | 11.4.2 Outline of a VLBI System | 278 |
| | 11.4.3 Amplitude Calibration | 280 |
| | 11.4.4 Delay and Phase Corrections; Fringe Fitting | 280 |
| | 11.4.5 Basic VLBI Analysis for Imaging | 282 |
| | 11.4.6 Geodesy and Astrometry | 283 |
| | 11.4.7 Methods | 283 |
| | 11.4.8 Astrometry | 285 |
| | 11.4.9 Geodesy | 286 |
| | 11.4.10 Space VLBI | 287 |
| | 11.4.11 VLBI Arrays | 288 |

| | | |
|---|---|---|
| 11.5 | Wide-Field Imaging | 288 |
| | 11.5.1 The *w*-term and its Effects | 289 |
| | 11.5.2 Effects of the Primary Beam | 291 |
| | 11.5.3 Peeling | 292 |
| | 11.5.4 Mosaicing | 292 |
| 11.6 | Low Frequency Imaging | 294 |
| | 11.6.1 The Challenges | 295 |
| | 11.6.2 Low Frequency Arrays | 297 |
| 11.7 | Further Reading | 298 |

**Part III  The Radio Cosmos**

**12  The Sun and the Planets** 301
| | | |
|---|---|---|
| 12.1 | Surface Brightness of the Quiet Sun | 301 |
| 12.2 | Solar Radio Bursts | 303 |
| 12.3 | Coronal Mass Ejection (CME) | 306 |
| 12.4 | The Planets | 307 |
| 12.5 | Further reading | 308 |

**13  Stars and Nebulae** 309
| | | |
|---|---|---|
| 13.1 | Thermal Radio Emission from Stars | 309 |
| 13.2 | Circumstellar Envelopes | 312 |
| 13.3 | Circumstellar Masers | 313 |
| | 13.3.1 Silicon Oxide | 314 |
| | 13.3.2 Methanol | 314 |
| | 13.3.3 Water | 315 |
| 13.4 | The Hydroxyl Masers | 315 |
| 13.5 | Classical Novae | 318 |
| 13.6 | Recurrent Novae | 320 |
| 13.7 | Non-Thermal Radiation from Binaries and Flare Stars | 324 |
| 13.8 | X-Ray Binaries and Microquasars | 324 |
| 13.9 | Superluminal Motion | 325 |
| 13.10 | H II Regions | 329 |
| 13.11 | Supernova Remnants | 331 |
| 13.12 | Further Reading | 337 |

**14  The Milky Way Galaxy** 338
| | | |
|---|---|---|
| 14.1 | The Structure of the Galaxy | 338 |
| 14.2 | Galactic Rotation: The Circular Approximation | 342 |
| 14.3 | Spiral Structure | 345 |
| 14.4 | The Galactic Centre Region | 349 |
| 14.5 | The Black Hole at the Galactic Centre | 351 |
| 14.6 | The Spectrum of the Galactic Continuum | 352 |
| 14.7 | Synchrotron Radiation: Emissivity | 355 |
| 14.8 | The Energy Spectrum of Cosmic Rays | 356 |
| 14.9 | Polarization of the Galactic Synchrotron Radiation | 357 |

| | | |
|---|---|---|
| 14.10 | Faraday Rotation: the Galactic Magnetic Field | 359 |
| 14.11 | Loops and Spurs | 363 |
| 14.12 | The Local Bubble | 365 |
| 14.13 | Further Reading | 366 |
| **15** | **Pulsars** | **367** |
| 15.1 | Neutron Stars | 367 |
| | 15.1.1 Neutron Star Structure | 368 |
| 15.2 | Rotational Slowdown | 369 |
| 15.3 | Magnetic Dipole Moments | 371 |
| 15.4 | Rotational Behaviour of the Crab Pulsar | 371 |
| 15.5 | Glitches in Other Pulsars | 373 |
| | 15.5.1 Superfluid Rotation | 373 |
| 15.6 | Radio, Optical, X-Ray, and Gamma-Ray Emission from the Magnetosphere | 374 |
| 15.7 | Polar Cap Radio Emission | 376 |
| | 15.7.1 Polarization | 377 |
| | 15.7.2 Individual and Integrated Pulses | 379 |
| | 15.7.3 Nulling, Moding, and Timing Noise | 380 |
| 15.8 | Magnetars | 381 |
| 15.9 | X-Ray Binaries and Millisecond Pulsars | 383 |
| 15.10 | Binary Millisecond Pulsars | 384 |
| | 15.10.1 Gamma-Ray Pulsars | 384 |
| 15.11 | The Population and Evolution of Pulsars | 386 |
| 15.12 | The Radiation Mechanism | 387 |
| 15.13 | Pulsar Timing | 388 |
| 15.14 | Distance and Proper Motion | 390 |
| 15.15 | Binary Radio Pulsars | 390 |
| | 15.15.1 The Analysis of Binary Orbits | 391 |
| | 15.15.2 Post-Keplerian Analysis | 392 |
| 15.16 | Searches and Surveys: The Constraints | 394 |
| 15.17 | Detecting Gravitational Waves | 395 |
| 15.18 | Further Reading | 396 |
| **16** | **Active Galaxies** | **397** |
| 16.1 | Star-Forming Galaxies | 397 |
| 16.2 | Active Galactic Nuclei | 401 |
| 16.3 | Radio-Loud AGN | 402 |
| | 16.3.1 Classification of Extended Radio Sources | 402 |
| | 16.3.2 FR II Sources | 403 |
| | 16.3.3 FR I Sources | 405 |
| | 16.3.4 Core–Jet Sources | 407 |
| | 16.3.5 Repeated Outbursts | 407 |
| | 16.3.6 Synchrotron Emission from Extended Sources | 409 |
| | 16.3.7 Sub-Galactic-Scale Radio Sources | 411 |

|  |  |  |  |
|---|---|---|---|
|  | 16.3.8 | Compact Radio Sources | 412 |
|  | 16.3.9 | Superluminal Motion | 414 |
|  | 16.3.10 | Brightness Temperatures in Compact Sources | 415 |
| 16.4 | | Other Properties of Radio-Loud AGN | 416 |
|  | 16.4.1 | Optical Emission Lines | 416 |
|  | 16.4.2 | Spectral Energy Distributions (SEDs) | 417 |
|  | 16.4.3 | Variability | 417 |
|  | 16.4.4 | Host Galaxies and SMBHs | 419 |
| 16.5 | | Unified Models of Radio-Loud AGN | 420 |
|  | 16.5.1 | Radiative-Mode AGN | 420 |
|  | 16.5.2 | Kinetic-Mode AGN | 426 |
| 16.6 | | Accretion Rates and Feedback | 427 |
| 16.7 | | Radio-Quiet AGN | 429 |
| 16.8 | | Summary of AGN Phenomenology | 430 |
| 16.9 | | Surveys, Source Counts, and Evolution | 431 |
|  | 16.9.1 | Early Source Counts | 431 |
|  | 16.9.2 | Modern Source Surveys | 433 |
|  | 16.9.3 | Modern Source Counts | 435 |
|  | 16.9.4 | Evolution of Source Populations | 437 |
|  | 16.9.5 | Future Continuum-Source Surveys | 438 |
| 16.10 | | Further Reading | 440 |

**17  The Radio Contributions to Cosmology** — **441**

| 17.1 | | The Expanding Cosmos | 441 |
|---|---|---|---|
| 17.2 | | A Brief History | 441 |
| 17.3 | | Geometry and Dynamics | 442 |
| 17.4 | | The Early Universe: The CMB | 445 |
| 17.5 | | The Cosmic Dipole: The Coordinate Frame of the Universe | 447 |
| 17.6 | | The Blackbody Spectrum of the CMB | 447 |
| 17.7 | | The Search for Structure | 448 |
|  | 17.7.1 | First Observations of Structure | 450 |
|  | 17.7.2 | Wilkinson Microwave Anisotropy Probe (WMAP) | 451 |
|  | 17.7.3 | Planck | 452 |
|  | 17.7.4 | South Pole Telescope | 453 |
| 17.8 | | The Derivation of Cosmological Quantities | 454 |
| 17.9 | | Polarization Structure of the CMB | 455 |
| 17.10 | | The Transition to the Era of Reionization | 457 |
| 17.11 | | The Sunyaev–Zel'dovich Effect | 458 |
| 17.12 | | Gravitational Lensing | 461 |
| 17.13 | | Ray Paths in a Gravitational Lens | 462 |
|  | 17.13.1 | Imaging by Extended Lenses | 464 |
| 17.14 | | Lensing Time Delay | 465 |
| 17.15 | | Weak Gravitational Imaging | 465 |
| 17.16 | | Further Reading | 466 |

*Appendix 1   Fourier Transforms*                                          467
   A1.1   Definitions                                       467
   A1.2   Convolution and Cross-Correlation                 470
   A1.3   Two or More Dimensions                            474
   A1.4   Further Reading                                   474
*Appendix 2   Celestial Coordinates and Time*                              475
   A2.1   The Celestial Coordinate System                   475
   A2.2   Time                                              477
   A2.3   Further Reading                                   479
*Appendix 3   Digitization*                                                480
   A3.1   Digitizing the Signal                             480
       A3.1.1   Amplitude Quantization   480
       A3.1.2   Time Quantization and the Nyquist Criterion   482
       A3.1.3   Aliassing                483
*Appendix 4   Calibrating Polarimeters*                                    487
   A4.1   Single-Dish Radio Telescopes                      487
   A4.2   Polarization in Interferometers                   488
   A4.3   Further Reading                                   490
*Appendix 5   Spherical Harmonics*                                         491

*References*                                                               493
*Index*                                                                    517

# Preface

Astronomy makes use of more than 20 decades of the electromagnetic spectrum, from gamma-rays to radio. The observing techniques vary so much over this enormous range that there are distinct disciplines of gamma-ray, X-ray, ultraviolet, optical, infrared, millimetre, and radio astronomy. Modern astrophysics depends on a synthesis of observations from the whole wavelength range, and the concentration on radio in this text needs some rationale. Apart from the history of the subject, which developed from radio communications rather than as a deliberate extension of conventional astronomy, there are two outstanding characteristics of radio astronomy which call for a special exposition.

First, the astrophysics. Radio is essential for observing:

- ionized atmospheres of stars and interstellar plasma, penetrating dust and gas which often obscure other wavelengths;
- the processes of star and planet formation;
- molecules in cold interstellar clouds;
- hydrogen, the fundamental element in the Universe;
- pulsars, the most accurate clocks in the Universe;
- cosmic magnetic fields;
- the structure of the early cosmos.

Second, the techniques: low energy radio photons can be treated as classical waves. Hence, in contrast with other regimes, they can be coherently amplified and manipulated in complex receiver systems. Coherent amplification enables one to take account of the phase as well as the intensity of incoming waves, allowing the development of interferometers with the highest angular resolution in astronomy and the development of aperture synthesis, now realized in powerful new interferometric radio telescopes such as LOFAR, LWA, MeerKAT, MWA, ASKAP and ALMA, with the first phase of the Square Kilometre Array (SKA) impending. The basic techniques follow well-established principles, but the advent of massive computer power and broadband fibre optic communications has only recently brought these impressive instruments within the range of possibility. At the same time, existing radio telescopes, now including the new five-hundred metre FAST, using new receiver technology continue front-line research in several astrophysical domains, such as pulsars, fast radio bursts, and large area surveys for low-brightness emission. Fundamental cosmology has been transformed by observations of the cosmic microwave background from spacecraft and from the ground.

In view of the developments in the last ten years, this fourth edition of the book has been completely revised and reorganized. The new generation of radio telescopes, with dramatically improved performance, and new generations of astronomers, require a presentation which is a combination of fundamental principles, an exposition of the basics of telescope techniques, and a survey of the radio cosmos. Our rewriting has followed the advice of many colleagues, and we have provided references to recent reviews as well as to papers which represent the current state of the art. We are also providing on-line supplementary material, presenting a wide range of colour images and other material to complement the text. It will be available at www.cambridge.org/ira4.

We are aiming particularly at a graduate student audience attracted by radio astronomy with its new observational capabilities – in particular SKA, which will grow in size and power over the next two decades.

For the reader wishing to progress further in the subject there is a growing list of books at a more advanced level. Of these we particularly recommend:

*Interferometry and Synthesis in Radio Astronomy*, A. R. Thompson, J. M. Moran, and G. W. Swenson (Springer).
*Essential Radio Astronomy*, J. J. Condon and S. M. Ransom (Princeton University Press).
*Tools of Radio Astronomy*, T. L. Wilson, K. Rohlfs, and S. Hüttemeister (Springer).

Our intention is to provide an introduction which is useful both to the observer and to the astrophysicist; perhaps it will appeal most to those who, like ourselves, enjoy the membership of both categories

## Acknowledgements

We have gained many insights from the three books listed in the preface. In our treatment of interferometry we have made extensive use of tutorial presentations in the proceedings of: (i) the NRAO synthesis imaging workshops; (ii) the European Radio Interferometry Schools (ERIS), (iii) the CSIRO astronomy and space science (CASS) radio astronomy schools, all of which are available on the web. The introductory presentations by Rick Perley were particularly helpful. We also acknowledge pedagogic insights from Tim Bastian, Wim Brouw, Ger de Bruyn, John Conway, Tim Cornwell, Darrell Emerson, André Fletcher, Mike Garrett, Simon Garrington, Neal Jackson, Hans-Rainer Klöckner, Robert Laing, Ray Norris, Tetsuo Sasao, and David Wilner. The videos by Aaron Parsons on the 'Astrobaki' website[1] are highly recommended. We are grateful to Clive Tadhunter for permission to reproduce Figure 16.16.

Colleagues at the University of Manchester's Jodrell Bank Centre for Astrophysics have read critically parts of the text, contributed original diagrams, or offered astronomical and technical insights – sometimes all three! They are: Adam Avison, Rob Beswick, Ian Browne, Clive Dickinson, Malcolm Gray, Neal Jackson, Scott Kay, Paddy Leahy,

---

[1] casper.berkeley.edu/astrobaki/index.php/Radio_Astronomy:_Tools_and_Techniques.

Ian Morison, Tom Muxlow, Anita Richards, Ralph Spencer, Peter Thomasson, Patrick Weltevrede, Althea Wilkinson, and Nick Wrigley. Finally we are indebted to Christine Jordan for guiding FG-S through the intricacies of his LaTeX set-up and for assembling the text of this book.

Bernard F. Burke
Francis Graham-Smith
Peter N. Wilkinson

# Part I

## The Emission, Propagation, and Detection of Radio Waves

# 1

# The Role of Radio Observations in Astronomy

## 1.1 The Discovery of Cosmic Radio Waves

The data give for the coordinates of the region from which the disturbance comes, a right ascension of 18 hours and declination of $-10°$.

– Karl G. Jansky 1933

Jansky's discovery of radio emission from the Milky Way is now seen as the birth of the new science of radio astronomy. Most astronomers remained unaware of this momentous event for at least the next decade, and its full significance only became apparent with the major discoveries in the 1950s and 1960s of the 21 cm hydrogen line, the evolution of distant radio sources, quasars, pulsars, and the cosmic microwave background.

Radio astronomy had revealed a previously unseen Universe and is now one of the prime observational tools available to astronomers. There are several fields of application in which it is especially, sometimes uniquely, useful, as follows.

**The cosmic microwave background (CMB)**  The early Universe is observable as a black body whose $\sim 2.7$ K temperature has maximum emissivity at millimetre wavelengths.

**High energy processes in galaxies and quasars**  These emit intense radio waves from charged particles, usually electrons, moving at relativistic velocities.

**Cosmic magnetic fields**  These are revealed in radio sources and in interstellar space by the polarization of radio waves.

**Astrochemistry**  Molecular constituents of clouds in the Milky Way and in distant galaxies are observable by radio spectroscopy.

**Star and planet formation**  Condensations of atoms and molecules are mapped by millimetre-wave synthesis arrays.

**Kinetics of galaxies**  Radio spectroscopy, especially of the 21 cm hydrogen line, reveals the dynamic structure of galaxies.

**Neutron stars**  The timing and structure of pulses from pulsars opens a wide field of research, from condensed matter in neutron star interiors to the gravitational interactions of binary star systems.

**General relativity**  Pulsars, the most accurate clocks in the Universe, are used to measure the geometry of space–time.

3

There are several reasons for radio astronomy's wide and diverse range of astrophysical impact. Radio waves penetrate dust and gas, which absorb and scatter radiation in most other wavebands, allowing us to see into galaxies and molecular clouds. Thermal emission from cold interstellar dust and the free–free emission from hot interstellar plasma are both best seen in radio. Intense non-thermal synchrotron emission is generated by relativistic electrons spiralling around magnetic fields. Synchrotron radiation, although it can be detected up to X-rays and beyond, is a particularly prominent long-wavelength phenomenon, giving radio astronomy a unique role in the investigation of some of the most energetic objects in the Universe. Finally, the development of aperture-synthesis imaging provides the means to produce the highest resolution images possible at any wavelength; even the region near the event horizon around the black hole in the centre of the Milky Way is accessible to study.

While in this text we have emphasized the particular contributions of radio astronomy, we stress that in many areas of study astrophysics has become a multiwavelength endeavour. Each regime contributes in its own unique way towards a greater understanding. Thus optically measured redshifts are vital to establish the cosmological distances of active galactic nuclei (AGN) and the components of gravitational lenses, whilst a combination of X-ray, gamma-ray, and radio observations is needed to obtain a coherent picture of AGN.

The importance of a multiwavelength approach is perhaps best exemplified in our own Galaxy, the Milky Way, which is the origin of the radio noise first observed by Jansky. The Galaxy is a complex assembly of stars of widely varying ages, embedded in an *interstellar medium*, or ISM, of ionized and neutral gas, itself displaying a great diversity and complexity throughout the electromagnetic spectrum. Most observations target the surfaces of the stars, or nearby gas ionized by those stars, where the temperatures bring thermal radiation naturally into the visible range. X-ray astronomy deals with much hotter regions, such as the million-degree ionized gas which is found in such diverse places as the solar corona and the centres of clusters of galaxies. Infrared astronomy studies relatively cool regions, where thermal radiation from the dust component of the ISM is a prominent feature; warmer regions are also studied, where the thermal radiation from star-forming regions is strong. In contrast, radio astronomy, using much longer wavelengths, addresses a broad range of both thermal and non-thermal phenomena, including the thermal radiation from the 21 cm line of neutral hydrogen in the ISM, and the thermal radiation from a wide variety of molecular lines coming from dense, extremely cold, gas concentrations that are found within the ISM. The radio noise discovered by Jansky comes from non-thermal synchrotron radiation from very high energy electrons circulating in the magnetic field that permeates the interstellar medium in the Galaxy; its polarization is a tracer of the magnetic fields of the ISM.

The methods of the radio astronomer are often quite different from those in other wavelength regimes. There is a fundamental physical reason for this. Whilst in principle the radio signals gathered by a telescope can be understood as sums of myriad radio quanta, these quanta have the lowest energies of all across the electromagnetic (EM) spectrum ($10^{-7}$ eV at 30 MHz to $10^{-3}$ eV at 300 GHz). This means that we never have to worry about their quantum statistical properties, and radio signals can be treated as classical waves up to the sub-mm regime (THz frequencies). As a consequence, before their power is measured (*detected* in radio terminology), radio signals can be turned into

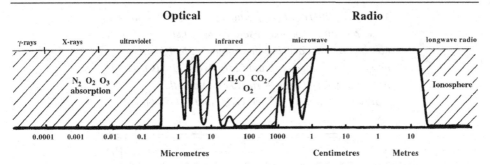

Figure 1.1. The electromagnetic spectrum, showing the wavelength range of the atmospheric 'windows'. The radio range is limited by the ionosphere at wavelengths greater than a few metres, and by molecular absorption in the sub-millimetre range.

complex voltages in the receiver. These voltages can be coherently manipulated, amplified, and split many ways and their frequencies changed, all the while maintaining the relative phases of the constituent waves.

While the methods may differ there are the same observational aims in the whole of astronomy, from the radio to the X-ray and gamma-ray domains. Nature presents us with a distribution of (frequency-dependent) brightness on the sky, and it is the task of the astronomer to deduce, from this brightness distribution of electromagnetic radiation, what the sources of emission are and what physical processes are acting.

To illustrate the relation between radio and other astronomies, the energy flux of electromagnetic radiation arriving at the Earth's surface from the cosmos is plotted in Figure 1.1. The wavelength scale runs from the AM radio broadcast band (hundreds of metres) to the gamma-ray region. The atmosphere is a barrier to all but two wavelength regions, radio and optical (including the near-infrared). Observations from spacecraft, clear of the atmosphere, now extend over the whole range of wavelengths but are necessarily limited to telescopes with small overall dimensions. The large ground-based telescopes on which radio astronomy depends use the whole of the radio 'window', from metre to millimetre wavelengths.

The atmospheric blockage at wavelengths short of the ultraviolet arises from a combination of nuclear interactions and electronic ionization and excitation, principally in ozone, oxygen, and nitrogen molecules. It is so complete that at ultraviolet and shorter wavelengths all observations must be carried out above the atmosphere. The optical–near-infrared window (0.3–1.1 microns) is relatively narrow; within this range the eye's sensitivity spans an even smaller region, 0.4–0.7 microns. In the infrared the atmospheric absorption arises from quantized vibrational transitions, principally in water vapour and carbon dioxide molecules. There are some atmospheric windows at infrared wavelengths, in particular 8–14 microns, through which the Earth's surface radiates heat energy into space. As we move into the sub-mm to mm waveband the absorption arises from quantized molecular rotations, and again there are some windows. Infrared and sub-mm observations can be made from high, dry, mountain sites or ultra-cold sites such as the South Pole, but for the most part observations must be taken from aircraft, balloons, or satellites.

Table 1.1. *International frequency band designations;*
*these remain in common use although a simpler A–M*
*naming system covering the range d.c. to 100 GHz is*
*now recommended. P-band (230–470 MHz) is used at*
*the Jansky Very Large Array (JVLA) and the*
*waveguide bands Q (33–50 GHz) and U (40–60 GHz)*
*are also in use.*

| Band | Frequency(GHz) |
|------|----------------|
| L    | 1–2            |
| S    | 2–4            |
| C    | 4–8            |
| X    | 8–12           |
| Ku   | 12–18          |
| K    | 18–27          |
| Ka   | 27–40          |
| V    | 40–75          |
| W    | 75–110         |

It is easy to see that there is a great stretch of the spectrum at the radio end (covering four orders of magnitude from millimetre to decametre waves) in which the atmosphere has much less effect. The bottom end is limited by the ionospheric plasma which, although variable, does not usually allow the passage of wavelengths longer than ~30 metres (10 MHz). Before Jansky's discovery there was no reason to expect much of interest in the radio spectrum; if stars were the principal sources of radiation, very little radio emission could be expected. The maximum thermal emission from even the coolest of the known stars falls at visible or infrared wavelengths, and their contribution to the radio end of the spectrum was regarded as almost negligible. The slow response to Jansky's discovery is therefore understandable in terms of both technical difficulty and lack of expectation.

The radio spectrum is often described by bands in frequency whose names are rooted in history, like the S, P, D states of atoms. Early names were: HF (high frequency, below 30 MHz); VHF (very high frequency, 30–300 MHz); UHF (ultra high frequency, 300–1000 MHz); microwaves (1000–30 000 MHz); and millimetre waves and sub-millimetre waves beyond that. A commonly used set of names covering the narrower bands is listed in Table 1.1.

## 1.2 The Origins of Radio Astronomy

The discoveries which opened the window of radio astronomy depended on advances in technique, and most of them arose unexpectedly. Here we summarize the major discoveries which introduced the wide scope of astrophysics to which radio now contributes.

**1935** When Karl Jansky, at the Bell Telephone Laboratories, discovered cosmic radio waves, he was investigating the background of sporadic radio noise which might have limited the usefulness of radio communications on what was then an unexplored short-wavelength band. He built a directional array working at 20 metres wavelength (frequency 15 MHz), and identified our Milky Way galaxy as the main source of the background of radio noise. The radiation originates not in the stars but in energetic electrons in the interstellar medium. The detailed structure of this radiation was investigated by large radio telescopes and by the WMAP and Planck satellites (Chapter 17).

**1937**   Grote Reber built the first reflector radio telescope, with which he mapped radio emission from the Milky Way galaxy (Reber 1944).

**1942**   The first detection of radio emission from the Sun, by James Hey during World War II, was again a surprise. The strong radio noise which was jamming metre-wavelength radar was found to be associated with sunspots and solar flares. This was emphatically non-thermal; a lower intensity thermal radiation was later found to originate in the solar corona (Chapter 12).

**1948**   Two of the most powerful discrete radio sources, Taurus A (the Crab Nebula) and Cassiopeia A, were found by John Bolton in Australia and by Martin Ryle and Graham Smith in Cambridge, UK, to be young remnants of supernova explosions (Chapters 13 and 14). Among these, the Crab Nebula has probably been the subject of more papers than any other object in the Milky Way galaxy.

**1951**   Radio spectroscopy began with the discovery of the 21 centimetre hydrogen line almost simultaneously in the USA, The Netherlands, and Australia. As predicted by Jan Oort and H. van der Hulst, spectroscopy transformed our understanding of the structure of the Milky Way galaxy and eventually of many other distant galaxies. Line radiation from many molecular species has become the new discipline of astrochemistry (Chapter 3).

**1954**   The development of interferometer techniques, initially in Australia and the UK, led to the accurate location and identification of radio galaxies, such as Cygnus A, at large extragalactic distances. Cygnus A was discovered by Hey in his early survey of the radio sky (Hey *et al.* 1946); its trace can even be discerned on Grote Reber's pioneering 1944 map (Reber 1944). It is, however, an inconspicuous object optically, and it was not identified until its position was known to an accuracy of 1 arcminute (Smith 1952; Baade and Minkowski 1954). A substantial proportion of these extragalactic sources were found later to have very small diameters; these became known as quasars (Chapter 16). The structure of these powerful radio emitters revealed the very energetic processes now known to be due to a black hole, with plasma jets streaming out to huge distances.

**1958**   The first suggestion that radio astronomy might contribute to cosmology was in 1958, when the large numbers of extragalactic sources discovered in the Cambridge surveys led Martin Ryle to suggest that their statistics appeared to contradict the steady-state cosmological theory. These sources are now numbered in millions, and their evolution traces the phases in the development of the Universe and its contents (Chapter 16).

**1962**   Interferometer techniques, notably that of aperture synthesis developed by Martin Ryle at Cambridge, and the radio links for long baselines, developed by Henry Palmer at Jodrell Bank, have become the foundations of modern radio telescopes covering the whole available radio spectrum (Chapters 8, 10, and 11). The design of the huge Square Kilometre Array (SKA) is derived directly from these early advances in telescope technology.

**1965**   The cosmic microwave background (CMB) was discovered by Arno Penzias and Bob Wilson, at the Bell Telephone Laboratories (Chapter 17). Like Jansky, they set out to measure the background against which radio communications must contend, but at the much shorter wavelength of 7.4 centimetres (4.1 GHz). This discovery, with the subsequent detailed evaluation of the structure of the CMB, marks the transformation of cosmology into a precise discipline.

**1966**   Pulsars were discovered by Jocelyn Bell and Antony Hewish in a survey of discrete radio sources which unusually was deliberately aimed at finding rapid fluctuations. Following the identification of pulsars as neutron stars, which can provide extremely accurate clocks, pulsar research has not only initiated the study of neutron stars, their physics, and their origin, but has provided the most accurate test of general relativity and demonstrated the existence of gravitational waves many years before their direct detection in 2016 (Chapter 15).

These major steps forward depended on the determination of pioneering observers to make the best possible use of the available technology. Receiver sensitivity, angular resolution, frequency coverage, spectral resolution, and time resolution have all achieved orders-of-magnitude improvements between Jansky and the measurement of the CMB structure. Radio telescopes grew, and continue to grow, bringing increased sensitivity and angular resolution. The advent of digital computing allowed the development of aperture synthesis, which led to the Square Kilometre Array, the largest telescope project in any part of the spectrum. There was also an element of good fortune in most of the discoveries listed above, but good fortune often comes to those who are prepared for it in their exploitation of a new technique. There is plenty of scope for new discoveries; the existing telescopes are continually being improved, and the SKA will provide such huge volumes of data that a new approach is already needed to the handling of such a flood of information.

### 1.3 Thermal and Non-Thermal Radiation Processes

During the pioneering stage of radio astronomy, a wide range of celestial objects turned out to be detectable, and two broad classes of emitter became clearly distinguished. At centimetric wavelengths the radio emission from the Sun could be understood as a thermal process, with an associated temperature. The term *temperature* implies that there is some approximation to an equilibrium, or quasi-equilibrium, condition in the emitting medium; in this case the medium is the ionized solar atmosphere. The mechanism of generation is electron–ion collisions, in which the radiation is known as *free–free emission* or *bremsstrahlung* (Chapter 2). At metre wavelengths, however, the outbursts of very powerful solar radiation observed by Hey could not be understood as the result of an equilibrium process. The distinction was therefore made between *thermal* and *non-thermal* processes, a distinction already familiar in other branches of physics. Many of the most dramatic sources of radio emission, such as the supernova remnants Cassiopeia A and the Crab Nebula, the radio galaxies M87 and Cygnus A, pulsars, and the metre-wave backgrounds from our own Milky Way galaxy, are non-thermal in nature. Nevertheless, for practical reasons the term 'temperature' was adopted in a variety of contexts, following practices that had been used widely in physics research during the 1940s. We return to this point later in this section and in Chapter 5.

The archetype of thermal sources is a black body, in which the radiation is in equilibrium with the emitting material, no matter what that is, and there is no need to specify any details of emission or absorption processes. The mathematical form of the spectrum is always the same and its intensity as a function of frequency or wavelength depends only on the temperature. The best example in radio astronomy, and indeed the best known anywhere, is the cosmic microwave background (CMB) radiation, which at wavelengths shorter than 20 cm becomes the predominant source of the sky brightness (except for a strip about 3 degrees wide along the galactic plane caused by radiation from the interstellar medium). The CMB spectrum is specified (almost) completely by the temperature 2.725 K. As just noted, no radiation process need be invoked in the calculation; the radiation was originally in equilibrium with matter in an early stage of cosmic evolution, and has preserved its blackbody spectrum in the subsequent expansion and cooling of the Universe. Thermal emission, being essentially a random process, exhibits no preferred direction at emission and hence is unpolarized. However, polarization can be subsequently imposed during the propagation path, just as sunlight from the sky is polarized by scattering in the Earth's atmosphere. We discuss the polarization of the CMB in Chapter 17.

Within a thermally emitting source, individual components can exhibit different temperatures. In a plasma excited by a strong radio frequency field, for example, the electron and ion components of the gas may each show velocity distributions that can be approximated by Maxwell–Boltzmann distributions but at quite different temperatures. Each component is in a state of approximate thermal equilibrium but the systems are weakly coupled and derive their excitation from different energy sources. One can speak, therefore, of two values of *kinetic temperature*, the *electron temperature* and the *ion temperature*.

A two-state system such as the ground state of the hydrogen atom, in which the magnetic moments of the proton and electron can be either parallel or antiparallel, can be used as a simple and illuminating example of how the temperature concept can be generalized. Given an ensemble of identical two-state systems (atoms or molecules) at temperature $T_s$, the mean relative population of the two states, $\langle n_2/n_1 \rangle$, is given by the Boltzmann distribution:

$$\langle n_2/n_1 \rangle = \exp(-\epsilon/kT_s), \tag{1.1}$$

where the energy separation $\epsilon$ corresponds to a photon energy $h\nu$. If the two states are degenerate, the statistical weights[1] $g_1$ and $g_2$ must be applied. The relationship can be inverted; for any given average ratio of populations, there is a corresponding value of the temperature, defined by the Boltzmann equation. This defines the *state temperature*, $T_s$.

---

[1] The number of states which have the same energy.

The state temperature need not be positive. When $\langle n_2/n_1 \rangle$ is greater than 1, Eqn 1.1 requires a negative state temperature, and this is precisely the condition for *maser* or *laser*[2] action to occur. Whereas a beam of photons with energy $\epsilon$ traversing a medium at a positive state temperature will suffer absorption, it will be coherently amplified as a result of stimulated emission if the state temperature is negative. Naturally occurring masers are common in astrophysics, particularly in star-forming regions and in the atmospheres of red giants. These are treated in Chapters 3 and 13. The non-thermal population inversion is maintained by a pumping mechanism, which can be either radiative or collisional; the action may be either to fill the upper state faster than the lower state or to populate both states, but with the lower state being drained of population more rapidly.

Atomic and molecular systems almost always have a large number of bound states, and one can associate a state temperature with each pair of states. If the system is in a state of thermal equilibrium, all these temperatures will be the same. Blackbody radiation is associated with a continuum of energy states.

In Chapter 2 we explain the idea of the *brightness temperature*, mentioned above. In brief, this assigns to an emitter of radiation at frequency $\nu$ the temperature that it would have to have if it were a black body. This need not, and usually does not, correspond to a physical temperature. For example, the sky at long radio wavelengths is far brighter than is expected from the thermal cosmic background at 2.73 K. Instead the Milky Way shines principally by the synchrotron mechanism and its brightness temperature at 10 m wavelength (30 MHz) can exceed 100 000 K (Chapter 14). In the extreme conditions found in quasars and active galactic nuclei, the emission, also boosted by bulk relativistic effects, can exhibit a brightness temperature exceeding $10^{12}$ K (Chapter 16). Synchrotron-emitting electrons have a distinctly non-Maxwellian energy distribution, which instead takes the form of a power law. As a result the emission spectrum is also a power law, rising towards lower frequencies, completely different from that of a black body. Since synchrotron emission involves a magnetic field, which has a preferred direction in space, the radiation is polarized.

In Chapter 2 we develop the theory behind the radiation mechanisms outlined in this section in more detail. We also introduce the concepts of radiative transfer within an emitting/absorbing medium; this leads to a brief exposition of maser action in Chapter 3, where we discuss the physics of radio spectral lines more generally. The various effects of propagation in the ionized and magnetized stellar medium are introduced in Chapter 4. These effects are the tools of radio astronomy, giving access to such diverse quantities as the dynamics of gas motions close to an active galactic nucleus and the configuration of the magnetic field of our Galaxy. We end Chapter 4 by considering the propagation effects in the Earth's atmosphere on the radiation, just before it arrives at the telescope.

## 1.4 Radio Observations

In Chapter 5 we consider the nature of the radio signal received by our telescopes and recognize that one is nearly always looking for weak signals in the presence of random

---

[2] These acronyms stand for Microwave (or Light) Amplification by the Stimulated Emission of Radiation.

noise from other sources. In that chapter we recognize that the use of thermodynamic concepts of temperature has more than a formal value. General features associated with measuring brightness temperatures and general theorems about antennas and receivers can be deduced from thermodynamic considerations.

The principles behind the design of basic types of radio astronomy receivers and how they are calibrated are developed in Chapter 6. The radio emission from many cosmic radio sources contains spectral lines, and may also be polarized. Spectrometers and polarimeters are described in Chapter 7.

Observations do not take place as an abstract process, and the diligent observer will have a knowledge of the characteristics of the instrument that is being used to take the data. The same observer will always be aware that the statistical significance of a result must be evaluated. With this familiarity, advantage can be taken of new and unexpected uses of an instrument, while due caution can be exercised in interpreting data that may contain instrumentally induced flaws.

The temperature concept introduced above is extended to the practice of receiver measurement. An ideal amplifier should add as little noise as possible to the system, although the laws of quantum mechanics prevent an amplifier from being entirely noise-free. The total excess noise is described as the *system-noise temperature*; the definition arises from the properties of a resistor as a noise generator. Every physical resistance generates noise, because of thermal fluctuations in the sea of conduction electrons, and the noise power per unit bandwidth that can be extracted from the resistor is proportional to its temperature. In effect the resistor acts like a one-dimensional black body. The excess noise observed with any radio astronomy receiver can be described by stating what temperature a resistive load would have to have, when connected to the input, to generate the observed noise. This turns out to be an entirely practical way to describe the system, because the faint continuum radio signals that one deals with are most conveniently calibrated by using as a reference the continuum noise generated by a hot (or cold) resistive load.

The actual radio receiver is but one part of the overall information collection system. The incoming radio energy first has to be collected, the more the better given its ultra-low levels. In Chapter 8 we therefore describe the principles behind single-aperture radio telescopes (including phased arrays), how to characterize their properties, and some practical issues involved with using them.

A large fraction of modern-day radio observations are taken with interferometer arrays of various types; one of the latest being ALMA, with the SKA to follow over the next decade. It is therefore vital for the student of radio astronomy to have a basic understanding of interferometers before moving on to more advanced texts for the subtler details. It is a widespread experience of lecturers that interferometry is the hardest part of any radio astronomy course. We have therefore split our exposition into two chapters. In Chapter 9 we develop the basic ideas and then in Chapter 10 move on to cover the essential details of modern-day interferometric imaging. In these two chapters we take care to differentiate between adding interferometers, such as phased arrays and aperture arrays, introduced in Chapter 8, and correlation interferometers, the dominant type. The SKA will encompass both types. In Chapter 11 we introduce some more advanced techniques of interferometry.

In discussing the principles of radiometric and spectroscopic receivers, of single-aperture telescopes, and of interferometer arrays, understanding the language of Fourier transforms is essential. Fourier transform methods have wide applications to nearly all fields of science and technology, including radiation processes, antenna theory and, especially, aperture-synthesis interferometry. Many readers will be familiar, to a greater or lesser extent, with the Fourier transform; as an aid to the memory, Appendix A1 summarizes its basic properties and some of its radio astronomical applications.

Chapters 12–16 describe how these various radio techniques have provided new insights into the astrophysics of stars, neutron stars, galaxies, and quasars. Observations of the cosmic microwave background (CMB), which is the subject of Chapter 17, have transformed our understanding of the Universe and have given access to some of the most fundamental aspects of cosmology. Chapter 16 also deals with the evolving population of radio galaxies and quasars within the Universe.

The rich history of radio astronomical discovery has, in significant part, set the agenda for astrophysics and cosmology since the 1960s. In recognition, the Nobel Prize in Physics has been awarded to radio astronomers on four occasions: in 1974 to Ryle and Hewish for the development of aperture synthesis and the discovery of pulsars; in 1978 to Penzias and Wilson for the discovery of the CMB; in 1993 to Taylor and Hulse for the indirect detection of gravitational radiation from a binary pulsar; in 2006 to Mather and Smoot for measurements of the spectrum and anisotropy of the CMB. We can be confident that these will not be the last.

## 1.5 The Challenge of Manmade Radio Signals

The principal challenge to all radio astronomical observations is that natural sources of radiation produce very weak signals, even from the most powerful cosmic sources and using the largest radio telescopes. For example, consider a radio observation of the supernova remnant Cassiopeia A (Cas A), the strongest radio source in the sky apart from the Sun. Now ask: what would be the energy collected by the 76 m Lovell Telescope (LT) at the Jodrell Bank Observatory (UK) over its entire operational lifetime if it had been pointed continuously at Cas A? Let our virtual observations occupy the long-term protected band 1400–1427 MHz around the 21 cm atomic hydrogen line. The calculation is then straightforward. The flux density (see Chapter 5) of Cas A at 1400 MHz is $\sim 2 \times 10^{-23}$ W m$^{-2}$ Hz$^{-1}$, thus, multiplying by the 2700 m$^2$ effective area of the LT (its geometrical area with a 60% aperture efficiency) and the 27 MHz bandwidth gives a rate of energy collection of $\sim 1.5 \times 10^{-12}$ W. Now multiplying by the number of seconds ($\sim 2 \times 10^9$) since the LT was commissioned in 1957 we obtain the total energy collected. The answer is $3 \times 10^{-3}$ joules, which is the same as that required to power a hand-torch bulb (3 V; 0.5 A) for 2 milliseconds!

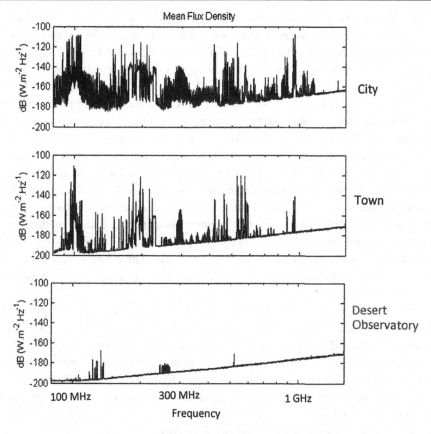

Figure 1.2. The radio frequency spectrum in sites of different population density. Top to bottom: city, town, desert.

Nevertheless, with modern low-noise amplifiers and high-gain but very stable receivers based on solid state devices, the ultra-low power levels associated with a wide range of cosmic radio sources are measured routinely. Early radio astronomers did not have these technological advantages. Their receivers used thermionic devices producing much higher intrinsic noise levels and suffering larger gain variations against which the natural signals had to be discerned. The origins and development of the subject need to be understood in this light.

The radio pioneers had one major advantage over their successors – the low occupation of the radio spectrum by manmade signals. The overriding challenge of present-day radio astronomy comes from the fact that most modern communication systems depend on radio transmissions; as a result the 'interference-free' space for radio astronomical observations is becoming ever more limited. Figure 1.2 illustrates the problem well. It shows the radio frequency spectrum, from $\sim 10\,\mathrm{MHz}$ to above $1\,\mathrm{GHz}$, in sites of different population density: in a city, a town, and a desert location. These spectra immediately make clear why the current generation of telescopes, built many decades ago in more or less convenient locations, have to operate in non-ideal environments and why the next-generation radio telescopes must be sited in remote regions well away from people. The Square Kilometre Array will therefore be built in the deserts of Australia and South Africa.

Figure 1.2 also shows just how powerful manmade signals are compared with natural ones. The strongest signals reach $10^{-11}$ W m$^{-2}$ Hz$^{-1}$ and hence are a million million times greater than those from the strong radio source Cas A. Sharing the spectrum between these ever-growing commercial transmissions and radio astronomy is a major task recognized by the International Astronomical Union (IAU) and the International Scientific Radio Union (URSI), working through the International Telecommunications Union (ITU), the body that allocates specific bands of the spectrum to the many and various users.[3] Parts of the spectrum are protected from powerful transmissions such as television, radio broadcasts, and radar; mobile phones, by using cellular networks, are confined to remarkably narrow bands. Satellite networks, particularly navigation systems (GPS, Glonass, Galileo), cannot be avoided anywhere in the world. One of the best protected radio astronomy bands is 1400 to 1427 MHz, covering the 21 cm hydrogen spectral line. Other bands are allocated with various degrees of protection at approximately octave intervals throughout the whole radio window shown in Figure 1.1.

There are, however, large advantages in using the much wider bandwidths provided by modern receiver techniques. In such wide bands unwanted signals, known as radio frequency interference (RFI), are inevitably picked up along with the wanted signals, and must be recognized and rejected. This is achieved by splitting the receiver band into thousands of separate channels, and rejecting those containing RFI. Interferometer techniques are also helpful, especially when the elementary radio antennas are sited far apart and are subject to different, uncorrelated, RFI. The narrower, specifically allocated, bands remain vitally important, especially for the most precise measurements.

### 1.6 Further Reading

*Supplementary Material* at www.cambridge.org/ira4.

Graham-Smith, F. 2014. *Unseen Cosmos*. Oxford University Press. A general account of radio astronomy.

Longair, M. 2006. *The Cosmic Century*. Cambridge University Press. A comprehensive history of astronomy up to 2000.

Sullivan, W. T. 2009. *Cosmic Noise*. Cambridge University Press. A history of early radio astronomy.

---

[3] The case for protection is made in Europe and South Africa by a joint Committee for Radio Astronomy Frequencies (CRAF), the corresponding body in the Americas is the Committee on Radio Frequencies (CORF) and in the Asia–Pacific region the Radio Astronomy Frequency Committee for the Asia–Pacific Region (RAFCAP).

# 2

# Emission and General Properties of Radio Waves

All branches of astronomy require an understanding of the radiation which a telescope collects and how this can be modified between emission and detection. In this chapter we outline the main emission mechanisms and the basic principles of radiative transfer.

Radio radiation can be treated mainly in terms of classical (non-quantized) electromagnetic waves. Only in the extreme of the cold CMB observed at millimetre wavelengths need we use the full spectrum of blackbody radiation. Photon energies are insignificant in most thermal sources such as planetary surfaces and interstellar dust, and in the non-thermal sources such as synchrotron radiation from the interstellar medium and quasars. Radio spectral lines, which we deal with in Chapter 3, do indeed arise from quantum processes but their observation need take no account of photons. Our account in this chapter of emission processes will therefore be mainly in terms of electrodynamics.

Material between the emitting region and the receiver can alter the intensity of the incoming radiation via the absorption and re-emission of energy. This can occur in ionized interstellar clouds or in the Earth's atmosphere. Refraction occurs in neutral astrophysical plasmas (Chapter 4); the protons have a negligible effect but the electrons can have a significant effect on the wave velocity, hence on the refractive index, and large-scale variations can alter the direction of propagation. The combined effect of an interstellar magnetic field and an ionized plasma can be observed as a Faraday rotation of the plane of polarization. Scattering away from the line of sight, by small-scale variations in refractive index, is not usually of importance except at the longest radio wavelengths.

## 2.1 Electromagnetic Waves

In free space, the propagation of electromagnetic waves is governed by the wave equation; for the electric field $\mathcal{E}$ this is

$$\nabla^2 \mathcal{E} = \frac{1}{c^2} \frac{\partial^2 \mathcal{E}}{\partial t^2}. \tag{2.1}$$

The magnetic field $\mathcal{B}$, obeying a similar equation, lies perpendicular to the electric field, with the cross-product $\mathcal{E} \times \mathcal{B}$ determining the direction of propagation, $\hat{n}$. (If a plasma is present, in which a current $j$ can flow, a term $(4\pi/c)\partial j/\partial t$ must be added to the right-hand

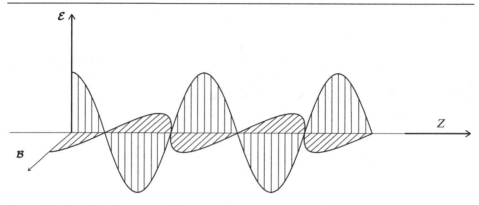

Figure 2.1. Electric and magnetic fields in a linearly polarized wave.

side of Eqn 2.1; this gives rise to interesting propagation phenomena, discussed later in this chapter.)

The electromagnetic wave equation has a rich set of solutions, and simplifying principles are needed. Near the source, Eqn 2.1 requires spherical waves, but astronomical sources are at great distances, such that the curvature of the wavefront at reception is negligible; this means that plane waves are an excellent approximation.[1] A monochromatic transverse wave, with $\mathcal{E}$ in the direction (arbitrary) of a unit vector $\hat{\mathbf{a}}$, and propagating in the $z$ direction with inverse wavelength $k = 1/\lambda$, frequency $\nu$, and initial phase $\phi$, will take the simple form

$$\mathcal{E}(x, y) = \hat{\mathbf{a}}\mathcal{E}_0 \cos[2\pi(\nu t - kz) + \phi] \tag{2.2}$$

or, in the more mathematically convenient complex notation,

$$\mathcal{E}(x, y) = \hat{\mathbf{a}}\mathcal{E}_0 \exp[i2\pi(\nu t - kz) + \phi]. \tag{2.3}$$

The vector $\hat{\mathbf{a}}$ lies in the $x$, $y$ plane, perpendicular to the direction of propagation $z$, and is parallel to $\mathcal{E}$, as illustrated in Figure 2.1. This defines the polarization direction: if $\hat{\mathbf{a}}$ has a constant orientation as the wave propagates, the wave is *linearly polarized*, with the plane of polarization defined by the electric field. The general polarization case is discussed in Section 2.2.

When space is not empty, and there is an index of refraction $n$, the plane-wave propagation of Eqn 2.3 becomes

$$\mathcal{E} = \hat{\mathbf{a}}\mathcal{E}_0 \exp[i2\pi(\nu t - knz)]. \tag{2.4}$$

When the index of refraction varies from place to place, an initially plane wave can develop a curved wavefront, and if the variations in the index of refraction are rapid, solving the

---

[1] The requirement that a plane wave should be a sufficiently good approximation to a spherical wave is important when a radio telescope of aperture dimension $d$ is to be calibrated by using a local transmitter of wavelength $\lambda$. In Section 8.3 we derive the so-called Rayleigh distance $R_{\mathrm{ff}} = 2d^2/\lambda$, beyond which the approximation holds.

propagation problem can become difficult unless suitable approximations can be used. When the curvature of the wavefront is sufficiently small, the *eikonal representation* is usually applicable. In this approximation the electromagnetic waves are described by ray paths perpendicular to the wavefront, a representation well known as geometrical optics.

The phenomenon of atmospheric scintillation is a particularly illuminating (and important) case to consider, both for optical and for millimetric radio observations. The Earth's atmosphere is inhomogeneous and time varying, and the quasi-plane light wave from a star undergoes distortions of two kinds. The wavefront is tilted locally, and this means that the apparent position of the star shifts; the star position jitters rapidly with time, degrading the image. The same phenomenon arises at long radio wavelengths (typically metres) owing to the ionosphere and affects interferometers such as LOFAR and SKA-low (see Section 11.6). When the wavefront distortion is limited, the eikonal approximation holds. It may happen, however, that the wavefront distortion across the aperture is so complex that focussing and interference effects come into play and a simple tilt is inadequate as a description. In this event, which is more severe the larger the aperture, a full wave description is needed. In radio astronomy, dramatic scintillation effects can be seen at longer wavelengths, where the ionospheric plasma, the interplanetary plasma, and the interstellar medium may all play a role. These effects are discussed more fully in Chapter 4.

The geometric optics approximation has, however, a wide range of applicability. In addition to its usefulness in describing the propagation of radiation through an inhomogeneous medium of gently varying refractive index, the approximation holds at interfaces provided that the interface is not sharply curved on the scale of a wavelength. For example, at the boundary of the receiving aperture of a radio telescope, or of any astronomical telescope, there is a sharp discontinuity, and the phenomenon of diffraction results. This presents a central problem for all radio telescopes, since the diffraction limitations of finite apertures determine the achievable angular resolution: in Chapter 8 we discuss the angular resolution issue in terms of wave optics.

## 2.2 Wave Polarization

Measuring the polarization state of the radiation is important in many radio astronomy observations. The outstanding example is radio emission from pulsars, which may be almost 100% polarized (Chapter 15), giving a remarkable diagnostic of the magnetic field geometry at the emitter. The Faraday rotation of linearly polarized pulsar signals provides a direct measurement of galactic magnetic fields, which may often be supplemented by observing the weaker polarization of interstellar synchrotron emission. The tiny degree of linear polarization in the cosmic microwave background (the CMB, see Chapter 17) provides vital clues to the earliest phases of the Universe.

A simple, linearly polarized, wave has orthogonal sinusoidally varying electric and magnetic fields $\mathcal{E}$ and $\mathcal{B}$ as illustrated in Figure 2.1. This description must be generalized,

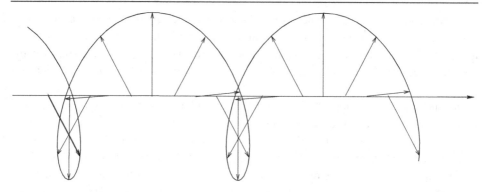

Figure 2.2. A right-hand circularly polarized wave at an instant of time. The tip of the electric vector rotates clockwise as seen from the source and counterclockwise when viewed with the wave approaching.

since cosmic radio sources are not monochromatic and can be fully or partially polarized, either linearly or circularly (see below); a systematic description of the polarization state is therefore necessary.[2] We start, however, by retaining the simplification of monochromatic waves.

### 2.2.1 The Polarization Ellipse

Equation 2.3 can be extended into a more general form, in which the total amplitude is the vector sum of independent complex amplitudes representing two waves linearly polarized along the orthogonal $x$ and $y$ axes. Thus, with the phases explicitly shown, we have:

$$\mathcal{E}(z,t) = (\hat{\mathbf{x}}\mathcal{E}_x e^{i\phi_1} + \hat{\mathbf{y}}\mathcal{E}_y e^{i\phi_2}) \exp[i2\pi(\nu t - kz)]. \tag{2.5}$$

The *polarization* of the resultant wave is determined by the relative amplitudes and phase $\Delta\phi = \phi_1 - \phi_2$. Let us first consider a specific case in which the amplitudes are the same, $\mathcal{E}_x = \mathcal{E}_y$, and $\Delta\phi = \pi/2$ (i.e. 90° or a quarter of a wavelength, with the phase of the $x$ component leading that of the $y$ component); the resultant wave vector rotates in the $x, y$ plane. The tip of the electric vector traces a circle, and the wave is said to be *circularly polarized*.

Figure 2.2 shows how such a circularly polarized wave develops in space; instantaneously the vector follows a spiral with a pitch of one wavelength. The choice of $\Delta\phi = \pi/2$ means that the wave in Figure 2.2 represents right-handed circular polarization. As viewed from along the $z$ axis in the direction of propagation the tip of the vector rotates clockwise,

---

[2] Polarization formalism can be algebraically involved and there are many detailed descriptions in textbooks; the classic texts by Kraus and by Born and Wolf are recommended. Our introductory treatment is brief and, as an aid to understanding, the reader is encouraged to look at the animations illustrating the various polarization states that are widely available on the web.

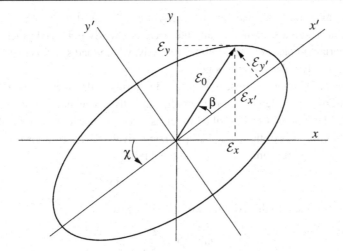

Figure 2.3. The polarization ellipse. The tip of the electric vector $\mathcal{E}$ traces an ellipse; it may be regarded as the sum of two quadrature phased components on the major and minor axes, or two components $\mathcal{E}_x$ and $\mathcal{E}_y$ with a different relative phase.

while an observer viewing from the opposite direction sees the the vector tip rotating counterclockwise.[3]

In the general case the projection of the electric vector $\mathcal{E}$ on a given $x, y$ plane erected at a fixed value of $z$ describes an ellipse in time whose shape depends on the ratio of $\mathcal{E}_x$ and $\mathcal{E}_y$ and their relative phase $\triangle\phi$; the propagation field amplitude describes an elliptical helical locus in the direction of propagation along the $z$ axis. The *polarization ellipse* is shown in Figure 2.3; this is the ellipse which characterizes the wave while the axes $x$ and $y$ are arbitrary directions on the sky, which may be the axes of an antenna feed system.

The limiting cases are the most widely used: *circular polarization*, with $\mathcal{E}_x = \mathcal{E}_y$ and hence with $\mathcal{E}$ rotating with a constant amplitude as in Figure 2.2; *linear polarization*, where $\triangle\phi = 0$ or $\pi$, the two components reach a maximum at the same time and the ellipse collapses to a straight line in a position angle determined by the ratio of $\mathcal{E}_x$ and $\mathcal{E}_y$. The general elliptical case is, of course, designated *elliptical polarization*.

Elliptical polarization may also be regarded as the sum of two circularly polarized waves with opposite handedness. The orientation of the major axis depends only on their relative

---

[3] Once there were differences in convention between physicists/astronomers and electrical engineers, but the terminology was finally regularized when the IAU adopted the engineers' convention that the plane of polarization is the plane of the electric field, and defined right-handed circular polarization as the case when the electric vector, in a fixed plane perpendicular to the ray, rotates in a clockwise direction when viewed in the direction of propagation. An early optical convention used the plane of $\mathcal{B}$ rather than $\mathcal{E}$ as the plane of polarization. The definition of the handedness of circular polarization also was opposite in optical and radio terminology. The convention is now agreed upon but care should be taken in reading the literature since the convention is still not universally followed (unfortunately, in pulsar astronomy the opposite convention is widely used; see Chapter 15). A useful operational definition is that a helical antenna, wound in the sense of a right-handed screw, radiates right-handed circular polarization away from the transmitter; a left-handed helix radiates left-handed circular polarization. Right-handed and left-handed helices also receive right- and left-handed circular polarization respectively.

phase, and the axial ratio depends on their relative amplitude; if the amplitudes are equal, the ellipse collapses to a straight line and the wave is linearly polarized. Such an analysis is often appropriate in observational radio astronomy, where feed systems may be arranged to measure circularly polarized components.

The analysis of the rotated ellipse in Figure 2.3 in terms of the linearly polarized component waves in Eqn 2.5 is elementary but not algebraically trivial and here we give only the salient points. The field can be described in terms of components $\mathcal{E}_{x'}$ and $\mathcal{E}_{y'}$ aligned with the major and minor axes of the rotated ellipse respectively. These components can then be written in terms of a single amplitude $\mathcal{E}_0$ and the auxiliary angle $\beta$:

$$\mathcal{E}_{x'} = \mathcal{E}_0 \cos \beta; \quad \mathcal{E}_{y'} = \mathcal{E}_0 \sin \beta. \tag{2.6}$$

The angle $\beta$ is related to the axial ratio of the ellipse by

$$\tan^{-1} \beta = \frac{\mathcal{E}_{y'}}{\mathcal{E}_{x'}} \tag{2.7}$$

and specifies the character of the polarization. A plane-polarized wave has $\beta = 0$ while circular polarization results when $\beta = \pm\pi/4$, where the positive sign defines right-handed polarization (corresponding to the phase convention that $\mathcal{E}_{x'}$ leads $\mathcal{E}_{y'}$ by $\Delta\phi = \pi/2$). The angle $\beta$, the orientation angle of the major axis $\chi$, and its amplitude $\mathcal{E}_0$ specify completely the state of a plane monochromatic wave.

A strictly monochromatic wave does not change its state of polarization, but the radiation from astronomical sources is far from this simplest case. Instead it is noise-like and is the resultant of many independent or incoherent emitters over a finite bandwidth $\Delta\nu$. At any instant a radio telescope therefore receives a superposition of statistically independent polarized waves with a variety of amplitudes and phases, all fluctuating on timescales $\sim 1/\Delta\nu$ (the coherence time, discussed in more detail in Chapter 6). The polarization ellipse therefore flickers about in time chaotically and hence is only valid at a particular instant.

Astrophysical radiation is often partially polarized, in the sense that there is more power in one polarization state than another and it can be regarded as a combination of a fully polarized part plus a completely unpolarized part. If the emission region is thermal (such as emission from a black body), there are no preferred directions – the source is randomly polarized and the random changes of direction result in an average polarization of zero. We say that the source is unpolarized, in which case its only measurable characteristic is the incoming energy flux (i.e. its power).

The polarization ellipse of the polarized part, whilst also constantly fluctuating, maintains its orientation, shape, and sense of rotation when averaged over an observing time that is long compared with the coherence time. One cannot, however, make use of the time average of the fluctuating polarization ellipse since it is cast in terms of unobservable electric field amplitudes. Thus, while an understanding of the polarization ellipse is necessary background for the radio astronomer and is helpful for understanding a polarimetric receiver's response, for practical measurements of partial polarization a different approach is necessary. We set out this analysis, in terms of the Stokes parameters, in Chapter 7.

## 2.3 Blackbody Radiation

Blackbody radiation and its quantitative understanding by Max Planck played a vital role in the development of quantum theory in the first decade of the twentieth century. This revolution in physics is well described in many standard textbooks. For our practical purposes in radio astronomy we need only to consider the emission from an idealized surface which has the property that it absorbs all the radiation falling upon it; this surface will also emit blackbody radiation corresponding to its temperature $T$. The emission from an ideal surface has a characteristic smoothly varying spectrum described by the Planck distribution:

$$I_\nu = \frac{2h\nu^3}{c^2} \frac{1}{e^{h\nu/kT} - 1}, \tag{2.8}$$

where $I_\nu$ is the *specific intensity*, defined as the power emitted at a specific frequency $\nu$, per unit area, per unit frequency interval, per unit solid angle; its link with telescope measurements will become clear in Section 2.4. The Planck distribution is the solid line in Figure 2.4 for $T = 2.725$ K. As $T$ increases the peak of the curve shifts to higher frequencies and the integral under the curve rises rapidly; this integral is the total power radiated per unit surface area, given by the Stefan–Boltzmann law

$$P_{\text{tot}} = \sigma T^4, \tag{2.9}$$

where $\sigma$ is Stefan's constant.[4]

The Planck distribution can also be defined in terms of wavelength:

$$I_\lambda = \frac{2hc^2}{\lambda^5} \frac{1}{e^{hc/\lambda kT}-1}, \tag{2.10}$$

and it is important to be clear about which form one is using in plots: intensity per unit frequency interval against frequency (as in Figure 2.4) or intensity per unit wavelength against wavelength. Obviously the power radiated per unit area per steradian between two points on the distributions, labelled either in terms of frequency or wavelength, must be equal, i.e.

$$\int_{\nu_1}^{\nu_2} I_\nu d\nu = \int_{\lambda_2}^{\lambda_1} I_\lambda d\lambda, \tag{2.11}$$

but the distributions look rather different. This is so because equal steps $d\lambda$ in the plot of $I_\lambda$ against $\lambda$ correspond to steps in frequency $d\nu$ that are stretched at the high frequency end and compressed at the low frequency end of the plot of $I_\nu$ against $\nu$, since $d\nu = (-c/\lambda^2)d\lambda$ with the minus sign implying that the ordinates increase in opposite directions. We illustrate this point in the radio astronomy context with the cosmic microwave background radiation (Figure 2.4 and Chapter 17), whose spectrum agrees beautifully with the full Planck law for $T_{\text{CMB}} = 2.726 \pm 0.010$ K. The maximum of Eqn 2.8 occurs at $\nu_{\text{max}} = 58.8T$ GHz, i.e. 160.3 GHz (corresponding to a wavelength of 1.87 mm), while the maximum of Eqn 2.10 occurs

---

[4] Recommended values of constants are: Boltzmann $k = 1.3807 \times 10^{-23}$ J K$^{-1}$; Planck $h = 6.626 \times 10^{-34}$ J; Stefan $\sigma = 5.670$ W m$^{-2}$ K$^{-4}$.

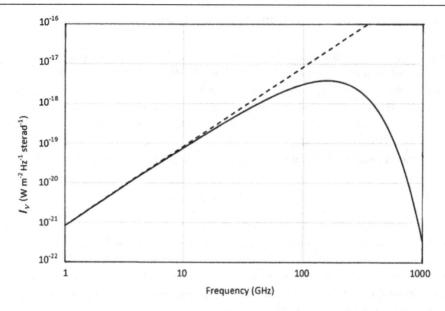

Figure 2.4. The Planck (solid line) and Rayleigh–Jeans (broken line) functions for a black body at the temperature of the cosmic microwave background 2.725 K; the peak of the Planck function is at 160.2 GHz.

at $\lambda_{\max} = 0.2898T^{-1}$ cm, i.e. 1.06 mm (corresponding to a frequency of 282.8 GHz). The apparent disagreement arises simply from the difference in how the distributions are defined.[5]

The full Planck formula is needed at short radio and millimetre wavelengths and at low temperatures but in most circumstances in radio astronomy we can use a simpler version. Equation 2.8 may be rewritten as

$$I_\nu = 2kT \frac{\nu^2}{c^2} \left( \frac{h\nu/kT}{e^{h\nu/kT} - 1} \right). \tag{2.12}$$

When $h\nu \ll kT$, i.e. at frequencies where quantum effects may be ignored, the term in parentheses tends to unity, leaving

$$I_\nu = \frac{2kT\nu^2}{c^2} = \frac{2kT}{\lambda^2}. \tag{2.13}$$

This is the Rayleigh–Jeans (RJ) approximation. The specific intensity is now proportional to temperature and to the square of the frequency, and a useful rule of thumb is that the approximation is valid at frequencies well below $\approx 20T_{\text{source}}$ GHz; this corresponds to the linear region of the Planck spectrum when it is plotted logarithmically, as in Figure 2.4. When dealing with flux densities (see Section 5.2), Eqn 2.13 is commonly used

---

[5] Helpful pedagogic discussions of the different forms of the Planck distribution are given by B. H. Soffer and D. K. Lynch (1999), *Am. J. Phys.*, **67**, 946 and J. M. Marr and F. P. Wilkin (2012), *Am. J. Phys.*, **80**, 399.

in the wavelength form given on the right, since it is simpler to remember. We shall use the RJ approximation freely except where explicit recognition of quantum effects is required. At much shorter wavelengths, with frequencies well above $kT/h$, the term in parentheses in Eqn 2.12 dominates, in which the flux of blackbody intensity falls exponentially with increasing frequency (a situation commonly met with in X-ray astronomy).

Before leaving this discussion one should note that in general thermal emitters behave as black bodies only over a restricted frequency range, outside which their radiation falls short by a factor $\epsilon$, the dimensionless *emissivity*, which is a function of frequency. The spectrum is therefore a combination of the Planck formula multiplied by a frequency-dependent factor. An example is the continuum spectrum of an interstellar dust cloud whose physical temperature, $\leq 50$ K, corresponds to a blackbody spectral peak at $\nu \leq 3 \times 10^3$ GHz (i.e. $\lambda \geq 0.1$ mm). The dust grains are typically much smaller than 0.1 mm and so do not radiate efficiently below the peak. At lower frequencies the cloud spectrum ($I_\nu \propto \nu^{3-4}$) therefore falls away more rapidly than the RJ approximation ($I_\nu \propto \nu^2$).

Thermal noise powers are small by everyday standards. Consider a radio telescope with a narrow beam pointed at the Moon and let the beamwidth be much smaller than the Moon's disk, so that sidelobe effects can be neglected. As an example, at $\lambda = 3$ cm a 100 m radio telescope has a primary beamwidth of just over 1 arcmin, about 30 times smaller than the disk (this is the *filled-beam* case discussed in Chapter 5). At this wavelength the Moon is nearly an ideal black body at a temperature of 226 K and the RJ approximation (Eqn 2.13) holds. The antenna temperature will therefore be 226 K and the power output equals that from a resistive load at 226 K; with a bandwidth of 1 MHz the thermal noise power (Eqn 2.16) will be 0.003 picowatts. For an interesting comparison, let a small transmitter generating power $P_{tr}$ be placed on the Moon, at a distance $D_{moon}$ from Earth, radiating its power through a small antenna with gain $G_{tr}$. A radio telescope of diameter $d_{tel}$ will receive an amount of power $P_{rec}$ given by

$$P_{rec} = P_{tr} \left( \frac{G_{tr}}{4\pi D_{moon}^2} \right) \frac{\pi d_{tel}^2}{4} = \frac{1}{16} P_{tr} G_{tr} \frac{d_{tel}^2}{D_{moon}^2}. \tag{2.14}$$

(Here it is assumed that the effective area of the telescope is its geometrical area, a goal approached but never reached in practice.)

As an example, take a one-watt transmitter, comparable to a mobile phone, radiating from an antenna that has a gain of 4 over an isotropic antenna, and set the Moon distance as 380 000 kilometres. With these numbers the power received by a 100 m telescope is almost ten times greater than the thermal noise power from the lunar surface.

The surfaces of several planets are useful as standard sources of blackbody radiation for calibrating millimetre-wave observations, especially from spacecraft. They are, however, not perfect black bodies, and their temperatures vary somewhat with wavelength. The CMB is an established standard with amplitude 2.725 K; it has a dipolar component (0.0035 K), due to the Earth's motion, which is used as a standard in measurements of small angular structure in the CMB (Chapter 17).

Blackbody radiation also plays a vital role in radio receivers. The corresponding derivation for the noise power flowing in a single-mode transmission line (or a resistor) connected to a black body at temperature $T$ leads to a one-dimensional analogue of the Planck law:

$$P_\nu = \frac{h\nu}{e^{h\nu/kT} - 1}.$$ (2.15)

Here $P_\nu$ is the *power spectral density*, which is the power per unit bandwidth flowing in each direction along the transmission line. In the RJ approximation this reduces to

$$P_\nu = kT,$$ (2.16)

showing that the power spectral density is proportional to temperature and that the total power within a band $\Delta\nu$ is $kT\Delta\nu$. The noise generated in a resistor $R$, known as *Johnson noise*, has the same rms voltage as in a matched transmission line with characteristic impedance $R$, giving $\langle v_n^2 \rangle = 4RkT\Delta\nu$. We will return to the topic of noise in radio receivers in Chapter 6.

## 2.4 Specific Intensity and Brightness

Whatever the emission mechanism, or mechanisms, at work, a radio telescope simply measures the power coming from a region of the sky over a band of frequencies; this band may be split up into many sub-bands for observing spectral lines. After leaving the source the radiation can be affected by various processes along its path. In the diagnosis of the mechanisms and processes at work, radio astronomers use a set of generic physical concepts and a simple picture of radiative transfer. These ideas, which permeate the whole of our subject, are explained in this section and the next. We start with the concepts of specific intensity and surface brightness.

The specific intensity $I_\nu$ of a source is a fundamental property related to the physics of the emission process; we first encountered it in the discussion of the Planck radiation law for black bodies in Section 2.3 but it is a general concept applicable to all emission mechanisms. $I_\nu$ is the energy flux (power) radiated per unit area into unit solid angle at a specific frequency $\nu$ over a range $d\nu$; its units are W m$^{-2}$ Hz$^{-1}$ sterad$^{-1}$.

In Figure 2.5 we illustrate an idealized observational situation in which an antenna (area $A_{ant}$) captures radiation from a uniform source of emission which extends beyond the antenna's reception beam (with solid angle $\Omega$). The reception beam intercepts an area of the source $A_{source}$. From the point of view of the source the solid angle subtended by the antenna is $\omega$ and thus the power emitted from an area $A_{source}$ into that solid angle is $dP_{em} = I_\nu d\nu A_{source}\omega$ watts.

At this stage the observer knows nothing about the specific intensity of the source. The antenna system simply measures the incoming flux $F$, i.e. the energy arrival rate $dE/dt$ (watts) per unit area at the collector, $F = (dE/dt)(1/A_{ant})$. We define the observed *brightness* $B$ of the source to be the flux divided by the solid angle of reception, i.e. $B = (dE/dtA_{ant})(1/\Omega)$; thus the observed *specific brightness* is $B_\nu = dE/(dtd\nu A_{ant}\Omega)$; notice that $B_\nu$ also has units W m$^{-2}$ Hz$^{-1}$ sterad$^{-1}$. The power captured by the antenna from within its reception beam is, therefore, $dP_{cap} = B_\nu d\nu A_{ant}\Omega$ watts.

If the emitting region is at a distance $r$ then, from the definition of solid angles, $dP_{em} = I_\nu d\nu A_{source}A_{ant}/r^2$ and $dP_{cap} = B_\nu d\nu A_{ant}A_{source}/r^2$ watts. Since $dP_{em} = dP_{cap}$ it follows

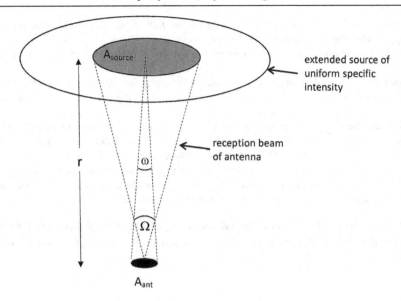

Figure 2.5. An extended source of uniform specific intensity $I_\nu$ emitting radio energy towards an antenna which subtends a solid angle $\omega$ at a point on the source. Radio energy arrives at the antenna from within the solid angle of the reception beam, $\Omega$.

that $B_\nu = I_\nu$. Stated another way, if the distance $r$ increases, the incoming power falls by $1/r^2$ but this is exactly cancelled out by the larger area of the source intercepted by the beam. Figure 2.5 also illustrates the fact that an antenna with a small collecting area and a consequently large beam solid angle receives the same power as one with a large collecting area and a small beam solid angle *as long as the emission region covers the larger beam*. The solid angle of the beam is proportional to $\lambda^2/A_{ant}$. The power collected is also proportional to $A_{ant}$ and hence for both antennas the power collected is proportional to $\lambda^2$. We will meet this argument in the more sophisticated form of the 'antenna equation' in Chapter 5. The terms $I_\nu$ and $B_\nu$ are frequently used interchangeably but we prefer to use $I_\nu$ for the value of the specific intensity emitted from the source and along the ray path. Under the circumstances described above, the observable $B_\nu$ therefore provides information on the physical conditions at the source.[6]

Since the observed brightness is linked to the physical conditions at emission it should be no surprise that in many circumstances it is conserved regardless of the distance to the source. However, when the source is *smaller than the reception beam* the $1/r^2$ fall-off is not cancelled by the larger source area; however, the emission physics does not change and the source does not become less bright. The flux decreases according to an inverse square law simply because the observed area of the source becomes smaller.

The equivalence of $I_\nu$ and $B_\nu$ does break down under a variety of circumstances.

---

[6] Infrared astronomers often use the same conventions as radio astronomers, using the specific brightness defined as a function of frequency. In optical astronomy the use of the specific brightness per wavelength interval continues in common usage.

(i) *Absorption and/or emission by matter along the ray path* If the intervening material absorbs
energy it must also re-emit energy (Kirchhoff's law);

(ii) *Scattering of energy out of the telescope beam* by small-scale variations in refractive
index;

(iii) *Cosmological effects* For an object at redshift $z$ the expansion of the Universe decreases the
energy arrival rate by $(1+z)^2$ and, together with a relativistic angular size effect, its true surface
brightness is reduced by a factor $(1+z)^4$;

(iv) *Diffraction* The $I_\nu \equiv B_\nu$ identity relies on geometrical optics, as drawn in Figure 2.5. In reality
the telescope beam has 'sidelobes' pointing towards regions of different brightness, e.g. the
ground around the telescope. Only if all the sidelobes are embedded in regions of the same
brightness can we claim to have measured the intrinsic $I_\nu$ from a measurement of $B_\nu$. Assessing
the effect of radiation entering via the sidelobes is a basic challenge when measuring the sky
brightness with a radio telescope. We return to this in Chapter 8.

Scattering is rarely an issue for radio astronomy and for the rest of this section we focus
on point (i): if energy is absorbed or emitted on its way to us, the specific intensity will not
be conserved.

## 2.5 Radiative Transfer

Figure 2.6 illustrates the basic situation from the point of view of the receiving antenna.
Radiation from a background source passes through an intervening medium, typically a
cloud of gas, which can both absorb incoming wave energy and also emit and absorb
its own energy. For the general case we can ignore the detailed microscopic physics of
the interactions and describe the effects in terms of macroscopic absorption and emission
coefficients.

The *specific emissivity* $j_\nu$ (the symbol $\epsilon_\nu$ is sometimes used) is the power emitted per
unit volume, per frequency interval, per steradian,

$$j_\nu = \frac{dE}{\Omega d\nu dt dV},$$
(2.17)

or equivalently the specific intensity added per unit distance through the medium (see
Figure 2.6); it has units W m$^{-3}$ Hz$^{-1}$ sterad$^{-1}$. The power added within the solid angle $\Omega$
is therefore $j_\nu d\nu d\sigma ds\Omega$. The diminution of the specific intensity depends upon the *linear
absorption coefficient* $\kappa_\nu$, which is the fraction of incident power lost while the radiation
passes through unit distance; its units are m$^{-1}$. The linear absorption coefficient is then
defined by

$$dI_\nu = -\kappa_\nu I_\nu ds.$$
(2.18)

The power removed from within the solid angle $\Omega$ is therefore $\kappa_\nu I_\nu d\nu d\sigma ds\Omega$. The specific
intensity $I_\nu$ entering the volume element emerges with its magnitude reduced by $dI_\nu$. The
change in specific intensity will be proportional to $I_\nu$ provided that the absorption process
is a small perturbation of the absorbing system. This implies that the system must be in

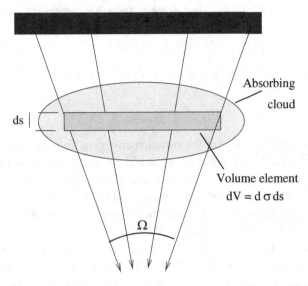

Emitting region, specific emissivity $I_\nu$

Figure 2.6. An emitting region seen through an intervening cloud. Material in the cloud can both absorb incoming energy and emit and absorb its own energy. Net absorption leads to a decrease in specific intensity along the path. Net emission leads to an increase. The volume element has cross-sectional area $d\sigma$ perpendicular to the page.

steady-state equilibrium, with a means of redistributing the absorbed energy. In terms of the increment in specific intensity $dI_\nu$ the power equation is

$$dI_\nu d\nu d\sigma \, \Omega = j_\nu d\nu d\sigma \, ds\Omega - \kappa_\nu I_\nu d\nu d\sigma \, ds\Omega. \tag{2.19}$$

Equation 2.19 easily simplifies to

$$\frac{dI_\nu}{ds} = j_\nu - \kappa_\nu I_\nu, \tag{2.20}$$

which describes the development of the specific intensity along a ray path; this is the *equation of radiative transfer*. Some elementary solutions can be written down immediately.

(i) A ray with specific intensity $I_\nu^0$ at the origin has, at a position $s$ along the ray path,

$$I_\nu(s) = I_\nu^0 + \int_0^s j_\nu(s')ds' \quad \text{(emission only)}, \tag{2.21}$$

$$I_\nu(s) = I_\nu^0 \exp\left(-\int_0^s \kappa_\nu(s')ds'\right) \quad \text{(absorption only)}. \tag{2.22}$$

The exponential term in Eqn 2.22 occurs frequently in radiative transfer problems, and the integral within it is called the *optical depth* $\tau_\nu$ (which is dimensionless):

$$\tau_\nu(s) \equiv \int_0^s \kappa_\nu(s')ds'. \qquad (2.23)$$

If $\kappa_\nu$ is constant within the absorbing region of depth $s$ then $\tau_\nu = \kappa_\nu s$ and Eqn 2.22 simplifies to

$$I_\nu = I_\nu^0 \exp(-\tau_\nu), \qquad (2.24)$$

where $I_\nu$ is the specific intensity of the radiation emerging from the absorbing region. If $\tau_\nu$ is large then this implies strong attenuation and the medium is termed *optically thick*; in the limit $\tau_\nu = \infty$ it is completely opaque and we only observe a surface layer. If $\tau_\nu$ is small then the medium is termed *optically thin* and in the limit $\tau_\nu = 0$ it is completely transparent. The case where $\tau_\nu = 1$ marks the transition between the optically thin and optically thick regimes, at which point $I_\nu = 0.368 I_\nu^0$.

(ii) In the case of a uniform medium in thermal equilibrium with a radiation field at a specific temperature, the specific intensity is determined only by the temperature of the source, and the rate of emission equals the rate of absorption, i.e. $dI_\nu/ds = 0$ and $I_\nu$ is constant along any ray path. These are the conditions met by blackbody radiation and $I_\nu$ must, therefore, take the form of a Planck function (Eqn 2.8 and, in the RJ approximation, Eqn 2.13). Since $dI_\nu = 0$ we have $j_\nu = \kappa_\nu I_\nu$ and

$$\frac{j_\nu}{\kappa_\nu} = I_\nu(T). \qquad (2.25)$$

Thus there is a universal relation between the emissivity and the absorption coefficient, known as Kirchhoff's radiation law. The same reasoning applies even if the medium is not in equilibrium with the incident radiation field *as long as it can be described by a single temperature*. These conditions constitute local thermodynamic equilibrium (LTE). For example, the Earth's troposphere is not in equilibrium with the incoming radiation from the Sun at 5800 K; instead its physical temperature is approximately constant at $260 \pm 10$ K. The troposphere is mainly heated by the infrared radiation from the Earth's surface and, since it absorbs only a fraction of this radiation, it is optically thin.

Dividing Eqn 2.20 through by $\kappa_\nu$ and recognizing that $d\tau_\nu = \kappa_\nu ds$, we obtain

$$\frac{dI_\nu}{d\tau} = \frac{j_\nu}{\kappa_\nu} - I_\nu. \qquad (2.26)$$

The term $j_\nu/\kappa_\nu$ is called the *source function* $S_\nu$; its units are those of specific intensity. The source function $S_\nu$ is a measure of how much energy from the original source is removed from the path and replaced by energy generated in the medium. It contains all the physics of the medium, e.g. scattering, stimulated emission, and non-uniformity as well as a mixture of absorbing and self-emitting processes. Equation 2.26 therefore has many solutions depending on the assumed physical conditions.

A simple case, met with commonly in radio astronomy, occurs when a distant source is observed through an isothermal cloud with temperature $T$ and optical depth $\tau_\nu$.

Table 2.1. *Expressions for $T_{obs}$ for various optical depths*

| $\tau$ | $T_{obs}$ |
|---|---|
| 0 | $T_{source}$ (cloud perfectly transparent) |
| 0.1 | $0.9T_{source} + 0.1T_{cloud}$ (source dominates, cloud seen in absorption) |
| 1 | $0.37T_{source} + 0.63T_{cloud}$ (cloud dominates, cloud seen in emission) |
| $\infty$ | $T_{cloud}$ (cloud perfectly opaque, only surface is seen) |

Equation 2.26 can then be solved by multiplying through with an integrating factor $e^{\tau_\nu}$ and integrating by parts. The solution is

$$I_\nu = I_\nu^0 \exp(-\tau_\nu) + S_\nu[1 - \exp(-\tau_\nu)], \tag{2.27}$$

which is the sum of the attenuated source radiation and the radiation from within the medium attenuated by the optical depth from that point to the point of reception (see Figure 2.6). The source function $S_\nu$ is the specific intensity of an optically thick cloud at temperature $T$; in local thermal equilibrium (LTE) conditions this is given by the Planck function for a black body.

In the RJ regime all specific intensities can be described by a brightness temperature, either the actual physical temperature of a black body or an equivalent blackbody temperature (Chapter 5). This enables Eqn 2.27 to be written for a particular frequency as

$$T_{obs} = T_{source}\exp(-\tau_\nu) + T_{cloud}[1 - \exp(-\tau_\nu)], \tag{2.28}$$

where $T_{obs}$ is the observed brightness temperature of a background source observed through a foreground cloud with uniform temperature and properties. Figure 2.7 shows Eqn 2.28 graphically while Table 2.1 lists some illustrative cases.

## 2.6 Free–Free Radiation

The brightness temperature of radiation from a thermally emitting source is related to the temperature of the source. The simplest case for radio emission is radiation from an ionized gas, in which the emission process is the acceleration of free electrons in elastic collisions with ions; this is known as free–free, or bremsstrahlung, radiation. It is broadband radiation, in contrast with the spectral lines of recombination radiation (Chapter 3).

In a free–free collision, the time over which most of the radiation is emitted is so brief that, particularly in the radio spectrum, the acceleration can be approximated by a delta-function, and the emissivity can be found from straightforward electromagnetic theory. A full calculation, however, requires a summation over all directions, impact parameters, and velocities. A summary of the principal results is given in Condon and Ransom (2016).

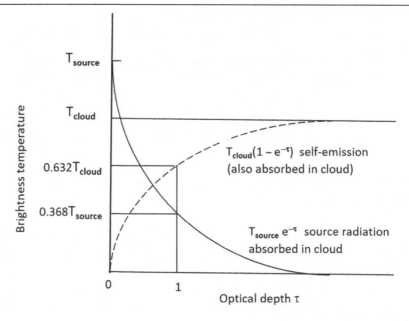

Figure 2.7. The effect on a background source of an isothermal foreground cloud in local thermo-dynamic equilibrium. When the optical depth $\tau = 1$, the source brightness temperature is reduced to 0.368 of its intrinsic value and the cloud emission has reached 0.632 of its optically thick value. See also Table 2.1.

For a completely ionized gas with electron density $n_e$ and ion density $n_i$, with charge number $Z$ per ion, the specific emissivity for free–free emission, $j_{ff}(\nu)$, is

$$ j_{ff}(\nu) = \frac{1}{2\pi} mc^2 \sigma_T^{3/2} \left(\frac{mc^2}{kT}\right)^{1/2} Z^2 n_e n_i \exp\left(-\frac{h\nu}{kT}\right) \bar{g}_{ff}(T, Z, \nu). \qquad (2.29) $$

Here, $\sigma_T$ is the classical Thomson cross-section of the electron and $\bar{g}_{ff}(T, Z, \nu)$ is the velocity-averaged *Gaunt factor*,[7] whose value varies between 4 and 5 over the radio range for temperatures of order $10^4$ K. Numerically,

$$ j_{ff}(\nu) = 0.54 \times 10^{-38} Z^2 n_e n_i T^{-1/2} \exp\left(-\frac{h\nu}{kT}\right) \bar{g}_{ff} \quad \text{erg cm}^{-3} \text{ sterad}^{-1}. \qquad (2.30) $$

A useful practical form of this equation, at radio wavelengths, for a temperature $T_4$ measured in units of $10^4$ K and with the frequency in GHz, is

$$ j_{ff}(\nu) \approx 3.15 \times 10^{-38} Z^2 n_e n_i T_4^{-1/2} \left[ 13.2 + \ln\left(\frac{T_4^{3/2}}{Z\nu_{GHz}}\right) \right] \quad \text{erg cm}^{-3} \text{ sterad}^{-1}. \qquad (2.31) $$

---

[7] The Gaunt factor is a quantum mechanics correction to the classical analysis. For X-rays in the kilovolt range, and for a million-volt plasma, the Gaunt factor is of order unity. A detailed plot can be found in Condon and Ransom (2016).

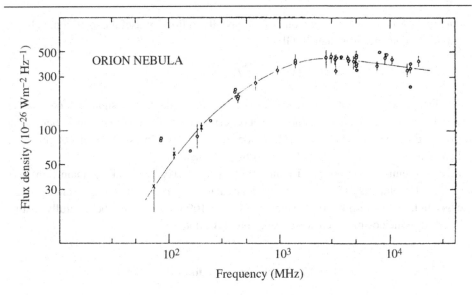

Figure 2.8. The spectrum of the observed radio emission from the Orion Nebula M42, showing the effect of increasing optical thickness at lower radio frequencies (after Terzian and Parrish 1970).

The exact form, Eqn 2.29 above, has an exponential cutoff at the high energy, short wavelength, end of the spectrum, expressing the fact that a colliding electron cannot emit more energy than it possesses. With the inclusion of this factor the expression is as valid for radiation in the X-ray domain as it is at radio wavelengths. When X-ray spectral observations of a source show this exponential behaviour, the term *soft spectrum* is used, to distinguish it from the power-law *hard spectrum* of synchrotron radiation.

With the emission spectrum for free–free radiation defined, Kirchhoff's law, Eqn 2.25, can be used to derive the linear absorption coefficient, $\kappa_{ff}$. For free–free radiation,

$$\kappa_{ff} = j_{ff}(\nu)\frac{c^2}{2kT\nu^2}. \tag{2.32}$$

Over a wide frequency range, an ionized gas cloud may be opaque (optically thick) at low frequencies and transparent (optically thin) at high frequencies. The absorption coefficient increases inversely as the square of the frequency. Hence, at low frequencies it will have a thermal spectrum corresponding to the electron temperature; a Rayleigh–Jeans spectrum, with the specific intensity increasing as the square of the frequency, is therefore observed. At a sufficiently high frequency, however, the plasma must become optically thin, and the spectrum bends over to become approximately flat, diminishing slowly under the influence of the Gaunt factor. A spectrum of this sort is shown in Figure 2.8, which shows the theoretical and observed continuum spectra of the Orion Nebula, M42.

The optical depth $\tau_\nu$ was defined in Section 2.5 as the integral of the absorption coefficient along the line of sight. For ionized hydrogen in the radio domain, where $Z = 1$

and $n_e = n_i$, Mezger and Henderson (1967) gave the following useful expression, valid to $\approx 5\%$ for frequencies less than 10 GHz:

$$\tau_\nu \approx 8.235 \times 10^{-2} T_e^{-1.35} \nu^{-2.1} \int n_e^2 dl. \tag{2.33}$$

The integral of the square of the electron density along the line of sight is known as the *emission measure*, and it is a commonly met observational parameter in plasmas such as the ionized hydrogen clouds known as H II regions. It is usually expressed in units of $cm^{-6}$ pc, as here, since these are the units of choice observationally.

Free–free emission is observed mainly within the Milky Way galaxy, mainly in H II regions and in planetary nebulae. On a wide angular scale, it contributes appreciably to the background, and at millimetre wavelengths (10 to 100 GHz) it must be carefully distinguished from the cosmic microwave background (Chapter 17).

### 2.7 Synchrotron Radiation

Synchrotron radiation is intense broad-bandwidth radiation produced by highly relativistic electrons gyrating around magnetic fields. The process by which electrons with large relativistic energy radiate when they are accelerated in a magnetic field first received attention when the first electron synchrotrons were built in the 1940s, when it was found to be an important mechanism of energy loss. The radiation has been most commonly called synchrotron radiation ever since (the alternative 'magnetobremsstrahlung' has largely fallen out of use). In the rest frame of the electron elementary gyroradiation is produced, and we present this treatment first. The remarkable (and intense) synchrotron radiation is then derived by transforming to the observer's frame by a Lorentz transformation.

The frequency of gyration of a non-relativistic electron circulating in a magnetic field $\mathcal{B}$ is known as the *cyclotron frequency*, $\nu_{cyc}$:

$$\nu_{cyc} = \frac{e\mathcal{B}}{2\pi mc}. \tag{2.34}$$

Note that this is in Gaussian units (commonly used in astrophysics; 1 gauss = $10^{-4}$ tesla), so

$$\nu_{cyc} = 2.80\mathcal{B} \text{ MHz.} \tag{2.35}$$

The cyclotron frequency is independent of the pitch angle (the inclination of the velocity vector to the magnetic field), so the electron may have a velocity component along the magnetic field direction.

When the energy $\gamma m_0 c^2$ of the electron is relativistic ($\gamma \gg 1$), the frequency of gyration is reduced (as the effective mass increases) and becomes the gyrofrequency $\nu_g$, which is no longer independent of energy:

$$\nu_g = \frac{\nu_{cyc}}{\gamma}. \tag{2.36}$$

Figure 2.9. The radiation pattern of an electron moving on a curved trajectory with relativistic velocity. The radiation is concentrated in a beamwidth of approximately $\gamma^{-1}$ radians.

For electrons with a velocity component along the magnetic field, moving with pitch angle $\alpha$, the radius $R$ of the helical trajectory is, for relativistic electrons,

$$R = \frac{\gamma m c^2 \sin \alpha}{eB} \qquad (\gamma \gg 1). \tag{2.37}$$

Thus, for a relativistic electron, with its energy $E_{\mathrm{GeV}}$ expressed in GeV,

$$R = \frac{1}{3} \times 10^5 \frac{E_{\mathrm{GeV}}}{B} \quad \mathrm{m}. \tag{2.38}$$

The scale of the phenomenon can be appreciated by noting that, for an electron with an energy of 10 GeV (corresponding to $\gamma \sim 20\,000$) and an interstellar magnetic field of 3 microgauss, its helical radius will be $10^{12}$ m or about 7 a.u., large compared with terrestrial phenomena, but considerably less than a parsec. The gyrofrequency will be $4.2 \times 10^{-4}$ Hz, i.e. the electron takes about 40 hours to complete one turn but, as we will see, the bulk of the radiation is produced at much higher frequencies.

A highly relativistic electron radiates in a narrow beam, with width of order $1/\gamma$ ($\sim 10$ arcsec for our 'standard' 10 GeV electron), in the direction of motion. Figure 2.9 shows the geometry of this headlight radiation as the electron proceeds along its helical path. The observer receives a short pulse each time the beam crosses the line of sight. The time transformation from the accelerated system to the observer's frame contracts the sweeping time by $1/\gamma^2$, so that the total time that elapses for the main radiation pattern to flit by the observer is of the order of

$$\delta t \approx \frac{1}{\gamma^3 \nu_g}. \tag{2.39}$$

This is a pulse which is enormously shorter than the inverse of the period $\nu_g^{-1}$. For the 10 GeV electron, $\delta t \approx 3 \times 10^{-10}$ s. If the magnetic field were absolutely uniform and the electron energy constant, the radiation would consist of periodic impulses, and a Fourier series would describe the spectrum as a sum of harmonics of the fundamental gyrofrequency. In real radio sources, the pulses are asynchronous, and the spectrum is the same as that of a single pulse. The spectrum is then concentrated at a characteristic frequency $\nu_0$, which is the inverse of the pulse duration:

$$\nu_0 \approx \gamma^3 \nu_g, \tag{2.40}$$

i.e. $\sim$ 3.4 GHz for the 10 GeV electron, or, in terms of the (non-relativistic) cyclotron frequency $\nu_{cyc}$,

$$\nu_0 \approx \gamma^2 \nu_{cyc}. \tag{2.41}$$

The full calculation of the spectrum is well documented (see particularly the review by Ginzburg and Syrovatskii (1969, correcting several errors in the earlier literature) and the textbook treatments by Rybicki and Lightman (1979) and by Condon and Ransom (2016)). The spectrum is expressed in terms of a natural parameter $x$, defined by

$$x \equiv \nu/\nu_{crit}, \tag{2.42}$$

where

$$\nu_{crit} = \tfrac{3}{2} \gamma^3 \nu_g \sin\alpha = \tfrac{3}{2}\gamma^2 \nu_{cyc} \sin\alpha. \tag{2.43}$$

The angle $\alpha$ is the pitch angle of the electron trajectory; without the trigonometric term, $\nu_{crit}$ is known as the *critical frequency*.

It is often convenient to consider the synchrotron radiation from an electron with energy with relativistic factor $\gamma$, or energy $E_{Gev}$, as though it were concentrated at $\nu_{crit}$. For a magnetic field $B\sin\alpha = B_\perp$ gauss, $\nu_{crit} = 4.2B_\perp\gamma^2 = 16 \times 10^6 B_\perp E_{GeV}^2$ MHz. For example, the electrons responsible for synchrotron radiation at 10 MHz from the Galaxy, where the field is a few microgauss (Chapter 14), have relativistic factors of order $\gamma \approx 10^3$, i.e. energies of order 1 GeV, while the radiation observed at 1 GHz is due to electrons with energies ten times higher.

The full analytical calculation gives the power spectrum of the radiation from a single electron in terms of the parameter $x$ as

$$P(\nu)d\nu = \sqrt{3}\frac{e^3 B \sin\alpha}{mc^2} F(x)d\nu, \tag{2.44}$$

where $F(x)$, which contains the shape of the spectrum, involves an integral of a modified Bessel function of order 5/3:

$$F(x) \equiv x \int_x^\infty K_{5/3}(\xi)d\xi. \tag{2.45}$$

The function $F$ is plotted in Figure 2.10, showing the concentration of the synchrotron emission spectrum at a frequency of the order of $\gamma^2$ times the cyclotron frequency (i.e. $\gamma^3$ times the gyrofrequency).

Synchrotron radiation is polarized, and it is convenient to resolve the radiation into components perpendicular and parallel to the magnetic field. This involves another function, $G(x)$:

$$G(x) \equiv x \int_x^\infty K_{2/3}(\xi)d\xi. \tag{2.46}$$

The power spectra of the two polarized components are

$$P_{\perp,\|}(\nu) = \frac{\sqrt{3}}{2}\frac{e^3 B \sin\alpha}{mc^2} \begin{cases} F(x) + G(x) & (\perp B), \\ F(x) - G(x) & (\| B). \end{cases} \tag{2.47}$$

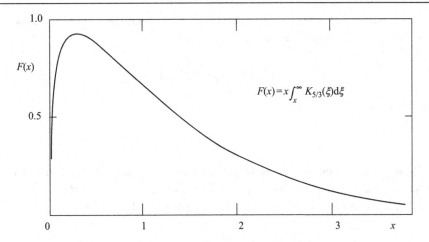

Figure 2.10. The function $F(x) \equiv x \int_x^\infty K_{5/3}(\xi)d\xi$, which appears in the spectrum of synchrotron radiation from a single electron.

The difference between $P_\perp$ and $P_\parallel$ implies that the synchrotron radiation is linearly polarized in the plane of motion of the electron, and the degree of linear polarization $\Pi(\nu)$ in the radiation is

$$\Pi(\nu) = \frac{P_\perp(\nu) - P_\parallel(\nu)}{P_\perp(\nu) + P_\parallel(\nu)} = \frac{G(x)}{F(x)}. \tag{2.48}$$

The degree of polarization is very little dependent on $\alpha$ and $\nu$. In most circumstances the degree of linear polarization is around 70%–75%. Strong linear polarization, therefore, is the hallmark of synchrotron radiation. The observation of strong polarization in the radiation from the Crab Nebula supernova remnant was the confirmation that the mechanism was synchrotron.

The total power radiated by a relativistic electron moving at an angle $\alpha$ to the direction of the field $\mathcal{B}$ is

$$P = \frac{2}{3} \frac{e^4 \gamma^2 \mathcal{B}^2 (\sin\alpha)^2}{m^2 c^3}. \tag{2.49}$$

An assemblage of electrons will have some distribution in pitch angle, so an average over all angles must be taken. If the electrons all have the same energy, but are distributed isotropically, the average of the factor $(\sin\alpha)^2$ is $2/3$.

The loss of energy by the electron may be considered as a collision process, in which an electron with the classical Thomson cross-section $\sigma_T$ encounters a magnetic field with energy density $U_\mathcal{B} = \mathcal{B}^2/8\pi$. The Thomson cross-section is $\sigma_T = (8\pi/3)r_0^2$ ($r_0$ is the classical electron radius $e^2/mc^2$). From Eqn 2.49, with an isotropic electron distribution a highly relativistic electron radiates a power

$$P = \frac{4}{3} \gamma^2 c \sigma_T U_\mathcal{B}. \tag{2.50}$$

From the energy loss rate, Eqn 2.50, the lifetime of a relativistic electron follows. For large $\gamma$ the electron energy decays by half in a time

$$t_{1/2} = mc^2 \left( \frac{4}{3} \gamma c \sigma_T U_B \right)^{-1} \tag{2.51}$$

or, in numerical terms,

$$t_{1/2} = \frac{16.4}{\mathcal{B}_\perp^2 \gamma} \text{ yr.} \tag{2.52}$$

Here $\mathcal{B}_\perp$ is in gauss. Note that $\mathcal{B}_\perp$ is the perpendicular component; if the field direction is randomly oriented for an assembly of electrons, the average lifetime is $25\mathcal{B}^{-2}\gamma^{-1}$ yr. Thus, a 10 GeV galactic cosmic ray electron in a 3 microgauss field has a lifetime of about $10^8$ yr.

### 2.7.1 A Power-Law Energy Distribution

Up to this point, we have considered the radiation from a single relativistic electron. We now consider a relativistic gas with a number-density distribution in energy $N(E)$; we assume it has an isotropic pitch-angle distribution. The energy distribution of cosmic rays, and apparently also of the radiating electrons in many other synchrotron sources, is in the form of a power law:

$$dN(E) = CE^{-p}dE, \tag{2.53}$$

where $p$ is the *spectral index* and C is a constant. The spectrum emitted by a single electron was given in Eqn 2.44, which can be convolved with the energy distribution. Following Ginzburg and Syrovatskii (1969), with a power-law distribution and an isotropic pitch angle distribution, the specific emissivity, i.e. the power emitted per unit volume per unit solid angle per unit frequency interval, can be written in closed form. A function $a(p)$ (given in Table 2.2) is usually used to simplify the power-law index dependency. For typical cases that are met with in radio astronomy, the function $a(p)$ is of the order of 0.1. The net result for the specific emissivity for a power-law electron distribution with spectral index $p$ and isotropic pitch-angle distribution is

$$j_\nu = \frac{e^3}{mc^2} \left( \frac{3e}{4\pi m^3 c^5} \right)^{(p-1)/2} C \mathcal{B}^{(p+1)/2} \nu^{-(p-1)/2} a(p). \tag{2.54}$$

For an optically thin source, the specific intensity follows on integrating along the line of sight (see Section 2.5 for this elementary solution of the equation of transfer). When numerical values are substituted, one obtains for uniform density and path length $L$ a specific intensity

$$I_\nu = 1.35 \times 10^{-25} C \mathcal{B}^{(p+1)/2} \left( \frac{6.26 \times 10^{18}}{\nu} \right)^{(p-1)/2} a(p)L \quad \text{W m}^{-2} \text{ sterad}^{-1} \text{ Hz}^{-1}. \tag{2.55}$$

Table 2.2. *The function a(p) as a function of energy spectral index p (Ginzburg and Syrovatskii 1969)*

| $p$ | 1 | 1.5 | 2 | 2.5 | 3 | 4 | 5 |
|---|---|---|---|---|---|---|---|
| $a(p)$ | 0.283 | 0.147 | 0.103 | 0.0852 | 0.0742 | 0.0725 | 0.0922 |

This shows that an ensemble of relativistic electrons with a power-law distribution in energy produces a synchrotron spectrum that is another power-law distribution:

$$dN(E) \approx E^{-p} \implies I(\nu) \approx \nu^{-(p-1)/2}. \tag{2.56}$$

Note that this result does not depend on a detailed integration of $F(x)$, but follows simply from the scaling given in Eqn 2.44; it may be derived by assuming simply that the synchrotron emission from an electron with relativistic factor $\gamma$ occurs only at frequency $\gamma^2 \nu_{\text{cyc}}$.

In later chapters we will find many examples where the observed flux density $S(\nu)$ of continuum radiation follows a power law $S(\nu) \propto \nu^{\alpha}$.[8] Equation 2.56 shows that $\alpha$ may be related to the electron energy spectrum by $\alpha = (1 - p)/2$. For most radio galaxies and quasars, $\alpha$ lies in the range $-2$ to $0$. Figure 2.11 shows examples, including for contrast the thermal spectra of the Sun and the Orion Nebula.

In Section 2.7.2 we will see that this power-law spectrum does not continue indefinitely to lower frequencies. At some point, depending on the physical conditions, the spectrum curves over as the source becomes opaque to its own radiation. We say that it has become self-absorbed.

When the relativistic electron gas is contained in an ordered magnetic field, the electrons in every energy range will produce linearly polarized radiation. The polarization analysis for a single electron, leading to Eqn 2.48, can now be extended for an energy power-law distribution of index $p$ by integrating over energy and pitch angle. The resulting polarization, for an isotropic pitch-angle distribution, is

$$\Pi = \frac{p+1}{p+7/3}. \tag{2.57}$$

For a typical spectral index of $-2.5$, the resulting degree of linear polarization will be 0.72, close to the value for a single electron. In practice the radiation observed is the sum over different regions with non-uniform magnetic fields; superposition effects along and across the line of sight within the observing beam can considerably reduce the degree of polarization.

---

[8] Note the sign convention for the spectral index $\alpha$; earlier work in radio astronomy, including the first edition of this text, used $S(\nu) \propto \nu^{-\alpha}$, similar to the convention for the cosmic ray power law. At long wavelengths most radio sources have synchrotron spectra which fall steeply with frequency; the earlier convention therefore gave the same sign for the cosmic ray and radio spectral indices. In contrast the positive convention is always used at millimetric and infrared wavelengths, where spectra that increase with frequency are often observed, and we recommend its use throughout the radio range.

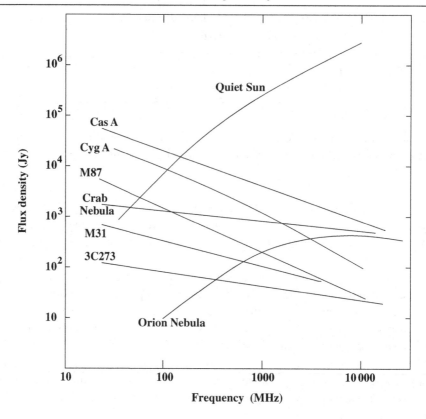

Figure 2.11. Spectra of synchrotron sources and two thermal sources (the Sun and the Orion Nebula).

We will see in later chapters that synchrotron radiation is important in radio astronomy wherever there are charged particles (usually electrons) with very high energies, moving in a magnetic field (which may be weak). The Milky Way galaxy is the prime example (Chapter 14), since it dominates the radio background at all but the shortest radio wavelengths. The most intense cosmic sources, the radio galaxies and quasars, emit by synchrotron radiation. Supernova remnants, such as the Crab Nebula, are powerful synchrotron sources, particularly when they are young, deriving their energy for a short time from the supernova explosion or for longer from a pulsar at their centres.

### 2.7.2 Synchrotron Self-Absorption

As we have seen in Sections 2.4 and 2.5, the specific intensity of radiation from a source in thermodynamic equilibrium cannot exceed a limit set by the temperature of the source; at this limit emission and absorption are balanced along the line of sight. This must apply to synchrotron radiation, although in this case the source 'temperature' must be related to the kinetic energy $E$ of the radiating electrons, which are not necessarily in thermal equilibrium with their surroundings. The temperature is then a 'kinetic temperature' $T_k$. As we show in

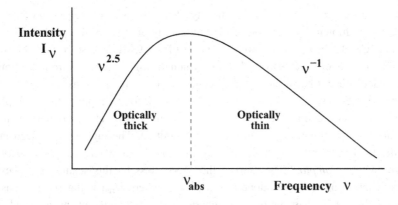

Figure 2.12. A radio spectrum, showing the effect of self-absorption. Below a turn-over frequency $\nu_{abs}$ the source becomes optically thick, and the spectral index approaches 2.5. In the optically thin region the index, shown here as $-1$, depends on the energy spectrum of the radiating electrons, as in Eqn 2.56.

Section 2.5, the emission must then be balanced by an inverse process of absorption, which in this case is known as *synchrotron self-absorption.*

The processes of the radiation and self-absorption of sychrotron radiation are spread over a range of energies and frequencies, as seen in the spectral function of a single electron (Fig. 2.10), but it is nevertheless a useful approximation to assign a single energy $E$ to each frequency, as in our discussion of $\nu_{crit}$ in relation to the power spectrum. Then the electrons radiating at frequency $\nu$ have the same energy, and equivalent temperature $T \propto \nu^{1/2}$. Since the spectrum of optically thick radiation from a source in full thermodynamic equilibrium is $S \propto T\nu^2$, the spectrum of a synchrotron source has an index of 2.5 instead of 2.0. (For detailed calculations see Ginzburg and Syrovatskii 1969.) Figure 2.12 shows the effect of self-absorption on a typical synchrotron spectrum with index $-1$, which becomes 2.5 below a turnover frequency $\nu_{abs}$.

Synchrotron self-absorption is met with in compact radio sources; this is discussed in Chapter 16. Such sources can be dominated by several emitting regions of different sizes, whose spectra become self-absorbed at different freqencies. These can add together to produce a relatively flat composite spectrum over a wide frequency range.

## 2.8 Inverse Compton Scattering

In classical Compton scattering, a high energy X-ray photon collides with a low energy electron, resulting in the transfer of energy from the photon to the particle and a consequent reduction in the photon energy. The opposite situation often occurs in astrophysics, when a relativistic electron (with Lorentz factor $\gamma$) transfers energy to a low energy photon; this is *inverse Compton scattering.* To first order the scattered photon is boosted up in frequency by a factor $\gamma^2$; thus for Lorentz factors of $10^{3-4}$ radio photons ($\nu \sim 10^{8-10}$ Hz) can be boosted into the ultraviolet and X-ray bands ($\nu \sim 10^{15-18}$ Hz).

This is an important process for cosmic ray electrons, whose energy may be limited by losses to the ambient background of the low energy photons of the cosmic microwave background (CMB). Inverse Compton scattering of the CMB is also responsible for the Sunyaev–Zel'dovich effect (Section 17.11), in which the high energy electrons concentrated within a galactic cluster boost the energy spectrum of the CMB photons.

Within a high energy source of synchrotron radiation, for example in the inner parts of the jet in AGN (see Sections 16.3.10 and 16.5) the relativistic electrons find themselves in a sea of photons which they themselves have emitted. The high energy electrons can then scatter the lower energy ('soft') photons, in a process now termed *synchro-Compton* or *synchroton self-Compton*. The ratio of the energy loss by this scattering process and that from synchrotron radiation alone is $U_{rad}/U_{mag}$, where $U_{rad}$ is the energy density of the photon field and $U_{mag}$ is that of the magnetic field in which the electrons spiral. When $U_{rad}/U_{mag} \geq 1$, synchrotron self-Compton becomes the dominant process in energetic radio sources to the extent that it places an upper limit to the brightness temperature of an emitting region. This can be qualitatively understood since bright sources must have high photon densities; the effect is further discussed in Section 16.3.10.

Analyses of inverse Compton radiation may be found, for example, in the books by Condon and Ransom (2016), and by Rybecki and Lightman (1979).

## 2.9 Further Reading

*Supplementary Material* at www.cambridge.org/ira4.
Condon, J. J., and Ransom, S. M. 2016. *Essential Radio Astronomy*. Princeton University Press.
Rybicki, G. B., and Lightman, A. P. 1979. *Radiative Processes in Astrophysics*. Wiley.

# 3

# Spectral Lines

Radio spectral lines are seen as distinct features at precise frequencies in the broad spectra of many radio sources. They are essentially quantum phenomena, in contrast with the classical electrodynamics of synchrotron and free–free radiation. Their frequencies reveal their origins in atoms and molecules, in both ionized and neutral gas clouds, in the interstellar medium, in stellar atmospheres, and in molecular clouds. Their widths, and the Doppler shifts in their centre frequencies, are diagnostic of local temperatures and dynamics. Furthermore, spectral lines are observed in some distant galaxies, some with large redshifts, providing insights into the early development of structure in the Universe.

The galactic continuum radiation discovered by Jansky and mapped by Reber aroused little interest in astronomers used to the rich information provided by visible spectral lines. When Reber's 1940 paper finally reached the Leiden Observatory in 1943, in war-torn Holland, the director, Jan Oort, realized its significance and more particularly recognized the potential impact of a radio line, if one existed. He asked his student, Henk van de Hulst, to see if there might exist an observable line. His response was that the most common element in the Universe, hydrogen, had such a possible line in its ground state, at a wavelength of 21 cm. The line was first detected at Harvard by Ewen and Purcell in 1951; Purcell knew that the Leiden group was on the verge of detecting the line and generously proposed that they should publish jointly. This was followed shortly afterwards by another successful detection by the group in Australia.

Predictions of other spectral lines at radio frequencies were made later by Townes (1957) and by Shklovskii (1960). In 1963 Weinreb *et al.* discovered spectral line emission from OH (hydroxyl) at 1665 and 1667 MHz. Lines from $H_2O$ (water), $NH_3$ (ammonia), and $H_2CO$ (formaldehyde) were discovered shortly afterwards, and when the Kitt Peak 11 m telescope of the US National Radio Astronomy Observatory (NRAO) began operations in 1970, a veritable flood of new discoveries came: at the time of writing (2018), approximately (two hundred) molecular species have been detected in the ISM, and the list is lengthening rapidly (Müller *et al.* 2005; Endres 2016).[1]

There are three main types of low energy transition between quantized energy levels in atomic and molecular gases, giving rise to photons in the radio spectrum. These are due to recombination (an extension to high quantum numbers of the familiar optical series),

---

[1] An up-to-date list is maintained at the Cologne Database of Molecular Spectroscopy
www.astro.uni-koeln.de/cdms/molecules.

hyperfine structure within atoms, and quantized rotation in molecules. Rotational transitions can occur within excited vibrational states, or there can even be combined (ro-vib) transitions at mm and shorter wavelengths. In 'open shell' molecules the phenomenon of lambda-doubling is often important (Section 3.4).

Spectral lines may be seen in emission or absorption. For atoms to be in a suitable ground state, or for molecules to exist, the gas temperature must be less than a few thousand kelvins. If in thermal equilibrium, this gives relatively low brightness temperatures; the narrow spectral widths further increase the observational threshold. This means that cm-wave thermal emission lines are usually only detected from regions large enough to fill an observing beam of a few arcsec or larger, e.g. a single dish or the VLA in a compact configuration. Since, for a given brightness temperature, the flux density is proportional to $\nu^2$, mm and sub-mm thermal lines can be imaged at higher resolution, but non-thermal emission will always be brighter. This can be maser emission (Section 3.7) or thermal line absorption against a non-thermal background, such as H I absorption against a bright source. Absorption can only be detected where there is a suitable background, so it does not give a complete picture of a target.

## 3.1  Radio Recombination Lines

Following the Bohr atomic model, a neutral hydrogen atom may be thought of as a single electron in orbit round a more massive proton, the radius of the orbit, and with it the atomic energy, being constrained by a simple quantum rule to a series of values designated by a quantum number, $n$; the closest orbit, with the lowest energy, is designated $n = 1$. A spectral line photon is emitted at an electronic transition from a higher to a lower energy orbit.

In ionized hydrogen there is a continuous process of recombination and detachment, in which electrons cascade down through quantized energy levels and are reionized by absorbing energy from collisions and photons. The familiar Lyman and Balmer series at optical wavelengths are lines originating in transitions between electron orbits close to the proton, i.e. with small quantum number. The recombination process starts from states of much higher quantum number $n$, generating an extensive set of radio lines, generically known as radio recombination lines (RRLs). Just as the Lyman series of transitions to the ground state is designated Ly$\alpha$, Ly$\beta$, etc. for $\Delta n = 1, 2 \ldots$, transitions ending on state $n$ are designated $n\alpha$, $n\beta$, etc. The formula for the frequency of hydrogen lines (frequency rather than wavelength is more appropriate for the radio spectrum) is, for transitions from state $n_2$ to $n_1$,

$$\nu = 3.289842 \times 10^{15} \left( \frac{1}{n_1^2} - \frac{1}{n_2^2} \right) \left( 1 + \frac{m_e}{M} \right)^{-1} \text{Hz.} \tag{3.1}$$

The numerical factor on the left is the *Rydberg constant* expressed in hertz. The factor in parentheses on the right allows for the finite mass of the proton; it is the *reduced-mass correction* for electron mass $m_e$ and total atomic mass $M$. The high quantum numbers involved in radio recombination lines correspond to orbits with remarkably large radii. The

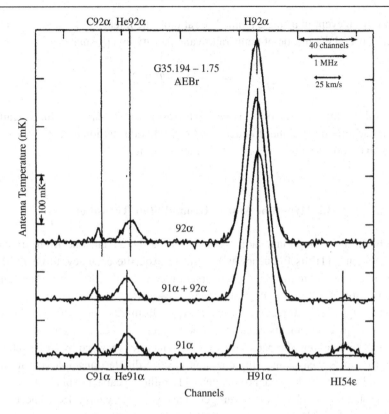

Figure 3.1. Radio recombination lines 91$\alpha$ and 92$\alpha$ from carbon, helium, and hydrogen in the galactic H II region G35.194-1.75. (From Quireza *et al.* 2006.)

mean radius for an $n = 1$ orbit is $\sim 0.5 \times 10^{-10}$ m; the radius varies as $n^2$, so that for $n \approx$ 100 the atom has a size of about three microns. The RRLs are generated by an electron so far removed from the nucleus that it is effectively a classical dipole radiator. These lines are only found in low density regions of the ISM, where collision rates are low enough for orbitals with low binding energy to survive.

The first detection of a recombination line, the 109$\alpha$ hydrogen line at 5009 MHz, was claimed by Dombrovskii and confirmed by Hoglund and Mezger (1965). As radio telescopes and receivers improved, the weaker lines of heavier hydrogen-like recombining atoms, also at frequencies given by Eqn 3.1 but shifted up from the hydrogen frequencies by the appropriate reduced-mass factor, were also detected. An example of adjacent recombination lines at 8.6 GHz in carbon, helium, and hydrogen, all three designated 92$\alpha$, is shown in Figure 3.1. The lines of hydrogen-like helium and carbon are well separated from the hydrogen lines. The adjacent 91$\alpha$ lines, also shown in Figure 3.1, have the same profiles.

Radio recombination lines and free–free emission may be observed from the same region. The two intensities have a different dependence on electron temperature $T_e$ and electron density $n_e$, since free–free is generated by collisions between ions and electrons rather than by an event in a single atom. The ratio of line intensity to free–free continuum intensity is a useful diagnostic of temperature and density in a plasma:

$$\frac{I_{\text{line}}}{I_{\text{continuum}}} \propto \nu^{2.1} n_e^{-1} T_e^{-1.15}. \tag{3.2}$$

As a result of the $\nu^{2.1}$ factor, it is easier to discern recombination lines against the continuum background at higher frequencies. For further discussion on RRLs see the book by Gordon and Sorochenko mentioned in Further Reading.

## 3.2 Hyperfine Atomic Ground-State Transitions

Historically the most important radio spectral line is the 21 cm line from ground-state atomic hydrogen (H I). Its frequency is known, to exquisite accuracy, to be 1420.40575 MHz. The line arises from the magnetic interaction between the electron and proton spins; a photon is emitted when the alignment flips between the two quantized states, parallel (higher energy) and antiparallel (lower energy). Both electron and proton have spin and possess magnetic moments; their magnetic moments combine to give a total angular momentum with quantum number $J$ either 0 or 1. The line is formed by a magnetic dipole transition $10^4$ times weaker than the common electric dipole transition, and its lifetime is correspondingly very long, approximately 11 million years. Despite this very small transition probability, and the low hydrogen density in the galaxy ISM, the long lines of sight through the Milky Way galaxy provide sufficient optical depth for the line to be observable in the whole of the galactic plane.

Despite the apparent simplicity of the transition the interpretation of H I observations is non-trivial because of the different ways in which an atom can be excited and de-excited, and the different phases in which the ISM gas is found (for an introduction see Kulkarni and Heiles 1988). In principle, by observing in optically thick directions the observed brightness gives a measure of the temperature of the ISM. This is so because in local thermal equilibrium (LTE) the number of transitions is proportional to the Boltzmann factor $e^{-\Delta E/kT}$, where $\Delta E$ is the difference in energy ($5.87 \times 10^{-6}$ eV) between the two spin states; the temperature thus derived is therefore called the *spin temperature*. It turns out that in the ISM the excitation to the upper state is mainly due to collisions between neutral atoms, which for any such atom typically occurs every few hundred years. This is sufficient to ensure that the spin temperature approximates to the kinetic temperature of the gas; however, the line may be observed from regions with different temperatures and densities (see e.g. Kalberla and Kerp (2009)) and this complicates the interpretation. The 21 cm line can also be seen in absorption against a bright continuum source but the signal will have a contribution from emission, so again interpretion is not always straightforward (see e.g. Dickey and Lockman (1990)). It is easier to use Doppler shifts to study the motion of the emitting gas, and 21 cm line studies have proven to be a powerful tool for studying galactic structure in general. This is described in Chapter 14 in the specific case of the Milky Way but many nearby galaxies can be imaged in H I and their kinematics investigated.

There are other examples of hydrogen-like hyperfine lines.[2] Deuterons have spin 1, and so deuterium can have either $J = 3/2$ or $J = 1/2$, giving a line at 327 MHz. This deuterium line was detected by Rogers *et al.* (2007), demonstrating that the cosmic abundance of deuterium is two orders of magnitude lower than the terrestrial abundance (see a discussion by Weinberg (2017) for the relation to the primordial abundance). Other hyperfine transitions are in principle observable in the ISM, but are far weaker; one hyperfine line, of singly ionized He[3] II, has been detected in highly excited H II regions (Rood *et al.* 1984).

The $^2P_{3/2} - ^2P_{1/2}$ fine structure line of singly ionized carbon at 157.74 μm, C II, is one of the main coolants of diffuse or heated gas found at, for example, the edges of molecular clouds. At high redshifts it can be imaged at (sub-)mm wavelengths, for example at $\sim 350$ GHz for $z \sim 4.5$, and it is an invaluable tracer of the dynamics and properties of early galaxies.

### 3.3 Rotational Lines

The energy of a spinning molecule must be quantized; transitions between these rotational energy levels may emit or absorb a photon in the radio range. Homonuclear diatomic molecules such as molecular hydrogen and nitrogen, together with other symmetric molecules such as carbon dioxide and methane, generally have neither electric nor magnetic dipole moments and therefore give rise to no observable lines in the radio spectrum.[3] Molecules that have a dipole moment undergo transitions from one rotational state to another, subject to the rules of quantum mechanics. Diatomic molecules such as carbon monoxide, CO, offer the simplest case. Classically the moment of inertia $I$ for a molecule with masses $m_1$ and $m_2$ is $I = \mu R^2$, where $R$ is the bond length and $\mu = m_1 m_2 / (m_1 + m_2)$ is the *reduced mass* (as in Section 3.1). Then the angular momentum $L = I\omega$, where $\omega$ is the rotational angular frequency, and the rotational energy is $\frac{1}{2}I\omega^2 = L^2/2I$. When quantized, the square of the total angular momentum of the molecule takes the value $J(J + 1)\hbar^2$, where $\hbar$ is Planck's constant divided by $2\pi$ and $J$ is a rotational quantum number. Thus the allowed rotational energy levels for a diatomic molecule are $E_J = J(J + 1)\hbar^2/2I_J$. More massive molecules rotate more slowly, and hence the gaps between energy levels are smaller, giving lines at lower frequencies.

The selection rule for a transition of a polar molecule radiating a photon having one unit of angular momentum is $\Delta J = \pm 1$ (thus conserving angular momentum). The transition

---

[2] See Wilson, Rohlfs, and Hüttemeister (2013), *Tools of Radio Astronomy.*
[3] Molecular oxygen ($O_2$) is an exception, since it has a magnetic dipole moment in its ground vibrational state. It has a rich spectrum around 60 GHz, and narrower lines such as at 119 GHz, which makes these spectral regions unobservable from the ground owing to the high $O_2$ abundance in the Earth's atmosphere, but it has proved surprisingly elusive even from satellite observatories, probably because its reactivity leads to a short lifetime in the gas phase. The molecule $O_2$ was detected for the first time outside the solar system, by the spacecraft Herschel, at 487, 774, and 1121 GHz, from the Orion Nebula (Goldsmith *et al.* 2011).

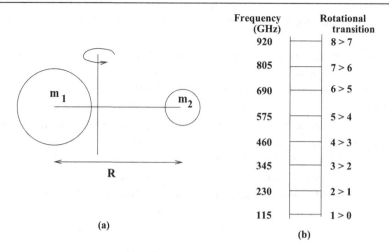

Figure 3.2.  (a) Diatomic molecule $^{12}C^{16}O$, with (b) a ladder of rotational lines.

from state $J$ to $J-1$ corresponds to an energy change $E_J \rightarrow E_{J-1} = BJ(J+1)-B(J-1)J = 2BJ$, where the rotational constant $B = \hbar^2/2I$ Hz.[4]

The successive transitions downward in $J$ and thus in frequency provide a ladder of transitions that, because of their different lifetimes, give information about the physical conditions in molecular clouds. The common molecule $^{12}C^{16}O$ exhibits such a ladder in the millimetre and submillimetre regions of the spectrum (see Figure 3.2). The lowest frequency transition, $J(1 \rightarrow 0)$ to the ground state occurs at approximately 115 GHz (2.6 mm), the $2 \rightarrow 1$ transition at 230 GHz, and so on. The less abundant isotopic variation, $^{13}C^{16}O$, has a larger moment of inertia and so the ground-state transition occurs at a lower frequency, 110 GHz. Linear molecules composed of more than two atoms, such as hydrogen cyanide, HCN, also are characterized by a single moment of inertia, and the dipolar formalism also applies.

There are rotational lines from many molecules that have more complicated non-linear structures. Symmetric rotor (or 'top') molecules such as ammonia, $NH_3$, have one moment of inertia for rotation about the symmetry axis and two separate but equal moments of inertia for rotations perpendicular to the symmetry axis. Thus for a state characterized by total angular momentum quantum number $J$ there are $|K| \leq J$ possible projections of the angular momentum along the symmetry axis, $2J + 1$ in all. A given projection $K\hbar$ can be in either direction so for $K \neq 0$ the proper state description is given by the symmetric and antisymmetric superpositions of the two possible orientations. The pair is called a $K$-doublet and in the absence of perturbations the states have the same energy. Ammonia is, however, subject to 'inversion doubling', which results in a slight splitting of these degenerate levels (see Section 3.4).

---

[4] A complication in the above theory is that rotation will stretch the atomic bond, and so the moment of inertia $I_J$ can depend on the rotational state. See the treatment by Wilson, Rohlfs, and Hüttemeister (2013), *Tools of Radio Astronomy*.

Most of the common non-linear molecules in the ISM are asymmetric rotors, with three different principal moments of inertia. The water molecule $H_2O$, with its boomerang shape, is a familiar example. Asymmetric rotors have only the total angular momentum $J$ as a good quantum number, although the levels can be characterized by an asymmetry parameter. For small asymmetry, each $K$-doublet is split into two separate energy levels. As the asymmetry parameter grows larger, there develops a continuous structure of these levels, so they are characterized by states designated $J_{kk}$.

For water, every $J$ state has $2J + 1$ substates, expressed in the form $J_{ab}$ where each $a$ and $b$ varies from 0 to $J$ (there is only one $J_{00}$ level). There are in addition two families of energy levels, ortho and para, based on the relative spins of the H atoms. Water in the Earth's atmosphere absorbs strongly at transitions between energy levels that correspond to a few hundred K or less. The lowest frequency transition well studied from Earth is at 22 GHz, in the ortho family, designated $6_{16}$–$5_{23}$, with an upper energy level of 643 K; its observation is possible because it is a strong maser (see Section 3.7). The high abundance of water in astronomical molecular enviroments provides dozens of detectable (sub-)mm thermal and maser lines; the lowest energy state detectable from the ground is the 183 GHz line (200 K).

## 3.4 Degeneracy Broken by Rotation

In many cases a molecule has two states that would have equal energy, but the degeneracy is broken by molecular rotation and a photon can be emitted in a transition between the states. The first molecule to be detected as a radio line was of this type, the hydroxyl radical OH, discovered in absorption at a wavelength of 18 cm against the radio continuum of Cas A (Weinreb *et al.* 1963). An OH molecule has an electric dipole moment and can rotate with angular momentum $J$. The angular momentum of its single unpaired electron, projected along the molecular axis, must be quantized. This projection is designated by $\Lambda$, with the value $\Lambda = 1$ for OH. There are two possible states, with the electron angular momentum either in the direction of the rotation or opposed; in the absence of rotation these have the same energy. The molecular rotation breaks the degeneracy; the resulting pair of states is called a *lambda doublet*. In addition, the magnetic moment of the electron couples with the proton's magnetic moment to give a pair of hyperfine states for both levels of the doublet, giving four lines at around 18 cm wavelength. In units of frequency, these are found at 1612, 1665, 1667, and 1720 MHz. In many H II regions these lines appear in emission, but greatly enhanced by maser action, as discussed in Section 3.7. Figure 3.3 shows the lowest four hyperfine levels for the $^2\Pi_{3/2}$ state.[5]

Another example in which degeneracy is broken by rotation is the oxygen molecule $O_2$. This molecule has an electron spin of 1, and hence has a net magnetic moment.[6] If the

---

[5] For an explanation of the spectroscopic notation and more details of the energy levels in OH, see Section 5.5 in Gray (2012).

[6] Contrary to what one might naively expect, the electrons do not pair each other off completely, so it is in a triplet state.

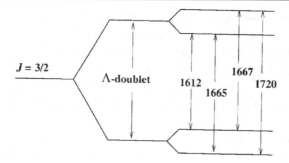

Figure 3.3. Energy levels of OH, showing $\Lambda$-doubling and hyperfine splitting. The transition frequencies are in MHz.

molecule could not rotate, its energy would be the same no matter which way along the molecular axis the magnetic dipole was pointing. This degeneracy is lifted when molecular rotation is taken into account. The actual frequency of the transition will depend upon the rotational state of the molecule. Thus, there is a rich cluster of lines in the region around 55 GHz (this makes the atmosphere opaque at around 60 GHz; see the diagram of atmospheric absorption in Figure 4.3).

The ammonia molecule, $NH_3$, presents another type of degeneracy lifting, called inversion doubling. The molecule is pyramidal in form and the nitrogen atom can tunnel through the triangle of hydrogen atoms, and so its ground-state vibration will consist of a continual inversion, in what has been described as an oil-can mode. As a result, the ground-state vibrational state can be regarded as a linear superposition of two states, one odd and one even and, because the rotation lifts the degeneracy, there is an electric-dipole transition between the two, at about 25 GHz; this was discovered by Cheung *et al.* (1969). The energy depends upon the rotational state, so, just as in the case of OH, there is a rich set of lines.[7] A somewhat different case is presented by methanol, $CH_3OH$. The OH group sits on top of the methyl pyramid, and the H atom angles off to the side. There is thus an allowed internal rotation, influenced by the three-fold potential well of the methyl group. Townes and Schawlow (1955) gives a more detailed description of the allowed states.

Ammonia is a good example of the complexity of the transitions of even a simple molecule and how this can provide information (Longmore *et al.* 2007). The total angular momentum of its component atoms $\sqrt{J(J+1)}\hbar$ can, but does not have to, lie parallel to the principal molecular rotation axis, and the component H spins can be antiparallel or parallel. The $K = J$ states (see Section 3.3) are known as metastable tops (shaped like a child's spinning top, or a shuttlecock) and the energy difference for inversion through the plane of the H atoms (the 'feathers') is small, giving a ladder of transitions starting around 24 GHz and separated by tens of MHz for $J = K = 1, 2, 3, 4, 5, \ldots$[8] These are collisionally excited, providing an indication of the gas kinetic temperature. Moreover,

---

[7] See Wilson, Rohlfs, and Hüttemeister (2013), *Tools of Radio Astronomy*.
[8] See Section 5.7 in Gray (2012).

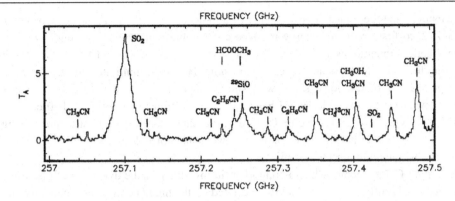

FREQUENCY (GHz)

FREQUENCY (GHz)

Figure 3.4. Spectrum of emission from a molecular cloud, showing many spectral lines emitted by molecules. (From Blake *et al.* 1986.)

the electric quadrupole moment of the larger N atom gives an easily resolved hyperfine structure of five components separated by just under 1 MHz ($\sim 10 \ km^{-1}$, much more than the Doppler width in a typical star-forming core); the central component is intrinsically brighter and any observed decrease in the ratio between it and the weaker hyperfine lines provides the optical depth of a gas cloud (in thermodynamic equilibrium).

## 3.5 Detected Lines

The wide variety of molecules that occur in the ISM offers many opportunities to determine local densities and temperatures. Astrochemistry has developed as the study of molecular lines has progressed, and powerful new instruments, notably the large synthesis array ALMA and several large single mm–submm telescopes have come into use.

The number of molecules that have now been identified through their radio spectral lines is continuously increasing; Figure 3.4 shows the rich spectrum typical of a dense molecular cloud. Many molecules have several detected lines, and in the stronger lines it is often possible to detect lines from isotopic species, such $^{13}CO$ or $C^{18}O$. A number of the simpler molecules with hydrogen are observed in deuterated form, such as HDO, DCN, and the deuterated forms of ammonia, especially $NDH_2$. The less abundant isotopic species can be especially valuable when, as for CO, the lines of the principal species are optically thick; also, when there is a question of whether local thermodynamic equilibrium is present, the lines of the isotopic species can provide a cross-check on this.

More than 200 molecular species have now been identified in the ISM and the solar system, even extending to structures as large as $C_{60}$, buckminsterfullerene (see Herbst and van Dishoeck 2009). Which species are observed and the relative intensities of their lines depends on the temperature of the emitting gas cloud. The rotational states are collisionally excited and the more kinetic energy in the collision, the higher the $J$-level which can be reached. In thermal equilibrium the molecular velocities follow a Maxwell–Boltzmann distribution, and the typical kinetic energy of a molecule near the peak of the distribution is $3kT/2$; to first order this energy must be comparable with that of the higher-$J$ level. The kinetic temperatures in molecular clouds are low, 10–30 K, and hence the slower rotating heavier molecules are the easier to excite. However, multi-atom molecules have a larger variety of rotational states with similar energies, so each individual transition may be weaker; for example $HC_3N$ has three detected transitions for $J = 1$–0 close to 9.1 GHz, six at $J = 2$–1 close to 18.2 GHz, etc. due to its hyperfine structure (giving different possibilities in spin–orbit coupling between the molecule as a whole and its atoms, governed by quantum mechanical rules).

The light CO molecule, which is important in investigating the dynamics of the Milky Way galaxy, is sufficiently abundant to be excited by the high energy tail of the Maxwell–Boltzmann distribution, even in low temperature clouds.

## 3.6 Linewidths

The intrinsic linewidths of most radio transitions are narrow, usually very much less than 0.1 Hz, and collisional broadening is negligible in typical low density astronomical environments (an exception is recombination radiation at very high $n$-numbers).[9] The line profiles are therefore determined by combinations of thermal and bulk velocities in the gas. For the non-relativistic velocities encountered in the ISM the simple Doppler formula $\Delta\nu/\nu_{em} = V/c$ is appropriate. Here $\Delta\nu = \nu_{em} - \nu_{obs}$ is the frequency shift away from the emitted or rest frequency $\nu_{em}$ and $V/c$ is the ratio of the gas velocity and the speed of light. By convention, a positive frequency shift, with $\nu_{em} > \nu_{obs}$, means that the velocity $V$ is away from the observer, so a redshift in the optical convention is a shift to lower frequencies. Line profiles are often described as a function of velocity, scaled simply from frequency; for the 21 cm hydrogen line at 1420 MHz the conversion is 4.74 kHz per $km\,s^{-1}$.

For an optically thin emitter, which is often the case in radio astronomy, the line profile $\phi(\nu)$ is determined only by the local velocity distribution of the emitting atoms or molecules, since it is the Doppler shift arising from the component of velocity along the line of sight that will shape the line (Figure 3.5 and Section 14.2). The line profile is usually written as a function normalized such that its integral over all frequencies is unity. For a source in thermal equilibrium at kinetic temperature $T$ the line profile is Gaussian and in one dimension can be written in terms of radial velocity $V_z$ and a width parameter $\sigma_t$:

$$\phi(V_z) = \frac{1}{\sigma_t\sqrt{\pi}} \exp\left[ - \left(\frac{V_z}{\sigma_t}\right)^2 \right].  \tag{3.3}$$

---

[9] Asymmetric molecules, where the larger atom has a large moment of inertia, may have 'blended hyperfine' splitting. In some cases, e.g. ammonia, this is larger than the typical linewidth in galactic sources (of order $1\,km\,s^{-1}$) but in others, e.g. AlO, multiple components for some transitions are blended to give the appearance of a single line spanning perhaps $10\,km\,s^{-1}$.

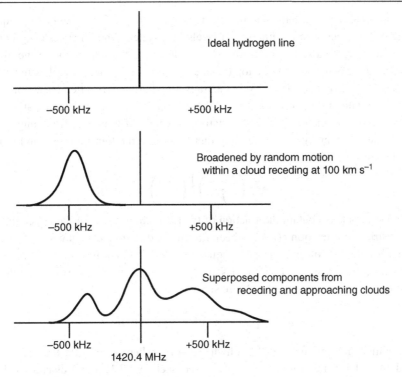

Figure 3.5. The shape and central frequency of a radio spectral line as observed from a cloud of gas is determined through the Doppler effect by the internal motions (here 40 km s$^{-1}$) and bulk velocity of the cloud.

The width parameter $\sigma_t = \sqrt{2kT/m}$ m s$^{-1}$, where $k$ is the Boltzmann constant and $m$ is the mass of the atom or molecule in kg. More practically, one can write $\sigma_t = 1.29\sqrt{T_{100}/m_H}$ km s$^{-1}$, where $m_H$ is the number of atomic mass units and $T_{100} = T/(100$ K$)$.

The thermal velocity distribution within the volume under observation may not determine the line profile, since there may be much larger turbulent velocities present and these may or may not have a Gaussian distribution. There may also be a systematic flow that results in an asymmetric or complex line shape, or there may be several different components centred on different velocities.

### 3.6.1 Line Emission and Absorption

The specific emissivity $j_\nu$ of an incoherent source, introduced in Chapter 2, determines the amount of energy $dE$ emitted into a solid angle $d\Omega$, in a direction $\hat{n}$, in a frequency interval $d\nu$, per time interval $dt$, from a volume element $dV$. It depends upon a variety of pseudo-temperature parameters. If the gas is reasonably close to kinetic equilibrium, the velocity distribution $f(v_z)$ will depend upon the kinetic temperature but the transition rates will depend upon the state temperature, discussed in Section 1.3 and defined by Eqn 1.1. If the transition is between the $i$th and $j$th levels of a system (the $i$th state being lower in energy), there may be a state temperature $T_i$ that describes the population of the $i$th state compared to the ground state and another state temperature $T_{ij}$ that describes the relative populations of the two states involved in the transition. Fortunately, often we are observing lines that terminate in the ground state, and there is then only one state temperature, $T_s$. If the system partition function[10] is $\mathcal{Z}$, the statistical weight of the upper (radiating) state is $g_i$, and if it lies at an energy $\epsilon_i$ above the ground state and the transition probability is $A_{ij}$, the specific emissivity is

$$j_\nu = \left[ \frac{g_i}{\mathcal{Z}} \exp\left( -\frac{h\nu_{ij}}{kT_s} \right) \right] \left( \frac{h\nu_{ij}}{4\pi} \right) A_{ij} n\phi(\nu). \tag{3.4}$$

The square brackets contain the fraction of the population that is in the upper state; the velocity distribution function $f(v_z)$ has been transformed to frequency units, so that $n\phi(\nu)$ is the number density of emitters per frequency interval. For a transition to the ground state, provided that the emitted photon has an energy that is small compared with the state temperature and when no higher levels are significantly populated, Eqn (3.4) simplifies to

$$j_\nu = n \left( \frac{g_1}{g_1 + g_0} \right) \left( \frac{h\nu}{4\pi} \right) A_{ij}\phi(\nu). \tag{3.5}$$

These conditions are met for the 21 cm hydrogen line, where $g_1 = 3$ for the upper $F = 1$ state and $g_0 = 1$ for the ground $F = 0$ state (see Section 3.7.1). For a discrete hydrogen cloud that is optically thin, the resulting brightness temperature follows from Eqn 2.21, which is one of the elementary solutions to the radiative transfer equation:

$$T_b = \frac{3}{32\pi} \left( \frac{hc\lambda}{k} \right) A_{10} N_\nu, \tag{3.6}$$

where $N_\nu$ is the surface density of hydrogen atoms per Hz along the line of sight (i.e. the integral of $n\phi(\nu)$ along the ray path). Note that the result is independent of state temperature for this approximation, since $h\nu \ll kT$.

When the medium is not optically thin, the optical depth $\tau_\nu$, defined by Eqn 2.23, must be calculated from the absorption coefficient $\kappa_\nu$. This follows from Kirchhoff's law,

$$\frac{j_\nu}{\kappa_\nu} = I_\nu(T), \tag{3.7}$$

so that, with state temperature $T_s$,

$$\tau_\nu = \int \kappa_\nu \, dz = \frac{\lambda^2}{2k} \int \frac{j_\nu}{T_s} \, dz. \tag{3.8}$$

---

[10] The partition function $\mathcal{Z}$ normalizes the probability for each state; it is the sum of the relative probabilities of finding a system in a particular state, taken over all posssible states of the system. Thus, if a system has degeneracy $g_i$ and energy $\epsilon_i$ above the ground state, $(g_i/\mathcal{Z}) \exp(\epsilon_i/kT)$ is the properly normalized probability of finding the system in that state.

The emissivity can be taken from Eqn 3.5 for any two-state system that satisfies its requirements. For the special case of the 21 cm hydrogen line, $g_1/(g_1 + g_0) = 3/4$, as noted above. The more general expression Eqn 3.4 is needed for most molecular systems in thermal equilibrium. This is seldom the case in the interstellar medium, where the state distribution frequently cannot be described in terms of a single state temperature. In such cases, due caution must be exercised.

For the 21 cm line, when a single state temperature holds along the ray path the optical depth is related to the population of hydrogen atoms by

$$\tau_v = \frac{N_v}{1.835 \times 10^{18} T_s}. \tag{3.9}$$

In this expression $N_v$ is the number of atoms per square cm, per km s$^{-1}$, along the line of sight. The approximation is good for estimation purposes, but in the interstellar medium the temperature is not uniform; if the density and/or the state temperature varies along the line of sight (in H atoms per cubic cm per km s$^{-1}$) and at velocity $v$ is $\rho_v$, the integral of $\rho_v/T_s$ must be taken along the line of sight.

## 3.7 Masers

The maser phenomenon was first observed in radio astronomy during the period 1963–1967, when studies of the 18 cm hydroxyl (OH) absorption lines towards the galactic plane showed a non-equilibrium ratio. Emission from near the W49 H II region, detected in 1965, had a specific intensity far too great to be generated by a thermal source, in which the quantized energy levels are populated according to the Boltzmann distribution. In thermal equilibrium the expectation is that the linear absorption coefficient $\kappa_\nu$ will be positive, giving the rate of attenuation of the specific intensity along a ray path. The probability of induced absorption is exactly the same as the probability of induced emission and, since the energy-level populations are greater for the lower-lying states in any system, the induced absorptions predominate. If a non-equilibrium distribution is introduced, however, this may no longer be the case since a population inversion (a negative state temperature) can occur and the induced emission transitions will be greater than the absorptions. The specific intensity will grow along the ray path and the 'absorption' coefficient will be negative. Such a system is known as a *maser* (an acronym for microwave amplification by stimulated emission of radiation; the equivalent for light is a *laser*).

Maser action requires a 'pump' to maintain the population of the upper energy level above the equilibrium thermal level (a population inversion). The pump may be a source of radiation, or the mechanism may be collisional, but in order to satisfy the second law of thermodynamics the equivalent temperature of the pumping process must be higher than the temperature corresponding to the energy difference of the energy-level pair that serves as the maser.

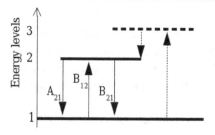

Figure 3.6. Masing can occur if level 2 becomes overpopulated with respect to level 1. Level 3 represents an additional level which provides a route for level 1 to be depopulated and/or level 2 to gain population. The transitions between levels 1 and 2 are labelled with their Einstein coefficients.

A two-level model of molecular energy levels can be used to explain an equilibrium system with populations $N_1$ and $N_2$ in the lower and upper states (see Figure 3.6). The photon energy corresponding to the energy difference will be $h\nu_{12}$ and the probabilities of transitions are given by the Einstein coefficients: for emission, $A_{21}$ for a spontaneous transition from the upper to the lower state and $B_{21}$ for a stimulated such transition; $B_{12}$ for absorption. Then $g_2 B_{21} = g_1 B_{12}$, where $g_1$, $g_2$ are the statistical weights of the lower and upper transitions. The equation of radiative transfer, given in its basic form in Eqn 2.21, can now be written, for the line centre frequency, in the form

$$\frac{dI}{ds} = h\nu_{12} \left[ (B_{12}N_1 - B_{21}N_2)\frac{I}{c} + \frac{A_{21}N_2}{4\pi} \right]. \tag{3.10}$$

In thermal equilibrium, $N_1 > N_2$ and $I \propto I_0 e^{-\kappa s}$, where the absorption coefficient $\kappa$ is positive so that $I < I_0$, i.e. absorption occurs, as explained in Section 7.1. Masing requires the inversion of this ratio via a third energy level, to increase the net population of level 2 with respect to level 1. Figure 3.6 represents this (in practice a complex series of transitions through more than three levels may be involved).

We consider a simple system in which the pumping mechanism populates the upper masing level at a rate $R$ producing a strong maser and at a low temperature such that the spontaneous transition rate is negligible in comparison, so we can ignore the $A_{21}$ term in Eqn 3.10. However, collisions between molecules can still induce transitions at a rate $C$ between states 1 and 2. In this approximation the equilibrium populations of states 1 and 2 will be determined by equating the rates of transition from the upper and lower states:

$$N_1(B_{12}I + C) \approx N_2(B_{21}I + C) - R. \tag{3.11}$$

Solving this expression for the population difference and substituting in Eqn 3.10 gives

$$\frac{dI}{ds} \propto -I \left( h\nu_{12}\frac{BR}{C + BI} \right), \tag{3.12}$$

where we assume that states 1 and 2 have identical statistical weights, so $B_{21} = B_{12} = B$. The solution of this equation gives the specific intensity along a ray path $s$:

$$I \propto I_0 e^{-\kappa_{21}s}, \tag{3.13}$$

where the negative absorption coefficient $\kappa_{21}$ is given by the expression in parentheses on the right-hand side of Eqn 3.12, so the net exponent is positive and $I > I_0$. If the pump rate $R$ is much greater than the stimulated emission rate then

$$\kappa_{21} \propto h\nu_{12}BR. \tag{3.14}$$

In this regime, the maser is said to be unsaturated. If the collision rate $C$ is much greater than the stimulated emission rate $BI$, it drives the state populations towards the thermal, non-inverted, state and the maser is quenched.

If the specific intensity grows to the point that the rate of induced radiative transitions approaches the pump rate then the right-hand side of Eqn 3.12 becomes constant. The specific intensity grows linearly, $I \propto I_0 R$, and the maser is *saturated*; the maser output is controlled by the rate at which the upper state is populated by the pump. Saturated emission is often very bright but if the pump rate is low then even faint masers can saturate.

The exponential nature of unsaturated maser amplification has several characteristic consequences:

– The intensity at the maser frequency can be amplified by many orders of magnitude, producing brightness temperatures far higher than any possible kinetic molecular gas temperature, $10^4$–$10^{15}$ K, occasionally even higher.
– The line centre is initially amplified more than the wings so the spectral profile is narrowed; if the emission becomes saturated the profile widens again to resemble a brighter version of the thermal profile.
– Interferometric images show that maser emission in an individual spectral channel can similarly be beamed into a very narrow spot, with a much smaller angular size than the emitting cloud, but emission from shocked regions may be amplified into a larger angle (not exceeding the cloud size).
– The pump routes involve various thermal transitions with different probabilities for different masing levels of the same molecule, so the relative brightness of different masers can diverge from the expected brightness ratios between lines in thermodynamic equilibrium.
– Masers can be very variable on timescales from hours to decades, owing both to the effects of small changes in temperature, radiation, etc. and to the effects of velocity changes or turbulence on directional beaming.

In the astrophysical context masers arise from warm molecular gas where there is sufficient velocity-coherent column density along the line of sight, subject to the presence of a disturbance providing a pump mechanism. They commonly arise around late-type stars and star-forming regions. Solar-system comets can mase and masers have recently been detected in protoplanetary disks. Shocked regions, occurring for example in supernova remnants and active galactic nuclei, often favour masing. The probability of stimulated emission is proportional to $\lambda^3$; cm-wave masers have been the most studied, but ALMA and the SOFIA observatory have detected masers down to $\lambda = 0.24$ mm.

Doppler shifts of maser lines provide a direct tool for study of the velocity fields in masering regions at very high spectral and angular resolution, thanks to maser beaming and spectacularly high brightness temperatures. There may also be a high degree of polarization in maser line emission, due to the Zeeman splitting of spectral lines in a magnetic field. The orientation of the magnetic field determines the polarization, which can be circular, elliptical, or linear. In recent years, the continuous wavelength coverage from cm to sub-mm wavelengths afforded by the upgraded JVLA, ATCA, ALMA, and so on, has allowed the imaging of multiple maser transitions of a given type of molecule with a wide range of excitation requirements and energies; notable among such molecules are water and methanol. Coupled with new models, this allows the reconstruction of physical conditions on scales an order of magnitude finer than is possible using thermal lines.

The review by Elitzur (1992) and the book by Gray (2012), *Maser Sources in Astrophysics*, may be consulted for further details and references.

### 3.7.1 Common Masers

Masing has been detected from a number of astrophysical molecules, and we summarize the commonest here. Other weak or rare masers are known, including CO in a few environments, hydrogen recombination lines, and electron cyclotron masers from stellar atmospheres.

The first-known maser, OH, has four ground-state transitions at 1612 and 1720 MHz (the satellite lines) and 1665 and 1667 MHz (the main lines). As mentioned in Section 3.4, the splitting of the $\Lambda$-doublet arises from the interaction of the electron's spin $J$ with that of the proton (hydrogen nucleus), $I$. A new quantum mumber, $F = J \pm I$, describes each pair of levels; $F$ does not change for the satellite lines whilst it does for the main lines.[2] All these lines are found in comets and star-forming regions. The three lower lines are common around evolved stars, whilst the 1720 MHz maser occurs in supernova remnants and the main lines are among the brightest extragalactic masers associated with active galactic nuclei. Hydroxyl, OH, lines around 4.8 and 6 GHz are also common in star-forming regions; a few detections are known at 8 and 13 GHz but no others are predicted below THz frequencies. The OH radical is mostly radiatively pumped by FIR emission from dust and traces of gas at relatively low kinetic temperatures, from $\sim$50 to a few hundred kelvins. It is a paramagnetic molecule with an unpaired electron and thus can show strong polarization, due to mG or even $\mu$G magnetic fields, often with Zeeman pairs showing a clear spectral separation of right- and left-hand circular emission peaks. If the linear polarization direction and strength are also measured, the magnetic field can be reconstructed in three dimensions.

Water, $H_2O$, masers were first identified in 1969, in the form of 22 GHz emission from the Orion Nebula. They are common in environments similar to OH but under denser and warmer conditions, for example in outflows from young stars where the 1.6 GHz OH and the 22 GHz water lines are found in the pre- and post-shock gas, respectively. The advent of terrestrial and space or airborne high frequency observatories, in very dry regions or above the troposphere, has revealed at least 20 more water masers up to THz frequencies, and over 100 are predicted. These are mostly collisionally pumped, although radiative pumping is possible at high temperatures. They cover a uniquely wide range of excitation energies, from 200 to several thousand kelvins, and of number densities and water abundances; this has the potential to constrain conditions on the scale of clouds as small as a few AU.

Silicon monoxide, SiO, masers are usually found just outside the stellar atmosphere of late-type stars, in gas temperatures around 2000 K. They are found in a series of rotational transitions, $J = 1$–$0$ around 43 GHz, $J = 2$–$1$ around 86 GHz, $J = 3$–$2$ around 129 GHz, etc. For each rotational transition masers may be detected in multiple vibrational transitions, usually strongest in $V = 1$. The most abundant isomer is $^{28}\text{Si}^{16}\text{O}$, but very strong masing is often seen in $^{29}\text{SiO}$, $^{30}\text{SiO}$, etc. The pumping method is complex but probably mainly radiative.

Methanol, $CH_3OH$, masers are almost exclusively found in the vicinity of star-forming regions. In general, the detection of a maser depends on the excitation conditions – non-detection is not proof of absence of the molecule – but the sufficient abundance of gas-phase methanol depends on its slow formation in the icy mantles of grains, followed by liberation as cloud collapse and protostellar evolution heats up the environment. They can be either radiatively pumped – Class 2, found close to young stellar objects (YSOs), including the best-known line at 6.7 GHz – or collisionally pumped, Class 1, found in more extended, cooler regions. Transitions have been detected up to a few hundred GHz with excitation energies up to a few hundred K.

Formaldehyde, HCHO, masers are rare; the 4.8 GHz transition is generally detected in star-forming regions that also support Class 2 methanol masers, but three OH megamaser galaxies also have formaldehyde masers. The pumping mechanism is complex and often involves very deep absorption, an 'anti-inversion'.

The lowest-frequency masing transition of hydrogen cyanide, HCN, is 89 GHz, and it has been less studied than the oxygen-rich maser species, but in recent years transitions have been detected up to around 800 GHz. It has the potential to open up high-resolution studies of carbon-rich environments, since these lines (from both the ground and vibrationally excited states) have excitation energies covering a wide range of temperatures similar to water, but more biased towards a few thousand kelvin.

### 3.8 Further Reading

*Supplementary Material* at www.cambridge.org/ira4.

Condon, J. J., and Ransom, S. M. 2016. *Essential Radio Astronomy*. Princeton University Press.

Gondon, M. A., and Sorochenko, R. I. 2007. *Radio Recombination Lines*. Springer.

Gray, M. 2012. *Maser Sources in Astrophysics*. Cambridge University Press.

Wilson, T. L., Rohlfs, K., and Hüttemeister, S. 2013. *Tools of Radio Astronomy*, 6th edn. Springer.

# 4

# Radio Wave Propagation

At the low frequency, low photon energy, end of the electromagnetic spectrum, the propagation of radio waves is subject to refraction in ionized gas, whether it be the terrestrial ionosphere or the interstellar medium. The refraction is strongly frequency dependent; it is also affected by an ambient magnetic field, which may cause birefringence. Radio waves are also refracted in the neutral terrestrial atmosphere, similarly to visible light; this is important in the pointing of narrow telescope beams, particularly at short radio wavelengths. Absorption and emission in the atmosphere are vitally important at millimetre wavelengths, providing severe limits on the sensitivity of observations.

In this chapter we set out the basic formulations of refraction in an ionized medium and of propagation in the terrestrial atmosphere.

## 4.1 Refractive Index

The upper atmosphere, at heights above 100 km, is partially ionized by solar ultraviolet radiation, forming the ionosphere. This has a major effect on the propagation of radio waves; ionospheric refraction affecting source positions by several arcminutes was observed in interferometric observations by Smith (1952) at 80 MHz. At low frequencies, below about 10 MHz, there may even be total reflection at vertical incidence. The Earth's magnetic field has a large influence on propagation through the ionosphere, causing birefringence, which may have complex effects on the polarization of reflected and refracted radio waves. Fortunately for the radio astronomer, the refractivity of the ionosphere decreases as $\nu^{-2}$, and at the frequencies of hundreds of MHz and above used in most observations the refractivity may be described comparatively simply. The same is true for the interstellar medium, where the very small refractivity can provide a useful probe of the density of the ionized gas and of the strength of the interstellar magnetic field.

The refractive index $n$ is most simply expressed in terms of the resonant frequency $\nu_p$ of the plasma:

$$n = (1 - \nu_p^2/\nu^2)^{1/2}, \tag{4.1}$$

where

$$\nu_p^2 = (1/2\pi\epsilon_0)^2(n_e^2/m) \quad \text{(MKS units)} \tag{4.2}$$

or

$$v_p^2 = (1/2\pi)^2 (n_e^2/mc) \quad \text{(Gaussian units)}. \tag{4.3}$$

To a good approximation $v_p = 9n_e^{1/2}$ Hz, where $n_e$ is the number density of electrons in $m^{-3}$; for interstellar space the electron density is usually quoted in $cm^{-3}$, giving the same numerical result in kHz. A typical interstellar electron density is $0.03$ $cm^{-3}$, corresponding to a plasma frequency of $1.56$ kHz; the refractivity $1 - n$ for a 100 MHz radio wave is then $1.2 \times 10^{-10}$.

The effects of phase delays and refraction in the ionosphere are important for low frequency synthesis telescope arrays such as LOFAR, whose observations extend to below 50 MHz and to baselines hundreds of kilometres long. Typical angles of refraction at 100 MHz are several arcminutes, often larger than the synthesized telescope beam. A more serious effect for LOFAR is differential phase changes at the separate telescope elements, which we discuss in Chapter 9 and 11.

### 4.1.1 Dispersion and Group Velocity

Since $n$ is less than unity, the velocity of the wave $c/n$ is greater than $c$; this is the *phase velocity* of an infinite monochromatic wave. Energy (and information) is carried at the *group velocity*, which is the modulation velocity of a group of waves with finite bandwidth. The frequency components of the wave group travel with different velocities, because of the *dispersion* in refractive index which appears in Eqn 4.1. The dispersion is such that the product of the group and the phase velocities is $c^2$; the group velocity is thus $cn$, which is less than $c$, resulting in a propagation delay.

The group velocity can be expressed as

$$v_g = cn = c\left(1 - v^2/v_p^2\right)^{1/2} \approx c(1 - \tfrac{1}{2}v^2/v_p^2 + \tfrac{1}{8}v^4/v_p^4 + \cdots) \tag{4.4}$$

and when $v \gg v_p$ the first term suffices. Over a path length $l$ the time delay $\Delta t$ relative to the light travel time in vacuo can then be written as

$$\Delta t = \int_0^l \frac{l}{v_g}\, dl - \frac{l}{c}, \tag{4.5}$$

and hence

$$\Delta t = \frac{e^2 c}{2\pi m v^2} \int n_e dl, \tag{4.6}$$

where the integral of $n_e$ along the line of sight is called the *dispersion measure* and is discussed more completely in Chapter 15. To give one example, a pulse observed at a frequency of 100 MHz from a pulsar at a distance of 100 light years, travelling through interstellar ionized gas with a density of $0.03$ $cm^{-3}$, will arrive 3.4 s later than the same pulse observed at a much higher frequency, say 1000 MHz.

In the solar corona the plasma density is high enough for the refractive index to approach zero for radio frequencies up to and beyond 100 MHz. The power flux along a ray path is given by the Poynting vector, $|S| = |\mathcal{E}|^2/Z = \sqrt{\epsilon/\mu}\,|\mathcal{E}|^2$ and is constant along the ray, so that when $\epsilon$ tends to zero, the field $\mathcal{E}$ becomes very large. This greatly increases the absorption, and the plasma may become opaque. The brightness temperature of the Sun at low frequencies is dominated by this effect. For example, at 100 MHz the brightness temperature is approximately $10^6$ K over an area which is much larger than the visible disk; this is the temperature of the outer parts of the corona. At progressively higher radio frequencies the corona and then the chromosphere become transparent, and the brightness temperature falls correspondingly towards the photospheric temperature (see Chapter 12).

## 4.2 Faraday Rotation

When a wave passes through a plasma, the electrons respond to the electric field by oscillating and reradiating at the wave frequency; this is the origin of the refractive index in Eqn 4.1. The response of the electrons is affected by any static magnetic field $B$, which forces an electron with velocity **v** into curved paths owing to the Lorentz force $e\mathbf{v} \times \mathbf{B}$. If the incident wave is circularly polarized, the electron response will be a circular motion whose amplitude depends on the direction of the static field. The addition of a static magnetic field $B$ in the direction of propagation therefore splits the refractive index into two components, and the plasma becomes *birefringent*, i.e. one of the two hands of circular polarization travels slightly faster than the other; which one depends on the direction of the static field.[1] The refractive index can be written as

$$n^2 = 1 - \frac{\nu_p^2}{\nu(\nu \pm \nu_B)}, \tag{4.7}$$

where $\nu_B$ is the gyrofrequency, the frequency at which the electron circulates around the magnetic field:

$$\nu_B = \frac{eB}{2\pi m} \text{ (MKS units)}; \quad \frac{eB}{2\pi mc} \text{ (Gaussian units)}. \tag{4.8}$$

The electron gyrofrequency in a field of 1 tesla is $2.8 \times 10^{10}$ Hz. Interstellar magnetic fields are usually quoted in microgauss, so that it is useful to remember that the gyrofrequency is 2.8 Hz per microgauss. (The same factor occurs in the Zeeman frequency splitting of the 21 cm hydrogen line.)

The analysis of propagation along the direction of the magnetic field is based on the 'transverse' case, the right-hand and left-hand circularly polarized waves being the normal modes. The general formalism applies equally well for large angles to the field, taking the component $B\cos\theta$ of the field along the wave normal. (This 'quasi-longitudinal' approximation is generally applicable in astrophysical circumstances, but may not be applicable when the wave frequency is comparable with the plasma frequency, as can occur for low

---

[1] For a wave travelling along the direction of the magnetic field the left-hand circular polarization has the lower phase velocity.

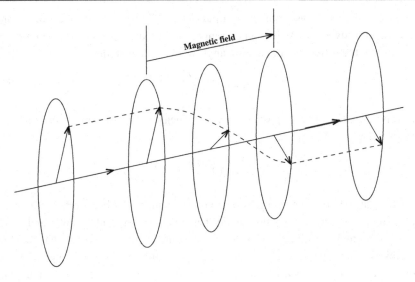

Figure 4.1. Faraday rotation. Propagation along a magnetic field in a plasma rotates the plane of linear polarization of a radio wave.

radio frequencies in the ionosphere.) In all cases, the ionized medium is found to be bire-fringent when the proper normal modes are identified.

A plane polarized wave may be regarded as the sum of two circularly polarized compo-nents of opposite handedness (Section 2.2). In a plasma with a static magnetic field these two components propagate with different phase velocities so after travelling some distance their relative phase has altered; the effect is to rotate the plane of polarization of the resultant when they are added together. This is *Faraday rotation*[2] (Figure 4.1). The phase difference is proportional to $NB_{||}$, where $B_{||}$ is the component of the magnetic field along the line of sight.

In radio observations the angular rotation around a ray path follows from Eqn 4.7 above:

$$\theta_R = R\lambda^2 \tag{4.9}$$

where $R$ is the *rotation measure* of that path, usually expressed in units of rad m$^{-2}$. Using the conventional astrophysical units, measuring distances in parsecs, $N$ in cm$^{-3}$, and $B$ in microgauss, the rotation measure is

$$R = 0.81 \int NB_{||} \, dl. \tag{4.10}$$

Thus for the typical galactic values of $N = 0.03$ cm$^{-3}$, $B_{||} = 3$ microgauss, and $l = 1$ kpc, $R \approx 73$ rad m$^{-2}$.

---

[2] The effect was first observed by Faraday in plane-polarized light in a glass block between the poles of an electromagnet.

For a radio source such as a quasar or a pulsar, which radiates with the same plane of linear polarization over a wide frequency range, the Faraday rotation can be empirically determined from the difference in position angle at two wavelengths:

$$R = \frac{\theta_1 - \theta_2}{\lambda_1^2 - \lambda_2^2}. \tag{4.11}$$

Alternatively, the difference in position angle $\theta$ (measured in radians) between adjacent frequencies $\nu$ and $\nu + \delta\nu$ MHz is

$$\Delta\theta_R = -2R \left(\frac{300}{\nu}\right)^2 \frac{\delta\nu}{\nu}. \tag{4.12}$$

Measurements at only two wavelengths or frequencies may well be insufficient to provide an unambiguous answer: the position angle $\theta$ is only known modulo $180°$ and a whole number of turns may be missed between the two measurements, particularly at longer wavelengths/lower frequencies. For example, if $R = 100$ rad m$^{-2}$, $\theta$ rotates by 9 rad or about one and a half turns at $\lambda = 30$ cm (1 GHz). Rotation measures can be ten or more times higher than this and, to counteract the $n\pi$ ambiguity, $\theta$ should be measured at three or more wavelengths.

The sign of $R$ is diagnostic of the direction of $B_{||}$, the component of the magnetic field along the line of sight. If the field direction is primarily towards the observer then the linear polarization rotates counterclockwise going from shorter to longer wavelengths (i.e. a positive slope in a plot of $\theta$ against $\lambda^2$). By convention this case corresponds to positive $R$. The converse is true if the field is primarily away from the observer. Extensive rotation measure surveys using a combination of extragalactic radio sources and pulsars have enabled a picture of the magnetic field morphology of the Milky Way to be built up, as will be discussed in Chapter 14.

Before leaving this section it must be noted that the above discussion refers to a so-called *Faraday screen* situation, where the region causing rotation lies in front of, and is distinct from, the source of emission. Significant deviations from a $\lambda^2$ behaviour can occur if emission and rotation occur together and the concept of the *Faraday depth* (Burn 1966) then becomes relevant, but this is beyond our present scope.

## 4.3 Scintillation

The plasmas of the ionosphere, of interplanetary space, and of the interstellar medium all contain random irregularities. Propagation through a medium with random fluctuations in refractive index initially corrugates the wavefront and then leads to the amplitude fluctuations which are familiar in the optical domain as the twinkling of stars. Scintillation is only observed when the source has a sufficiently small angular diameter. In the radio domain the most important example is provided by the propagation through interstellar space of radio pulses from pulsars.

A simple presentation of the theory by Scheuer (1968) in terms of random phase changes imposed on a plane wavefront by density irregularities, with typical dimension $a$ and with

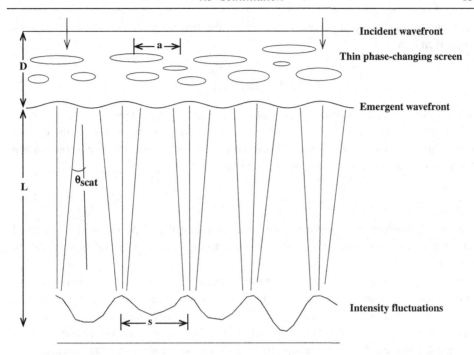

Figure 4.2. Scintillation: fluctuations in refractive index corrugate a plane wavefront, which then develops amplitude irregularities.

differences in refractive index due to density fluctuations $\Delta n_e$, contained in a slab with thickness $D$ leads to phase perturbations $\Delta\phi$ across the wavefront:

$$\Delta\phi = D^{1/2}a^{1/2}r_e\lambda\Delta n_e. \tag{4.13}$$

The angle of scattering is given by

$$\theta_{\text{scat}} = \left(\frac{\Delta\phi}{2\pi}\right)\frac{\lambda}{a} = \frac{1}{2}\left(\frac{D}{a}\right)^{1/2}r_e\Delta n_e\lambda^2. \tag{4.14}$$

This is the apparent angular size of a distant point source observed through the screen formed by the slab.

As the wave progresses beyond the screen to distances $L$ greater than $L\theta_{\text{scat}}$ the rays cross and an interference pattern develops (Figure 4.2). The ray paths will differ by $\approx (1/2)\,\theta_{\text{scat}}^2 L$, which will usually be many wavelengths. Interference between the rays will therefore depend on the wavelength; the resultant amplitude will vary over a wavelength difference $\delta\lambda$ given by

$$\frac{\delta\lambda}{\lambda} = \frac{2\lambda}{\theta_{\text{scat}}^2 L}. \tag{4.15}$$

The model may be extended to the practical situation, where the screen fills the whole line of sight over a distance $L$, by setting $L = D$; only a small numerical factor is involved in the results for scattering angle and interference. The frequency bandwidth of scintillation $\Delta\nu_{scat}$ then becomes

$$\Delta\nu_{scat} = \frac{8\pi^2 ac}{D^2(\Delta n_e)^2 r_e^2 \lambda^4}. \qquad (4.16)$$

The $\lambda^{-4}$ behaviour is characteristic of scattering in pulsar observations.

The lateral scale $s$ of the interference pattern is the same as the scale $a$ of the density irregularities provided that $\Delta\phi \ll 1$. This is the case of weak scattering (see below). When $\Delta\phi \gg 1$, i.e. for strong scattering, the lateral scale $s$ becomes

$$s = \frac{a}{\Delta\phi} = \frac{\lambda}{2\pi\theta_{scat}}. \qquad (4.17)$$

The lateral structure moves past the observer in a way that depends on the combined velocities of the source and the observer relative to the screen, and the source is seen to scintillate.

In the important case of scintillation in the turbulent interstellar medium, there is a very wide spectrum, of random size and amplitude of density irregularities, to be considered. Two distinct regimes of scintillation may be observed, corresponding to large and small scales of turbulence (see the review Rickett 1990). On the large scale, scintillation may be regarded as the effect of individual ionized clouds which act as prisms or lenses; this is referred to as refractive scintillation and is responsible for some of the slow changes in the observed intensities of pulsars (Chapter 15) and quasars (Chapter 16). The more usual small-scale scintillation effects are understood as due to random diffraction.

A wavefront with randomly distributed fluctuations of phase can be treated by diffraction theory; the wave spreads in angle as it propagates, with an angular spectrum that is determined from the phase distribution by a Fourier transform. (The treatment is closely related to the diffraction theory of an antenna aperture, discussed in Chapter 8.) A lateral scale $a$ of the phase structure $\Delta\phi$ corresponds to a scattered wave at angle $(\Delta\phi/2\pi)(\lambda/a)$. If the total phase irregularities are less than a radian, there is an unscattered plane wave component; in this case, known as weak scattering, the appearance of a point source will be that of an unscattered point surrounded by a scattered halo. For strong scattering there is no unscattered wave, and the total amplitude of the randomly scintillating wave varies according to a Rayleigh distribution.

The effects of scintillation can be used as a tool for determining the angular sizes of radio sources. For the same reason that in the Earth's atmosphere stars scintillate but planets do not, only small radio sources scintillate. We note two examples here. First, the development of VLBI was encouraged by the observation of scintillations of compact radio sources in the solar wind, which indicated that there were compact sources smaller than ten milliarcseconds in size. Antony Hewish and his colleagues set out to exploit this difference when they built their large low frequency array at Cambridge – this was the array with which they discovered pulsars (Chapter 15). Second, the ionized interstellar medium will cause scintillations in radio sources such as pulsars whose angular size is of order 1 microarcsecond or smaller; notably the radio afterglow of the gamma-ray bursters shows scintillations which damp out as the radio source expands, showing that the angular diameter of the afterglow was less than 3 microarcseconds during the first month of its expansion (Frail *et al.* 1997).

Resolving the angular size of the diffracted image of a pulsar requires an interferometer with a very long baseline. This has been achieved by the space observatory RadioAstron (Section 11.4.10). Combined with ground-based radio telescopes, baselines up to 330 000 km, using a 1 metre wavelength (324 MHz) have been used to observe the pulsar B0329+54 (Popov *et al.* 2017). The diffracted image diameter was found to be $4.8 \pm 0.8$ milliarcseconds (mas), containing sub-structure at a smaller angular scale. These observations provide a new probe of the structure of the interstellar medium.

## 4.4 Propagation in the Earth's Atmosphere

The refractive index of the neutral dry atmosphere is almost the same at optical and radio wavelengths. The effect of water vapour is, however, more than 20 times greater at radio wavelengths; this is due to the permanent dipole moment of water molecules. The refractive index $n$ of air at radio wavelengths and at temperatures encountered in the atmosphere is conveniently quoted in terms of a refractivity, $N = 10^6(n - 1)$. For most purposes this is given adequately by the Smith and Weintraub (1953) formula,

$$N = 77.6T^{-1}(P_\mathrm{d} + 4810P_\mathrm{v}T^{-1}), \tag{4.18}$$

where $P_\mathrm{d}$ is the partial pressure of the dry air and $P_\mathrm{v}$ is the partial pressure of the water vapour, both in millibars (1 mb = 100 pascals; 1 atmosphere = 1013 mb). At frequencies above 100 GHz the component due to water vapour is slightly greater because of the effect of the very broad infrared transitions in water vapour (Hill and Clifford 1981).

Refraction in the atmosphere affects the direction of arrival of a wavefront at a radio telescope in much the same way as for an optical telescope, except that radio observations are more often made close to the horizon. If the atmosphere can be treated as a flat slab, with refractive index $n$ at ground level, the apparent increase $\Delta z$ in elevation at zenith angle $z$ is given by

$$\Delta z = (n - 1)\tan z. \tag{4.19}$$

This first-order approximation must be elaborated near the horizon by the results of ray tracing, taking account of Earth's curvature. It is useful to remember, however, that $n - 1$ is approximately $3 \times 10^{-4}$, so that the refraction near the horizon is approximately 20 arcmin. This must be taken into account in the pointing model for large radio telescopes.

The refractivity of the atmosphere is also important in interferometric observations because of the extra path length which is introduced. A plane parallel slab atmosphere affects all telescope elements of an interferometer array in the same way, so that the only important changes in path length are those due to differences in the atmosphere above the individual telescope sites; this is important for interferometers with baselines longer than a few km, where there may be large differences in water vapour content at the sites. It is especially true in VLBI, where there may also be important differences in path length due to Earth curvature, so that the elevation of a source above the horizon may differ greatly between the various contributing sites. Differences in path length may be estimated from measurements of partial pressures at the telescope sites, but since the excess path lengths may be large (more than 10 m for zenith distances greater than 80°), correction for refraction may be a serious limitation. The effects on path length are obviously more serious at shorter wavelengths; for millimetre-wavelength arrays such as ALMA they may prove to be the limiting factor in the overall performance. We discuss this problem at greater length in Chapter 11.

At radio frequencies below about 10 GHz atmospheric absorption is small (less than a few per cent) and often negligible except for the most precise measurements. At metre wavelengths even rain might not matter much, except for the effects the liquid water might have on the electrical components of the receiver. The attenuation at higher frequencies arises from molecular resonances of oxygen, ozone, and water vapour. These resonances occur within the radio band from 22 GHz upwards. The wings of their pressure-broadened absorption bands affect atmospheric transmission at lower frequencies and even extend into the infrared in the case of water vapour. Figure 4.3, from Smith (1982), shows the

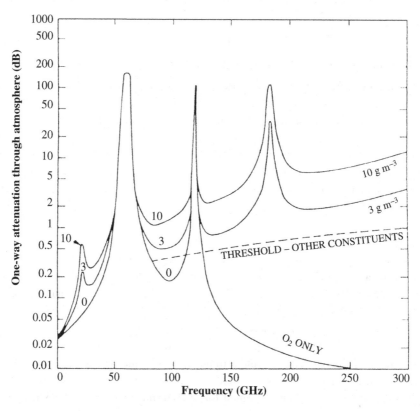

Figure 4.3. Atmospheric absorption as a function of frequency. The total attenuation at the zenith is shown for the oxygen content alone and with surface water vapour densities of 3 g m$^{-3}$ and 6 g m$^{-3}$ (2 km scale height), equivalent to precipitable water vapour values (PWVs) of 6 mm and 20 mm respectively.

attenuation at the zenith for frequencies up to 300 GHz for a standard atmosphere with low and moderate water vapour content.[3]

Atmospheric absorption has a second effect on radio astronomical observations. If a radio source is observed through the atmosphere with temperature $T_{atmos}$ and optical depth $\tau$ the observed temperature is

$$T_{obs} = T_{source} \exp(-\tau) + T_{atmos}[1 - \exp(-\tau)] \qquad (4.20)$$

as in Eqn 2.28. Thus not only is the signal from above the atmosphere reduced but there is also an added component of thermal radiation from the atmosphere. For a typical value[4] of $T_{atmos} \approx 270$ K and absorption of 1 dB (corresponding to a transmission of 0.794 and hence an optical depth $\tau = 0.23$) the atmospheric noise contribution is $\sim 56$ K. This may double the system temperature in a mm-wave receiving system, which is a more serious effect than the attenuation of the signal itself (see a further discussion in Section 11.3).

The contribution of the atmospheric emission can be measured by monitoring the system temperature as the telescope is tipped away from the zenith. This manoeuvre is sometimes called a 'sky dip'. As the zenith angle $z$ increases, the path length through the atmosphere and hence the optical depth increase:

$$\tau_z = \frac{\tau_{zenith}}{\cos z} = \tau_{zenith} \sec z. \qquad (4.21)$$

At $z = 60°$ (or elevation 30°) the path length and the optical depth are doubled and hence the increase in the system temperature compared with that at the zenith directly provides the zenith contribution.

The optical depth is a function of the partial pressures of the absorbing gases and these fall off exponentially with altitude (Figure 4.4). The scale height is that at which the pressure of a constituent falls to 1/e of its sea level value. For oxygen this height is $\sim 8.5$ km and for water vapour $\sim 2$ km; hence at an altitude $h$ km the water vapour optical depth falls to $e^{-h/2}$ of its sea-level value. High mountain sites are therefore favoured for millimetre-wave observations; for example at $h = 5000$ m the optical depth due to water vapour is only $\sim 8\%$ of its sea-level value. At balloon altitudes (30–40 km) water vapour can be ignored. The effects of water vapour for millimetre-wave observations are sufficiently serious to drive the site choice to otherwise inconvenient locations. Hence ALMA is located on the Chajnantor plateau in northern Chile (altitude 5000 m with an annual median precipitable water vapour less than 2 mm), while many specialized CMB experiments are carried out in Antarctica. The South Pole site has an excellent combination of altitude ($\sim 2800$ m) and extreme cold. As a result the annual median of the precipitable water vapour is $< 0.5$ mm.

---

[3] The water vapour content is often quoted as millimetres of precipitable water vapour (PWV); this is the depth of liquid water if all the water vapour from the surface to the top of the atmosphere were to be condensed. For a 2 km scale height, a density of 1 g m$^{-3}$ at the surface is equivalent to 2 mm of precipitable water vapour. In temperate latitudes the precipitable water vapour typically ranges from $\sim 10$ mm in winter to $\sim 40$ mm in summer, with an annual mean of $\sim 20$ mm.

[4] The temperature of the troposphere at a particular altitude can be estimated using an average lapse rate of $-6.5$ K per km. Thus for a 2 km scale height the temperature is typically 13 K colder than that at the surface.

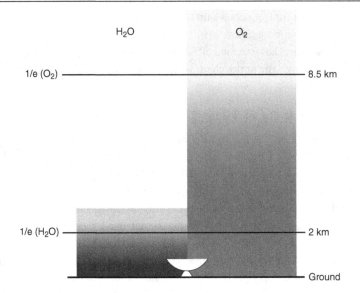

Figure 4.4. Scale heights of water vapour and oxygen in the atmosphere (courtesy of S. Lowe).

The frequency-dependent optical depth at a particular site can be calculated using computer models such as the '*am* atmospheric model' (Paine 2017)[5] and ATM (Pardo *et al.* 2001).

## 4.5 Further Reading

*Supplementary Material* at www.cambridge.org/ira4.

---

[5] http://doi.org/10.5281/zenodo.438726.

# 5

# The Nature of the Received Radio Signal

The vast majority of astronomical observing starts with a telescope intercepting incoming electromagnetic signals carrying information about the Universe; these signals have the characteristics of random noise. For the radio astronomer, the incoming radiation can be treated as a superposition of classical electromagnetic waves and the signals have the characteristics of Gaussian noise, the result of an assemblage of many independent oscillators with random frequency and phase. As one moves to shorter wavelengths, through the infrared and into the optical, ultraviolet, and X-ray bands, the discrete character of photons becomes increasingly dominant, and the random noise, sometimes called shot noise, obeys Poisson statistics.

This chapter therefore begins with an introduction to the properties of Gaussian random noise. We then move on to consider the generic aspects of reception at a radio telescope, in particular the measurement of the sky *brightness temperature* and how it is calibrated with respect to accessible sources of blackbody radiation. An important practical concept in this discussion is that of the *antenna temperature*. Brightness temperatures can only be measured when the emitting source is comparable with, or larger than, the reception beam of the telescope. When, however, the source is much smaller than the beam then only a more limited parameter, the *flux density*, can be measured.

One difference in terminology should be noted: radio telescopes are sometimes called radio antennas. Here, the terms radio telescope and radio antenna will be used interchangeably; in general usage, the term antenna is used when the angle of reception is large, as it is for television or communication antennas. When the angle of reception is small, as it is for steerable paraboloids (which are, indeed, analogous to optical telescopes) the term radio telescope is used.

## 5.1 Gaussian Random Noise

All the natural signals with which we are dealing have been generated by random processes of one form or another and, when incident on a telescope, exhibit random variations of the electric field in both amplitude and phase. In the receiver these electric field variations are turned into voltages. The signal voltages are indistinguishable from (and usually much smaller than) the voltages generated by thermal fluctuations in the resistive components of the receiver itself and from other non-astronomical sources of noise incident on the telescope.

The fundamental accuracy of radio measurements depends upon the statistical properties of the voltage amplitude $V(t)$. The noise power $P(t)$ is proportional to the square of the amplitude and, for much of this discussion, the units can be chosen[1] so that $P = V^2$. Although the average power from natural sources can vary with time, in most cases (man-made interference excepted) the variation is much slower than the coherence time of the radiation. For present purposes the average power at the output of the receiver can be taken to be stable,[2] and the combination $V(t)$ of these voltages can be treated as a *stationary random variable*; its statistical properties will not vary over time, but will depend on the particular characteristics of the system.

The amplitude expectation for $V(t)$ is described by a Gaussian distribution; *Gaussian random noise* is the most common form met with in practice[3] and the simplest to deal with quantitatively. The probability that the value of $V(t)$ will fall between $V$ and $V+dV$ is determined by the *probability density function* $\mathcal{P}(V)$; this must be normalized so that its integral is unity. (Note the distinction from the *probability distribution function*, which is the probability that $V(t)$ will take on a value *less* than $V$.) For a given probability density function, three quantities are of particular interest: the mean value $\langle V \rangle$, the mean square value $\langle V^2 \rangle$ of $V(t)$ over a given interval, and the autocorrelation $R(\tau)$ of $V(t)$ (see Appendix 1).

For Gaussian random noise, the probability density function is

$$\mathcal{P}(V) = \frac{1}{\sigma\sqrt{2\pi}} \exp\left(-\frac{V^2}{2\sigma^2}\right). \tag{5.1}$$

The amplitude has an equal probability of being positive or negative, and its mean value is zero. The only stable, and hence measurable, quantity at a single point in space is the average power $P(t) = V(t)^2 = \sigma^2$, the variance of the random variable for the given probability density.

The power spectrum of a random noise voltage may be obtained from the Fourier transform of the autocorrelation of its amplitude (see, in Appendix 1, the Wiener–Khinchin theorem). In the real world, the observer can only estimate the signal power or its autocorrelation by evaluation over a finite timespan $\mathcal{T}$, so one arrives at an *estimate* of the power spectrum; the estimate becomes more precise as $\mathcal{T}$ increases. The autocorrelation function $R(\tau)$ is defined by the product of $V(t)$ with itself but shifted by a lag $\tau$, integrated over all time; the function is then defined for a range of lags (see also Section 7.2). For the present finite-time case,

$$R_{t,\mathcal{T}}(\tau) = \int_{t-\mathcal{T}/2}^{t+\mathcal{T}/2} V(t)'V(t'+\tau)\mathrm{d}t', \tag{5.2}$$

---

[1] The power dissipated is proportional to $V^2/R$, where $R$ is the resistive term in the impedance. Since $R$ is constant throughout a 'matched' receiver it can be ignored when considering relative power levels.

[2] In Chapter 6 we will discuss the gain variations which arise in practical receivers and result in so-called '$1/f$ noise'; this can dominate the Gaussian noise on long timescales and steps must be taken to combat it.

[3] This is a consequence of the central limit theorem: the sum or average of samples drawn from *any* distribution with a finite mean and standard deviation will tend towards Gaussian, with the approximation improving as the samples get bigger.

where the subscripts $t, \mathcal{T}$ indicate that for each lag the integration extends over an interval $\mathcal{T}$ centred on time $t$. The estimated power spectrum, $S_{t,\mathcal{T}}(\nu)$, is the Fourier transform

$$S_{t,\mathcal{T}}(\nu) \overset{\text{FT}}{\Longleftrightarrow} R_{t,\mathcal{T}}(\tau) \tag{5.3}$$

(see also Chapter 7). In the limit where the averaging time $\mathcal{T}$ goes to infinity, the relation converges to

$$S(\nu) \overset{\text{FT}}{\Longleftrightarrow} R(\tau). \tag{5.4}$$

The spectrum of Gaussian noise is evaluated by inspecting Eqn 5.3 for the cases of zero and non-zero time lag. When $\tau$ is non-zero the integral goes to zero, since for white noise,[4] i.e. with no bandwidth limitation, $V(t)$ is completely uncorrelated from one instant to the next. When $\tau = 0$, however, the autocorrelation (Eqn. 5.2) becomes the mean square voltage $\langle V^2 \rangle$; i.e. it is the variance of the Gaussian noise distribution. For a sufficiently long integration time, therefore, the autocorrelation for a Gaussian random signal is a Dirac delta function (technically, a distribution with an infinite spike at $t = 0$, zero elsewhere, but with an integral whose value is 1). Let $t = 0$; for stationary noise the result should converge to the same answer for any epoch:

$$\lim_{\mathcal{T} \to \infty} R(t) = V^2 \delta(t) \quad \text{(for } v(t) \text{ Gaussian).} \tag{5.5}$$

The Fourier transform of the delta function is a constant, from which it follows that Gaussian noise has a flat power spectrum with a power spectral density (the power per unit bandwidth) independent of frequency and equal to the square of the standard deviation $\sigma$ of the probability density $V(t)$.

Two points are worth noting before we leave this discussion. First, since one sample is as good as any other sample, random noise has no redundancy and thus cannot be compressed with algorithms such as MPEG and JPEG. Second, while Gaussian processes are of fundamental importance in radio astronomy – and indeed in science as a whole[5] – they do not describe all important phenomena; for example, turbulence and many other phenomena exhibit power-law distributions (see the review Clauset *et al.* (2009) mentioned in the Further Reading section).

## 5.2 Brightness Temperature and Flux Density

Before seeing how measurements of the power in random noise signals are used to measure properties of the radio sky, we must be aware of the limitations imposed at reception by the antenna or telescope.

An antenna can be treated either as a receiving device, gathering the noise power from an incoming radiation field and conducting the electrical signals to the output terminals,

---

[4] A flat power spectrum is often called a 'white' spectrum, from the colour analogy, and so Gaussian noise is termed 'white noise'. Its quality is familiar as the hiss from an AM radio that is not tuned to a station; the hiss contains all audio frequencies, with no noticeable tonality.

[5] A classic text is Davenport and Root (1958); see Further Reading.

or as a transmitting system, launching electromagnetic waves outward. The two cases are equivalent because of time reversibility: the solutions of Maxwell's equations are valid when time is reversed. As a transmitter the antenna produces a beam of radiation whose solid angle is determined by the size of the aperture: the larger the aperture, the narrower is the beam and the greater is the maximum power flux at the centre of the beam. The concept of the *power gain* of an antenna, which arises in transmission, is therefore closely related to that of its *effective area*, which applies to reception.

We will return to the power gain in Chapter 8. For the moment we consider a telescope used in the receiving mode; here it is natural to think of it as a receiving area, intercepting a power flux $S$. As was first introduced in Section 2.8, a radio telescope intercepts a power flux $S$ that is the rate at which energy $dE$ crosses a unit area $dA$ perpendicular to the direction of propagation:

$$S \equiv dE/dAdt. \tag{5.6}$$

When integrated over the *effective area* $A_{\text{eff}}$ of the telescope we obtain the incoming power $P_{\text{rec}}$ in units of W m$^{-2}$. The effective area $A_{\text{eff}}$ is a function of direction measured with respect to the antenna axis (generally the direction of maximum response) and the received power (in watts) is

$$P_{\text{rec}} = A_{\text{eff}}S. \tag{5.7}$$

The incoming power flux is distributed over a finite receiving band, and it is usually a function of frequency $\nu$. For this reason, we introduce the concept of a *spectrum*, described by the power per unit bandwidth $P_\nu$, derived from the *specific flux* $S_\nu$ or *flux density*, which has units of W m$^{-2}$ Hz$^{-1}$:

$$S_\nu \equiv dE/dAdtd\nu. \tag{5.8}$$

The observed power flux is equal to the flux density integrated over the receiving bandwidth:

$$S = \int S_\nu d\nu. \tag{5.9}$$

The power per unit frequency interval $P_\nu = A_{\text{eff}}S_\nu$ is the *specific power*, or *power spectral density* (PSD), which has units of W Hz$^{-1}$.

The range of directions for which the effective area is large is the *antenna beamwidth*. From the laws of diffraction the beamwidth of an antenna with characteristic size $d$ is of order $\lambda/d$ radians. The detailed power gain over all angles is known, from radio engineering, as its *polar diagram*[6] and its integral is the antenna-beam solid angle $\Omega_A$. The polar diagram, in addition to having a principal lobe (or main beam), exhibits the complete diffraction pattern, and the response outside the principal beam is referred to as the *sidelobe response*. These concepts are explained and illustrated in more detail in Chapter 8.

Most astronomical sources are not black bodies but it is convenient to use a black body as a fundamental reference. We therefore define a source to have *brightness temperature*

---

[6] In optical terminology the polar diagram is usually called the *point spread function*.

$T_B$ such that a hypothetical blackbody at a physical temperature $T_B$ would produce the same specific intensity $I_\nu$, which can often be equated to the specific brightness $B_\nu$, as was explained in Section 2.4. The brightness temperature $T_B$ should not be taken to be a physical temperature in the usual sense although in some cases (see below) it can be.

The specific brightness is expressed as a brightness temperature using the Rayleigh–Jeans approximation:

$$B_\nu = 2kT_B/\lambda^2 \tag{5.10}$$

with units of W m$^{-2}$ Hz$^{-1}$sterad$^{-1}$. This is Eqn 2.13 expressed in its most-often-used form. Then referring back to Section 2.4 we see that the specific brightness and the flux density are related by

$$B_\nu = S_\nu/\Omega_{\text{source}}. \tag{5.11}$$

Alternatively, the flux density is the integral of the specific brightness over the solid angle of the source:

$$S_\nu = \int\int_{\text{source}} B_\nu(\theta,\phi)d\Omega \tag{5.12}$$

with units of W m$^{-2}$ Hz$^{-1}$. In the Rayleigh–Jeans regime we therefore obtain the useful equation

$$S_\nu = 2kT_B\Omega_{\text{source}}/\lambda^2, \tag{5.13}$$

which can also be rearranged to give the brightness temperature for a source of a given flux density.

The distinction between the observables $S_\nu$ and $B_\nu$ becomes clear when we consider how the quantity measured by a radio telescope depends on the solid angle subtended by a point on the source, $\Omega_{\text{source}}$, compared with the solid angle of the reception beam, $\Omega_A$. The limiting situations are illustrated in Figure 5.1 (Here we are ignoring the effect of the sidelobes; we discuss this issue in Chapter 8.)

When the emitting region is larger than the beam the brightness of the region covered by the beam can be measured. The receiver detects the incoming power and, since we know the frequency and bandwidth of the observations, the collecting area of the antenna and the solid angle intercepted (here defined directly by $\Omega_A$ itself) we can measure a quantity with units W m$^{-2}$ Hz$^{-1}$ sterad$^{-1}$ which is the specific brightness. However, when the source is much smaller than the telescope beam its solid angle cannot be determined – we say that the source is unresolved. As a result we cannot measure its brightness and only the more limited information, the flux density in W m$^{-2}$ Hz$^{-1}$, is available. A convenient unit of flux density has been designated the *jansky* (Jy):

$$1 \text{ Jy} = 10^{-26} \text{ W m}^{-2} \text{ Hz}^{-1}.$$

The Moon is one of the strongest radio sources in the sky and we can use it as a quantitative example relating brightness temperature to flux density. The Moon's blackbody temperature is 233 K at 1.4 GHz, its angular diameter is $\approx 0.5°$ and hence its solid angle is $\sim 6\times10^{-5}$ sterad; the Moon's flux density at 1.4 GHz is therefore $\sim 872$ Jy. In comparison the

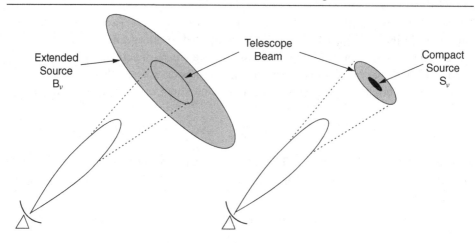

Figure 5.1. Extended and compact sources: the distinction between the observables $S_\nu$ and $B_\nu$. The dotted lines show the solid angle of the reception beam, $\Omega_a$, which is defined by the characteristics of the antenna. (Note that no sidelobes are shown.)

Table 5.1. *Flux densities of a source with visual*
*magnitude* $m_v = 0$

| Band name | Peak wavelength (microns) | $S_\nu$ (janskys) |
|-----------|---------------------------|-------------------|
| V(visual) | 0.556  | 3540 |
| J         | 1.215  | 1630 |
| H         | 1.654  | 1050 |
| K         | 2.179  | 655  |
| L         | 3.547  | 276  |
| M         | 4.769  | 160  |
| N         | 10.472 | 35   |

flux density of Cas A, the strongest source after the Sun, is $\sim 1700$ Jy at 1.4 GHz. Deep surveys with current interferometer arrays such as eMERLIN and the VLA are now sensitive to discrete sources with flux densities down to $\sim 1\,\mu$Jy. The SKA will probe the nJy regime.

The jansky is also widely used in the infrared spectrum; for example the Spitzer Space Telescope and the Herschel Space Observatory have been calibrated over a wide band of wavelengths in terms of janskys. The commonly encountered infrared bands J, H, K, L, M, and N are determined by the Earth's atmosphere, which has numerous absorbing bands of water and carbon dioxide that block observations between the bands. The wavelength equivalents of these bands are shown in Table 5.1, which also shows the flux density in these bands for a star with visual magnitude $m_v = 0$.

The magnitude scale of optical astronomy, which astronomers have used since the time of Hipparchus, is now defined as a logarithmic scale where a magnitude difference of 5 corresponds to a factor of 100 in flux ratio. The scale is fixed by measurements of the bright star Vega, an A0 star with a spectrum close to blackbody. Separate calibrations are necessary for the individual photometric bands conventionally used in optical photometry.[7] The flux densities for the infrared bands in Table 5.1 are quoted in *Allen's Astrophysical Quantities* (Allen 2000).

Radio astronomers, following the physicists and electrical engineers who founded the new science, use a logarithmic scale of power to express the wide range of power encountered in radio engineering, the *decibel*, abbreviated dB and defined as ten times the $\log_{10}$ of the power ratio. A power gain of 10 is therefore 10 dB, of 100 it is 20 dB, and so forth. An optical ratio of 1 magnitude is exactly 4 dB, but no decibel scale for fluxes has been adopted.

### 5.2.1 Brightness Temperatures of Astronomical Sources

The many types of cosmic radio source discussed in Chapters 12 to 17 exhibit a wide range of brightness temperatures. To set the scene we give some examples. They are split into somewhat arbitrary divisions, which are in practice useful aids to memory.

(i) 'Cold', associated with actual blackbody or thermal line radiation from regions with physical temperatures of a few K to about 100 K. Sources include:

  – the cosmic microwave background, $T_B = 2.726$ K;
  – interstellar dust, $T_B = 10$ K to 30 K;
  – molecular clouds, $T_B = 10$ K to 20 K, the source of molecular spectral lines;
  – atomic hydrogen clouds, $T_B = 50$ K to 100 K, the source of the 21 cm spin-flip line.

(ii) 'Warm', associated with diffuse free–free emission from astrophysical plasmas in higher energy environments; $T_B$ is typically 5000 to 15 000 K and is similar to the physical temperature. Sources include:

  – stellar photospheres;
  – tenuous atmospheres around stars (stellar winds);
  – planetary nebulae;
  – gaseous nebulae ionized by hot stars (H II regions).

(iii) 'Hot', associated with non-thermal synchrotron radiation. If $T_B > 10^5$ K, one can assume that the radiation is non-thermal even if the spectrum is not known. Typical synchrotron sources include:

  – supernovae;
  – pulsars;
  – gamma-ray bursters (super-powerful supernovae);
  – active galactic nuclei.

(iv) 'Hot', associated with non-thermal maser sources, which exhibit $T_B$ values from $\sim 10^9$ to $> 10^{12}$ K, and with coherent emission from pulsars, in which $T_B$ may exceed $10^{30}$ K.

---

[7] The flux equivalents for commonly used visual bands were presented by Gray (1998); these are centred on the Johnson and Stromgren V bands at 5480 Å, where the flux of an $m_v = 0$ star is given as $F_\lambda = 3.68 \times 10^9 \frac{5480}{\lambda}^2$ erg s$^{-1}$cm$^{-2}$ Å$^{-1}$; this is equivalent to 3590 Jy at wavelength 0.556 microns.

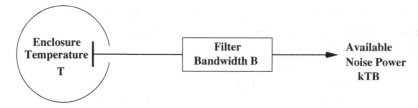

Figure 5.2. An antenna enclosed by a black body at temperature $T$.

## 5.3 Antenna Temperature

The action of a radio antenna is to couple the power flux in the incident radiation field to the receiver system for measurement. Figure 5.2 shows an antenna that is enclosed in a black body, i.e. an enclosure filled with radiation in thermal equilibrium with the walls of the enclosure, at temperature $T$. The output terminals of the antenna are connected to a transmission line that in the radio case is usually a coaxial cable or a waveguide. The signal, with or without amplification, passes through a filter that determines the bandwidth $\Delta\nu$ over which the signal is averaged. The transmission line is terminated by a matched load, a resistor having the same impedance as the line.

If the transmission line is thus terminated by a matched load resistor which is thermally isolated from the outside world, the second law of thermodynamics requires that the resistor will reach an equilibrium temperature $T$ that is identical to the blackbody temperature. Furthermore, the same radio frequency power density $P_\nu = kT\Delta\nu$ (the Johnson noise derived in Section 2.3) must flow in both directions along the transmission line.[8] This applies whatever the termination of the line, whether it is a resistor, another antenna, or an infinite transmission line at temperature $T$. We return to the subject of matching to maximize power transfer in Section 6.1.1.

If we imagine that the antenna in Figure 5.2 is replaced by a radio telescope with a narrow receiving pattern that only gathers the radiation from a small solid angle in the blackbody, the result is the same. This leads to the conclusion that, if the telescope is viewing a blackbody surface, at temperature $T$, that is larger than the receiving pattern of the telescope then a matched load will still reach the temperature $T$ after it reaches thermal equilibrium with the incoming radiation from the blackbody surface. This circumstance is called the *filled beam* case and is the situation represented in Figure 5.1. It is necessarily an approximation, because all the sidelobes of the telescope beam pattern must also be within the angle subtended by the black body, but the sidelobe patterns fall rapidly enough for most telescopes to make the approximation a useful one, except for highly exact measurements such as those of the CMB (Chapter 17). Another circumstance where the full sidelobe pattern must be taken into account arises when one is viewing an extended object in a

---

[8] The Johnson noise is independent of frequency, which seems inconsistent with the specific intensity of the incoming blackbody radiation, $I_\nu = 2kT\nu^2/c^2 = 2kT/\lambda^2$ (Eqn 2.18); however, the antenna beamwidth is proportional to $(\lambda/d)^2$ and so the wavelength dependencies exactly cancel.

region of low surface brightness and there is a region of high surface brightness nearby. For example, this is necessary when studying the faint hydrogen emission in the outer parts of galaxies, where the more intense emission from the inner parts can be picked up by the sidelobes and result in a false signal being added to the signal under study.

Returning to the concept of power gain, when transmitting, the antenna has a direction-dependent power gain $G(\theta,\phi)$ (referring to the polar diagram) relative to an idealized isotropic antenna; $G(\theta,\phi)$ is therefore unitless. The formal relationships defining polar diagrams and antenna beams in general will be derived in Section 8.3. For our present purposes it is sufficient to know that the maximum power gain over an isotropic antenna (i.e. the on-axis power gain) can be written as

$$G = \frac{4\pi}{\Omega_A},$$
(5.14)

where $\Omega_A$ is the antenna beam solid angle, which includes both the main lobe and the sidelobes.

Equation 5.14 provides a simple way to think about the on-axis power gain, as the number of times the antenna beam solid angle $\Omega_A$ can be fitted into the $4\pi$ steradians of a sphere.[9] We can now relate the gain and the effective area of the antenna. In Figure 5.2 let the antenna have an isotropic gain pattern, i.e. it has gain equal to unity in all directions and its effective area is therefore constant and independent of direction (while these idealizations are impossible to achieve in practice, the argument remains valid). The antenna is bathed in a radiation field of specific intensity $I_\nu$ and the power density that it gathers in and delivers to its output will be given by, in the Rayleigh–Jeans regime,

$$P_\nu d\nu = \frac{1}{2}\int_{4\pi} I_\nu A_{\mathrm{eff}} d\Omega d\nu = 2\pi A_{\mathrm{eff}}\left(\frac{2kT}{\lambda^2}\right)d\nu.$$
(5.15)

The factor $1/2$ arises from the fact that an antenna can deliver power from only one component of polarization to a given output port.

This, by the second law of thermodynamics, must equal the power flow in one direction in the transmission line from the matched load at temperature $T$ to the antenna port. Equating this to the Johnson noise from the matched load we obtain

$$kT d\nu = \frac{4\pi}{\lambda^2} kT A_{\mathrm{eff}} d\nu$$
(5.16)

and, since the gain of the isotropic antenna is unity, it follows that, for any antenna,

$$A_{\mathrm{eff}} = \frac{\lambda^2}{4\pi} G.$$
(5.17)

This relation is valid for any off-axis direction. An often-used corollary of Eqn 5.17 is obtained by substituting for $G$ using Eqn 5.14, yielding

$$A_{\mathrm{eff}}\Omega_A = \lambda^2.$$
(5.18)

---

[9] Strictly this number is the *directivity*; the power gain takes into account losses in the antenna and transmission line but these are often small and it is common to refer only to the gain.

The above considerations lead to the convenient concept of an *antenna temperature* $T_A$, which is the temperature that a resistor would have if it were to generate the same power density at frequency $\nu$ that is observed to be coming from the antenna output port. The definition is not restricted to blackbody conditions: one considers only the narrow band of frequencies in which the measurement is made.

In Section 2.4 we defined the *brightness temperature* $T_B$ as the temperature at which a black body would have to be in order to duplicate the observed specific intensity of the emitting source. In the particular situation shown on the left in Figure 5.1, where the entire telescope beam is filled with uniform radiation of brightness temperature $T_B$, the antenna temperature $T_A$ is equal to $T_B$, and the antenna plus receiver combination acts like a 'radio thermometer' or *radiometer*.

We should also consider the situation where the source is smaller than the beam, as on the right-hand side of Figure 5.1. The expected antenna temperature can then be calculated if the flux density or the solid angle of the source is known. Taking these in turn: the power density received by an antenna of effective area $A_{eff}$ observing an unpolarized source of flux density $S_\nu$ is $S_\nu A_{eff}/2$ (the factor 2 is for one polarization per receiver). Equating this to the power density $kT_A$ from a matched load at temperature $T_A$ and rearranging gives

$$T_A = \frac{S_\nu A_{eff}}{2k}. \tag{5.19}$$

Alternatively, with an antenna beam $\Omega_A$ the antenna temperature from a source of brightness temperature $T_{source}$ and solid angle $\Omega_{source}$ is

$$T_A = T_{source}\left(\frac{\Omega_{source}}{\Omega_A}\right), \tag{5.20}$$

where the ratio of the solid angles is the *beam dilution factor*. It is left as an exercise for the reader to show that Eqns 5.19 and 5.20 are equivalent.[10]

To recapitulate: the noise power per Hz received from a source with brightness temperature $T_B$ (its effective blackbody temperature) filling the entire antenna beam solid angle $\Omega_A$ is independent of frequency and equal to the Johnson noise power $kT_B$ per Hz from a matched resistive load at physical temperature $T_B$. Temperature is therefore a convenient way to characterize the amounts of power in a radio astronomy system and hence we use the concepts of the brightness temperature, the antenna temperature, and (in the next chapter) the receiver noise temperature. These temperatures can be conveniently calibrated with respect to a source of blackbody radiation, as we now describe.

### 5.3.1 Adding Noise Powers

As we discuss in Chapter 6, independent noise signals are combined in the receiver, which also generates its own noise. The amplified voltages are square-law detected and averaged

---

[10] Use Eqn 5.18 and $S_\nu$ = average brightness $\times \Omega_{source}$.

over many coherence times. The form of the combination is surprisingly simple. For a pair of noise voltages we have

$$\langle [V_1(t) + V_2(t)]^2 \rangle = \langle V_1(t)^2 \rangle + \langle V_1(t)^2 \rangle + \langle 2V_1(t)V_2(t) \rangle. \tag{5.21}$$

Since the voltages are from independent sources of radiation they are 'uncorrelated', i.e. they have no long-term phase relationships with each other. As a result the average of the product term $\langle 2V_1(t)V_2(t) \rangle$ tends to zero and we are left with

$$\langle [V_1(t) + V_2(t)^2] \rangle = \langle V_1(t)^2 \rangle + \langle V_1(t)^2 \rangle = T_1 + T_2. \tag{5.22}$$

In other words the noise powers, not the voltages, add. The antenna temperature $T_A$ is in practice the sum of various contributions, as follows.

### 5.3.2 Sources of Antenna Noise

The antenna temperature $T_A$ corresponds to the noise generated by the total antenna system and presented to the radiometer. The contributors are:

– radiation from the source or region under study, $T_s$;
– radiation from the cosmic microwave background, $T_{cmb}$;
– radiation from the Milky Way galaxy, $T_{gal}$;
– radiation from the atmosphere, $T_{atm}$;
– radiation picked up in the sidelobes of the beam, in most cases principally thermal noise from the ground, $T_{sl}$;
– thermal noise associated with losses in the antenna itself, in the transmission line, or in any passive components in the signal path before the low-noise amplifier, $T_{loss}$.

The antenna temperature is the following sum:

$$T_A = T_s + T_{cmb} + T_{gal} + T_{atm} + T_{sl} + T_{loss}. \tag{5.23}$$

If a discrete celestial source is being studied, its noise contribution is often negligibly small, but a bright source can contribute significant noise $T_s$ if the telescope collecting area is large enough.

Figure 5.3 summarizes the main contributions to the antenna temperature as a function of frequency. At low frequencies, the intense galactic background noise dominates; as the frequency increases this background temperature decreases ($T_{gal} \propto \nu^{-2.8}$). At centimetre wavelengths, radiation from the atmosphere and stray radiation from the ground are the dominant noise sources unless one is observing an intense source such as a nearby supernova remnant or the Sun.

Low-noise amplifiers for the first stages of receivers at 10 GHz and above are usually InP high electron mobility transistors (HEMTs) cooled to 20 K, providing receiver temperatures comparable with or below the sky background (see Section 6.4.1). For later amplifier stages, and for the first stage of receivers at lower frequencies, GaAs metal-oxide–semiconductor field effect transistors (MOSFETs) are also used; these have the advantage of giving a more linear performance.

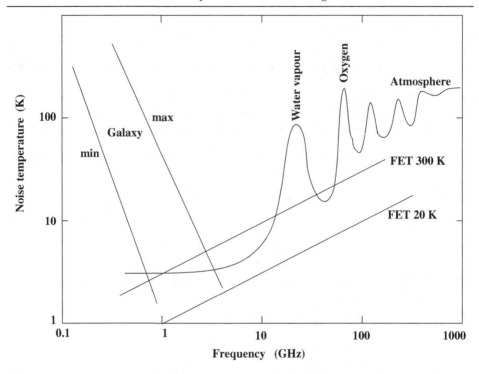

Figure 5.3. Typical sky and receiver temperatures. The steep spectrum of galactic radiation dominates at frequencies below 1 GHz. Atmospheric absorption dominates above 20 GHz. FET 300 K and FET 20 K refer to room-temperature and cooled field effect transistors.

At millimetre wavelengths, radiation from water vapour and oxygen in the atmosphere causes the antenna temperature to rise dramatically (note the effects at line frequencies such as that of the water line at 22.3 GHz and the extensive oxygen complex at 55–60 GHz). At submillimetre wavelengths, absorption by water vapour becomes increasingly severe, and high, dry, observing sites are required (see also the discussions in Chapters 9 and 11).

### 5.3.3 Measuring the Antenna Temperature

We will take operational advantage of the definitions of antenna temperature and thence brightness temperature, since black bodies and resistive loads at a well-determined temperature are the measurement standards. The process of measurement is schematized in Figure 5.4, which shows a simplified radiometric system that compares the antenna temperature to the temperature $T_{ref}$ of a reference resistor whose physical temperature must be very carefully monitored and controlled. An alternative to a resistor is a sheet of highly absorbent material in front of an input feed horn.

The antenna temperature obviously depends on the brightness temperature in the beam, which may vary across it. This may have a large effect on the system noise, as, for example, if the Sun were to occupy part of the telescope beam. The Sun's brightness temperature

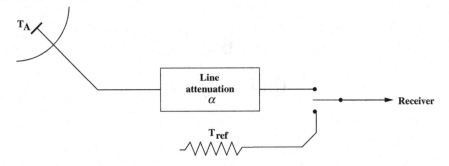

Figure 5.4. A simple radiometer, showing the effect of line attenuation. The receiver can be switched from the antenna to a resistive load with effective noise temperature $T_{ref}$.

can range from 6000 K (short-centimetre and millimetre wavelengths) to more than $10^6$ K (metre wavelengths), so a serious degradation of sensitivity may occur even if the Sun is only in the sidelobes of the antenna pattern. As a result, sensitive total-power measurements using ground-based telescopes may be restricted to night-time hours while spacecraft measuring the CMB always point away from the Sun and the Earth and have radiation shields around the reflector dish.

Another practical consideration is loss in the transmission line to the receiver. This is directly akin to radiative transfer, considered in Section 2.5. If the line is at temperature $T_{line}$, and the fraction of the signal transmitted is $\alpha$, then in the ideal case of matched antenna and line the antenna temperature becomes

$$T_A = \alpha T_B + (1 - \alpha)T_{line}, \tag{5.24}$$

which is an application of Eqn 2.28. In practice one quotes the line loss in dB (decibels); thus, for a loss of 0.5 dB, $\alpha = 10^{-0.05} = 0.891$ (an 11% loss of signal) and if $T_{line} = 290$ K (ambient temperature) then the unwanted addition to the noise is $290(1 - 0.891) \approx 32$ K. Loss in front of the receiver is a double blow to performance.

In Chapter 6 we will trace the various noise contributions into and through a telescope and receiver system and describe how the overall system is calibrated.

## 5.4 Further Reading

*Supplementary Material* at www.cambridge.org/ira4.

Davenport, W. D., and Root, W. L. 1958. *An Introduction to the Theory of Random Signals and Noise*. McGraw Hill.

Clauset, A., Shalizi, C. R., and Newman, M. E. J. 2009. Power law distributions in empirical data. *SIAM Review*, **51**, 661–703.

# 6

# Radiometers

A radio telescope intercepts a portion of the randomly fluctuating incident radiation field and directs it into the receiver, where it is turned into a fluctuating noise voltage. If the noise is purely random, with no correlation from one sample to the next, then the only measurable quantity is the intensity, or total power, measured with a *radiometer*; the frequency spectrum is flat, without features. If the frequency spectrum contains discrete line components, there will be statistical correlations between samples. A receiver which measures the signal power as a function of frequency is a *spectrometer* (Chapter 7). In Chapter 2 we saw that radiation generated by thermal processes is unpolarized but that non-thermal synchrotron radiation in particular can be significantly polarized, with the state of polarization containing information about the emitting region and the medium between the source and the telescope. A receiver which is sensitive to the polarization state is a *polarimeter* (Chapter 7). It is possible to combine all three capabilities in one receiver.

In Chapter 5 the receiver took the simplest form of a matched resistor whose equilibrium temperature reflected the incoming power. But natural radio signals are very weak and must be amplified by at least $\sim 100$ decibels (a factor $\sim 10^{10}$ in power, see Section 6.8) before they can be detected at a convenient signal level. The use of amplifiers for the first stage of a receiver defines the boundary between radio and mm and sub-mm wave astronomy. At wavelengths longer than a few mm (below $\sim 100$ GHz) low-noise transistor amplifiers (LNAs) are used; however, the laws of quantum mechanics require any amplifier to add noise to the signal, and at the shorter mm and sub-mm wavelengths transistor amplifiers generate too much self-noise. They have been replaced by superconductor–insulator–superconductor (SIS) mixers, which do not provide gain but are well suited for spectroscopy and interferometry. However, SIS mixers also add noise, so, for the most sensitive mm and sub-mm wave radiometry and polarimetry, the energy of the incoming stream of photons is measured directly with superconducting *bolometers*.

No matter what the wavelength band, there will always be extraneous noise signals that corrupt the observations and, almost always, the cosmic radio signal is far weaker than the extraneous noise. In this chapter we focus on the challenges faced by the radio astronomer in extracting the total power. In Chapter 7 we deal with measurements of the frequency spectrum and the polarization state of the incoming signal.

## 6.1 The Basic Radiometer

In its simplest form, a receiver can be represented as a combination of an *ideal amplifier* and a lossless *passive filter* that attenuates frequencies outside the bandpass, followed by a square-law power detector. This basic radiometer is illustrated in Figure 6.1.

An ideal receiver should be *linear*, which means that it amplifies signals within the bandpass faithfully, with no distortion. In practice a series of amplifiers are needed to provide the $\sim 100$ dB gain required and there is often a translation to a lower frequency before the last amplification stages (Section 6.8).

### 6.1.1 Impedance Matching and Power Transfer

The efficient transfer of power collected by the antenna into the receiver is critical given the extreme faintness of natural signals. It is a characteristic of any type of travelling wave that an abrupt change in propagation conditions results in power being reflected back at the interface, with a corresponding loss in transmission efficiency. Maximum power transfer between two components in the receiver chain therefore occurs when their (in general complex) impedances are matched, i.e. when one is the complex conjugate of the other. Ordinarily, one tries to obtain impedances that are purely resistive, or at least have as small a reactive component as possible.

The schematic antenna system shown in Figure 6.1 is a combination of a parabolic reflector with a 'feed', usually a waveguide horn (see Chapter 8), at its focus. The first task is therefore to couple radiation travelling in free space (whose characteristic impedance is 377 ohms) into the waveguide transmission line, whose impedance is invariably 50 ohms.[1] Impedance matching can be achieved by shaping the horn, most simply by reducing the cross-section from the mouth to the throat where it meets the transmission waveguide (see Chapter 8).

The horn's electrical impedance is dominated by its *radiation resistance* – a concept most easily grasped in transmission mode. Power flows down the transmission line to the horn where it is launched into space as EM radiation. If the impedance presented by the horn is matched to the transmission line then the power transfer is maximized – but note that as much power is lost in the transmitter as is transferred into space.[2] Since power is being

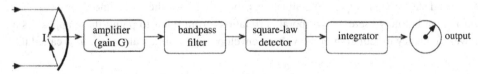

Figure 6.1. The basic components of a linear radiometric receiver.

---

[1] This standard emerged in the 1940s for transmission lines used in high-power transmitters as a compromise between power handling and low loss.

[2] The effects of impedance matching and power sharing are exhibited by a simple d.c. circuit in which a battery of voltage $V$ and internal resistance $R_{int}$ is connected to an external resistor $R_{ext}$. By applying Ohm's law one

dissipated as radiation the horn can be regarded as having a radiation resistance $R_r$. Note that power will also be lost as heat in the ohmic resistance in the material of the horn; this should be minimized for the reasons laid out in Section 5.3.3. When the antenna is in receive mode it absorbs power and sets up currents $i(t)$ in the transmission line corresponding to a power $\langle i(t)^2 \rangle R_r$. Impedance matching 'downstream' of the antenna ensures maximum power transfer to the receiver but again only half the power is transferred; an equal amount is reradiated from the horn (if it is lossless).

### 6.1.2 Power Amplification

Returning to Figure 6.1, the input signal amplitude $v_s(t)$ from the antenna is amplified to become $v_0(t)$ at the output of the amplifier and filter (the signal amplitude is designated by its associated voltage but the current could equally well be used, as above). The power gain of the amplifier is frequency dependent but extends over a larger range than that of the bandpass-defining filter $\Delta\nu$ and for our purposes can be represented by a single value $G$, yielding an output amplitude $v_0(t) = \sqrt{G}v_s(t)$.

All the incident power is transferred from the source to the amplifier, with no reflection, if the input impedance of the amplifier is matched to the source impedance. It is, however, often the case that the noise generated in the amplifier is dependent on the impedance match and, consequently, the optimum ratio of signal to noise is obtained with a considerable mismatch.[3] As bandwidths have grown in modern-day radiometers, achieving an optimal impedance match across the operating band can be an engineering challenge. Since power is voltage squared divided by the resistance, it follows that the square of the ratio of output and input voltages will give the power gain only when the matched impedances at input and output are equal. More generally, if the receiver has an input impedance $Z_i$ and an output load $Z_0$,

$$G = \frac{v_0^2}{v_i^2} \frac{Z_i}{Z_0}. \tag{6.1}$$

### 6.1.3 Bandwidth and Coherence

Before the bandpass filter the voltage has the form of white noise, with rapid fluctuations and zero mean. Figure 6.2 illustrates that after the filter the time series still has zero mean but now there are temporal correlations from point to point; these appear as a lower frequency modulation of the voltage envelope whose timescale $\sim 1/\Delta\nu$ is called the

---

can easily show that the maximum power dissipated in the external resistor occurs when $R_{ext} = R_{int} = R$ and is equal to $V^2/4R$. The same amount of power is dissipated in the internal resistance of the battery.

[3] This seems counterintuitive since signals from the sky will be reflected back. But one of the subtleties of low-noise amplifier (LNA) design is that the noise performance and the input match are not directly related (Pospieszalski 1989). Whilst matching enables more signal power to pass into the LNA, the noise power will also increase in such a way that the signal-to-noise ratio may be degraded.

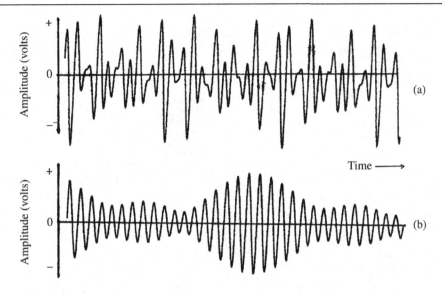

Figure 6.2. Temporal coherence and frequency bandwidth in random noise: (a) broad bandwidth, $\Delta\nu = \nu_0/2$; (b) narrow bandwidth, $\Delta\nu = \nu_0/10$.

*coherence time* $\tau_{\mathrm{coh}}$. This behaviour can be regarded as a statistical extension of the concept of 'beats' between two sinusoids (see also Appendix 1) but more generally as an elementary case of the effect of a receiver *transfer function*. Transfer functions are considered in more detail in Section 7.2.

Two examples of filtered white noise are shown in Figure 6.2, the first being relatively broadband, with a bandwidth of half the centre frequency $\nu_0$, while the second is for a bandwidth only one-tenth of $\nu_0$. As the bandwidth is reduced the voltage variations within a coherence time increasingly take on the form of quasi-sinusoids of frequency $\nu_0$, and the number of zero crossings ($\sim \nu_0/\Delta\nu$) within $\tau_{\mathrm{coh}}$ grows. For example, if $\nu_0$ is 1 GHz and $\Delta\nu$ is 100 MHz the envelope varies on timescales of 10 nanoseconds and typically contains 10 zero crossings. These considerations become important when we consider the theory behind interferometers in Chapter 9.

## 6.2 Detection and Integration

Although it now has some persistence in time, the band-limited noise signal is still a Gaussian random variable; to measure the power in this fluctuating signal we need to turn it into a steady non-zero voltage. The simplest way to do this is to square the noise voltage and average the result for a long period $\mathcal{T}$.

Since power is proportional to the square of the voltage amplitude the signal must be multiplied by itself. This is conventionally done in a *square-law device*, usually known

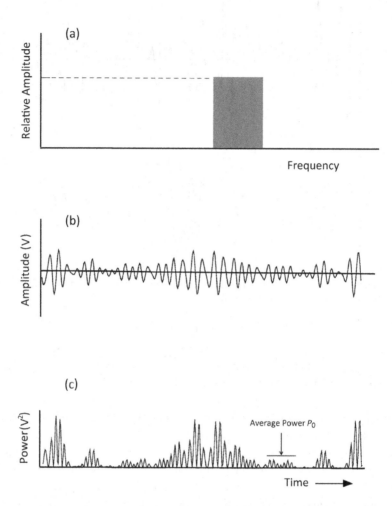

Figure 6.3.  Detection of band-limited noise. (a) A bandpass filter centred at $\nu_0$ passes a frequency range $\Delta\nu$. (b) The zero crossings of the filtered random noise are at typical time intervals $\Delta t \sim 1/\nu_0$. The waveform is quasi-sinusoidal within a modulation envelope exhibiting a coherence time $\tau_{\mathrm{coh}} \sim 1/\Delta\nu$. (c) The same wave after square-law detection.

as a *detector*. The effect is shown in Figure 6.3. The detector can be a simple diode, with associated filters to prevent the radio frequency signal from proceeding on and possibly interfering with the following circuitry. The average power may be written as

$$\langle P \rangle = \int_{t-T/2}^{t+T/2} p_{\mathrm{d}}(t)\mathrm{d}t = \int_{t-T/2}^{t+T/2} [V_o(t)]^2 \mathrm{d}t. \tag{6.2}$$

In this expression the instantaneous power $p_d(t)$ measured by the detector is the square of the amplitude $V_o(t)$ at the output port of the receiver. Since $\mathcal{T}$ is finite, $\langle P \rangle$ remains only an estimate of the true value, with an associated uncertainty.

The averaging circuit can take many forms. In earlier radiometers (and even today in some instances), an analogue integrator was used, typically an $RC$ circuit acting as a low-pass filter:

$$\langle P \rangle = p_d(t)(1 - e^{-t/RC}) \tag{6.3}$$

and the accumulated charge on the capacitor is effectively an average over the time constant $RC$ of the circuit.

In present practice the signal is almost always digital, the integral in Eqn 6.2 being taken repeatedly over the averaging time $\mathcal{T}$. These integrations are stored, and can be summed later into a longer-term average if desired.

### 6.3 Post-Detection Signals

#### 6.3.1 Time Series

What is the form of the signal after square-law detection and averaging? As a first illustration consider the case of a single sinusoidal voltage component of the noise, $V_o(t) = A \cos(2\pi V_o t)$, which is squared to yield[4] $p_d(t) = 0.5A^2\{1 + \cos[2(2\pi V_o t)]\}$; this is a double-frequency sine wave with minimum zero and peak value $A^2$. After averaging, the oscillating component tends to zero and we are left with the average power $\langle P \rangle = A^2/2$.

When the band-limited noise signal is squared the individual sine waves behave in the above manner but they have a range of amplitudes, frequencies, and phases. The time series of the fluctuating power $p_d(t)$ is shown in Figure 6.3(c).The large peaks in the envelope of $p_d(t)$ are associated with the maxima in the voltage envelope, while within the envelope there are sinusoidal fluctuations at $2V_o$. The averaged output $\langle P \rangle$ from the integrator has a constant component, $P_0$, which is the average power that would be measured for an infinite integration time, and a randomly varying component $\Delta P(\mathcal{T})$. Thus the detector power can be written

$$\langle P \rangle = P_0 + \Delta P(\mathcal{T}). \tag{6.4}$$

It is clear from Figure 6.3(c) that the distribution of $p_d(t)$ is far from Gaussian; in particular its rms deviation from the mean is much larger than that for the original noise voltage distribution.[5] The probability distribution of $p_d(t)$ is vitally important in the analysis of radar systems with a range of input return signal strengths. With no additional signals and only Gaussian noise into the detector the probability follows a chi-squared distribution for one degree of freedom; when a signal is added the probability function is a Rice distribution.

---

[4] Using the trigonometrical identity $\cos^2 \theta = (1 + \cos 2\theta)/2$.
[5] As an elementary illustration of the effect, take the integers 1 to 5; their mean is 3.0 with rms deviation 1.6; after squaring, the mean is 11.0 and the rms deviation is 9.7.

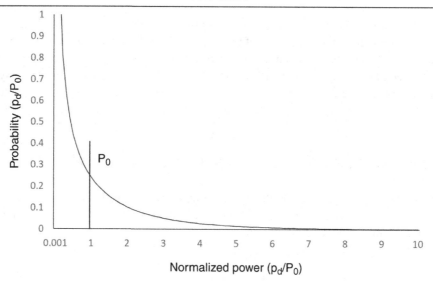

Figure 6.4. The probability distribution of post-detection noise.

The noise-only case is shown in Figure 6.4, where the $y$ axis is $p_d(t)$, normalized to the steady mean $P_0$. The mean value of chi-squared for $k$ degrees of freedom is $k$ and its variance is $2k$; hence for $k = 1$ the mean $P_0$ in Figure 6.4 is at $(p_d(t)/P_0) = 1$ and the rms deviation about the mean is $\Delta P(\mathcal{T}) = \sqrt{2}P_0$.

The uncertainty in the final measurement can now be specified. A detailed analysis is given in Thompson, Moran, and Swenson (2017, Appendix 1.1), but we can employ a heuristic argument as follows. For an averaging time $\mathcal{T}$ there are $\mathcal{T}/\tau_{coh}$ coherence times in each integration and to capture all the information in the $p_d(t)$ time series we require two samples per coherence time (according to the Nyquist–Shannon theorem – see Appendix 3); hence, there are $N = 2\mathcal{T}\Delta\nu$ independent samples. Even though the probability distribution of $p_d(t)$ is non-Gaussian, for many samples we can invoke the central limit theorem and expect the error on the mean value to be the rms deviation divided by $\sqrt{N}$. We have already seen that the rms deviation $\Delta P(\mathcal{T}) = \sqrt{2}P_0$, so the error on the mean is $\sqrt{2}P_0/(2\mathcal{T}\Delta\nu)^{1/2} = P_0/(\mathcal{T}\Delta\nu)^{1/2}$.

In the Rayleigh–Jeans regime the power spectral density $P = kT_{sys}$ where $T_{sys}$ is the summed total of noise from the antenna and self-noise from the receiver; $T_{sys}$ is called the *system noise temperature*. We therefore obtain a relation for the rms fluctuations in the output from the integrator:

$$\Delta T_{rms} = \frac{T_{sys}}{\sqrt{\Delta\nu\mathcal{T}}}. \tag{6.5}$$

This is the *radiometer equation*, one of the most important in radio astronomy. For an ideal square bandwidth[6] and averaging time, the equation is exact provided that $T_{sys}$ is uniform

---

[6] The effective bandwidths of filters with various-shaped frequency responses are tabulated in Bracewell (1962).

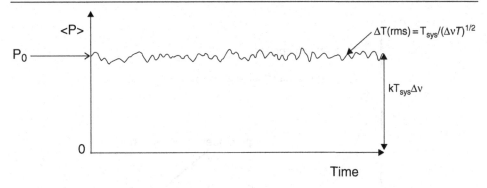

Figure 6.5. The averaged post-detector output.

over the band. For a digitized signal, a correction factor, dependent on the number of bits per sample, is required. Figure 6.5 shows a pictorial representation of the power output after square-law detection and averaging.

The receiver measures the total power, of constant value $P_0 = kT_{sys}\Delta\nu$, calibrated in terms of the system temperature $T_{sys}$, but this is still fluctuating. Increasing $\Delta\nu$ increases the power output and decreases the level of fluctuations relative to $T_{sys}$. Increasing $\mathcal{T}$ also decreases the relative fluctuations. The effect of both is, however, only proportional to their square root and hence both wide bandwidths and long integration times are required to detect faint signals.

### 6.3.2 Spectrum

A complementary way of thinking about the signal after detection is via its frequency spectrum. The full treatment is algebraically involved, but the main steps can be understood heuristically. The input white-noise voltage can be represented as a sum of sine waves with different amplitudes, frequencies, and phases, i.e.

$$V(t) = \sum_i a_i \cos(2\pi\nu_i t + \phi_i). \tag{6.6}$$

Square-law detection involves multiplying the signal by itself, which is a non-linear process (compare with mixing in Section 6.6), so different input frequencies combine to produce new frequencies in the post-detection output. A simplification arises from the fact that all the input phase information is lost since squaring implies forming the product $V(t)V^*(t)$, where $V^*(t)$ is the complex conjugate. After the detector we can therefore represent the power as

$$p_d(t) = V^2(t) = \sum_i a_i^2 \cos^2(2\pi\nu_i t) + \sum_i \sum_j a_i a_j \cos(2\pi\nu_i t) \cos(2\pi\nu_j t). \tag{6.7}$$

The first term on the right-hand side represents the individual input frequencies multiplied by themselves; the second term represents the frequency cross-products. Equation 6.7 can

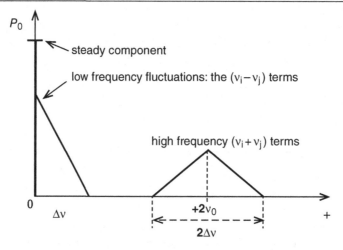

Figure 6.6. Spectrum of post-detection noise.

be written as[7]

$$p_d(t) = \frac{1}{2} \sum_i a_i^2 + \frac{1}{2} \sum_i \sum_j (a_i a_j)\{\cos[2\pi(\nu_i - \nu_j)t] + \cos[2\pi(\nu_i + \nu_j)t]\}. \quad (6.8)$$

The first term on the right-hand side represents the constant term $P_0$. The second term contains both low frequency, $\nu_i - \nu_j$, and high frequency, $\nu_i + \nu_j$, components. The latter lie around twice the centre frequency of the bandpass while the former represent the envelope modulation. There are many ways to produce low modulation frequencies by choosing $\nu_i$ and $\nu_j$ close together within the bandpass $-\Delta\nu/2$ to $+\Delta\nu/2$. But as $\nu_i - \nu_j$ increases there are fewer ways to choose appropriate values of $\nu_i$ and $\nu_j$, while for $\nu_i - \nu_j = \Delta\nu$ there is only one way to choose. These considerations underlay the characteristic triangular shapes in the post-detection spectrum of band-limited noise shown in Figure 6.6.[8] To measure the constant term $P_0$ as accurately as possible the high frequency spectral components must be suppressed by averaging. This has the effect of a low-pass filter, rejecting frequencies above $\sim 1/\mathcal{T}$ and retaining only spectral components with frequencies close to zero.

### 6.3.3 Recognizing a Weak Source

To illustrate what we have discussed so far, Figure 6.7 shows a schematic radio telescope record of the receiver output power after averaging over a timescale $\mathcal{T} \sim 1$ second. Note that

---

[7] Following the reasoning in Section 6.3.1 and using the identity $\cos A \cos B = [\cos(A - B) + \cos(A + B)]/2$.
[8] An equivalent way to arrive at this power spectrum is by autocorrelating the rectangular voltage spectrum in Figure 6.3, with the addition of a symmetric band of negative frequencies, since the voltages are real.

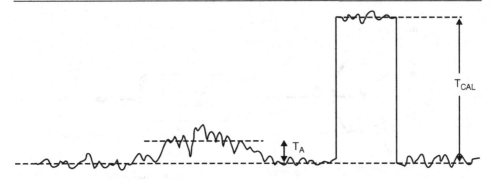

Figure 6.7. A weak radio source passing through a telescope beam, followed by an injected calibration signal. Note that the whole recording sits on top of the system noise $T_{sys}$.

the large constant component of $T_{sys}$ shown in Figure 6.5 has been suppressed to emphasize the fluctuations. The telescope beam scans across a weak source, taking some tens of seconds to pass; the source gives rise to a peak antenna temperature $T_A$ several times larger than the rms fluctuations $\Delta T$ in Eqn 6.5. This variation can readily be detected. However, if for a weaker source $T_A = 1$ millikelvin, $T_{sys} = 100$ K and $\Delta\nu = 100$ MHz, then even a one-sigma variation in the output would require $\mathcal{T} \sim 100$ seconds and the source would not be recognized. To make recognition almost certain (at five sigma), averaging over 2500 seconds would be needed.

Such long integration times place severe requirements on the gain stability of the system, since tiny (e.g. one part in $10^4$ to $10^5$) gain fluctuations in the receiver can mask real sources. In Section 6.5 we consider the temperature calibration of the system, which sets the magnitude of the thermal noise fluctuations $\Delta T$, and in Section 6.8 we show how receiver gain variations can be corrected with changes to the design of the radiometer.

## 6.4 System Noise Temperature

The sensitivity of a telescope may be expressed as a single system temperature $T_{sys}$, in which the various noise contributions from antenna and receiver are added. The sources of antenna noise were discussed in Section 5.3.1.

### 6.4.1 Receiver Temperature

Sometimes $T_{loss}$, associated with attenuation in several passive components in front of the low-noise amplifier (LNA), is included with the receiver temperature. We have chosen to include $T_{loss}$ with the antenna temperature and so start our discussion of the receiver temperature $T_{rec}$ with the contribution of the first stage.

As we will see in Section 6.7, the noise temperature of a well-designed receiver is completely dominated by the Johnson noise generated by its first stage. In most radio

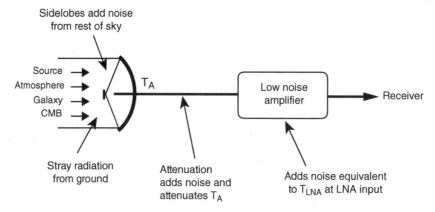

Figure 6.8. Schematic of noise contributions.

astronomy receivers this is a low-noise amplifier based on high electron mobility transistors (HEMTs) with an excess noise temperature contribution $T_{LNA}$. Obviously the LNA design is aimed at adding as little noise as possible, so cooling the first stages of receivers typically to 20 K, using closed cycle helium gas refrigerators, is standard practice for frequencies from 1 GHz to above 100 GHz. The cooled *cryostat* section (the so-called *front end*) may often include a filter to reduce interference from strong radio signals in adjacent frequency bands. These filters may use superconducting materials, which introduce an almost negligible loss in the observing frequency band. The lowest receiver noise is available from indium phosphide HEMTs. Other types of field effect transistors (FETs) are also used, some of which have lower $1/f$ noise and better linearity in response to large signals, an advantage in situations where large interfering radio signals may occur.

Even with cooling and the best transistors, an absolute lower limit, $T_{LNA} > h\nu/k$, is imposed by the laws of quantum mechanics. The so-called *quantum limit* $\approx 0.05$ K per GHz; hence $T_{LNA}$ cannot be less than 0.5 K at 10 GHz. In practical amplifiers it can be up to ten times the quantum limit although state-of-the art systems achieve 4–5 times $h\nu/k$ at frequencies up to 100 GHz and beyond. The current state of the art is summarized in Figure 5.3, where receiver temperatures are shown in relation to typical antenna temperatures.

At frequencies significantly below $\sim 1$ GHz, the front end is usually not cooled, with significant cost savings. Although $T_{LNA}$ is typically 9–10 times higher than when cooled to 20 K this becomes less and less important as the galactic background noise increases. At frequencies below $\sim 200$ MHz down to tens of MHz, where wide-field arrays such as LOFAR, MWA, and LWA operate, the galactic contribution completely dominates $T_{sys}$. In this case the receiver designer stresses the linearity of the front end in the face of high intensity signals including manmade ones (Tillman *et al.* 2016).

Figure 6.8 is a schematic showing the various contributions to noise in a radiometer. Table 6.1 shows some typical system noise budgets for a range of receivers operated at sea

Table 6.1. *Typical system-noise budgets (in kelvins) for radio astronomy receivers at sea level, operating at ambient temperature (300 K, (a)) and cryogenic temperature (20 K, (c)).*

| Source | 400 MHz (a) | 1.4 GHz (c) | 10 GHz(c) | 30 GHz (c) |
|---|---|---|---|---|
| CMB | 3 | 3 | 3 | 3 |
| Galaxy | 20 | 2 | < 1 | < 1 |
| Atmosphere | < 1 | 2 | 4 | 13 |
| Sidelobes | 7–10 | 7–10 | 7–10 | 7–10 |
| Loss | 5 | 3 | 3 | 3 |
| LNA | 20 | 4 | 5 | 10–15 |
| Total | 55–58 | 21–24 | 22–25 | 33–41 |

Notes: (i) the galactic contribution corresponds to regions well away from the plane of the Milky Way, and for $\nu <$ 400 MHz it rapidly becomes dominant (see Figure 5.3); (ii) the loss contributions are strongly system dependent and can vary considerably from the quoted values; (iii) the sidelobe contributions are typical for parabolic dishes and can be reduced at the expense of $A_{\text{eff}}$ (see Chapter 8); (iv) at 1.4 GHz, low-noise amplifiers at ambient temperature have $T_{\text{LNA}} \leq 30$ K and can produce cost-effective systems, while at higher frequencies cryogenic cooling is always required; (v) at 30 GHz the atmospheric contribution is for sea level in clear-sky conditions with precipitable water vapour (PWV) = 20 mm.

level. Further information on low-noise amplifiers for radio astronomy techniques may be found in Bryerton *et al.* (2013).

### 6.4.2 Receivers for Millimetre and Sub-Millimetre Waves

At millimetre and shorter wavelengths the noise performance of HEMT-based amplifiers becomes uncompetitive. If the receiver is to be used for spectrometry or interferometry it should, like those based on transistor LNAs, operate coherently. This means it should maintain the phase relationships of the incident radiation to allow for temporal comparisons within the receiving band (autocorrelation, see Chapter 7) or comparison with signals from other spatially separated telescopes (cross-correlation, see Chapters 9 and 10).

For coherent operation at mm and sub-mm wavelengths, SIS (superconductor–insulator–superconductor) diodes give the best performance for the first stage, heterodyning (see Section 6.6) the signal down to a lower frequency. Since there is a loss of signal, rather than amplification, in the mixing process it is vital that the preamplifier that follows, typically a transistor LNA, has the lowest possible noise temperature. For frequencies up to $\sim 900$ GHz (wavelength 0.35 mm), SIS-mixer receivers have been developed for use in the ALMA synthesis telescope and several single-aperture telescopes. Typical values of $T_{\text{rec}}$ range from 65 K at 200 GHz to 230 K at 900 GHz, corresponding to $\sim 5$ times the quantum limit.

When the highest radiometric sensitivity is required, in particular for studies associated with the cosmic microwave background, incoherent bolometric detectors are used. A bolometer is a total power detector; energy in the incoming photon stream is converted to heat in the detector (essentially a resistor), whose temperature rise is proportional to the incoming power. In that sense bolometers are like the ideal radiometer depicted in Figure 6.1. A recent technology is the transition edge (TES) bolometer, whose resistors are biassed to operate right on the edge of the superconducting condition, at which point the resistance is a very sharp function of temperature (e.g. May *et al.* 2012). The energy of an incoming photon drives the detector into the non-superconducting state and changes the current flow through the device.

While bolometers offer a great advantage in terms of absolute signal-to-noise ratio this comes with several practical disadvantages.

(i) They are broadband power detectors and hence suffer from 'power loading' from any unwanted radiation which arrives at the detector. The bandwidth of a bolometer has therefore to be set by quasi-optical interference filters incorporated into the signal path.

(ii) They are not suitable for spectroscopy since they operate in a total-power mode, i.e. incoherently.

(iii) They can only detect one state of polarization per detector. This contrasts with coherent receivers, which can be sensitive to the $Q$ and $U$ Stokes parameters from the output of a single feed horn (see Section 7.7). A polarimeter based on bolometers therefore requires a filter to select the desired polarization state for each detector and thus requires two antennas plus detectors to measure the Stokes $Q$ and $U$.

(iv) They need to be cooled close to the absolute zero of temperature (to $\sim 0.1$ K in the case of the Planck high frequency instrument (HFI) detectors, Chapter 17).

Despite all these caveats, bolometers are the detectors of choice for short mm and sub-mm wave radiometry. Their overriding advantages are very low Johnson noise and broad bandwidths. The system sensitivity is usually set by photon noise due to the thermal loading from the atmosphere and components in the signal path. Ground-based systems are severely limited by atmospheric emission but this can be minimized by operation from sites such as the South Pole or from high altitude balloons. In space the thermal loading from within the system may be $\sim 20$ K across a wide range of mm and sub-mm wavebands. The bottom line is that, for radiometry and polarimetry above $\sim 100$ GHz, bolometric sensitivity can be several times better than can be achieved with any other detector. This sensitivity advantage is multiplied for wide-area survey work now that multi-pixel arrays with inbuilt antennas for each detector have been developed.

### 6.4.3 System Equivalent Flux Density (SEFD)

When studying discrete sources it is useful to define the *system equivalent flux density* (SEFD); this is the flux density of a source which would give the same output noise power as the receiving system. Using Eqn 5.19, the SEFD is given by

$$S_{\text{sys}} = \frac{2kT_{\text{sys}}}{A_{\text{eff}}}. \tag{6.9}$$

For example, a 25 m diameter parabolic dish with an aperture efficiency of 56% (typical of dishes in the JVLA and eMERLIN) has $A_{\text{eff}} = 275$ m$^2$ and thus $2k/A_{\text{eff}} = 1.00 \times 10^{-25}$ W m$^{-2}$ Hz$^{-1}$; recalling that 1 Jy $= 10^{-26}$ W m$^{-2}$ Hz$^{-1}$ we see that $2k/A_{\text{eff}} = 10$ Jy K$^{-1}$. Thus if $T_{\text{sys}} = 30$ K then $S_{\text{sys}} = 300$ Jy.

## 6.5 Calibration of the System Noise

The system-noise temperature is the sum of independent components, one of which is known to very high accuracy ($T_{cmb}$), two of which may be measured to high ($< 1$ K) accuracy ($T_{loss}$ and $T_{LNA}$) in the laboratory, and the others of which ($T_{gal}$, $T_{atm}$, and $T_{sl}$) can be assessed with reasonable accuracy (1–2 K at short wavelengths), depending on the dimensions of the telescope beam and the relative contribution of the sidelobes.

### 6.5.1 Receiver Noise Calibration

The *receiver noise temperature* is the sum of the emission from passive components in front of the receiver and the receiver system itself, i.e. $T_{loss}$ and $T_{LNA}$. The calibration of the receiver system is achieved by injecting a noise signal, identical in character to the noise from the antenna, into the input. This is best done directly by covering the antenna or the antenna feed completely with absorbing material at a known temperature. By this means the effect of loss in the antenna and passive components is automatically included.

Alternatively one can use a well-matched resistive load in a thermal bath at a well-determined temperature as the source of noise power, substituting it for the antenna. However, depending where in the system the noise is injected, it may not be possible to determine where all the passive losses occur. This may not matter except for very precise work.

Whichever primary calibration scheme is adopted, it is important that both the antenna and the thermal load should be well matched to the receiver, presenting the same impedance – as was stressed in Section 6.1.1. Loads at two input temperatures, 'hot' and 'cold', are needed. For example, if the antenna system employs a feed horn (see Section 8.1) the standard approach is to place a sheet of microwave absorber in front of the horn. The absorber is first used at ambient temperature, typically $\sim 290$ K, as the hot load and then is saturated with liquid nitrogen, whose boiling point at normal atmospheric pressure is 77 K, to form the cold load. In each case the output power of the receiver is measured. If the receiver power gain is $G$, the output noise powers are: $P_{cold} = Gk(T_{rec} + T_{cold})\Delta\nu$ and $P_{hot} = Gk(T_{rec} + T_{hot})\Delta\nu$. The *Y-factor* is defined as

$$Y = \frac{P_{hot}}{P_{cold}} = \frac{T_{rec} + T_{hot}}{T_{rec} + T_{cold}} \qquad (6.10)$$

and the receiver noise temperature can then be calculated as

$$T_{rec} = \frac{T_{hot} - YT_{cold}}{Y - 1}. \qquad (6.11)$$

For example, in a particular test with loads at 290 K and 77 K the $Y$ factor was found to be 3.0 (measured as a difference of 5 dB). Using Eqn 6.11 we obtain $T_{rec} \approx 30$ K. It is important to note that this basic method depends on the receiver's responding linearly to a wide range of input powers, from 77 K to 290 K. To test for non-linearity it is necessary to use loads at three temperatures; one option is to use the sky as the third datum temperature.

For large horns or for wire antennas, such as dipoles of various types (see Section 8.1.3), it is impractical to use a microwave absorber as the load. In these cases resistors are used in place of the antenna[9] with the cold temperature provided by a bath of liquid nitrogen or maintaining the resistor in an active cryostat. For precise measurements of ultra-low-noise receivers, as on the low frequency instrument (LFI) on the Planck spacecraft ($T_{rec} = 8$ K), a hot load at 290 K provides too much power and so the load is kept in the cryostat with its temperature raised by a few degrees by electrical heating.

For less accurate measurements, at lower frequencies, the hot load can be a resistor at ambient temperature with the estimated brightness temperature of the sky defining the temperature of the cold load. For example, at 1.4 GHz with the antenna pointing to the zenith the combination of $T_{cmb} + T_{gal} + T_{atm}$ is in the range 7–9 K. This method relies on shielding the antenna sidelobes (responsible for $T_{sl}$) from the ground and instead turning them towards the cold sky, by placing the system inside a reflective 'ground screen' whose walls are angled at 45° to the zenith.

### 6.5.2  Secondary Methods

Hot and cold load tests are usually carried out infrequently to establish the receiver noise temperature on an absolute scale. To provide continuous checks on the receiver performance a more easily controlled noise source is required; most commonly this is a back-biased Zener diode. This is a *secondary* source, providing a much higher noise power corresponding to an effective noise temperature of the order of 10 000 K. A coupler circuit in the receiver input is used to inject a small proportion of the calibration signal without significantly attenuating the signal from the antenna.

The noise injection is usually accomplished by using a 20 dB or 30 dB *directional coupler*, thus leaking only 1% or 0.1% of the calibration signal into the receiver. Frequently, there is an adjustable attenuator as well, which allows one to adjust the calibration signal to the most convenient value. A small noise signal injected in this way can be used frequently, during an observation, to monitor the gain of the receiver system. As illustrated in Figure 6.7 a measurement might consist of moving on and off a celestial source, noting the change in $T_{sys}$, calibrating the change by turning the secondary calibrator on and off and, at some point, calibrating the secondary calibrator with respect to a primary blackbody source (e.g. Baars *et al.* 1977). Celestial sources of known flux density can also be used for calibration and monitoring of the system performance; their flux densities will, at some earlier time, have been determined with reference to a thermal load.

For both the noise injection and celestial source, methods of calibration noise fluctuations contribute to the measurement uncertainty. A good measurement usually requires an integration time long enough to give a signal-to-noise ratio of at least $5\sigma$, where $\sigma$ is the root mean square (rms) noise in the output; sometimes a $10\sigma$ result may be required.

---

[9]  As mentioned above, this approach does not measure the losses in the antenna itself.

When the rms fluctuations given in terms of temperature by Eqn 6.5 are expressed in terms of flux density, the limiting ($1\sigma$) flux sensitivity of a point source, $\Delta S$, is

$$\Delta S(\text{rms}) = \frac{2kT_{\text{sys}}}{A_{\text{eff}}\sqrt{\Delta\nu\mathcal{T}}} \tag{6.12}$$

or alternatively

$$\Delta S(\text{rms}) = \frac{S_{\text{sys}}}{\sqrt{\Delta\nu\mathcal{T}}}. \tag{6.13}$$

The strongest celestial sources produce antenna temperatures well above the system noise even with modest collecting areas. For example, following Section 6.4.3, a source with a flux density of 2000 Jy (similar to Cas A at 1.4 GHz) will produce a rise in antenna temperature of $\Delta T_A \sim 200\,\text{K}$ (i.e. 9.5 dB if $T_{\text{rec}} = 25$ K) on a 25 m diameter dish and $\sim 30\,\text{K}$ on a 10 m dish. The temperature rise $\Delta T_A$ is measured relative to one or more 'off source' positions with an offset several times larger than the beam size. A problem can arise when measuring $\Delta T_a$ with small-diameter systems owing to their relatively larger beam sizes; the brightness of the galactic background is not constant and will depend on where the 'off source' data are taken.

### 6.5.3 Relative and Absolute Calibration

Measurements of discrete sources are basically straightforward, since they involve the difference between the receiver output when the telescope is directed on and off the source; many terms in the extraneous noise budget change very little for the different pointings. But when an extended source such as the galactic background is being measured, absolute measurements of antenna temperature must be made over a wide range of telescope positions and over the duration of the observations. Sources of systematic error then make achieving high accuracy more difficult. For example, the atmospheric and particularly the sidelobe contributions will vary with time and with the beam position; the calibration loads (of which the reference could be a microwave absorber) may not present precisely the same impedance to the receiver and there may be minor variations in the receiver performance with time. As a result essentially all radio intensity maps of the galactic background made with ground-based telescopes contain low-level calibration-related artefacts and have absolute brightness temperature offsets (see for example the much used 408 MHz all-sky map in the Supplementary Material, which has an absolute offset level of $\sim 3$ K.) Measurements of very low sky temperatures are especially difficult and demanding; the ultimate challenge is the measurement of the cosmic microwave background (CMB), described in Chapter 17.

## 6.6 Heterodyne Receivers

Thus far we have assumed that all the amplification and filtering is carried out at and around the observing frequency. This is not generally the case. The receiver may incorporate a *mixer* or *frequency converter* that shifts the initial frequency band (centred at $\nu_0$) up or

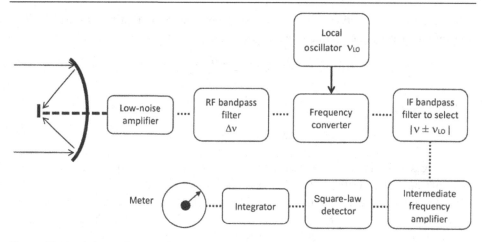

Figure 6.9. Basic heterodyne receiver.

down in frequency while otherwise preserving the signal; a *local oscillator* (LO) injects a pure sinusoid at frequency $\nu_{LO}$ and the resulting *intermediate frequency* band is at either the sum or difference frequency. A receiver of this type, which translates the input signal band to an intermediate frequency, is called a *heterodyne receiver* and its essential features are shown in Figure 6.9.

The first-stage low-noise amplifier usually passes a wide range of frequencies and hence may well pass unwanted signals, especially manmade ones, that are away from the desired observing band. These signals will affect the astronomical signal during and after the mixing process. Usually, therefore, a narrower range of frequencies $\Delta\nu$ is selected by a bandpass filter immediately after the first amplifier stage and centred at the desired observing frequency. If the unwanted signals are large enough to drive the amplifier out of its linear range, the filter has to be placed in front of the LNA; since filters have loss this has a doubly deleterious impact on the system performance – see Section 6.7.

The action of the frequency mixer circuit is illustrated in Figure 6.10. The product of two sinusoids at frequencies $\nu_0$ and $\nu_{LO}$ creates upper and lower sidebands, at frequencies $\nu_0 \pm \nu_{LO}$. The products are generated by the non-linear behaviour of the mixer, in which the instantaneous input level of one signal affects the output level of the other. There are many types but in a basic diode mixer the powerful LO signal (typically several milliwatts) at $\nu_{LO}$ modifies the diode's conductance, forcing it open or closed at different times during the LO cycle. The low-level astronomy input signal $\nu_0$ is therefore 'chopped' and the resulting output waveform contains a wide range of harmonics and sums and differences of input frequency components (usually plus a low-level remnant of $\nu_{LO}$); these are not shown in Figure 6.10, which is highly idealized.

The desired output is an intermediate frequency centred at $\nu_{IF} = \nu_0 \pm \nu_{LO}$; the choice is made by another bandpass filter after the mixer. Usually, since it is easier and cheaper to carry out a significant part of the amplification and most of the signal manipulation at a lower frequency, the difference frequency $\nu_{IF} = \nu_0 - \nu_{LO}$ is chosen.

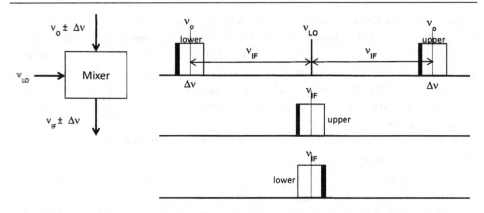

Figure 6.10. The action of a mixer. (left) The local oscillator (LO) converts the signal band, centred on $\nu_0$, to a higher or (more usually) a lower frequency band centred on $\nu_{IF}$. (right) The LO freqency can be chosen to lie above or below the astronomy band centred on $\nu_0$; here both options are shown, the thick solid lines indicating the low end of the signal spectrum. When $\nu_{LO} > \nu_0$ the lower sideband is translated to $\nu_{IF}$, and the order of the IF frequencies is reversed.

There is always a loss in signal power through a passive mixer, called its *conversion loss*. The power at the selected IF frequency can never be better than half the input RF power since the latter is split between the two IF sidebands. Power is also lost into a range of unwanted harmonics and in resistive losses. Typically $< 25\%$ of the input power is translated into the selected IF output; hence it is usual to assume that the conversion loss will be 6–10 dB.

The system shown in Figure 6.9 is simplified. Often there is more than one stage of frequency conversion and more than one intermediate frequency. The final conversion can even be to a band whose lower edge is close to zero frequency, when it is generally known as *video* or *baseband*. This is routinely done when the following stage is an autocorrelation spectrometer (see Section 7.2), while for broadband radiometry the final intermediate frequency can be at tens or hundreds of MHz. There is usually an adjustable attenuator somewhere in the system to set the output to an optimum level before it passes to the detector.

For radiometry in bands where there is no radio frequency interference (RFI), typically bands at mm and sub-mm wavelengths, the bandpass selecting filter after the first stage (which is typically an SIS mixer) may be omitted. In this case, the power at frequencies both above and below $\nu_{LO}$ is translated to the IF band, where the upper and lower sidebands are superimposed. This is called *double sideband* operation. It offers signal-to-noise advantages for observations of broadband continuum sources since radiation in two independent passbands is collected. For spectroscopy single sideband operation is usually preferred.

Despite its reliance on non-linear mixing, a well-designed heterodyne receiver (with good frequency filtering) retains the relative coherence of the input frequency components. The input frequencies are not jumbled up in the output; instead a well-defined band of IF frequencies is selected which share the same frequency offset for all the input frequency components. If the receiver gain is uniform across the passband then the relative amplitudes remain the same and the phase response of the bandpass-defining filters is often a linear function of frequency. This coherent behaviour of the receiver is vital for both spectrometry and interferometry. To maintain coherence for spectroscopy and interferometry the LO frequency must be stable, and in practice it is generated by a crystal oscillator or by a frequency synthesizer that uses a reference source of high stability, such as an atomic frequency standard.

The increased complexity of a heterodyne receiver is usually more than outweighed by several practical advantages.

(i) It translates the astronomy band to a more convenient intermediate (usually lower) frequency for further amplification and signal manipulation.

(ii) Amplification of the signal at different frequencies avoids the tendency for oscillation in the receiver as a result of the enormous amplification ($\sim 100$ dB) required for radio astronomy. A small amount of unwanted positive feedback in the receiver chain can give rise to oscillation if all the gain is at the same frequency.

(iii) The ability to vary the LO frequency enables the centre frequency of the band under study to be changed; this is very useful for spectroscopic observations, as will be discussed in Chapter 7.

## 6.7 Tracing Noise Power through a Receiver

We are now in a position to trace the flow of noise power through a heterodyne receiver and assess the relative contributions of each stage. The starting points are:

(i) an amplifier acts as if it had a resistor of temperature $T_{amp}$ at its input whose Johnson noise power $kT_{amp}\Delta\nu$ is amplified by the power gain $G = P_{out}/P_{in}$ of the amplifier;

(ii) the noise temperature of a lossy component (a filter or a mixer) is the same as its physical temperature; it has a power gain less than unity.

Consider first a chain of three amplifiers $A_1$, $A_2$, and $A_3$ with noise temperatures $T_1$, $T_2$, and $T_3$ and power gains $G_1$, $G_2$, and $G_3$. Dropping the $k\Delta\nu$ terms for clarity, the noise power output from $A_1$ is $G_1T_1$, from $A_2$ it is $G_2(G_1T_1 + T_2)$, and from $A_3$ is $G_3[G_2(G_1T_1 + T_2) + T_3]$. Now treating the entire chain as a single amplifier of noise temperature $T_{total}$ and gain $G_{total} = G_1G_2G_3$ we can write the output power as

$$G_{total}T_{total} = (G_3G_2G_1)T_1 + (G_3G_2)T_2 + G_3T_3 \qquad (6.14)$$

and, dividing by $G_{total}$, we obtain

$$T_{total} = T_1 + T_2/G_1 + T_3/G_1G_2. \qquad (6.15)$$

The argument can be continued for longer chains. We can see immediately that if $G_1$ is high enough then the noise from the later stages can be swamped by the noise from the first amplifier. A specific example shows the effect of varying $G_1$. A simple receiver chain consisting of an RF amplifier, a mixer, and an IF amplifier has the following parameters:

– first amplifier, $T_{LNA} = 20$ K, $G_{LNA} = 100$ (20 dB)
– mixer, $T_{mix} = 290$ K, $G_{mix} = 0.25$ (−6 dB)
– IF amplifier, $T_{IF} = 290$ K; $G_{IF} = 100$ (20 dB)

Following Eqn 6.15 we obtain $T_{total} = 20 + 290/100 + 290/(100 \times 0.25) = 34.5\,\mathrm{K}$. With these relatively low gain values, notice the effect of the lossy mixer and hence the significant noise contributions of the stages after the first amplifier. If we simply increase $G_{LNA}$ to 1000 (30 dB) and leave the other parameters the same then the second two terms together contribute only 1.5 K and so $T_{total} = 21.5\,\mathrm{K}$, an improvement of 60%.

The lesson is clear: if the gain of the first stage amplifier is $>$ 30 dB then the receiver noise temperature is essentially determined by the noise temperature of the first stage.

## 6.8 Gain Variations and Their Correction

A practical radiometer system must overcome the problem of instabilities in receiver gain as well as accounting for all sources of noise. Given the degree of amplification required, the performance of all receivers becomes dominated by gain variations rather than thermal noise as the required integration times increase. If the gain fluctuates significantly on the measurement timescale, for example during the time taken to move the radio telescope on and off a weak source, a meaningful measurement may be impossible.

It is instructive to calculate, step by step, the gain required in a typical cooled receiver. The receiver operates at 21 cm (1.4 GHz) and has a system temperature $T_{sys} = 25\,\mathrm{K}$ and a bandwidth $\Delta\nu = 100\,\mathrm{MHz}$ (probably limited by RFI). The total system noise power $kT\Delta\nu \approx 3.5 \times 10^{-14}\,\mathrm{W}$, or $\sim 105$ dB below the standard of one milliwatt (dBm). For square-law operation one wants to present a power level of $\sim 0.01$ mW ($-20$ dBm) to the detector so the receiver must have a gain of $\sim 85$ dB; this would be followed by a further 20–30 dB of low frequency (video) amplification to provide a sufficiently strong signal for the power-level measurement. If, however, the system is entirely digital and bypasses the square-law detector (which is typical for spectrometry) then, with power levels of several mW required to drive the analogue-to-digital converter, the RF + IF receiver gain must be $\sim 105$ dB. Typically, there might be 40–50 dB gain before the first mixer (with its $\sim 6$ dB loss) and another 60–70 dB gain in the remaining stages, although these figures are highly design dependent.

Now let us calculate how gain and thermal noise fluctuations combine. The noise power out of a receiver can be written as a constant term associated with a fixed gain $G$ plus a varying term associated with the gain variations $G(t)$:

$$P_0 + \Delta P(t) = [G + \Delta G(t)]kT_{sys}\Delta\nu. \tag{6.16}$$

A fractional gain variation then produces an output fluctuation

$$\Delta T_{gain} = \frac{\Delta G(t)}{G}T_{sys}. \tag{6.17}$$

Fractional gain fluctuations can span a wide range but are typically $10^{-4}$ to $10^{-5}$ on timescales of interest (see below). Thus, faint celestial sources producing antenna temperatures only one ten thousandth or one hundred thousandth of $T_{sys}$ will be lost or at best have their strength incorrectly estimated.

The output fluctuations due to gain variations and thermal noise are independent and so their effects add quadratically:

$$\Delta T_{total}^2 = \Delta T_{thermal}^2 + \Delta T_{gain}^2. \tag{6.18}$$

Substituting from Eqns 6.5 and 6.17 we obtain

$$\Delta T_{total}^2 = \frac{T_{sys}^2}{\Delta \nu \mathcal{T}} + T_{sys}^2 \left( \frac{\Delta G(t)}{G} \right)^2, \tag{6.19}$$

and hence

$$\Delta T_{total} = T_{sys} \left[ \frac{1}{\Delta \nu \mathcal{T}} + \left( \frac{\Delta G(t)}{G} \right)^2 \right]^{1/2}. \tag{6.20}$$

A simulated time series of the receiver output power is shown in Figure 6.11. The short-term fluctuations correspond to white thermal noise. The larger fluctuations arise from gain fluctuations termed 'flicker noise' or more commonly '1/f noise' because a Fourier transform of the time series reveals a frequency spectrum with a slope usually close to $-1$ (see Figure 6.12).

The term '1/f noise' is confusing at first sight. It is not noise in the sense of thermal noise, which has a flat output spectrum; instead the receiver gain tends to deviate away from an initial value with a deviation that is statistically proportional to the time elapsed; this is the inverse of the output frequency, hence the terminology. The gain drifts produce larger fluctuations in the output power on longer timescales and these eventually come to dominate on timescales longer than $1/f_{knee}$ (Figure 6.12), which is system dependent.

The gain fluctuations arise from a combination of effects. On timescales less than a second they are due to solid state physics phenomena in the transistors of the first-stage amplifier; on longer timescales, variations in the supply voltage, in the physical temper-

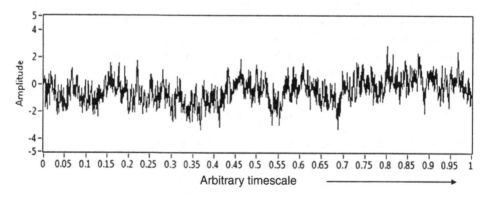

Figure 6.11. Simulated white plus $1/f$ noise.

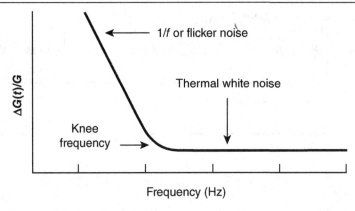

Figure 6.12. Spectrum of white plus $1/f$ noise.

atures of the receiver, and in atmospheric propagation conditions come into play.[10] The $1/f_{\text{knee}}$ timescale depends on the level of the white noise. For the very-low-noise (8 K) and broadband (6 GHz) 30 GHz receiver on the Planck spacecraft, $f_{\text{knee}} \sim 100$ Hz and the transistor gain fluctuations were larger than the thermal noise for timescales longer than $\sim 10$ ms. For uncooled receivers with smaller bandwidths, gain fluctuations begin to dominate on timescales of seconds or more.

Without correction the gain fluctuations would greatly curtail the effectiveness of radio astronomy receivers. Given a total system temperature of 25 K and a bandwidth of 100 MHz, an averaging time of 1 s would result in an rms uncertainty of 2.5 mK for a given measurement, by the fundamental relation Eqn 6.5. This is only one part in $10^4$ of the system noise, so the fractional gain stability needs to be much better, despite the total gain of $> 100$ dB ($> 10^{10}$). This is difficult to achieve without incorporating some type of calibration scheme into the radiometer itself. The switching strategy in the *Dicke radiometer*, named after its inventor, Robert Dicke, was one of the early types and remains in use to this day.

### 6.8.1 Dicke Switched Radiometer

A simplified block diagram of a Dicke radiometer is given in Figure 6.13. The change compared with Figure 6.1 is the comparison switch interposed in front of the receiver, switching rapidly between the usual signal path from the antenna and a path to a reference load which generates a noise signal equivalent to a black body at nearly the same physical temperature. The switched waveform drives the front-end RF switch and a lock-in amplifier, also known as a synchronous detector, which is sensitive not only to the magnitude of the input power but also to which phase of the switching waveform that power level applies.

---

[10] Time variations in many phenomena in the natural world exhibit $1/f$ spectra; for a pedagogical review see http://arxiv.org/ftp/physics/papers/0204/0204033.pdf.

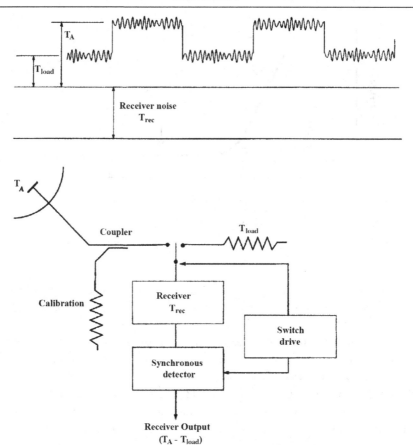

Figure 6.13. The Dicke switched radiometer, showing the switched signal as a function of time.

There are a number of ways of constructing such a device, but functionally one simply takes the product of the detected signal with the reference square wave that drives the comparison switch, as shown in Figure 6.13. These processes and measurements are now more easily carried out digitally.

The essence of how a Dicke switched receiver works can be described simply. It continuously takes the difference between the total power output of the receiver when it is switched to the antenna, proportional to $T_A + T_{rec}$, against the output when it is switched to the stable resistor load, proportional to $T_{load} + T_{rec}$. The gain variations affect each path, but on short timescales $t_{switch} \ll 1/f_{knee}$ they are very small and so the gain will be almost the same for each path even though the data are not taken at exactly the same time. The time average of the difference will therefore be constant and indeed will be zero (perfectly balanced) if one can arrange that $T_{load} = T_A$. In that case, only when $T_A$ changes, for example owing to the entry of a source into the beam, will the output move away from zero. Further illustrations of the signal flow are in the Supplementary Material. The Dicke switched system offers important advantages over a non-switched radiometer, as follows.

(i) The large constant offset in the output associated with $T_{rec}$ is automatically subtracted out.

(ii) On the switching timescale or longer, the gain variations are smaller compared with an unswitched receiver by the ratio $(T_{load} - T_A)/(T_{rec} + T_A)$. A perfectly balanced receiver is completely insensitive to gain variations on these timescales; only changes in $T_A$ are then reflected in the output. If it is not possible to make $T_{load}$ and $T_A$ closely equal, the software in the digital signal processing system can apply gain modulation factors to balance the radiometer.

While gain fluctuations can be almost eliminated in a switched receiver, thermal noise fluctuations will still be present as an unavoidable limit to the accuracy of measurement. The disadvantage of the Dicke receiver is an enhancement of the thermal noise level by a factor 2 compared with an unswitched radiometer. This arises because the antenna temperature and the reference load temperature are measured for only half of the time. This means that the relative accuracy of measuring either temperature will be degraded by a factor of $\sqrt{2}$ and, since the difference between two uncorrelated random variables is being taken, there is a further loss in accuracy of $\sqrt{2}$. The net fractional accuracy of the measurement, therefore, is

$$\Delta T = \frac{2T_{sys}}{\sqrt{\Delta \nu \mathcal{T}}}. \tag{6.21}$$

Despite the loss in accuracy by a factor 2 compared with the radiometer Eqn 6.5, the greater stability of measurement is often worth the price.[11]

One practical problem with the Dicke switched receiver concerns the RF switch itself. This has to operate continuously at high rates, with potential reliability issues, and always has some loss. Since the switch comes before the low-noise amplifier this brings the usual twofold degradation in system performance through signal loss and increased thermal noise.

### 6.8.2 Correlation Radiometers

An alternative to the Dicke switch takes advantage of the coherent nature of radio signals before detection. The basic components are shown in Figure 6.14. Here the antenna and the reference load are connected to a 'hybrid splitter' circuit whose outputs pass through amplifiers with power gains $G_1$ and $G_2$ and hence voltage gains $\sqrt{G_1}$ and $\sqrt{G_2}$. The splitter divides the *power* of each input signal equally between the two outputs (hence the '3 dB hybrid' in Figure 6.14) and also reverses the phase of one input *voltage* in one of the outputs.

Consider the noise voltages from the antenna $v_a$ and the reference load $v_{load}$ and ignore the amplifier noise for simplicity. The sum and difference outputs (due to the phase reversal) from the splitter after amplification are

$$\Sigma = \sqrt{G_1}(V_a + V_{load})/\sqrt{2} \quad \text{and} \quad \Delta = \sqrt{G_2}(V_a - V_{load})/\sqrt{2}.$$

---

[11] Dicke switched receivers are used in many applications which are not thermal noise limited, for example in the ALMA water vapour radiometers at 183 GHz.

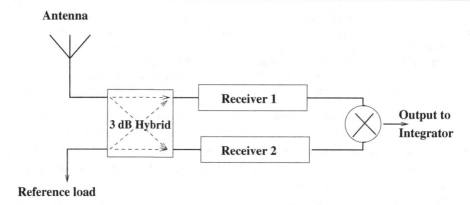

Figure 6.14. Basic differential radiometer, using a hybrid splitter circuit and multiplier, which measures the difference between the noise power from an antenna and that from a reference source.

After multiplication and time averaging (using a mixer plus low-pass filter or by digital processing) the incoherent cross-terms, which have no sustained phase relationships, average to zero. We are left with the self-coherent terms; thus

$$V_{\text{out}} = \sqrt{G_1}\sqrt{G_2}(\langle V_{\text{load}}^2\rangle - \langle V_A^2\rangle)/2,$$

which we can write as

$$V_{\text{out}} \propto \sqrt{G_1}\sqrt{G_2}(T_{\text{load}} - T_A)/2. \tag{6.22}$$

The signals from the load and the antenna each pass though both receivers and hence any gain variations affect them equally. The output therefore responds to the difference between the powers from, hence the temperatures of, the load and the antenna; when these are balanced, the effects of gain fluctuations are minimized, as in the Dicke switched system. However, correlation receivers have several advantages over Dicke systems.

- The appropriate radiometer equation is $\Delta T = \sqrt{2}T_{\text{sys}}/\sqrt{\Delta\nu\mathcal{T}}$, which is $\sqrt{2}$ better than the Dicke switched system. The reason is that the antenna signal is connected to the receiver all the time rather than only half the time, as in the Dicke system.
- Radio frequency losses in hybrid splitters are less than in switches, so $T_{\text{sys}}$ can be lower than in a Dicke system.
- The system is effectively continuously switching between the inputs at the RF frequency, rather than at the switch frequency as in the Dicke system. This is always higher than $f_{\text{knee}}$.

In practice correlation radiometers are more complicated than is shown in Figure 6.14, with additional hybrids and phase switches (the general design considerations for correlation receivers are discussed in Harris (2005)). The CMB satellites WMAP and Planck (see Chapter 17) used correlation radiometers (Figure 6.15), which also included a switching system that further reduced the effect of any gain fluctuations in the amplifiers outside the two hybrids. The WMAP radiometers compared the signals from two oppositely facing antennas, with orthomode transducers separating the signals from the two polarizations, $A, A'$ for one antenna and $B, B'$ for the other. The Planck radiometer measured differences

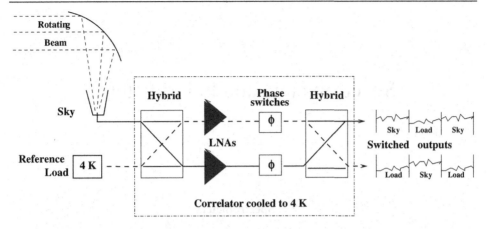

Figure 6.15. A simplified diagram of the low frequency radiometers on the Planck spacecraft, measuring the difference between the sky and reference load temperatures. Each of the frequency channels had two such radiometers, one for each polarization. The signals are added in the first hybrid, the sum and difference are amplified by 60 db in the low noise amplifiers, and the signals are separated in the second hybrid. A phase reversal of 180° interchanges the outputs. Not shown are further amplifiers after the second hybrid. The whole spacecraft rotates to scan the beam round the sky. For further details see Mennella *et al.* (2010).

between the signal from a single larger antenna and a cold load at a temperature (4 K) close to that of the CMB.

## 6.9 Digital Techniques

Many technical problems which arise in signal handling, and especially in multi-channel receivers, correlators, and the delay units commonly found in spectrometers, polarimeters and interferometers, are greatly eased by the use of digital rather than analogue electronic techniques. These processes are encountered in the later stages of signal processing, where the signal frequencies have been reduced by a heterodyne process to an intermediate or video frequency band where the signal may be sampled and digitally encoded (Appendix 3). The subsequent signal processing can then be free of distortion.

## 6.10 Further Reading

*Supplementary Material* at www.cambridge.org/ira4.
Lyons, R. G. 2011. *Understanding Digital Signal Processing*, 3rd edn. Prentice Hall.
Marven, C., and Ewers, G. 1996. *A Simple Approach to Digital Signal Processing*. Wiley Interscience.

# 7

# Spectrometers and Polarimeters

## 7.1 Spectrometers

Spectral lines are powerful diagnostics of physical and chemical conditions in space, and radio astronomers want to maximize the signal-to-noise ratio of the lines with respect to the off-line noise. A range of factors contributes, from the physics of the emission process and the distance to the source to the performance of the radio telescope and its receiver. We will describe several different types of spectrometer which can measure the power in multiple contiguous channels within the receiver band. This band must be wider than the spectral line, and the channel width $\delta\nu_{ch}$ of the filters must be considerably narrower than the linewidth $\delta\nu_{line}$. If $\delta\nu_{ch} \sim \delta\nu_{line}$ or larger then the signal-to-noise ratio falls. This is not obvious at first sight since wider channels have lower thermal noise, by the radiometer equation 6.5. However, the effect of diluting the beam power over a channel is greater, as we shall now show. The total power flux captured from a line of centre frequency $\nu_0$ is

$$P_{tot} = \int_{\nu_0 - \delta\nu_{ch}/2}^{\nu_0 + \delta\nu_{ch}/2} S(\nu)d\nu \quad \text{W m}^{-2}, \tag{7.1}$$

where $S(\nu)$ is the frequency-dependent flux density of the natural line. In the case where $\delta\nu_{line} \ll \delta\nu_{ch}$, this power is spread out across a channel and hence the apparent flux density of the line depends inversely on the channel width:

$$S'(\nu) = P_{tot}/\delta\nu_{ch} \quad \text{W m}^{-2} \text{ Hz}^{-1}. \tag{7.2}$$

From Eqn 6.9, if the receiving telescope has an effective area $A_{eff}$ then the antenna temperature produced by the line is

$$T_{line} = \frac{S'(\nu)A_{eff}}{2k} = \frac{P_{tot}A_{eff}}{2k\delta\nu_{ch}}. \tag{7.3}$$

This is the signal. Using Eqn 6.5 the rms noise in each channel for an integration time $\mathcal{T}$ is

$$\Delta T = \frac{T_{sys}}{\sqrt{\delta\nu_{ch}\mathcal{T}}}, \tag{7.4}$$

hence the signal-to-noise ratio for a line within a single channel is

$$\frac{T_{line}}{\Delta T} = \frac{P_{tot}A_{eff}\sqrt{\mathcal{T}}}{2kT_{sys}\sqrt{\delta\nu_{ch}}}. \tag{7.5}$$

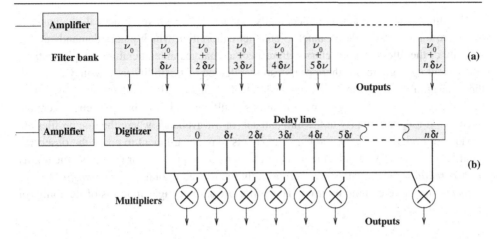

Figure 7.1. Radio spectrometers using: (a) a bank of narrowband filters; (b) a digital autocorrelation system. The longest delay $n\delta t$ is approximately equivalent to the spectral resolution, $(\delta\nu)^{-1}$.

This has the expected dependencies on the system parameters but is also inversely proportional to the square root of the channel width. To optimize the detection of a spectral line, one should therefore reduce the channel width so as concentrate the line's power into as few channels as possible, even though the noise fluctuations in each channel increase.

To simplify the argument we have not included the effect of the natural linewidth, but in practice for the optimum detection of a weak line one wants to have a small number of channels across the natural width. Of course if the line is strong (depending on $P_{tot}$ and $A_{eff}$) its detailed substructure can be measured by deploying a greater number of channels across its width. Since linewidths are dominated by Doppler broadening due to differential velocities in the emitting gas (see Chapter 3) it is usually possible to estimate the likely linewidth from prior knowledge of the physical conditions at the source.

### 7.1.1 Filter-Bank Spectrometers

The most direct approach to a multichannel system, schematized in Figure 7.1(a), is to split the signal after the IF amplifier stage by means of a large number of narrowband filters in parallel, each with its own detector, measuring the flux in discrete bins. The individual outputs can be accumulated in integrators that are read by the data-acquisition computer, which then stores and organizes the data. Usually there is a buffer amplifier in front of each filter to guarantee that there will be no cross-talk between channels.

In Figure 7.1(a) each filter is tuned to a separate frequency, but alternatively the filters may be identical, with each channel having its own local oscillator and mixer. In this more flexible type of multichannel receiver one can reconfigure the filter array, for example tuning groups of filters to individual lines in the spectrum or spacing the channels more broadly when searching for a line whose frequency is only approximately known.

The filter-bank approach is conceptually simple. However, the practical challenge is to ensure uniformity of the channel-to-channel calibration, which places demands on the stability of the filters and detectors with respect to changes in temperature, supply voltage, component ageing, etc. Furthermore, despite a degree of flexibility in how they are used, the total number of channels and the individual channel width are fixed at construction time. Despite these challenges and limitations, multichannel filter-bank spectrometers are in common use, particularly in the mm and sub-mm wavebands where the line profiles are broad and a relatively small number of channels may be sufficient to cover the observing band.[1] At longer wavelengths the number of channels required is larger and so the digital *autocorrelation spectrometer*, and increasingly the related digital *Fourier transform spectrometer*, are more commonly used. They offer great advantages in terms of flexibility of operation.

### 7.2 Autocorrelation Spectrometers

Autocorrelation (see also Section 5.1) involves multiplying a signal by a shifted version of itself (the shifts being either positive or negative) and integrating the result over a time period $T$. Each shift $\tau$ results in one point on the autocorrelation function (ACF) $R(\tau)$, whose extent therefore depends on the number of shifts. Each point on the ACF is a measure of how predictable the function is at a time shift $\tau$ when compared with the function at time $t$. This is its *temporal correlation* (see also Section 9.6), which is clearly related to its frequency content. A closely related system is the well-known Michelson spectral interferometer, in which two copies of the input signal are shifted relative to each other by physically changing the length of one of the arms. The spectrum is obtained from a Fourier transform of the output ACF, as formalized in the Wiener–Khinchin theorem (see Appendix 1 and Section 5.1).

Before considering how an autocorrelation function (ACF) can be measured, it is useful to gain a heuristic understanding of the Wiener–Khinchin theorem by considering the autocorrelation of a pure sine wave and then a sum of sine waves. The autocorrelation of a sine wave $A \sin(\omega \tau + \phi)$ is

$$R(\tau) = \frac{1}{T} \int_0^T A \sin(\omega t + \phi) A \sin[\omega(t + \tau) + \phi] \, dt, \qquad (7.6)$$

thus[2]

$$R(\tau) = \frac{A^2}{2T} \int_0^T [\cos \omega \tau - \cos(2\omega t + \omega \tau + 2\phi)] \, dt. \qquad (7.7)$$

The first term inside the integral is constant for a given value of $\tau$ and $\omega$ and therefore adds up over time. The second term is, however, oscillatory and does not add up. We therefore

---

[1] Complete filter-bank spectrometers with many hundreds of channels are now being fabricated on a chip.
[2] Using the relation $\sin A \sin B = [\cos(A - B) - \cos(A + B)]/2$.

obtain, for a given shift,

$$R(\tau) = \frac{A^2}{2T} \cos \omega \tau . T = \frac{A^2}{2} \cos \omega \tau. \tag{7.8}$$

The ACF is established by repeating the process for many values of $\tau$. Equation 7.8 shows that:

(i) The ACF is periodic, with the same period as the sine wave; the maximum positive value occurs for time shifts $\tau$ of zero and integer periods and the maximum negative value occurs for half-period shifts.
(ii) The sine wave defines its own starting point, so, regardless of its phase with respect to an arbitrary point in time, the ACF of a sinusoid always takes the form of a cosine wave with its first maximum at zero shift.
(iii) At the maxima the ACF is proportional to the magnitude squared, i.e. to the power in the sine wave.

In general a signal can be described by a sum of many sine waves of different amplitudes, frequencies, and phases, but the composite ACF remains simple. The individual frequency components are not mixed up, there are no harmonics, and the phases have vanished. Using the arguments above, readers can convince themselves of this behaviour for a two-component voltage signal $A \sin(\omega_1 \tau + \phi_1) + B \sin(\omega_2 \tau + \phi_2)$. Picking out the constant terms and dismissing the oscillatory ones, we obtain

$$A \sin(\omega_1 \tau + \phi_1) \rightarrow \frac{A^2}{2} \cos \omega_1 \tau \tag{7.9}$$

and

$$B \sin(\omega_2 \tau + \phi_2) \rightarrow \frac{B^2}{2} \cos \omega_2 \tau. \tag{7.10}$$

The ACF for this simple situation is shown in Figure 7.2.

Thus, for a set of sine waves the ACF picks out the power in each separate frequency component. One can immediately see that the power spectrum of the signal can be constructed from the Fourier transform of the ACF, which is the Wiener–Khinchin theorem.

Real data sets in radio astronomy are invariably dominated by noise. Figure 7.3 shows how the ACF can pick out a sine wave with a low signal-to-noise ratio. The ACF is a digitized sine wave, whose Fourier transform is the spectral line. In this idealized case the fall-off ('decorrelation') in the ACF magnitude is due to the finite length of the simulated data set. As one keeps on shifting, the overlap between the function and the shifted version tends to zero and hence so does the ACF.

Analogue spectrometer techniques are widely used at mm and sub-mm wavelengths, but at longer radio wavelengths digital autocorrelation spectrometers, introduced in 1963 by Weinreb (MIT Technical Report 412), provide superior performance. Digital autocorrelation spectrometers have the desirable characteristic that the entire band is analysed at the same time, with the practical advantage of using flexible and convenient digital components. There are special features of digitization, however, of which the user must be aware (see Appendix 3).

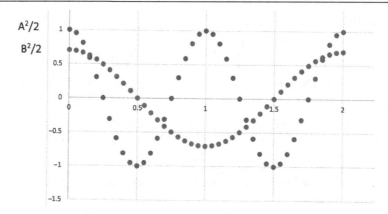

Figure 7.2. The discrete positive-going ACF $R(\tau)$ of two sampled cosine waves with different amplitudes ($A$ and $B$) and different frequencies and initial phases. Notice that they appear in the ACF as separate sinusoids, symmetrical about zero shift.

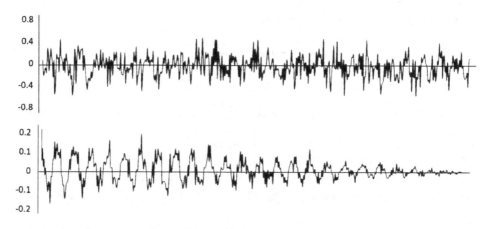

Figure 7.3. How a sine wave is picked out from noise by autocorrelation. (upper) Five hundred equispaced samples of Gaussian noise (zero mean, standard deviation 0.2) added to a sine wave of amplitude 0.1 and period 20 samples; (lower) the normalized ACF for 500 positive unit shifts (lags). The ACF for negative shifts is identical when reflected through zero lag; note that the unity correlation at zero lag is not shown. The sampled sine wave can clearly be recognized.

### 7.2.1 Linewidth and the Window Function

Figure 7.3 shows a fall-off in amplitude of the ACF with increasing lag, owing to the finite length of the data set. The same effect of *decorrelation* occurs in long data sets owing to the finite natural linewidth of the signal, and is a vital diagnostic of spectral linewidth. In the

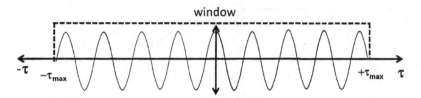

Figure 7.4. The ACF for a pure sine wave. Since the ACF is an even function (symmetric about zero shift) the values for negative shifts can be filled in automatically. If a range of shifts is limited to $\tau_{max}$ then the ACF has effectively been multiplied by a window function (in this case uniform) out to $\pm\tau_{max}$; this is the transfer function weighting the input frequency components in the ACF.

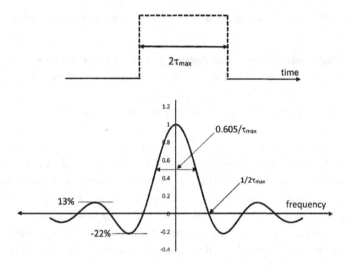

Figure 7.5. The length $2\tau_{max}$ of the sampling function (including both positive and negative lags) is Fourier transformed into the frequency response of the spectrometer. The square sampling function transforms into the sinc function $\sin(\pi\nu 2\tau_{max})/(\pi\nu 2\tau_{max})$ with which the spectrum is convolved.

receiver the voltage time series associated with a narrow spectral line is highly correlated from sample to sample (tending to infinity for $\delta\nu_{line} \to 0$) but, over a timescale $1/\delta\nu_{line}$, the time series less and less resembles a shifted copy of itself. If there is significant natural decorrelation within the 'window', i.e. the range of shifts over which the ACF is measured, then the line is resolved and its width can be measured.

Figures 7.4 and 7.5 illustrate the effect of a finite range of shifts. The maximum shift is $\tau_{max}$ but the ACF can automatically be extended to $-\tau_{max}$ owing to its even symmetry. The effect is to multiply the ACF by a 'window function', which acts as a transfer function weighting the ACF frequency components. In this particular case the weighting is uniform within the window. The convolution theorem (see Appendix 1) gives us

$$[R(\tau) \times W(\tau)] \overset{\text{FT}}{\Longleftrightarrow} P(\nu) \star W(\nu), \tag{7.11}$$

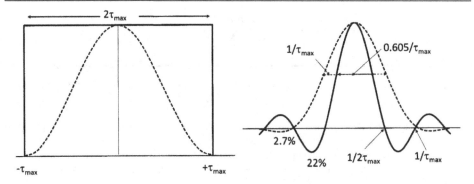

Figure 7.6.   (left) Rectangular and Hann windows; (right) their normalized impulse responses.

where the left-hand side is in the time domain and $W(\tau)$ is the window function extending between $\pm\tau_{max}$. The right-hand side is in the frequency domain; $P(\nu)$ is the power spectrum of the input signal and $W(\nu)$, the Fourier transform of the window function, is the frequency response to an impulse. The convolution with the impulse response sets the spectral resolution.

For the rectangular window in Figure 7.4 the impulse response is a sinc function. The half-width of the sinc function is $0.605/\tau_{max}$; if this width is less than $\delta\nu_{line}$ then the line will be resolved. There is a price to pay for uniform weighting and hence high spectral resolution, since a sharp-edged window produces unwanted 'ringing' in the frequency domain; in the case of a sinc function the amplitude of the first negative is ~22% of the peak. Although all the information in the spectrum is present, the appearance of such a spectrum can be confusing; a sidelobe of a strong spectral line falling at the same frequency as a nearby weak line will distort the weak line or even cause it to be missed entirely.

The solution is to smooth off the sharp edges of the window function with a weighting function.[3] The effect is to sacrifice spectral resolution for reduced sidelobes. Many different functions have been devised; one in common use was invented by J. von Hann and hence is called 'Hanning' weighting.[4] It takes the form

$$w_H(\tau) = \left[1 + \cos(\pi\tau/\tau_{max})\right]/2 \qquad (7.12)$$

and hence is also called raised cosine weighting. This function and its corresponding impulse response are shown in Figure 7.6. The first sidelobe is now almost a factor ten lower than in the sinc function but the spectral width is significantly greater. The first zero is at $1/\tau_{max}$, twice that of the sinc function, and the half-width is $\sim 1/\tau_{max}$, i.e. about 40% larger. The reduction in the sidelobe level proves, in practice, to be more important than the resolution penalty.

---

[3]  This is similar to the problem of excessive sidelobe levels in a radio telescope, where the solution is to taper the illumination of the aperture (Chapter 8).

[4]  A feature of the Hanning function is that it reaches zero at the end of the window. An extensive library of alternative weighting functions and their effects on the resolution and sidelobe levels can be found on the web.

## 7.3 Digital Autocorrelation Spectrometers

Autocorrelation spectrometers based on these principles but working with digitized data have been workhorse instruments at many radio observatories from the 1970s to the present day. One can imagine two different approaches to their design. Conceptually the simpler would be to implement a digital version of the classical Michelson spectral interferometer. Assume for simplicity that the signal band has been heterodyned down to baseband, i.e. from 0 to $\nu_{max}$ Hz. The digitizer should sample the band at greater than the Nyquist rate $2\nu_{max}$, hence the samples of the signal $V(t)$ are spaced at intervals $\delta t < 1/2\nu_{max}$ seconds. For simplicity let us also assume that only single-bit digitization is involved. To produce an ACF corresponding to an overall lag $\tau_{max}$ and an integration time $T$, one could imagine storing two sets of $M = T/\delta t$ digital samples of the signal voltage $V(t)$ in hardware shift registers. One point in the ACF is then generated by comparing samples, using a set of $M$ comparators (multipliers), in one copy of $V(t)$ with samples in the other copy which have been shifted by a particular lag. If both inputs are the same (either 0 or 1) then a comparator produces a 1; if they are different it produces a 0. The outputs of the comparators are summed to produce the value of the ACF for that lag and the process is repeated $n + 1$ times for lags ranging from zero to $n = \tau_{max}/\delta t$. This is a discretized version of Eqn 5.2 and exemplified in Section 7.2. The problem is that, for practical bandwidths, $M$ becomes very large and the hardware demands render this approach impracticable.

A sequential approach, developed by S. Weinreb in 1963 (MIT Technical Report 412) and schematized in Figure 7.1(b), enabled practical spectrometers to be constructed. Now only $n = \tau_{max}/\Delta t$ samples of $V(t)$ are stored in a single shift register. At a given time the state of each sample is compared to that of the first (zero-shift) sample. With $n$ separate lags covered by the shift register, $\delta t, \delta 2t, \ldots, n\delta t$, and including the zero-shift comparison, there are a total of $n + 1$ outputs; each is an estimate of one point in the ACF. Since only one sample of $V(t)$ is compared with the stored range of $V(t)$, the signal-to-noise ratio of these estimates is very low. The solution is to accumulate many estimates, and thus on the next clock pulse all samples move along by one and the process is repeated. The outputs from the comparators are read out at the clock rate (the Nyquist rate) and the results are accumulated in separate counters. The counters are read out at a more leisurely rate by a computer that carries out the longer-term accumulation. From a set of $M$ samples of the signal, and for an integration time $T = M\delta t$, one thereby assembles a set of $n+1$ estimates of the discrete ACF $R_n$:

$$R_n = \frac{\sum_{j-1}^{M} V(t_j)V(t_j + n\delta t)}{M};$$ (7.13)

note that division by $M$ is necessary to normalize $R_n$. The power spectrum is then produced by a discrete Fourier transform of $R_n$.

It is straightforward to calculate the number $n$ of lags required for a given spectral resolution $\delta\nu_{ch}$. Continuing with our earlier assumptions let us now assume that Hanning weighting has been applied to the ACF. The spectral resolution is

$$\delta\nu_{ch} \sim 1/\tau_{max} = 1/(n\delta t);$$ (7.14)

hence

$$n = 1/(\delta\nu_{\text{ch}}\delta t) \tag{7.15}$$

and thus

$$n > 2\nu_{\text{max}}/\delta\nu_{\text{ch}}. \tag{7.16}$$

As explained in Appendix 3 (on digitization), if the spectrum is not at baseband then $\nu_{\text{max}}$ is replaced by the bandwidth $\Delta\nu$. A numerical example helps: if $\nu_{\text{max}}$ or $\Delta\nu$ is 100 MHz and we require $\delta\nu_{\text{ch}} = 100$ kHz then the spectometer must allow $n = 2000$ lags. In practice $n$ would be rounded up to the nearest power of 2 (2048 in this case) for application of the fast Fourier transform.

The concept of determining a power spectrum by autocorrelating the signal amplitude, while it is a powerful and still utilized technique, has to be understood more thoroughly than straightforward multichannel spectrometry. The radio astronomer must beware, in particular, of the complications introduced by non-Gaussian signals, such as manmade interference. One must also consider the choices of quantization and window weighting and their effects on the signal-to-noise ratio of the final spectrum. Nevertheless the ubiquity of digital autocorrelation spectrometers in the world's radio observatories is a testament to their advantages.

In contrast with analogue systems, digital spectrometers can be used for long integration times without drift; the signal-to-noise ratio therefore improves with the square root of time for integration times of many hours. The computationally intensive Fourier transform is performed only after the autocorrelation data have been averaged up for a considerable time using simple digital logic. However, digital autocorrelation spectrometers require large amounts of custom-designed electronics, in particular specialized integrated circuits, to carry out the required logic processes (shift registers, multiplications, additions). As a result they are increasingly being supplanted by a more direct approach to forming the spectrum which can use industry-standard components.

### 7.3.1 Fast Fourier Transform Spectrometers

Many non-astronomical applications require the rapid calculation of a Fourier transform; as a result industry is continuously developing high speed multi-bit analogue-to-digital converters (ADCs) which can feed into increasingly powerful field programmable gate arrays (FPGAs), as in Figure 7.7. The latter enable the direct calculation of a many-channel spectrum from an input time series at speeds which were not available a few decades ago. Such fast Fourier transform spectrometers (FFTSs) are therefore increasingly the norm.

It is beyond the scope of this book to describe the inner workings of such a spectrometer but it is worthwhile to itemize the sorts of calculation it has to perform. Let us place a requirement of producing a 16 384 (i.e. $2^{14}$) channel spectrum over a bandwidth of 1 GHz. These are cost-limited parameters and are not fundamental.

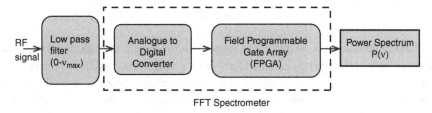

FFT Spectrometer

Figure 7.7. Schematic of a digital FFT spectrometer.

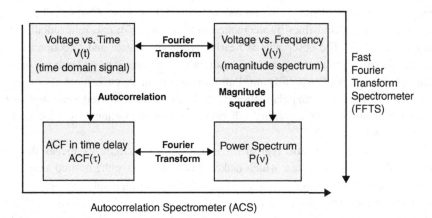

Autocorrelation Spectrometer (ACS)

Figure 7.8. The two types of digital spectrometer (bold arrows) exploit the Wiener–Khinchin theorem in different ways (see also Appendix 1).

- The channel width $\delta\nu_{ch} = 10^9/16\,384 = 61\,035$ Hz.
- Nyquist rate sampling at $> 2\nu_{max}$ requires $> 2$ giga samples per second, thus the time interval between samples $\Delta T < 0.5 \times 10^{-9}$ s.
- Data are taken over a total time $T$ consistent with the uncertainty relation $T \sim 1/\delta\nu_{ch}$ (here we are ignoring details of windowing for simplicity). Thus $T \sim 1/61\,035 = 1.6384 \times 10^{-5}$ s, corresponding to $1.6384 \times 10^{-5}/0.5 \times 10^{-9} = 32\,768$ data points (twice the number of channels).
- The spectrum is calculated using an FFT within the FPGA for 32 768 points and repeated every $T = 1.6384 \times 10^{-5}$ s, i.e. 61 035 times per second, to form $V(\nu)$, the voltage spectrum. The magnitude of each spectral frequency component is squared to produce $P(\nu)$, the power spectrum, and the successive power spectra are integrated up.

One can therefore see how the number of calculations required builds up. To avoid 'dead time' the calculations need to keep up with the data flow and hence everything needs to be done at high speed. These requirements are now attainable without having to resort to custom-designed integrated circuits.

To round off our presentation of digital spectrometry we show in Figure 7.8 the two complementary sequences for obtaining the power spectrum $P(\nu)$ from the input signal $V(t)$, by autocorrelation and by fast Fourier transform (see also Appendix 1).

## 7.4 Polarimetry

The polarization state of a cosmic radio signal is a diagnostic of both the astrophysical emission processes (Chapter 2) and the intervening propagation conditions (Chapter 4). A measurable polarization automatically implies significant anisotropy or a preferred direction in the source or intervening medium. Celestial targets include the scattered thermal emission from the cosmic microwave background (Chapter 17); synchrotron radiation from radio galaxies and quasars (Chapter 16) and nearby galaxies including the Milky Way (Chapter 14); scattered thermal-dust emission in the Milky Way (Chapter 17); Zeeman effects induced in spectral lines by magnetic fields (Chapter 3); coherent radiation from pulsars (Chapter 15). The observable is nearly always the degree and direction of linear polarization, the exceptions being cyclotron radiation from some stars and pulsars, in which circular polarization is often significant. Pulsars (and manmade emissions) also stand out because of their high fractional polarization (up to 100%) whereas for most celestial sources the polarized intensity of the incoming radiation is a small proportion of its total intensity, often only a few per cent.

The low signal-to-noise ratio typical of polarization measurements means that wide receiver bandwidths are required, which makes it hard to achieve the desired antenna and receiver characteristics across the band and hence to maintain high polarization purity. The resultant spurious instrumental contributions add to other unwanted effects associated with the reception beam itself. Separating these effects from the weak astronomical signal is a challenge – one which must be met with a combination of high quality radio engineering and astronomical experience. In what follows we focus on essential features of the single-dish polarimeters in common use; the calibration of such polarimeters is described in Appendix 4.

## 7.5 Stokes Parameters

In the mid-nineteenth century Sir George Stokes developed a practical way to characterize partially polarized light, in terms of linear combinations of average powers measured in different directions using polarizing filters. Measurement of the state of polarization of the wave, whether partial, plane, or circular, is fully specified by the four Stokes parameters $I$, $Q$, $U$, and $V$. The approach was largely ignored for almost a century until Subrahmanyan Chandrasekhar brought the formalism into astronomy in its present form during his study of radiative transfer in stellar atmospheres.[5]

Before considering partially polarized astronomical sources let us return, for simplicity, to the case of a fully polarized wave. Consider a radio telescope measuring two orthogonal electric field components on axes $x$ and $y$; it might, for example have separate receivers connected to crossed dipoles at the focus of a telescope.[6] The total power $\mathcal{E}_0^2$ in the received

---

[5]  S. Chandrasekhar 1981, *Radiative Transfer*, Dover.

[6]  We recall from Chapter 2 the fact that a single receiver is sensitive to only one of the orthogonal polarization components.

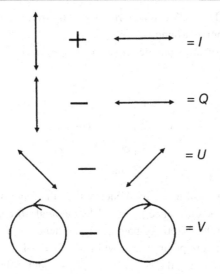

Figure 7.9. A graphical representation of the Stokes parameters for an approaching wave.

wave with any polarization is simply the average of the sum of the outputs of the two receivers detecting the squared amplitudes $\mathcal{E}_x^2$ and $\mathcal{E}_y^2$. The Stokes parameters can be written in a variety of ways but one of the simplest is based purely on practical operational definitions rather than on electromagnetic theory and the polarization ellipse:

$$
\begin{aligned}
I &= \mathcal{E}_x^2 + \mathcal{E}_y^2, \\
Q &= \mathcal{E}_x^2 - \mathcal{E}_y^2, \\
U &= \mathcal{E}_a^2 - \mathcal{E}_b^2, \\
V &= \mathcal{E}_r^2 - \mathcal{E}_l^2.
\end{aligned}
\tag{7.17}
$$

These relationships are illustrated in Figure 7.9.

In Eqns 7.17, $I$ is the total power (i.e. $\mathcal{E}_0^2$) and $Q$ and $U$ characterize the linear polarization; $Q$ is the difference between the powers measured in the $x$ and $y$ directions and $U$ is the difference between the powers measured along two orthogonal axes $a$ and $b$ rotated at 45° counterclockwise to $x$ and $y$. This additional power difference is required since the position angle of linear polarization cannot be unambiguously determined from only one set of axes. This measurement can be realized in practice simply by turning the crossed dipoles through 45°, and indeed some early polarization measurements were carried out in this manner. Note that $I$ could equally well be defined as $\mathcal{E}_a^2 + \mathcal{E}_b^2$. The circularly polarized component $V$ cannot be defined in terms of the difference between the powers on orthogonal linear axes no matter what their orientation, since it requires a measurement of the relative phase difference $\Delta\phi$ between $\mathcal{E}_x$ and $\mathcal{E}_y$. The component $V$ is therefore written as the difference between the powers measured with left and right circular polarization filters which require $\pi/2$ phase shifts between $\mathcal{E}_x$ and $\mathcal{E}_y$ to realize. Again, $I$ could be defined as $\mathcal{E}_l^2 + \mathcal{E}_r^2$. Note that equations similar to Eqn 7.17 can be written starting from $\mathcal{E}_a$ and $\mathcal{E}_b$ or $\mathcal{E}_l$ and $\mathcal{E}_r$, with the algebraic forms swapping around.

We do not show the derivation, but the Stokes parameters can be related to the parameters of the polarization ellipse:

$$I \equiv \mathcal{E}_0^2,$$
$$Q \equiv \mathcal{E}_0^2 \cos 2\beta \cos 2\chi,$$
$$U \equiv \mathcal{E}_0^2 \cos 2\beta \sin 2\chi,$$
$$V \equiv \mathcal{E}_0^2 \sin 2\beta.$$

(7.18)

It is clear that $I$ is proportional to the total energy flux, the other parameters being related to it by trigonometrical factors. Since in Eqn 7.18 the angle $\beta$ determines the axial ratio of the ellipse, $V$ must be the ellipticity parameter and hence $V = 0$ describes the case of entirely linear or fully random polarization, while $V/\mathcal{E}_0^2 = +1$ describes full right-handed polarization. The angles defining the polarization ellipse can be obtained thus:

$$\sin 2\beta = V/I,$$
$$\tan 2\chi = U/Q.$$

(7.19)

The ratio $U/Q$ therefore specifies the orientation of the ellipse; in practice this defines the position angle of linear polarization on the sky.

The Stokes parameters can also be expressed in terms of the real amplitudes $\mathcal{E}_x$ and $\mathcal{E}_y$ with relative phase $\Delta\phi$:

$$I = \mathcal{E}_x^2 + \mathcal{E}_y^2,$$
$$Q = \mathcal{E}_x^2 - \mathcal{E}_y^2,$$
$$U = 2\mathcal{E}_x\mathcal{E}_y \cos \Delta\phi,$$
$$V = 2\mathcal{E}_x\mathcal{E}_y \sin \Delta\phi.$$

(7.20)

As we shall see in Section 7.5.1, this form and related ones depending on the choice of the principal axes are the most common and provide a starting point for the design of polarimeters. In particular Eqns 7.20 refer to measurements on fixed axes rather than requiring an additional, and practically inconvenient, 45° rotation. Since $U$ and $V$ are defined in terms of products of electric fields and a phase angle rather than individual powers, the link with Eqns 7.17 is not immediately obvious.[7] However, for the specific case of a pure linearly polarized wave the link can be simply understood. The projections of $\mathcal{E}_x$ and $\mathcal{E}_y$ on the $a, b$ axes are: $\mathcal{E}_a = \mathcal{E}_y \sin 45° + \mathcal{E}_x \cos 45°$ and $\mathcal{E}_b = \mathcal{E}_y \cos 45° - \mathcal{E}_x \sin 45°$. Thus $\mathcal{E}_a = (\mathcal{E}_y + \mathcal{E}_x)/\sqrt{2}$, and $\mathcal{E}_b = (\mathcal{E}_y - \mathcal{E}_x)/\sqrt{2}$, and after squaring and subtracting we obtain $\mathcal{E}_a^2 - \mathcal{E}_b^2 = U = 2\mathcal{E}_x\mathcal{E}_y$. For pure linear polarization, $\cos \Delta\phi = 1$ while $\sin \Delta\phi = 0$ and hence $V = 0$. In the general case, where non-zero $\Delta\phi$ introduces a circularly polarized component, the trigonometrical terms cannot be neglected.

---

[7] See also E. Hecht (1970) Note on an operational definition of the Stokes parameters. *Am. J. Phys.* **38**, 1156.

Since the wave is described by two component amplitudes and one relative phase angle, three parameters in all, the four Stokes parameters of a monochromatic (fully polarized) wave must be related. The relation is

$$I^2 = Q^2 + U^2 + V^2. \tag{7.21}$$

Cosmic radio sources generally emit or are observed with a finite bandwidth $\Delta\nu$, so the observed amplitudes and relative phases fluctuate in a characteristic time $1/\Delta\nu$. For radio astronomers, the definitions of the Stokes parameters must therefore be generalized as time averages and are then written as

$$
\begin{aligned}
I &\equiv \langle \mathcal{E}_x^2 + \mathcal{E}_y^2 \rangle, \\
Q &\equiv \langle \mathcal{E}_x^2 - \mathcal{E}_y^2 \rangle, \\
U &\equiv \langle 2\mathcal{E}_x\mathcal{E}_y \cos\Delta\phi \rangle, \\
V &\equiv \langle 2\mathcal{E}_x\mathcal{E}_y \sin\Delta\phi \rangle.
\end{aligned}
\tag{7.22}
$$

The time-averaged Stokes parameters can also be written in terms of circular polarizations:

$$
\begin{aligned}
I &\equiv \langle \mathcal{E}_r^2 + \mathcal{E}_l^2 \rangle, \\
Q &\equiv \langle 2\mathcal{E}_r\mathcal{E}_l \cos \Delta\phi \rangle, \\
U &\equiv \langle 2\mathcal{E}_r\mathcal{E}_l \sin \Delta\phi \rangle, \\
V &\equiv \langle \mathcal{E}_r^2 - \mathcal{E}_l^2 \rangle.
\end{aligned}
\tag{7.23}
$$

As we shall see in Section 7.5.1, Eqns 7.22 and 7.23 are both exploited in practical polarimeters.

The total intensity $I$ can be partially polarized, with linear, circular, or elliptical partial polarization, and the four Stokes parameters provide a necessary and sufficient description in all cases. The polarized components are contained in $Q$, $U$, and $V$ and so for completely unpolarized radiation $Q = U = V = 0$, while for partially polarized radiation $I > (Q^2 + U^2 + V^2)^{1/2}$.

Most often we are concerned with the linear polarization (implicitly taking $V = 0$), in which case the polarized flux density is $p = (Q^2 + U^2)^{1/2}$ and the fractional degree of polarization is $m = p/I$. In terms of $Q$ and $U$ we have

$$Q = p \cos 2\chi, \tag{7.24}$$
$$U = p \sin 2\chi, \tag{7.25}$$

and the position angle of the electric vector on the sky is

$$\chi = (1/2) \tan^{-1}(U/Q). \tag{7.26}$$

In polarization interferometry (Section 11.2) we have to deal with the complex linear polarization

$$P = p e^{2i\chi} = m I e^{2i\chi} = Q + iU, \tag{7.27}$$

whose amplitude is the fractional polarization $m$ and whose phase is twice the position angle $\chi$ of the plane of polarization of the $\mathcal{E}$-vector (see also Figure 11.3).

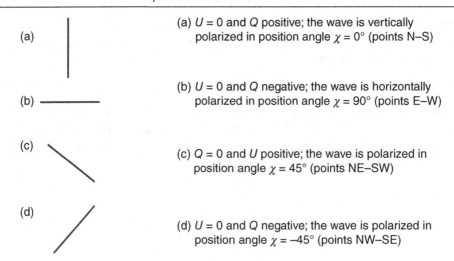

(a) $U = 0$ and $Q$ positive; the wave is vertically polarized in position angle $\chi = 0°$ (points N–S)

(b) $U = 0$ and $Q$ negative; the wave is horizontally polarized in position angle $\chi = 90°$ (points E–W)

(c) $Q = 0$ and $U$ positive; the wave is polarized in position angle $\chi = 45°$ (points NE–SW)

(d) $U = 0$ and $Q$ negative; the wave is polarized in position angle $\chi = -45°$ (points NW–SE)

Figure 7.10.   The relation between the Stokes parameters $Q$ and $U$ and the position angle of linear polarization.

Two receivers using orthogonal polarizations, either linear or circular, in a single telescope or in an interferometer can produce all four Stokes parameters simultaneously. Cosmic radio sources can, therefore, routinely be studied in polarization as well as intensity, which is a big advantage of radio astronomy. These sources frequently have complicated polarization properties and $V$ can be significant for pulsars and for radio bursts from the Sun and from Jupiter. However, for distributed emission, $V$ is invariably close to zero; thus such a source is only linearly polarized. In this case the polarization distribution can be characterized by maps of $Q$ and $U$ or by maps of the magnitude of the fractional polarization drawn in position angle $\chi = (1/2)\tan^{-1}(U/Q)$. These are commonly referred to as polarization vectors, although they are not strictly vectors since $\chi$ is ambiguous modulo $180°$; a better term is 'orientations'. Figure 7.10 illustrates the relationship between $Q$ and $U$ and the position angle of linear polarization. In the IAU definition this angle increases from north through east, in other words it increases counterclockwise when looking at the source.[8]

### 7.5.1 Choice of Orthogonal Polarizations

The challenge in radio polarimetry is to combine wide bandwidth with good impedance matching, low loss, and high isolation ($>30$ dB) between the orthogonal polarizations. Earlier in this section we saw that, by collecting two such components (invariably either

---

[8] Note, however, that the convention for the polarization angle adopted by the CMB polarization community (increasing clockwise when looking at the source) is opposite to the IAU definition and corresponds to a change of sign of the Stokes $U$ parameter.

linear, $X$, $Y$, or circular, $L$, $R$) one can derive the complete polarization state of radiation in terms of the four Stokes parameters.

In the standard situation in radio astronomy, the radiation is collected by a waveguide feed at the focus of a parabolic dish (see Section 8.6). In order for orthogonal polarizations to propagate into the receiver the cross-section of the feed system must be symmetrical; this usually means a circular waveguide, although square sections can be used. It is then a free choice as to which orthogonal pair, $X$, $Y$ or $L$, $R$, to start with. Since there is only one radiation field there should, in principle, be no difference in the outcome. In reality, though, there is a range of reasons to prefer one over the other depending both on astronomical requirements and practical receiver design and construction issues.

The most frequent requirement is to measure linear polarization from targets where the natural level of circular polarization is vestigial. Since $V \approx 0$ we need only to measure the Stokes $I$, $Q$, and $U$. The most obvious route would seem to be a receiver based on $X$, $Y$ but this is often not the favoured choice. A linear system derives $U$ from the cross-correlation of $X$ and $Y$; since linear polarizations are usually small the correlation will be weak and as a result any receiver gain fluctuations between the $X$ and $Y$ channels will have little effect. On the other hand, $Q$ is derived from the difference between the powers in the $X$ and $Y$ channels; these are large numbers dominated by the receiver noise and hence $Q$ is fully subject to receiver gain fluctuations. The alternative is to use a receiver based on the $L$, $R$ pair. Equations 7.23 show that in this case both $Q$ and $U$ are derived from cross-correlation measurements and hence are relatively immune from gain fluctuations. The Stokes $I$ is simply obtained from the sum of powers in the $L$ and $R$ channels.

It might seem, therefore, that polarimeters based on circular polarization would be the norm – indeed they are the commonest, and for reasons over and above the gain immunity for $Q$ and $U$. In particular the orientation of an $X$, $Y$ feed on a telescope with an alt-azimuth mount (see Chapter 8 and Section 11.2) is not fixed relative to the sky. It varies with the parallactic angle and hence depends on the celestial position of the source, the sidereal time, and the latitude of the telescope. The $X$, $Y$ feed orientations in the various telescopes in a VLBI array may, therefore, be very different and so the recorded data will not correlate correctly.[9]

The oft-quoted rule-of-thumb is *'to measure linear polarization correlate circulars, while to measure circular polarization correlate linears'*; however, $X$, $Y$ types are easier to build, to some extent to calibrate, and often, tellingly, can achieve wider fractional bandwidths compared with their $L$, $R$ counterparts. And, for measuring Stokes $V$, an $L$, $R$ polarimeter will suffer the same problem of gain fluctuations as does an $X$, $Y$ when measuring Stokes $Q$. As a result, polarimetry on some single dishes (and short-baseline interferometers) is based on linearly polarized inputs. Careful design and calibration can make either choice work well.

We start our outline of single-dish polarimeters by describing the ways in which orthogonal polarization modes, either linear or the hands of circular polarization, are extracted from the radiation flowing into a waveguide feed.

---

[9] See Section 11.2. The variation with parallactic angle is very useful for calibrating a single-dish polarimeter – see Appendix 4.

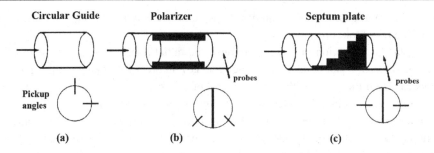

Figure 7.11. Orthomode transducers. (a) Two probes receive orthogonal linear polarizations. (b) A 90° polarizer before the probes converts orthogonal circular input modes into linear output modes. (c) A septum plate in a circular guide directly separates the two circular modes.

## 7.6 Polarized Waveguide Feeds

At long wavelengths, where dipoles are used as antenna feeds, a pair of crossed dipoles naturally provides two orthogonal linear polarizations. At shorter wavelengths, where symmetrical horn feeds are used, two orthogonal signals are received in a single waveguide and these must be separated and fed to a pair of mutually isolated amplifiers, one for each polarization. This action is carried out by an *orthomode transducer* (OMT), for which there are a multiplicity of designs to meet the ever-changing demands of bandwidth, cross-polar purity, and ease of fabrication. Most broadband high purity OMTs use waveguides (e.g. Bøifot 1991) but at longer wavelengths these can become unwieldy.

A simple OMT design, shown in Figure 7.11(a), consists of two coaxial probes into a symmetrical circular waveguide; each probe naturally picks up a linear polarization. However, as we have seen, a common requirement is to separate the circular modes. To do this we recall from Chapter 2 that the hands of circular polarization can be regarded as the sums of orthogonally polarized linear components with $\pm 90°$ phase shifts between them. To achieve separation one inserts a polarizer section of waveguide before the probes and orientated at 45° as indicated in Figure 7.11(b). This form of polarizer consists of posts or ridges which change the phase velocity of one linear component so as to introduce a 90° phase difference between the two orthogonal components. The frequency range over which the shift stays close to 90° defines the bandwidth. If the probes define axes $a$ and $b$ and the polarizer defines one of the axes $x$ or $y$ then, following the angle transformation in Section 7.5 and before considering any phase shift, $\mathcal{E}_a = (\mathcal{E}_y + \mathcal{E}_x)/\sqrt{2}$ and $\mathcal{E}_b = (\mathcal{E}_y - \mathcal{E}_x)/\sqrt{2} \equiv [\mathcal{E}_y + (\mathcal{E}_x + 180°)]/\sqrt{2}$, where $\mathcal{E}_x + 180°$ means 'rotate $\mathcal{E}_x$ by 180°'. Now if the polarizer changes the phase of $\mathcal{E}_x$ by $+90°$ we have $\mathcal{E}_a = [\mathcal{E}_y + (\mathcal{E}_x + 90°)]/\sqrt{2}$ and $\mathcal{E}_b = [\mathcal{E}_y + (\mathcal{E}_x + 270°)]/\sqrt{2} \equiv \mathcal{E}_y + (\mathcal{E}_x - 90°)]/\sqrt{2}$. Thus $\mathcal{E}_a$ and $\mathcal{E}_b$ correspond to the two hands of circular polarisation.[10]

---

[10] By analogy with the optical component which carries out the same mathematical function this waveguide polarizer is sometimes referred to as a quarter-wave plate (QWP).

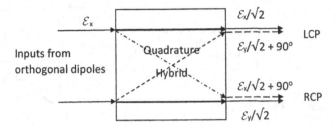

Figure 7.12. A hybrid circuit transforming linear to circular polarization.

The septum plate shown in Figure 7.11(c) combines the functions of OMT and polarizer in a single waveguide section. A metal plate, tapered at an angle of around 30° or, more often, in steps as shown in the figure, divides the circular guide into two semicircular halves at the output end. The two narrow output sections sustain only one linear polarization; the conversion to circular occurs in the tapered section, in which this linear mode is progressively coupled to the orthogonal linear mode. The phase of the orthogonal linear mode is arranged to be in quadrature, combining to give a circular mode at the input. The phase of the coupled linear mode is opposite on either side of the septum, so that the two circular modes are coupled separately to the two probes. (As seen from the two probes, the two sections of waveguide separated by the septum plate are mirror images.)

Septum polarizers are more compact than the linear OMT plus polarizer system shown in Figure 7.11(b), but since their operation depends on a subtle balance of electrical lengths in the waveguide they tend to offer narrower bandwidths.

### 7.6.1 Linear Feeds and Quadrature Hybrids

At frequencies lower than 1 GHz ($\lambda > 30$ cm), where waveguides become too large to deploy at dish foci, polarization-sensitive feeds are based on crossed dipoles, albeit often with other structures around them (see Section 8.1). Circular polarization can be produced by feeding the orthogonal dipoles into a quadrature or 90° hybrid coupler (Figure 7.12). This has two input ports and two output ports. A signal applied to either input will result in equal amplitude signals (reduced by $\sqrt{2}$) in the output ports that are 90° out of phase with each other, as indicated graphically in Figure 7.12. With the $X$, $Y$ dipole voltages as inputs the outputs are $L$, $R$.

## 7.7 A Basic Polarimeter

A schematic of a polarimeter based on $L$, $R$ inputs is shown in Figure 7.13. Looking back to the form of the Stokes parameters in Eqns 7.17, the outputs can be understood as follows:

- Output 1: total power from $R$
- Output 2: correlation of $R$ and $L$

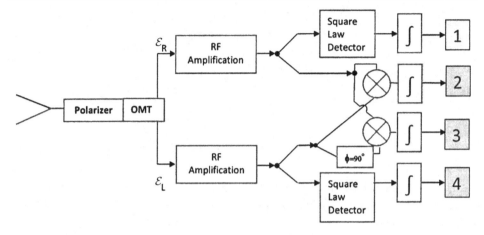

Figure 7.13. Schematic polarimeter based on circularly polarized inputs.

- Output 3: correlation of $R$ and ($L$ delayed by $90°$)
- Output 4: total power from $L$

Combinations of outputs 1 to 4 are related to the Stokes parameters thus:

- $I \equiv \langle \mathcal{E}_r^2 \rangle + \langle \mathcal{E}_l^2 \rangle \propto (1+4)$
- $V \equiv \langle \mathcal{E}_r^2 \rangle - \langle \mathcal{E}_l^2 \rangle \propto (1-4)$
- $Q \equiv 2\,\mathrm{Re}\langle \mathcal{E}_r \mathcal{E}_l \rangle \propto 2$
- $U \equiv 2\,\mathrm{Im}\langle \mathcal{E}_r \mathcal{E}_l \rangle \propto 3$

A similar scheme using orthogonal linear polarizations might use a circular waveguide feed with no polarizer, as in Figure 7.11(a), providing $\mathcal{E}_x$ and $\mathcal{E}_y$ to the two receiver inputs

### 7.8 Practical Considerations

Figure 7.13 is a highly idealized and simplified version of a real polarimeter system; these are many and varied in their arrangements. In practice the analogue sections of the two channels are likely to be of the heterodyne type with several amplifiers, mixers, filters, and the same local oscillator (LO) system to maintain phase coherence between the channels. Maintaining the precisely correct relative phase delays within the receiver is vital for accurate results, since they affect either the $U$ and $V$ Stokes parameters (for linear inputs) or the $Q$ and $U$ Stokes parameters (for circular inputs). The amplifier gains will vary, and to combat this various Dicke-like switching arrangements can be made. Finally, much of the post-IF data handling will be done digitally. In Figure 7.13, digital auto (to give total power) and cross-correlation and integration based on FPGAs can replace all the 'components' to the right of 'RF Amplification', while achieving bandwidths $>1$ GHz. To calibrate the polarimeter, measurements of known sources are needed, as discussed in Appendix 4 and in Section 11.2 in the case of interferometers.

## 7.9  Further Reading

*Supplementary Material* at www.cambridge.org/ira4.

Bracewell, R. N., 1962. Radio astronomy techniques, In: Flugge, S. (ed.), *Handbuch der Physik*, vol. 54, p. 42. Springer.

Heiles, C. 2012. A heuristic introduction to radioastronomical polarization. In: Astronomical Society of the Pacific Conference Series, vol. 278, p. 131.

Webber, J. C., and Pospieszalski, M. W. 2002. Microwave instrumentation for radio astronomy. *IEEE Trans. Microwave Theory and Techniques*, **50**.

Wilson, T. L., Rohlfs, K., and Hüttemeister, S. 2013. *Tools of Radio Astronomy*, 6th edn. Springer-Verlag.

# Part II

# Radio Telescopes and Aperture Synthesis

# 8

# Single-Aperture Radio Telescopes

A radio telescope intercepts the radiation coming from celestial sources, usually separating it into two polarization components. The radio telescope must meet two basic requirements, sensitivity and angular resolution. Sensitivity depends on having the largest possible collecting area while minimizing the contributions of extraneous noise; this depends upon both the telescope design and on the quality of the receiving system. Angular resolution is determined by the overall dimensions of the telescope.

In this chapter we consider single-aperture telescopes, for which large area and high angular resolution go together. It is economically impossible to get the highest angular resolution by extending the size of a single aperture indefinitely, so it is necessary to use widely spaced single-aperture telescopes in an *aperture-synthesis array*. Such arrays are discussed in Chapters 9–11.

The method of construction of the single aperture depends on wavelength. Antennas made from metal wires or rods, i.e. dipoles, Yagis, log periodics, and the like, are the norm at longer wavelengths (frequencies below ∼ 500 MHz) and connected arrays of such elements are employed in instruments such as LOFAR, the MWA, and the LWA (see Chapter 11) to produce large instantaneous collecting areas in a *phased array*. The individual elements each observe a patch of sky defining the *field of view* (fov), which can be large at low frequencies and within which narrower beams can be steered electronically (see Section 8.2).

Above ∼ 500 MHz, mechanically steerable parabolic dish reflectors with waveguide horns as the 'feeds' are the norm. Their fov can be increased with a *focal plane array* of independent receptors (often horns) at the focus which provides multiple simultaneous, but well-separated, beams on the sky. Densely packed electronically connected *phased array feeds* (PAFs) can yield a larger number of simultaneous beams, which can be arranged to overlap and hence completely cover a specific patch of sky (see Section 8.7).

## 8.1 Fundamentals: Dipoles and Horns

Because of reciprocity, a radio telescope can be analyzed either as a transmitter or as a receiver. For the transmitting case, the radiated electric field is generally described by the power gain $G$ (proportional to the square of the field strength) and for the receiving case by

131

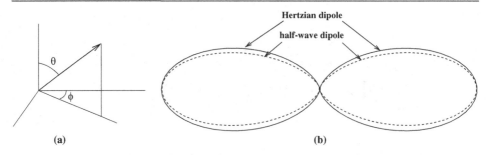

Figure 8.1. (a) Coordinate system, (b) radiation patterns (polar diagrams) of a Hertzian dipole (full line) and a half-wavelength dipole (broken line) in free space.

the effective area, $A_{\text{eff}}$. Both are functions of the direction of reception or transmission, and they are related by the fundamental relation between gain and effective area:

$$A_{\text{eff}} = (G/4\pi)\lambda^2. \tag{8.1}$$

The simplest case to analyse is the Hertzian dipole, an infinitesimal length $\Delta l$ of current varying sinusoidally at angular frequency $\omega$. The geometry and the radiation pattern are shown in Figure 8.1 for a dipole axis at $\theta = 0°$. Note that the coordinates $\theta, \phi$ are convenient for an isolated dipole, because of the azimuthal symmetry, as a consequence of which the radiation pattern does not depend on $\phi$.

The radiation pattern for the electric and magnetic fields has two parts, with different dependence on both the radial coordinate $r$ and the polar angle $\theta$. The first part is called the *near field* or *induction field*; it has two terms, diminishing inversely as the second and third powers of $r$. The second part is the radiation field, or *far field*, which varies inversely with distance; it has an electric field component only in the $\theta$ direction, and a magnetic field component[1] in the $\phi$ direction. These two field components are related by

$$\mathcal{E} = Z_0 \mathcal{H} \times \hat{\mathbf{n}}, \tag{8.2}$$

where $Z_0 = 377$ ohms is the impedance of free space and $\hat{\mathbf{n}}$ is the unit vector in the $r$ direction.

The electric field in the far-field region, for a short sinusoidally varying current element of amplitude $I_0$ and length $\Delta l$, is[2]

$$\mathcal{E}_0 = (Z_0/4\pi)(I_0\Delta l)k^2 \sin\theta \, \exp[j(\omega t - kr)]/(jkr), \tag{8.3}$$

where $k = 2\pi/\lambda$. The electric field is an expanding spherical wave and is accompanied by an orthogonal magnetic field. Both fields vary as $1/r$, and so the average value of the Poynting vector is $\langle \mathbf{S} \rangle = \frac{1}{2}\mathcal{E} \times \mathcal{H}$ (including a factor $\frac{1}{2}$ for the time average):

$$\langle \mathbf{S} \rangle = \frac{1}{8}Z_0(I_0\Delta l/\lambda)^2 \sin^2\theta/r^2. \tag{8.4}$$

---

[1] Note the use of $\mathcal{H}$ in antenna theory, whereas $\mathcal{B}$ is used in astrophysics and radiation theory.
[2] For the imaginary quantity we use either j or i, depending on the context: j for electrical engineering usage, and i otherwise.

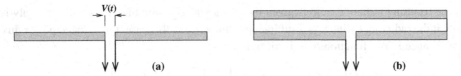

Figure 8.2. (a) Dipole antenna connected to a transmission line. (b) A folded dipole.

Integrating over all solid angles gives the average total power radiated:

$$\langle P \rangle = (\pi/3)Z_0 I_0^2 (\Delta l/\lambda)^2. \tag{8.5}$$

This reminds one of the average power dissipated in a resistor, $\langle P \rangle = \frac{1}{2}I_0^2 R$, and leads to the introduction of another physical quantity, the *radiation resistance* $R_r$ (see Section 6.1.1), since the power radiated by the Hertzian dipole in Eqn 8.5 is equivalent to the power dissipated in a radiation resistance $R_r$ given by

$$R_r = (2\pi/3)Z_0(\Delta l/\lambda)^2 \quad (= 789(\Delta l/\lambda)^2 \text{ ohms}). \tag{8.6}$$

It is immediately clear that if the dipole is very short compared to a wavelength then its radiation resistance will be very small, and hence it will be an inefficient antenna.

The power gain of an antenna is defined, for a given direction, as the ratio of the power radiated to that radiated by an isotropic radiator. Equation 8.4 leads to the result that the power gain of a Hertzian dipole is $G = (3/2)\sin^2\theta$ and so the maximum gain, in the direction perpendicular to the dipole, is 3/2. Thus, from Eqn 8.1, the maximum effective area of the Hertzian dipole, perpendicular to the dipole at $\theta = 90°$, is

$$A_{\text{eff}} = (3/8\pi)\lambda^2. \tag{8.7}$$

It may seem paradoxical that a very small wire should have an effective area that is an appreciable fraction of a square wavelength, but it should be remembered that the displacement current in electromagnetic theory acts like a real current, and the near-field components, while playing a negligible role in the far-field properties of the antenna, are real and are important over distances of the order of a wavelength.

Short dipoles are sometimes met with in radio astronomy, but the *half-wavelength dipole* is more commonly met in practice. When it is excited by a current that is at a frequency whose wavelength is twice the length of the dipole, it is said to be excited at its resonant wavelength.

The general case of radiation from a cylindrical conductor carrying a current is treated in the standard textbooks. The dipole has to be fed by a transmission line, as illustrated in Figure 8.2, and so the dipole has to be broken at the centre, where the voltage appears. The solution for the radiated electromagnetic field depends on matching the boundary conditions. These are: (i) the integral of the electric field across the gap (assumed to be negligibly small) must equal the voltage; (ii) the tangential component of the electric field must go to zero at the surface of the cylinder; and (iii) the current must go to zero at the end of the cylinder. For the half-wave dipole, the current along the dipole is distributed like a half-sinusoid, zero at the ends and maximum at the centre.

The resulting radiation field is given by an integral equation that can be solved analytically if the radiating cylinder is a sufficiently thin wire. In this case, the radiated power flux at a distance $r$, as a function of $\theta$, is given by

$$\langle S(\theta) \rangle = Z_0/(8\pi^2)(I_0/r)^2 \left[ \cos\left(\frac{\pi}{2}\cos\theta\right) / \sin\theta \right]^2. \tag{8.8}$$

The power gain and radiation resistance of the half-wavelength dipole at its resonant wavelength can be calculated by the same methods as those used above for the Hertzian dipole. The gain $G(\theta)$ is

$$G(\theta) = 1.64 \left[ \cos\left(\frac{\pi}{2}\cos\theta\right) / \sin\theta \right]^2, \tag{8.9}$$

so that the gain normal to the dipole, 1.64 (2.15 dB), is slightly greater than the maximum gain, 1.5 (1.76 dB), of the Hertzian dipole. The polar diagram $G(\theta)$ for the half-wave dipole is shown in Figure 8.1(b), where it is seen to be narrower (FWHM $= 78°$) than for the Hertzian dipole (FWHM $= 90°$). All directions perpendicular to the dipole axis give the same gain, and hence its three-dimensional radiation pattern is a torus.

At resonance, the radiation resistance of the ideal half-wave dipole is $R_r = 73$ ohms. In practice the actual antenna impedance $Z_A$ will be different, partly because of the finite thickness of the dipole, but mainly because it may need to be used over a wide bandwidth. The antenna impedance will be frequency dependent and will not be purely resistive; it will be a complex quantity $Z_A = R_A + jX_A$. Off resonance, the antenna impedance changes rapidly and the reactive term $X_A$ becomes significant. For thicker dipoles, resonance occurs at a lower frequency and $X_A$ varies more slowly with frequency, and so in most practical cases a fat dipole is used, thus increasing its usable bandwidth without introducing a severe impedance mismatch between receiving system and antenna. The folded dipole (Figure 8.2(b)) also has a comparatively wide bandwidth. At resonance it has a radiation resistance of 292 ohms, four times that of the single dipole (this arises because the parallel element doubles the current for a given voltage at the transmission line, and four times the power is radiated).

A dipole produces a *balanced* signal, with two voltages in antiphase. It may be connected to a receiver via a twin-wire balanced transmission line, or the connection may be via a coaxial cable (unbalanced) transmission line, in which the outer connector is connected to earth. In the latter case a transformer known as a *balun* is required.

Figure 8.3 shows four types of balun. The simplest, shown in Figure 8.3(a), consists of two mutually coupled coils, one balanced and the other unbalanced with one side grounded. This is impractical for most radio astronomy purposes, since it can only be realized at low frequencies. A second version, Figure 8.3(b), in use at frequencies up to 10 GHz (and probably realizable up to the highest frequencies at which transistor low-noise amplifiers can be used), has an amplifier with a balanced input, using a pair of input transistors back to back. Internal circuitry transforms the initially balanced transistor configuration to unbalanced circuitry with an unbalanced output, usually a coaxial cable. This should be realizable up to the highest frequencies at which transistor circuits can be fabricated.

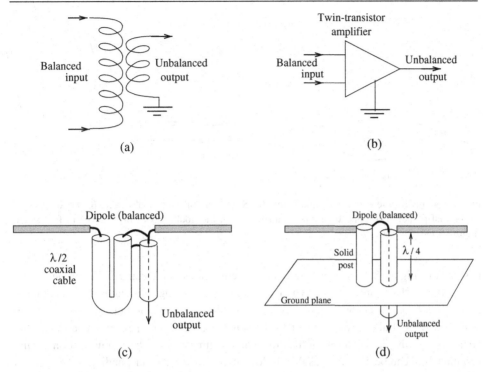

Figure 8.3. Balanced to unbalanced transformers, known as baluns: (a) coupled coils, suitable for low frequencies; (b) paired transistors; (c) half-wave coaxial; (d) split coaxial above a ground plane.

Figure 8.3(c) shows a simple passive balun, in which a half-wavelength coaxial cable provides a phase shift of 180° for one half of the dipole. It also transforms the dipole impedance downward by a factor 4. Thus, a folded dipole with an impedance of 280 ohms would conveniently present an impedance of 70 ohms to the coaxial cable. The fourth example, Figure 8.3(d), shows a robust structure that has been useful in radio astronomy. Here, the coaxial line is split, with the central conductor attached to one side of the dipole while the other side of the dipole is attached to the opposite half of the split coaxial line. For a fat dipole, this gives a good impedance match to a 50 ohm cable.

One should note that passive baluns work only over a finite bandwidth. The version in Figure 8.3(a) is limited by the impracticability of making coupled coils at high frequencies; while both the versions in Figures 8.3(c) and 8.3(d) have tuned transmission lines. This is not always a problem, since the dipole itself has a finite bandwidth; however, the need for a very wide bandwidth in modern arrays, such as LOFAR and MWA (Chapter 11), has led to an increasing use of the paired transistor balun (Figure 8.3(b)).

### 8.1.1 Ground Planes

Except for the case of dipoles attached to spacecraft, dipoles are usually found in combination with other structures, often a ground plane. A dipole above a conducting ground plane

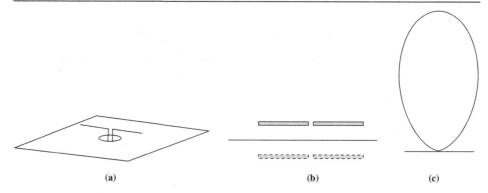

Figure 8.4. (a) Dipole above a ground plane. (b) Schematic of the virtual image. (c) Polar diagram (equatorial plane). The polar diagram is narrower than the double lobes of a dipole in free space (Figure 8.1).

is shown in Figure 8.4. To meet the boundary conditions, the electric field parallel to the conducting surface must be zero. The resulting solution is logically equivalent to a dipole with a reflection, a second virtual dipole, below the ground plane but reversed in phase. The electric field above the reflector (to be squared for the power pattern $G(\theta, \phi)$) is the vector sum of the field patterns of the dipole and its image. As discussed in Section 8.2 the resultant field can be described as the product of an 'array factor' depending on the pattern of isotropic radiators at the positions of the antenna elements[3] multiplied by the pattern of an individual element. The most obvious change from the free-space symmetry is that the lobe in the direction of the ground plane is absent. The specific power patterns depend on the spacing between the dipole and the ground plane; a single lobe, giving the maximum forward gain, occurs when the spacing is a quarter of a wavelength. The pattern equivalent to Figure 8.1(b) is shown in Figure 8.4(c). In this direction the FWHM ($\sim 73°$) is only a little smaller than for the free-space case. However, in the free-space case all directions perpendicular to the dipole give the same gain (hence the toroidal pattern) but the array factor concentrates that power into a pattern of FWHM $= 120°$. The overall forward gain is a factor 5.56 (7.45 dB), roughly a factor 4 greater than the free-space case.

### 8.1.2 The Horn Antenna

We noted the important role of the displacement current in the radiation from a short dipole; a *horn antenna* is an example of a system where all the current at the aperture is displacement current. Figure 8.5 shows a rectangular, tapered horn connected to a waveguide, into which protrudes the central conductor of a coaxial cable. The probe is positioned where the electric field strength is a maximum, typically about a quarter of a wavelength from the end.

---

[3] For a dipole $h$ degrees above the reflector (i.e. $h = 90°$ for a height $\lambda/4$) the array factor is $2 \sin(h \cos \theta)$, where $\theta$ is the angle from the normal to the dipole axis.

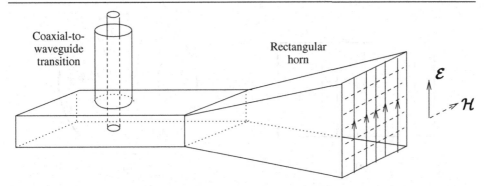

Figure 8.5. A rectangular horn antenna, with a transition to a coaxial cable.

A received wave propagates down the waveguide, its electric field generates a current in the coaxial probe, and the signal is propagated down the coaxial cable to the receiving device. The horn must be gently tapered,[4] since the wavefront in the horn will be curved and furthermore unwanted higher modes can be excited.[5] The combination of circular waveguide and cone-shaped horn is the set-up most commonly used in radio astronomy; the symmetry makes such horns well suited for feeding parabolic dishes and for achieving good polarization performance.

The mode that is excited in the waveguide approximates the field at the aperture of the horn if the taper is not too abrupt, thus avoiding a wavefront that is too strongly curved (although some curvature will be present, and this must be accounted for in calculating the actual gain pattern of the horn). The lowest modes in a rectangular and in a circular waveguide, the principal modes, form the two most important simple cases. The field configurations for these two cases are illustrated in Figures 8.6(a) and (b).

In both cases the electric field is transverse, with the magnetic field in the orthogonal direction, in loops along the axis of the waveguide. The fundamental mode in a rectangular waveguide, Figure 8.6(a), is designated $TE_{01}$, with a uniform electric field stretching from bottom to top of the waveguide but varying in intensity sinusoidally across the guide, peaking in the centre. The nomenclature $TE_{mn}$ describes a transverse electric mode, where the electric field is orthogonal to direction of the propagation of power down the guide. For a rectangular guide, $m$ is the number of half-wave patterns across the guide width and $n$ the number across the height. The configuration is independent of the proportions of the guide. The analogous case for a circular waveguide, shown in Figure 8.6(b), is called $TE_{11}$[6] since in polar coordinates there is one node both radially and azimuthally. If a waveguide is sufficiently large, it can support more than one mode; for example, a square waveguide

---

[4] Tapering also gradually matches the impedance in the guide to that of free space, thus maximizing the transfer of power.

[5] The boundary conditions only allow certain patterns of electric and magnetic fields to propagate within the waveguide; a *mode* is a particular pattern of electric and magnetic fields confined within the waveguide.

[6] The subscripts for the modes in a circular waveguide differ from those for the rectangular guide; $m$ is the number of full-wave variations of the radial component of the electric field around the circumference and $n$ is the number of half-wave variations across the diameter.

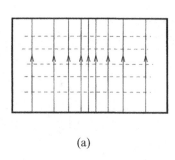
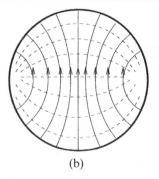

(a)                                                        (b)

Figure 8.6. The field pattern for transverse electric modes at the aperture of (a) a rectangular ($TE_{01}$) and (b) a circular ($TE_{11}$) horn. The solid lines show the $\mathcal{E}$-field, the broken lines the $\mathcal{H}$-field.

supports both the $TE_{01}$ and the orthogonal $TE_{10}$ mode and the same holds true for the circular case. In general, different modes propagate at different velocities.

The size of the horn aperture, measured in wavelengths, determines the angular dimensions of its radiation pattern. In most radio astronomy applications an axially symmetric gain pattern is desired, and this is determined by the proportions of the aperture. A horn fed in the $TE_{01}$ mode does not have a symmetrical pattern, with either a square or a circular aperture. For the square and circular horns, however, it is easy to see that because of their symmetry a $TE_{01}$ mode can be excited in any orientation. If two separate modes with the same phase and orthogonal to one another are excited then, because of the linearity of the field, their sum is a $TE_{01}$ mode at $45°$. If, on the other hand, they differ in phase by $90°$ then their superposition is a circularly polarized mode, and the resulting phase pattern must be symmetric. For this reason, circular polarization is frequently met with in feeds for paraboloidal antennas (Section 7.6 and Figure 8.25).The polar pattern of simple conical horns is, however, not symmetric in orthogonal directions. To symmetrize such a pattern more modes must be excited, and this is often achieved by periodic corrugations in the walls, as discussed in Section 8.6.

### 8.1.3  Wide-Band Antennas

Dipoles and horns have characteristic dimensions, centred on a given wavelength, and this limits their useful bandwidth. However, the bow-tie antenna shown in Figure 8.7(c) can offer a usable bandwidth of more than 3 : 1; it may be regarded either as a very fat dipole or as a scalar antenna (see below). Figure 8.9 shows the pair of bow-tie dipoles, mounted over a reflecting sheet, which, with minor differences, forms a basic element of both the LOFAR-HBA and the MWA, as further described in Section 8.2 and Chapter 11.

The exponential horn, shown in Figure 8.7(a), is an example of an antenna that has no characteristic size and hence a radiation pattern that in principle is independent of

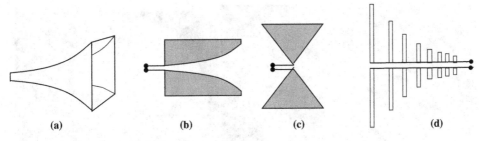

Figure 8.7. Broadband antennas: (a) exponential horn; (b) Vivaldi exponentially tapered slot; (c) bow-tie; (d) log-periodic.

wavelength (in practice, of course, this is not possible, since the horn has to have a finite size).

The Vivaldi dipole, shown in Figure 8.7(b), is an exponentially tapered slot cut in a conducting plane, and fed by a balanced twin transmission line. As in the exponential horn, a guided wave pattern propagates along the slot until it reaches a slot width of about half a wavelength, when it detaches and is radiated.

These tapered systems are examples of a class of antenna whose properties are described entirely by ratios; they are known as *scalar antennas*. The earliest example, the log-periodic array shown in Figure 8.7(d), was introduced by Rumsey (1966). The array has to be terminated at an upper and a lower size, so its wavelength range is still finite, but it turns out that the end effects do not seriously disturb the antenna properties. A pyramidal form of the scalar array is in use as a feed for the elements of the Allen Telescope Array, giving a bandwidth of 20 : 1 (from 0.5 to 12 GHz), the broadest-band feed in existence (Welch *et al.* 2017). It is shown in Figure 8.8. A drawback of this design is, however, that the phase centre varies as a function of frequency, which is inconvenient when it is operating as a feed at the focus of a parabolic dish. Recent developments in horn design offer bandwidths of about 5 : 1 without the problem of a varying phase centre. One approach is a development (Akgiray and Weinreb 2012) of the 'quad-ridge' horn design, in which the horn contains pairs of ridges closely resembling Vivaldis; another is the so-called 'Circular Eleven Antenna' developed at Chalmers University (Yin *et al.* 2013), which uses a flat plate of dipole elements resembling log-periodic arrays (Figure 8.8).

## 8.2 Phased Arrays of Elementary Antennas

At the longest wavelengths used in radio astronomy, when it is impractical to use large reflectors, it is possible instead to use flat arrays of dipole elements distributed over the ground, covering a larger area than can be achieved with a steerable reflector. Such an array was used by Antony Hewish and Jocelyn Bell in the discovery of pulsars (Chapter 15), at a wavelength of 3.7 m. The UTR-2 array at Kharkov is currently being developed into the Giant Ukranian Radio Telescope (GURT) (Konovalenko *et al.* 2016), which will be the single-site array with the largest collecting area (150 000 m²). The UTR-2 dipoles are

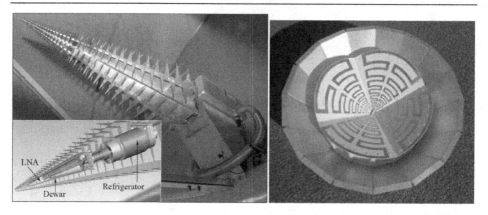

Figure 8.8. (left) The broadband feed used in the Allen Telescope Array; it consists of extended cross-polarized log-periodic elements covering 500 MHz to 11 GHz (Welch *et al.* 2017). (right) The Chalmers Circular Eleven broadband feed; each dipole element is log-periodic (Chalmers University, courtesy of J. Yang). For further information on wideband single-pixel feeds see https://www.skatelescope.org/wbspf/.

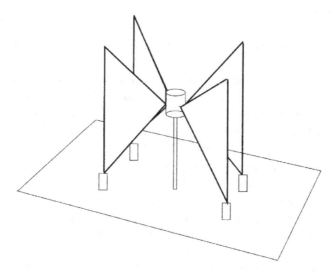

Figure 8.9. An orthogonal pair of bow-tie dipoles mounted over a reflecting sheet. In the basic element of the MWA telescope array, each dipole is an open framework, with the long side approximately one metre across, i.e. one half-wavelength at the centre of the band 80 to 300 MHz.

made of mesh, and are 'fat' in order to broaden their usable bandwidth, covering the range 8–33 MHz. We will discuss the latest generation of low frequency arrays at the end of this section and in Section 11.6.

The behaviour of such an array can be visualized most easily by considering it as a transmitting array. Consider an input signal, fed into a transmission line that splits symmetrically into other transmission lines, each splitting again; eventually each branch is connected to a dipole. This arrangement, illustrated in Figure 8.10, is called a Christmas tree network.

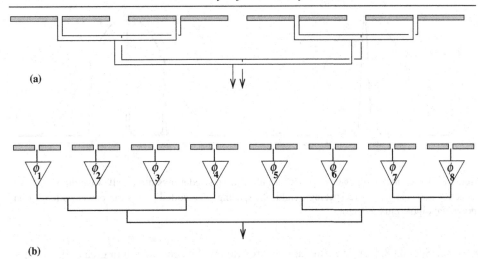

(a)

(b)

Figure 8.10. A phased array: (a) a 'Christmas tree' network feeds an array of dipoles in phase; (b) phase shifters (and amplifiers) allow a linear gradient of phase, which swings the angle of the radiated beam.

Such a fixed analogue splitting–adding network is the simplest to understand, and variants were used in all early phased arrays such as the Cambridge pulsar array. However, in modern practice, the data from individual antenna elements, or groups of elements, are digitized and the combinations are carried out in digital hardware under software control. The amplitude radiation pattern, $F_R(\theta)$, will be the product of the pattern of the dipole above a ground plane, $F_D(\theta)$ (the 'element factor'), and the vector sum of the radiation field from each element (the 'array factor'):

$$F_R(\theta) = F_D(\theta)\Sigma i_n \exp(i\phi(n)). \tag{8.10}$$

For simplicity, consider an even number of dipoles, fed by a Christmas tree array, with spacing between the dipoles equal to $\lambda/2$ (a common case for dense arrays). All the dipoles will have the same current excitation, which we take as unity, since it is only the relative pattern that is being studied. In the far field, the relative phases of a pair of adjacent dipoles will be $\phi = (\pi/2)\sin\theta$ (where $\sin\theta$ can be replaced by $\theta$ if the angle is small). The power gain will be proportional to the square of the amplitude gain of Eqn 8.10 above; a Christmas tree network has an even number of dipoles, so, with the restriction that $N$ is even, the resulting gain as a function of $\theta$ will be

$$G(\theta) = \frac{G_D}{N} \left[ \sin\left(\frac{N\pi}{2}\sin\theta\right) \Big/ \sin\left(\frac{\pi}{2}\sin\theta\right) \right]^2. \tag{8.11}$$

The gain, as shown in Eqn 8.11, is at its maximum $N$ times as great as the gain $G_D$ of a single dipole above a ground plane. It follows, then, that the larger the number of dipoles, the narrower the principal beam. Note that an array of $N$ dipoles, spaced by $\lambda/2$, has a principal beam with angular width of order $2/N$; this is to be expected from standard diffraction theory, which gives, for an aperture of size $D$, an angular resolution of order

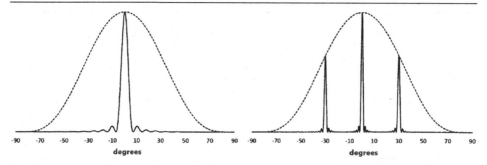

Figure 8.11. Normalized power patterns for uniformly phased arrays of 16 half-wave dipoles above reflecting screens: (left) $\lambda/2$ spacing; (right) $2\lambda$ spacing. The dotted line is the power pattern of an individual dipole plus reflector.

$\lambda/D$ (see Section 8.3.2). This fundamental relation will be considered in greater detail later in this chapter.

If the spacing is smaller than $\sim \lambda/2$, the elementary collecting areas overlap and there is mutual coupling between the elements; as a result the array gain is reduced. The spacing can be larger than $\lambda/2$, with a consequent sharpening of the beam, but if the spacing exceeds $\lambda$, the periodic nature of Eqn 8.11 will mean that other maxima will appear. Such arrays are called sparse arrays. Their total collecting area is maximized, each element delivering its full gain free from mutual coupling, but at the expense of higher sidelobes. Two cases of a uniformly spaced phased array are illustrated in Figure 8.11. The one-dimensional case has been used as an illustration, but the same formalism applies for two-dimensional arrays. Note that these 'adding' radio arrays, with all elements connected to a single receiver, are the direct analogue of diffraction gratings in optics.

Densely packed phased arrays intercept all the radiation falling upon them and hence mimic a filled aperture such as a dish. Like a filled aperture, they form an instantaneous beam and hence are *direct imaging* instruments. Such arrays are also called *aperture arrays* since they constitute the whole receiving aperture without the aid of reflecting surfaces. An example of a basic aperture array of Yagi antennas is shown in Figure 8.12.

Phase-stable amplifiers can be inserted to overcome the losses in the transmission lines from the individual elements of the array. Furthermore, the beam can be steered by inserting electronic phase shifters that give a progressive phase delay across the array, as shown in Figure 8.13, which illustrates the equivalence of physical-optics and electronic beam forming. Figure 8.13 shows an electronically steerable phased array with the same beamwidth as a mechanically steerable parabolic dish. Note that all the array elements (not just two as shown) are co-added with an appropriate delay gradient across the array for the chosen pointing direction. Different pointing directions can be simultaneously achieved electrically using multiple phasing networks with different phase gradients. Note, however, that the geometrical area presented by the array reduces as the pointing direction moves away from the perpendicular to its surface plane. A great advantage of electronically steered arrays is that, with multiple phasing networks, many simultaneous beams can be formed pointing in different directions; the number achievable depends on the cost of the hardware. In principle

Figure 8.12. A basic aperture array, part of the Manchester University Student Telescope (MUST).

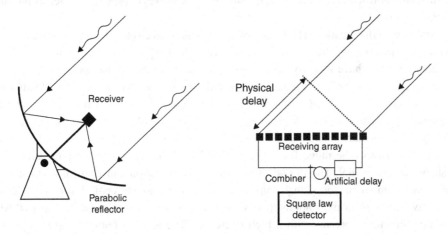

Figure 8.13. The equivalence of physical-optics and electronic beam forming. (left) A mechanically steerable parabolic dish; (right) an electronically steerable phased array.

sufficient *array beams* can be formed to fill the reception pattern of the primary elements, as illustrated schematically in Figure 8.14. Here the outer circle is the field of view (angular width $\sim \lambda/d$) set by the reception pattern of individual array elements each of characteristic dimension $d$. The inner circles represent multiple array beams from the combination of the elements with different phase gradients in two dimensions. Their angular width is $\sim \lambda/D$, where $D$ is the overall extent of the array.

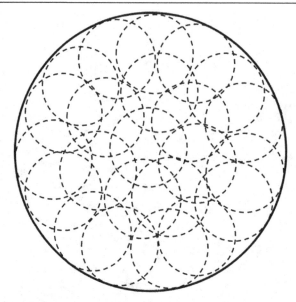

Figure 8.14. Filling the reception pattern with array beams.


Because of their instantaneous response, phased arrays are well suited for certain types of astronomical research programme, in particular for surveying the sky to detect pulsars and transient sources.

In addition to the Kharkov UTR-2 array, several large electronically phased-array radio telescopes, operating in the relatively unexplored low frequency range below 300 MHz (wavelength 1 m), have recently been constructed and are in routine operation (see also Chapter 11). One is the Low Frequency Array (LOFAR), centred in the Netherlands and extending through many European countries. The low band array (LBA), covering the frequency range from 30 to 80 MHz, is based on droopy dipoles which are inconspicuous tent-like wire structures. The heavily used FM radio band is omitted and the high band array (HBA) covers the range from 115 to 240 MHz; this band uses bow-tie antennas, as shown in Figure 8.9. The digital electronics at each station can form more than one tied array beam; then, the data having been transmitted to the central site, equivalent tied array beams from the stations across the network are cross-correlated. The LOFAR telescope therefore operates with a high degree of flexibility in both adding and cross-correlation (aperture synthesis) interferometer modes (see Chapters 9–11). Another recently constructed array is the MWA (Murchison Widefield Array), in Western Australia, which is designed to cover the band 80–300 MHz. Currently MWA is a single 'station' which, as in LOFAR-HBA, is electrically made up of 'tiles' each composed of 4 × 4 wide bandwidth bow-tie antennas on a ground screen. Each of the orthogonal bow-tie dipoles has its own individual transistor amplifier with a balanced input and an unbalanced output, as in Figure 8.3(b). The signals from the 16 antennas in a tile are summed in an analogue beam-forming network which has delay elements appropriate to form a pointable 'tied array' beam. The data from these beams are then cross-correlated to form an aperture-synthesis array. The SKA Low Frequency Aperture Array (LFAA), to be constructed in Western Australia, will be the largest phased array for astronomy. Its design is being informed by the experience gained with LOFAR and the MWA. Among many objectives all these new low frequency arrays will attempt observations of the greatly redshifted 21 cm hydrogen line, from the era after the Big Bang and before reionization (Chapter 17).

## 8.3 Antenna Beams

In this section we address the subject of antenna beams more generally. We have seen that in the specific case of a phased array most of the power is collected over a limited range of angles. This is true of antennas of any type, from the smallest to the largest including parabolic dishes, which we will meet later in this chapter. The width of the main, or primary, beam is usually defined between the half power points (full width half maximum, FWHM or $\theta_{1/2}$) on either side of the straight-ahead or boresight direction. Figure 8.15 shows a schematic power polar diagram of a low-gain antenna; the power relative to the on-axis peak is plotted logarithmically as a function of angle in order to reveal the low-level sidelobes clearly.

The main beamwidth is shown by the broken lines. The FWHM is typically $\sim \lambda/D$ radians, where $D$ is the dimension of the antenna aperture. Power also enters via the minor lobes or sidelobes, whose structure depends on the antenna design and is invariably more

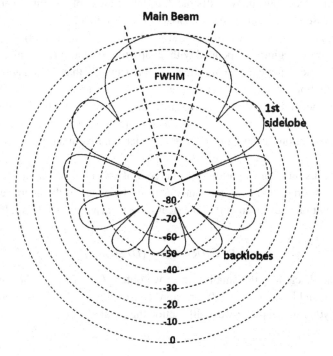

Figure 8.15. Schematic power polar diagram (logarithmic scale to show the sidelobes).

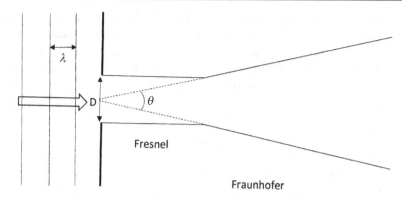

Figure 8.16. Fresnel and Fraunhofer regimes.

complicated than shown here. A small amount of radiation always leaks in from directly behind the antenna.

### 8.3.1 Aperture Distributions and Beam Patterns

We now set out more precisely the relation between the geometry of the telescope aperture (which may be a dipole array, or a horn, or a reflector) and the size and shape of the power beam. This is governed by classical diffraction theory, and the diffraction of plane waves at an aperture is treated in many textbooks.[7] In the context of antenna beams and their calibration the essential features are summarized in Figure 8.16.

Figure 8.16 gives a schematic illustration of the Fresnel (near-field) and Fraunhofer (far-field) diffraction regimes associated with an aperture of width $D$ illuminated by plane waves. In the Fresnel regime the beam extent is approximately that of the geometrical aperture; in the Fraunhofer regime the beam spreads out over an angle $\sim \lambda/D$. The boundary between the Fresnel and Fraunhofer regimes is to an extent arbitrary. In classical diffraction theory it is the Fresnel distance but in antenna contexts it is often called the Rayleigh distance. The derivation of the latter is easy to understand from simple geometrical considerations.

Figure 8.17 illustrates the situation of a source $S$ of spherical waves incident on a telescope aperture of dimension $D$ at a distance $R$ from the edges of the aperture. Across the aperture the maximum deviation of the spherical wave from a plane wave is $s$. In the triangle SOA we have

$$R^2 = (R - s)^2 + (D/2)^2. \tag{8.12}$$

Simplifying and discarding the small term $s^2$ we obtain $R = D^2/8s$. The usual criterion to define the far field is that $s = \lambda/8$ or, more conservatively, $s = \lambda/16$, and hence the distance $R_{\rm ff}$ beyond which plane-wave diffraction theory applies is

$$R_{\rm ff} = 2D^2/\lambda. \tag{8.13}$$

---

[7] For an elegant discussion see *Modern Classical Physics* by Thorne and Blandford (2018) Chapter 8.

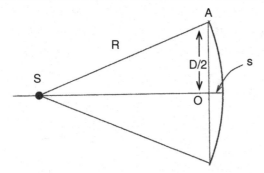

Figure 8.17. The far-field domain. If $R > 2D^2/\lambda$, the wavefront is plane within $\lambda/16$.

Celestial sources are always in the far field, but for calibrating the antenna both the near-field and far-field beams may be relevant. For a large paraboloid the tropospheric water vapour (mean height $\sim$ 1 km) lies within the near field and hence the appropriate incoming beam is approximately cylindrical. To measure the far-field beam one ideally uses an indoor or outdoor antenna test range, although these are of limited size. For antennas made from wires or rods (i.e. dipoles and their relatives) or waveguide horns, with effective apertures $D \leq 1$ m, $R_{ff}$ ranges from a few metres up to tens of metres and so beam measurements in the far field can readily be made.[8] But, for large parabolic telescopes or phased arrays, calibrating the far-field beam poses a severe challenge. For example, $R_{ff} = 2D^2/\lambda = 200$ km for a 100 m telescope operating at $\lambda = 10$ cm; this is close to the height of a low Earth-orbiting satellite. Even for smaller dishes it is often hard to find a transmitting site far enough away, unless there is a conveniently located and distant hill.[9] Geostationary satellite transmissions can be used for telescope calibration but only at the wavelength being transmitted.

### 8.3.2 Fraunhofer Diffraction at an Aperture

A heuristic picture of far-field beam formation can be gained by imagining an aperture to be built up out of a superposition of many elementary diffraction gratings with different *spatial wavelengths* $\xi_i$ (in metres) whose transmission varies sinusoidally from zero to 100%. When illuminated by plane waves, each grating produces three diffracted waves in the far field: one straight through and (using the small-angle approximation $\sin\theta \approx \theta$) one on either side at angles $\pm\theta_i$ radians, where $\theta_i = \lambda/\xi_i$. Figure 8.18 shows two such gratings, with their diffraction patterns.

This inverse relationship between grating spacing and beam angle immediately suggests that the angular (beam) and spatial (aperture) patterns constitute a Fourier pair (see Appendix 1). Equivalently, the aperture can be described in terms of a sum of *spatial*

---

[8] If $R_{ff}$ becomes inconveniently large then data can be taken in the near-field regime and mathematically transformed into the far field.
[9] The distance to the horizon is $\sim 3.6 \times \sqrt{h}$ km, where $h$ is the observer's height in metres above the ground.

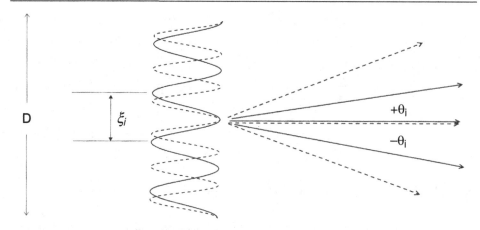

Figure 8.18. An aperture distribution of size $D$ can be imagined as comprising many sinusoidal diffraction gratings (only two shown here). When illuminated by plane waves each grating produces three diffracted waves in the far field.

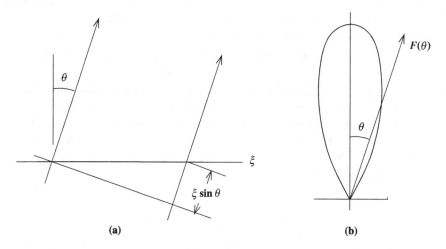

Figure 8.19. (a) A linear aperture and (b) its radiation pattern. The excitation distribution $i(\xi)$ and the radiation pattern $F(\theta)$ are related by a Fourier transform.

*frequency* components $u = \xi_i/\lambda$ cycles per metre with $\theta_i = 1/u_i$. The combined effect of the elementary gratings is to produce the complex angular pattern of electric field variations $F(\theta)$. As the aperture size $D$ increases, longer spatial wavelengths $\xi_i$ (lower spatial frequencies $u_i$) can be fitted in and thus more power is radiated at small angles; the beam therefore becomes narrower.

The alternative (Huygens) approach to beam formation is to sum the contributions of appropriately phased small elements. The simplest aperture consists of a line distribution $i(\xi)$ of excitation currents along an axis $\xi$ at a single wavelength $\lambda$. This is illustrated in Figure 8.19. An approximate derivation, following the Fraunhofer approximation, illustrates the basic principle. At a large distance from the aperture, in a direction $\theta$ to the normal, the contribution of each element $i(\xi)d\xi$ to the radiation field depends on the phase introduced by the path difference $\xi \sin\theta$; omitting normalizing factors and making the small-angle approximation $\sin\theta \simeq \theta$, the radiation pattern $F(\theta)$ (referring either to the electric or magnetic field) is

$$F(\theta) = \int \exp[-j(2\pi\xi\theta/\lambda)]\,i(\xi)\,d\xi. \qquad (8.14)$$

It is convenient from here on to measure lengths in wavelengths, for example setting $u = \xi/\lambda$.

The generalization to a two-dimensional aperture follows naturally. The distribution of current density across the aperture is often referred to as a *grading*, designated $g(u,v)$, giving

$$F(\theta,\phi) = F_e(\theta,\phi) \iint_{4\pi} g(u,v)\exp[-j2\pi(u\theta + v\phi)]\,du\,dv, \qquad (8.15)$$

where $\theta,\phi$ are assumed to be small angles, measured on a plane tangential to the sky, and $F_e(\theta,\phi)$ is the radiation pattern of a current element in the surface. Equation 8.15 sets out the basic relationship: *the voltage radiation pattern is the Fourier transform of the aperture current distribution.*

At this point one should note that $F(\theta,\phi)$ is a function of time; it varies at the frequency of the incoming radio waves and hence is not measurable; only the power is measured by the receiver.[10] This leads to a corollary relating the current distribution and the power gain $G(\theta,\phi)$. The power is proportional to the square of the field strength, $FF^*$, the autocorrelation theorem (Appendix 1) then yields the result

$$G(\theta,\phi) \overset{\text{FT}}{\Longleftrightarrow} g \otimes g. \qquad (8.16)$$

Thus, the second basic relationship is: *the power gain pattern is the Fourier transform of the autocorrelation of the aperture current density distribution.*

Fourier analysis therefore provides a simple key to the theory of telescope beam shapes, both in transmission and reception. It is even more important in the understanding of interferometers and aperture synthesis (Chapters 9 and 10); we therefore set out the basic theory of Fourier integrals and transforms in Appendix 1, where precise mathematical formulations can be found. The Fourier transform has already been referred to in Section 7.2, where the dualism between time and frequency is obvious in the analysis of electrical signals, and Eqns 8.15 and 8.16 show that there is the same dualism between the angular pattern of plane waves and the aperture function distribution. Some results which are familiar in this time and frequency domain can be applied directly to the geometrical relation between apertures and beams; for example the impulse function $\delta(t)$ has a particularly simple Fourier transform, which is a flat spectral distribution of amplitude unity. The corresponding relation in antenna theory shows that a point current source (although

---

[10] This is the same situation as when a laser beam is shone through an aperture and the pattern is observed on a screen with your eye, which also detects only power (intensity).

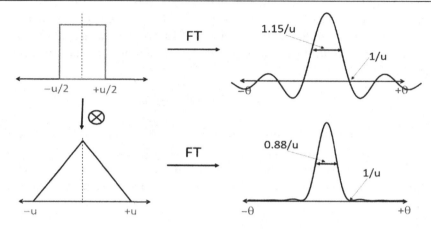

Figure 8.20. Voltage and power beams. (left) A one-dimensional uniform radiator of overall extent $u$ wavelengths and its triangular autocorrelation function; (right) the corresponding voltage beam $F(\theta)$ and power beam $G(\theta)$.

not physically realizable) has an isotropic radiation pattern, and similarly that an infinitely narrow radiation pattern can only be produced by an infinitely long array or aperture.

We now turn to the practical implications of the Fourier formalism using the small-angle approximation. First, consider the simple case of a one-dimensional radiator of length $u$ (measured in wavelengths), uniformly excited. For $i = \text{const} = 1$, Eqn 8.14 gives the radiation pattern (often called the *voltage beam*)

$$F(\theta) = (\sin \pi u\theta)/(\pi u\theta). \tag{8.17}$$

This is recognizable as the sinc function $(\sin x)/x = \text{sinc}(x)$ and the power beam is therefore a $\text{sinc}^2$ function (Figure 8.20). (see Appendix 1). Clearly this beam is the same in each dimension for a square aperture. The uniformly illuminated circular aperture of diameter $u$ gives a related result: the two-dimensional integration is a Fourier–Bessel transform yielding, for an aperture of diameter $u$,

$$F(\theta) = 2J_1(\pi u \sin \theta)/(\pi u \sin \theta). \tag{8.18}$$

The square of this function is the *Airy function*, shown in Figure 8.21. In two dimensions the Airy pattern consists of a series of rings (the sidelobes) centred on the dominant main beam.

Some specific beam parameters for uniformly illuminated square and circular apertures are given in Table 8.1. As noted in Section 8.2, the width (FWHM) of the principal lobe of a filled aperture is always of the order of $1/u = \lambda/D$ for apertures of diameter $D$, with corrections for the individual geometry. The first nulls also occur around $\pm\lambda/D$ away from the boresight direction. One further point to note is that the power beam is insensitive to the relative phase relationships between the spatial and the temporal variations in the angular field pattern $F(\theta, \phi, t)$. For example, in Figure 8.21 the odd-

Table 8.1. *Parameters of the power beams of uniformly illuminated symmetric apertures with diameter-u wavelengths.*

|  | Square | Circular |
|---|---|---|
| Main beam FWHM | 0.88*u* | 1.02*u* |
| First null | 1.0*u* | 1.22*u* |
| First sidelobe | 4.7% (−13.3 dB) | 1.74% (−17.6 dB) |

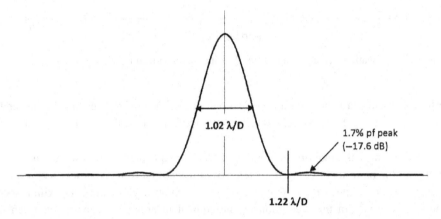

Figure 8.21. A cut through the Airy function: the power beam pattern for a circular aperture with uniform distribution of aperture current.

numbered sidelobes are $\pi$ out of phase with the main lobe while the even-numbered ones are in phase.

### 8.3.3 Effective Area

In Section 5.3 we introduced the generic concept of the direction-dependent antenna power gain $G(\theta, \phi)$, commonly known as the antenna beam pattern, and derived relationships (Eqns 5.14 and 5.15) between the peak (on-axis) gain $G$, the on-axis effective area $A_{\text{eff}}$, and the antenna beam solid angle $\Omega_A$. In this section we are concerned with the entire beam pattern, normalized to unity on the axis and written as $G_n(\theta, \phi)$. The antenna beam solid angle is defined as

$$\Omega_A = \int_{4\pi} G_n(\theta, \phi) d\Omega, \qquad (8.19)$$

where the integral is taken over the entire $4\pi$ steradians of the sphere centred on the antenna. It is simplest to envisage $\Omega_A$ when the antenna is transmitting. It is that solid angle into which all the power would be transmitted through a notional beam pattern which is constant as a function of angle and equal to the value of $G_n(\theta, \phi)$ on the axis. In polar

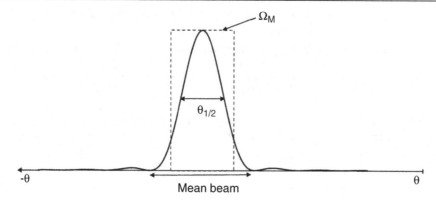

Figure 8.22. Schematic antennä beam pattern, showing the main-beam solid angle.

coordinates, picture a cone with a spherical cap and with no sidelobes, all the power in the sidelobes having been concentrated into the uniform conical lobe. From Eqn 5.18, $\Omega_A = \lambda^2/A_{\text{eff}}$.

The definition of the main-beam solid angle $\Omega_M$ is equivalent, but now the integral is restricted to the main lobe. Using Cartesian coordinates, Figure 8.22 shows a cut through a generic beam pattern $G_n(\theta, \phi)$ at constant $\phi$ with the power plotted linearly with respect to the on-axis peak. In these coordinates the equivalent area of the main beam can be shown with a rectangle. The main beamwidth $\theta_{1/2}$ depends on the feed taper (see Section 8.8), but in two dimensions, for a perfect Gaussian (which, in practice, is representative),

$$\Omega_M = 1.13\theta_{1/2}^2. \tag{8.20}$$

The main-beam efficiency $\epsilon$ is given by

$$\epsilon = \frac{\Omega_M}{\Omega_A} \tag{8.21}$$

and hence the fraction of the beam solid angle arising from the sidelobes is $(1 - \epsilon)$.

In the far field the power output from the antenna when observing a distribution of specific brightness $B_\nu(\theta, \phi)$ (W m$^{-2}$ Hz$^{-1}$ sterad$^{-1}$) is the effective area $A_{\text{eff}}$ (m$^2$) weighted by the antenna beam pattern (sterad):

$$P_\nu = \frac{1}{2}A_{\text{eff}} \int_{4\pi} B_\nu(\theta, \phi)G_n(\theta, \phi)d\Omega \quad \text{W Hz}^{-1}; \tag{8.22}$$

the factor 1/2 arises because there is a single polarization per receiver. Using Eqn 5.15, the power output can be expressed in terms of a distribution of brightness temperature $T_B(\theta, \phi)$:

$$P_\nu = \frac{A_{\text{eff}}k}{\lambda^2} \int_{4\pi} T_B(\theta, \phi)G_n(\theta, \phi)d\Omega \quad \text{W Hz}^{-1} \tag{8.23}$$

and then, since $P_\nu = kT_A$, as an antenna temperature $T_A$:

$$T_A = \frac{A_{\text{eff}}}{\lambda^2} \int_{4\pi} T_B(\theta, \phi)G_n(\theta, \phi)d\Omega \quad \text{K} \tag{8.24}$$

or alternatively, using Equation 5.15,

$$T_A = \frac{1}{\Omega_A} \int_{4\pi} T_B(\theta, \phi) G_n(\theta, \phi) d\Omega \quad \text{K.} \tag{8.25}$$

This is logical since, for a uniform distribution of brightness temperature, we have

$$T_A = \frac{T_B}{\Omega_A} \int_{4\pi} G_n(\theta, \phi) d\Omega \quad \text{K} \tag{8.26}$$

and, since the integral gives $\Omega_A$, the antenna temperature is the same as the surrounding brightness temperature for any lossless antenna. For a uniform temperature distribution Eqn 5.15 simplifies to

$$A_{\text{eff}} \int_{4\pi} G_n(\theta, \phi) d\Omega = \lambda^2, \tag{8.27}$$

which yields Eqn 5.18. Thus, for any lossless antenna the product of the on-axis effective area and the antenna beam solid angle is a constant and equal to the square of the wavelength. As discussed in Section 5.3, in a constant temperature field any lossless antenna, large or small, will yield the same power output. What differs is the direction from which the power is collected, i.e the balance between $\Omega_A$ and $\Omega_M$. In a large antenna the power mostly enters via the narrow main beam; in a small antenna the power collection is more uniformly distributed in angle.

## 8.4 Partially Steerable Telescopes

It is obviously difficult to construct an enormous collecting area which can be physically steered in all directions. In the early days of radio astronomy, the compromise solution was to have transit instruments that could be steered in elevation only, allowing the sky to drift by as the Earth rotates.[11] A particularly fruitful early approach was to construct cylindrical paraboloids, steerable in elevation, with a linear antenna or *line feed* along the focal line. They suffer from complexity of the feed arrangement, and can be hard to calibrate, but the reduced cost of the reflector area compared with that for circular paraboloids has kept them competitive for some applications such as the CHIME array (Section 17.10).

A most successful radio telescope design, which is steerable while avoiding the necessity of moving a very large reflector surface, is used in the 500 Metre Aperture Spherical Telescope (FAST) and the 1000 foot Arecibo telescope, described in Section 8.10 and shown in the Supplementary Material. In these telescopes the surface is part of a sphere, and the beam is steered by moving the feed system to illuminate different parts of the spherical surface. The surface introduces a very large spherical aberration, which is corrected either by actively deforming the surface (FAST) or by a large Gregorian secondary reflector (Arecibo).

---

[11] For a discussion of a wide range of radio telescope designs the interested reader is referred to the book by Christiansen and Hogböm (1985, see Further Reading); the books by Baars (2007) and Baars and Kärcher (2017), listed in Further Reading, are unsurpassed.

## 8.5 Steerable Telescopes

Ray diagrams of the optics of paraboloidal telescopes are shown in Figure 8.23; some of the largest telescopes are shown in the Supplementary Material (see also Section 8.11). All large telescopes of this type, including large optical telescopes, are supported by a structure that points the instrument in azimuth and elevation (colloquially, an Az–El mount): the telescope moves in elevation about its elevation axis, and this in turn is supported by the alidade which rotates in azimuth.[12] The Az–El mount has significant structural advantages, at the cost of two complications: first, following a source continuously through the zenith would require an instantaneous movement of 180° in azimuth; second, at the focus, the image of the sky rotates, i.e. the parallactic angle (Section 11.2.2) varies with position. The first complication is met by accepting a dead zone near the zenith, while the second

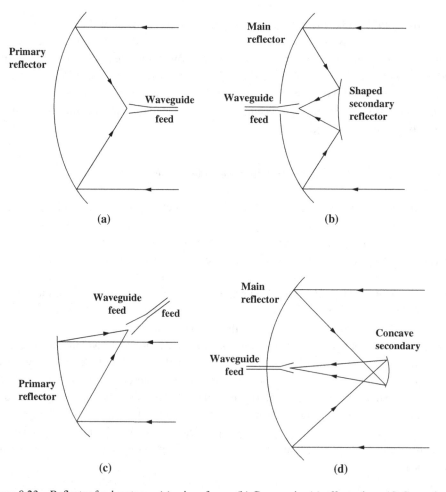

Figure 8.23. Reflector feed systems: (a) prime-focus; (b) Cassegrain; (c) offset prime; (d) Gregorian.

[12] Most old optical telescopes, and some older radio telescopes, have equatorial mounts.

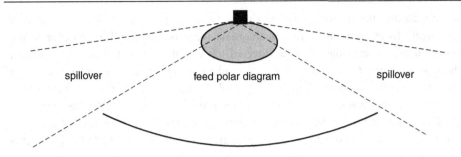

Figure 8.24. Spillover in a prime focus dish.

complication requires a rotating-focal-plane receiver or, in the case of a dual polarization radio receiver, a later correction in the software (Appendix 4).

The steerable paraboloid is commonly used in present-day radio astronomy, and its radiation pattern is of vital interest. The aperture efficiency must be maximized, and the sidelobe levels minimized, to the maximum feasible extent. The gain pattern of the telescope is given by the prescription in Eqn 8.16: the autocorrelation of the field amplitude across the aperture is the Fourier transform of the antenna power gain. The field amplitude across the aperture is determined, in the case of a prime-focus feed, by the radiation pattern of the feed; if the telescope is a Cassegrain or Gregorian, the secondary reflector is illuminated by a primary feed and it is the gain pattern of that combination that determines the aperture distribution.

Maximizing the power fed into the receiver implies collecting radiation from the edge of the dish with the same efficiency as from its centre. In practice this is never achieved, nor is it desirable; it is impossible to arrange a sharp transition from full illumination to zero at the edge of the aperture without generating large sidelobes. Unwanted sidelobes can pick up unwanted signals, whether astronomical or manmade. In order to reduce the sidelobes the power picked up by the feed is therefore tapered towards the edge of the reflector (for more details see Section 8.6). This has the effect of diminishing the aperture efficiency and, inevitably, some radiation from the feed will 'spill over' the edge of the reflector. In the receiving mode, the spillover part of the pattern will pick up radiation from the ground. Since the ground is at a physical temperature of 280–290 K, this can contribute significantly to the noise at the input of the low-noise receiver, degrading the system performance.

The geometry of prime-focus optics illustrated in Figure 8.23(a) makes such telescopes particularly prone to the effects of ground radiation from spillover. Figure 8.24 shows a feed illumination pattern at the prime focus. At the edge of the dish the power pattern is typically designed to fall by 12–15 dB (see Section 8.6) but there is inevitably spillover past the edge of the dish. When the dish points to the zenith (as illustrated) the spillover region intercepts ground radiation; when the dish points towards the horizon, half the spillover intercepts radiation from the sky and half radiation from the ground.

For this reason, the Cassegrain configuration, Figure 8.23(b), is more commonly met with (Baars 2007, Chapter 2). Here there is a secondary hyperboloidal reflector with one focus coincident with the focus of the primary paraboloid, while the feed is located at the conjugate focus, near the vertex of the main reflector surface. The spillover in the illumination of the secondary is now directed towards the sky rather than the ground; at short wavelengths this usually gives a considerable reduction in antenna noise. The Gregorian geometry, shown in Figure 8.23(d), places an elliptical secondary beyond the focal point, with its conjugate focal point near the primary surface as in the Cassegrain.

The feed at the secondary focus in a Cassegrain system requires a narrower beam than a primary feed, and correspondingly a larger aperture, normally around 5 to 8 wavelengths across, which may be inconveniently large at long wavelengths. Cassegrain systems are therefore usually used only at wavelengths $\sim 21$ cm or shorter. A practical advantage of secondary mirror optics is that the feed and receiver system are situated in a location which is easier to access than the prime focus.

A variant of the Cassegrain configuration, the shaped-surface Cassegrain, is in common use. The secondary has a perturbed surface profile, designed to throw the inner regions of the feed pattern further out on the primary reflector. This increases the radiation field in the outer parts of the primary, reducing the taper and increasing the antenna efficiency on-axis. In order to meet the optical condition of a plane wave across the aperture, the primary has to have a compensating deviation (typically only at the cm level) from a paraboloid to equalize the lengths of the ray paths. The gain in on-axis aperture efficiency is significant: it is difficult to achieve much better than 60% efficiency with a pure Cassegrain, while the shaped Cassegrain can achieve an efficiency of up to 85%. There are, however, compensating disadvantages to shaping compared with classical Cassegrain optics. The reduced taper means that the near-in sidelobes are higher. Furthermore, optimizing the optics for on-axis gain results in a more rapid fall-off of the gain away from the axis. The number of simultaneous beams which can be generated by various forms of focal plane array (see Section 8.7) is, therefore, much reduced. Shaped reflectors are therefore best suited for observations dominated by individual discrete sources (including deep space tracking) but are not well suited for wide-area sky surveys, particularly of diffuse emission.

There are two variants of the Cassegrain configuration that are in use: the Nasmyth and Coudé systems. In the Nasmyth configuration, a tertiary reflector intercepts the beam and sends it along the elevation axis. In this configuration the receivers are located on the alidade arm, outside the elevation bearing; since they only move in azimuth, they do not tilt as the telescope moves, a design advantage at millimetre wavelengths where the receivers may need to be cooled to liquid helium temperatures. The Coudé configuration takes the beam at the Nasmyth focus and uses a series of mirrors to transport the beam down the azimuth axis to the base of the telescope. This means that the receivers are fixed to the ground, and can be accessed even when the telescope is moving. Both radio and optical telescopes have used these configurations.

## 8.6 Feed Systems

As mentioned in the last section the telescope feed must have a pattern that optimizes the illumination across the aperture and also operates over as large a frequency bandwidth as possible. For low frequency ($\leq$ 1 GHz) operation on prime-focus telescopes, dipole systems are normally used. A simple dipole above a ground plane will not do, because it has an asymmetric pattern, narrower in the $\mathcal{E}$-plane than in the $\mathcal{H}$-plane. Examples of practical plane-polarized systems that have an approximately symmetrical pattern, suited to prime-focus use on paraboloids with an f-ratio of around 0.4, are shown in Figure 8.25. A pair of dipoles, suitably spaced, can do this (Figure 8.25(a)), although a broader-band system that is a hybrid of a dipole and a waveguide (Figure 8.25(b)) is often used; this can accommodate crossed dipoles for both planes of polarization. More frequently, especially on Cassegrain telescopes at higher frequencies, circular horns (previously discussed in Section 5.1.3) are used. One common design, the corrugated circular horn, illustrated in Figure 8.25(c), accepts both polarization modes.

The surface corrugations in the horn feed in Figure 8.25(c) are designed to control the aperture distribution and to produce symmetric beams in orthogonal directions for good polarization performance. At the aperture the corrugations are $\lambda/4$ deep, providing a high surface impedance which allows the wave to detach from the wall, while at the junction with the circular waveguide they are $\lambda/2$ deep, providing continuity with the conducting wall of the guide. Clarricoats and Olver (1984) designed compact versions in which the profile of the cone and the spacing of the corrugations are varied; these are physically smaller, so they can be installed more conveniently but nevertheless allow operation over a wide bandwidth. In Chapter 7 we described the different ways in which signals can be extracted from circular horns and used in polarimetric observations.

Some telescopes have separate receivers that can be plugged in when the receiving band is changed. There are feed designs that work at two separate frequencies, and three or even four bands are possible but are seldom seen in use, particularly for prime-focus telescopes. Frequency changing is often achieved by mounting an assemblage of feeds on a carousel at the focal plane, or, as in the case of the VLA, an off-axis secondary reflector can be rotated (nutated), bringing the feed for the desired band onto the optical axis. The Haystack telescope mounts the shortest-wavelength feed horn on-axis, with the longer-wavelength horns displaced from the axis but within the acceptable off-axis range.

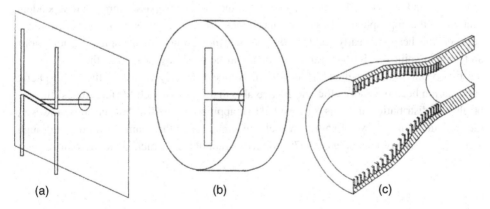

(a)                              (b)                              (c)

Figure 8.25. Telescope feed systems: (a) a pair of dipoles; (b) a hybrid system, dipole with waveguide; (c) the compact corrugated horn design used in many radio telescopes.

### 8.6.1 Twin-Beam Radiometry

The balanced receiver architectures outlined in Section 6.8 can be used to minimize the effects of the troposphere and ground spillover on single-dish radiometry. Here we consider only the simplest case, that of a twin-beam system operating in clear-sky conditions. The two receiver inputs are taken from a pair of closely spaced feeds at the focus. On a blank area of sky and with a static atmosphere the antenna temperatures for the two beams will be very similar and the balanced receiver produces a close-to-null output. But a discrete source in one beam will unbalance the receiver; in effect the other beam acts as a comparison load. In practice, however, the wind blows patches of water vapour across the field of view, causing fluctuations in both the atmospheric attenuation and emission (Section 4.4) which perturb the measurement.

This technique involves two different reception regimes: the far field (Fraunhofer regime) for the celestial source and the near field (Fresnel regime) for the troposphere (Section 8.3.1). In the far field there are two primary beams, each of diameter $\theta_{1/2} = (1.15 \pm 0.05)\lambda/D$, separated in angle by $\phi = x/f$ radians, where $D$ is the dish diameter, $f$ is its focal length, and $x$ is the separation between the feed centres (limited by their outer diameters). If, as is likely, the feeds are at one of the secondary foci then their diameters will be $5\lambda$–$8\lambda$ and the focal length will be many times that at the primary focus (see Section 8.6). An example of such a system is the 30 GHz ($\lambda = 1$ cm) OCRA-p radiometer (Lowe *et al.* 2007) mounted on the 32 m dish of the Torun Centre for Astronomy.[13] With $x = 8.7$ cm and $f = 97$ m, $\phi = 3.1$ arcmin; thus the primary beams ($\theta_{1/2} = 1.2$ arcmin) are well separated on the sky. Many groups have used similar techniques at short centimetric and millimetre wavelengths.

For large antennas ($D > 15$ m) the mean altitude $h$ of the tropospheric water vapour, typically about 1 km, is well below the Rayleigh distance. The near-field beams look like two cylinders, each with the diameter of the dish and separated in angle by $\phi$ (Figure 8.26; left). Within the troposphere the two cylinders largely overlap; the maximum offset $\delta = h\phi$ and it is easy to show that the area of overlap divided by the total area covered by the beams (Figure 8.26; right) is $(D - \delta)/(D + \delta)$. For OCRA-p with $h = 1$ km, $D = 32$ m, and $\phi = 9 \times 10^{-4}$ rad, only $\sim 5\%$ of the two beams passes through independent sections of the troposphere. Thus the balanced receiver, by continously taking the difference between the beams, greatly reduces the additive noise due to tropospheric fluctuations and ground spillover. Further improvements can be made by switching the positions of the two beams (e.g. Readhead *et al.* 1989). Notice that as $D$ increases the tropospheric cancellation becomes more efficient. Notice also that this approach 'differentiates' the sky brightness distribution and thus is not suited to mapping smooth distributions. It is, however, ideally suited for surveys of sources which are smaller than the beam separation, although techniques exist (Emerson, *et al.* 1979) to reconstruct objects which are somewhat larger.

---

[13]  web.astro.umk.pl/?lang=en.

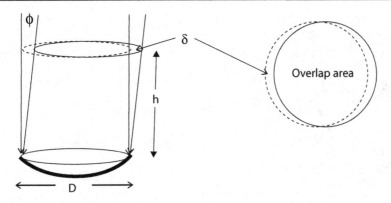

Figure 8.26.   The beams of a twin-beam radiometer overlap in the Fresnel region close to the telescope.

## 8.7 Focal Plane Arrays and Phased Array Feeds

Multiple feed systems, packed closely side by side in the focal plane, can be used to allow simultaneous observations on a mosaic of adjacent telescope beams. These systems are called *focal plane arrays* and are coming into general use. They are useful in reducing the time taken in large-scale surveys; for example the 63 metre Parkes radio telescope is equipped with an array of 13 dual polarization feeds on the 21 cm wavelength for hydrogen line surveys and in the search for pulsars (Staveley-Smith *et al.* 1996). At millimetre wavelengths an array of feeds, all contained in a single cryogenic package, is often used, particularly for spectral-line observations where the sources are extended objects with interesting structure.

Figure 8.27(a) shows the arrangement of the 13 feed horns at the focal plane of the Parkes telescope. It is possible to add more feed horns; a feed for FAST achieves 19 by filling the gaps in Figure 8.27(a). In such systems the outer beams inevitably suffer from aberration; furthermore, the feed horns cannot overlap and consequently the far-field beams do not fill the whole wide field of view, so that several separate telescope pointings are needed to observe the wide field. Both these limitations can be overcome by filling the focal plane with smaller elements, each with its own individual low-noise amplifier, and combining the signals from a group of elements to form a virtual feed at any point of the focal plane (Figure 8.27(b)). Each group forms its own beam; adjacent beams can overlap, giving a complete coverage of the wide field and avoiding the gaps inherent in the focal plane array. Within each separate beam, the signals from each element are combined with appropriate phase and amplitude weighting; the aperture distribution in the outer beams may be adjusted to compensate for aberration. This system might be termed a synthetic focal plane array, although by analogy with the primary beam-forming arrays already in use for low frequency radio astronomy (Section 8.2) it is usually known as a *phased array feed* (PAF).

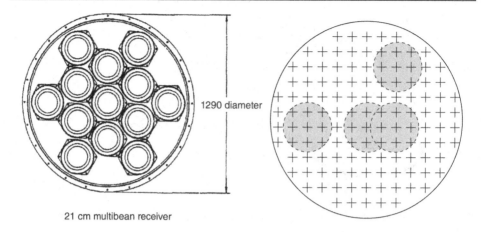

21 cm multibean receiver

Figure 8.27. (a) The array of 13 horn feeds used at the focal plane of the Parkes radio telescope. (b) The phased array feed. Signals from independent elements in the focal plane are combined in groups to form any number of telescope beams, including overlapping beams which can cover the whole field of view.

The design and use of PAFs is complicated by the inevitable coupling between adjacent closely spaced elements, owing to which the noise generated in one receiver element can leak into another. Combination of the elements with complete flexibility of amplitude and phase is complex and achievable only after digitization at each receiver. Broadband elements may be used, giving systems with bandwidths of 2 : 1 or more. The ASKAP PAF, developed in Australia, uses a planar array of 94 fat dipoles, each with dual polarization, while APERTIF, developed at ASTRON in The Netherlands uses a box-like structure of close-packed Vivaldi tapered slot dipoles (see the Supplementary Material). Landon *et al.* (2010) gives a basic discussion and describes the performance of a 19-element focal plane array on a 20 m telescope at Green Bank Observatory. All these systems are centred on the hydrogen line at 21 cm wavelength.

## 8.8 Antenna Efficiency

For an antenna in free space, the *antenna efficiency* $\eta$ is the fraction of power, from an on-axis plane wave impinging on the geometrical aperture, which is actually delivered to the antenna terminals. Three main factors reduce the efficiency of a paraboloidal dish for collecting incoming radiation and hence reduce its effective area $A_{\text{eff}}$. We will discuss these in turn and then bring them all together in Section 8.8.4.

### 8.8.1 Aperture Illumination

As discussed in Section 8.5 the polar pattern of the feed reception pattern is always tapered – it has a maximum near the centre and diminishes towards the edge of the aperture.[14]

---

[14] The corresponding procedure in optics is termed *apodization*. The telescope's power gain pattern also has an equivalent term in optical practice; it is called the *point-spread function*.

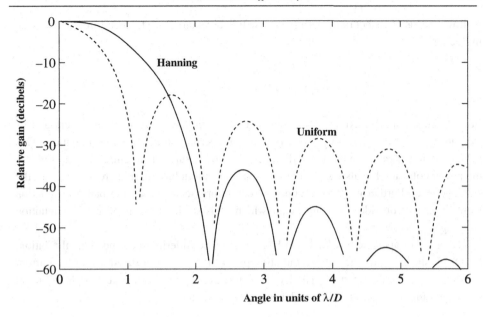

Figure 8.28. The power beam patterns for a circular aperture with uniform and Hanning distributions of aperture current.

This reduces the effective size and area of the aperture. A mathematically convenient form of tapering function is the Hanning taper (as also used in spectrometry, Section 7.2.1), which has the form of a raised cosine and can be written $\cos^2(\pi r/D), r < D/2$. A circular aperture, with Hanning taper in the aperture excitation $i(r)$, yields the power gain pattern shown in Figure 8.28. Note that the beamwidth is noticeably broader than for the uniformly illuminated aperture – hence the gain is reduced – but also note that the sidelobe level is much lower.

A tapered aperture distribution obviously implies that the effective area $A_{\text{eff}}$ in the direction of maximum gain is less than the geometric area. The reduction is called the *illumination efficiency*. A detailed discussion of feed tapers and illumination efficiencies for parabolic dishes is given in the book by Baars (2007). For low-noise receivers a compromise between gain, sidelobe level, and the effect of spillover on the system noise is usually reached when the taper is between $-12$ and $-15$ dB at the edge. This taper is a combination of the feed polar pattern and the 'free-space taper' simply due to the fact that the edge of the dish is further from the focus.[15] For a Gaussian feed pattern and a combined edge taper of $-12$ dB the illumination efficiency is 0.866; for a $-15$ dB taper the illumination efficiency is 0.808.

That part of the feed polar pattern which 'misses' the dish can be thought of in two ways. In transmission mode a fraction of the radiated power is lost; the reduction in performance is characterized as *spillover efficiency*. In reception mode it is more useful to think of the

---

[15] The free-space taper is $\sim 3$ dB for a prime-focus dish with focal ratio $f/D = 0.4$. The free-space taper for a Cassegrain is small owing to the much greater effective focal length. For more details see Baars (2007).

spillover not as a loss but rather as an increase in the system-noise temperature, as discussed in Section 6.4.

### 8.8.2 Blockage of the Aperture

In both the simple (on-axis) prime-focus and Cassegrain systems the feed system and its supporting structure inevitably block part of the aperture from incoming radiation. This brings a double penalty in reduced efficiency. The focus-support structure not only reduces the power collected but also, since the illumination pattern has missing areas, increases the sidelobe level. Further more, the support structure provides a scattering path for radiation from directions outside the main beam which adds to the amount of ground radiation entering the receiver.

The effect of aperture blockage can be treated by considering the fundamental relation, Eqn 8.15, showing that the voltage radiation pattern is the Fourier transform of the current distribution across the aperture. The feed structure causes a gap in the current distribution, so the grading function (the current density) can be written

$$g_{\text{eff}}(\xi, \eta) = g_0 - g_f - g_s, \tag{8.28}$$

where the original grading $g_0$ is reduced to an effective grading $g_{\text{eff}}$ by subtraction of the current distribution in the areas that are blocked by the feed and by the legs of the feed support. If these represent a fraction $\alpha$ of the (weighted) area, the gain will be reduced by a factor $2\alpha$, since the gain is proportional to the square of the current distribution. The geometric blockage of on-axis radio parabolas typically ranges from 4% to 7% (Lockman 1989) and hence the loss in gain is in the range 8% to 14%. The efficiency due to blockage, $\eta_{\text{block}}$, is therefore in the range 0.86 to 0.92.

The lost gain appears in unwanted sidelobes, whose pattern is given by the Fourier transform of $g_f + g_s$. The scale of the sidelobes due to a feed structure whose diameter is a fraction $\beta$ of the dish diameter will be like the radiation pattern of a roughly circular aperture extended in angle by a factor $1/\beta$ compared with the principal pattern. The sidelobes caused by the supports will be narrow fan beams; these are the familiar diffraction spikes seen in bright star images taken by optical telescopes. The patterns in Figure 8.29 and in the Supplementary Material illustrate the points raised in this and the previous section.

The problems of aperture blockage, resulting in higher sidelobe level and reduced antenna efficiency, can be eliminated if the primary reflector is an off-axis paraboloid (or shaped paraboloid), with the secondary outside the aperture and the focal plane outside the main reflector (see Figure 8.23(c)). Home satellite TV dishes are predominantly of this design. For large radio telescopes, however, there are penalties associated with this concept; the non-circular symmetry of the primary introduces extra complexity and cost, while the supporting structure for the secondary reflector adds an extra structural complication. Nevertheless, there are three examples in use: the GBT 100 m telescope at Green Bank, West Virginia, the elements of the Allen Telescope Array at Hat Creek in California (see Section 8.1.3), and the 13.5 m antennas of the South African MeerKAT array. The 15 m dishes of the SKA-MID are also off-axis paraboloids.

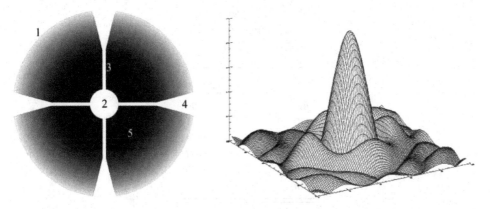

Figure 8.29. Aperture blocking and tapering. Simulated two-dimensional aperture distribution and three-dimensional voltage beam pattern for a dish telescope: 1, edge of aperture; 2, feed blockage; 3, support pylon; 4, pylon shadow; 5, tapered illumination. (Courtesy of N. Wrigley.)

### *8.8.3  Reflection at the Surface*

There is no difficulty in obtaining high reflectivity in the steel or aluminum surface of a reflector radio telescope. (Typically, the modulus of the surface impedance of a metal sheet, or even a wire mesh, is less than one ohm, which is to be compared with 377 ohms, the impedance of free space.) The thickness of the sheet is also unimportant, as the penetration depth is small compared with thicknesses needed for mechanical stability (the skin depth is a few microns for most metals).

A wire-mesh surface has an upper frequency limit, since an electromagnetic wave with a wavelength smaller than or comparable to the mesh size can leak through. Even if the leakage is small, extraneous radiation from the ground, leaking through the mesh, degrades the noise performance of the telescope. In practice, these effects are small if the mesh size is finer than one-twentieth a wavelength. Christiansen and Hogböm (1985) (see Further Reading) gives more detail on the RF transparency of wire mesh.

The practical problem is to maintain the correct profile over a large area, since deviations from the design profile cause losses in efficiency. A plane wavefront strikes the reflecting surface but the electric fields received at the focus from different parts of the surface do not all arrive at the same time. The phase differences mean that the vector sum of the fields is less than for a perfect surface. For a quick feeling for the scale of the deviations which give a significant gain loss consider two parts of the surface, one of which is $\lambda/16$ above and the other of which is $\lambda/16$ below the perfect profile. At the focus the phase errors will be $\pm\lambda/8$, i.e. $\pm 45°$. The resultant of the two vectors is therefore down by 0.707 in voltage or 0.5 in power.

The surface of a large steerable telescope is usually built up from separate panels mounted on a deep structural frame. Deformations may be caused by gravitational or wind forces, or by differential thermal expansion. The largest effects are gravitational deformations of the backing structure. One might think, naively, that this could be minimized by making the support structure stronger, but this cannot be achieved by

Table 8.2. *Reflectivity*
$\eta_{surf}$ *as a function of rms*
*surface irregularities $\delta$*

| $\delta$ | $\eta_{surf}$ |
|---|---|
| $\lambda/60$ | 0.96 |
| $\lambda/40$ | 0.91 |
| $\lambda/30$ | 0.84 |
| $\lambda/20$ | 0.67 |
| $\lambda/16$ | 0.50 |
| $\lambda/4\pi$ | 0.37 |

increasing the cross-section of the members since this would increase the weight and the increased strength would be cancelled by the increased gravitational torque. The deflection, for a given structural geometry, is independent of the member cross-sections. Significant strengthening could be obtained only by resorting to exotic composites, such as bonded boron fibers, but the expense would be too great for large radio telescopes.

The structural design can minimize the effect of these gravitational deformations by allowing the reflector to deform, but as far as possible constraining the deformations so that the surface remains close to a paraboloid. When the elevation of the telescope changes, the feed system can then be moved to compensate for any change in axis and focal length. This principle of *homologous design*, or *conformal deformation*, first proposed by von Hoerner (1966), was introduced with notable success in the 100 m Effelsberg radio telescope. In practice no structural system is perfect, and the remaining corrections can be taken out if the individual surface panels are mounted on motor-driven jacks, a system now used on many radio telescopes.

The mechanical setting of the surface may require a series of repeated measurements of the shape; these are best achieved not by conventional survey methods but, rather, by *radio holography* (see the Supplementary Material). The amplitude pattern $F(\theta, \phi)$ is measured by making interferometer observations of a point source, in which a secondary antenna is kept directed at the source while the antenna under survey is scanned across it. This gives full knowledge of phase as well as amplitude in the beam; a Fourier transformation (Eqn 8.15) then yields the surface current distribution and thus the surface figure.

Even if the large-scale surface profile is perfect, small-scale irregularities in the panels themselves will limit the final surface profile. The effects of any remaining deformations or inaccuracies in the surface may be estimated by Fourier analysis of the consequent errors in phase across the aperture. Phase imperfections take power from the main beam and transfer it to sidelobes. Irregularities on a large linear scale transform into sidelobes on a small angular scale, close to the main beam, while irregularities on a small scale throw power into far sidelobes.

The amount of power taken from the main beam and thrown into the sidelobes by surface irregularities represents a loss of efficiency. This loss is easily estimated for a random Gaussian distribution of phase error across the telescope aperture. A portion of wavefront with a small phase error of $\phi$ radians reduces the power in the main beam by a fraction $1 - \phi^2$, or more precisely $e^{-\phi^2}$, and the contributions from the whole surface add randomly. The phase error at a reflector with surface error $\delta$ is $4\pi\delta/\lambda$ for normal reflection. Although normal reflection only applies strictly on the axis of a reflector, the whole error is often quoted as a single rms error $\epsilon$, which is related to the surface efficiency $\eta_{surf}$ by the Ruze formula (1966)

$$\eta_{surf} = e^{-(4\pi\epsilon/\lambda)^2}. \qquad (8.29)$$

Some representative values are given in Table 8.2. An rms surface error of $\lambda/20$, for example, results in a surface efficiency of 67%.

The surface panels of large radio telescopes such as the GBT have an rms surface accuracy of the order of 0.02 mm, but the effects of gravity limit the overall accuracy to 1 mm before any active correction is applied to the surface. Programmed jacks can reduce the errors by a factor of five, to $\sim 200$ microns rms. This means that an uncorrected 100 m telescope can operate with reasonably high efficiency at wavelengths longer than $\sim 2$ cm (15 GHz) and down to $\sim 3$ mm (90 GHz) with active correction.

### 8.8.4 Summary

We have seen that the antenna efficiency $\eta$ is the combination of three principal factors:

$$\eta = \eta_{illum}\eta_{block}\eta_{surf}. \qquad (8.30)$$

Other factors could be included[16] but these are usually comparatively small and for our present purposes can be ignored. The antenna efficiency defines the effective area $A_{eff}$:

$$\eta = \frac{A_{eff}}{A_{geom}}. \qquad (8.31)$$

Substituting some typical values,

$$\eta_{illum} = 0.81 \quad \text{(for } -15 \text{ dB combined edge taper),}$$
$$\eta_{block} = 0.89 \quad \text{(typical value for on-axis optics),}$$
$$\eta_{surf} = 0.91 \quad \text{(for } \lambda/40 \text{ rms irregularities),}$$

gives $\eta = 0.66$. In practice anything over 0.6 is regarded as satisfactory for an on-axis paraboloid.

The trade-offs are as follows.

– *Illumination closer to uniform*  The antenna collects more power and hence $\eta$ rises but the sharper transition at the edge produces higher sidelobes; thus the relative amount of power in the main beam falls and the main beam efficiency $\epsilon$ falls.
– *Illumination more tapered*  The antenna collects less power and hence $\eta$ falls but the smoother transition at the edge produces lower sidelobes and the main beam efficiency $\epsilon$ rises.

---

[16] For example, losses in the antenna and cross-polarization losses. Atmospheric losses, or losses due to rain on the structure, which are time variable and frequency dependent, have to be applied as separate corrections.

The antenna beam parameters are typically:[17]

> half-power beamwidth $\quad \theta_{1/2} = (1.15 \pm 0.05)\lambda/D$;
>
> equivalent area of main beam $\quad \Omega_M = (1.15 \pm 0.05)\theta_{1/2}^2$.

In Section 8.9 we will describe how these factors affect radio astronomical observations.

## 8.9 The Response to a Sky Brightness Distribution

Single radio telescopes are often general purpose instruments used for a wide range of observational programmes. Over the course of time they may: survey regions of the sky for discrete, sometimes time-variable, sources (Chapters 15 and 16); map large areas of sky for diffuse extended emission in both continuum and spectral line modes (Chapters 14 and 16); study the time variability of targeted sources (Chapter 15); take part in VLBI arrays (Chapter 11).

The angular resolution of the telescope is, in practice, limited to $\sim \lambda/D$, roughly the half-power beamwidth. For studies of low-brightness extended emission over large areas and for time-variable sources (e.g. pulsars) this relatively low resolution (typically a few arcminutes to tens of arcminutes) is no handicap. If the source is detected with a high signal-to-noise ratio, the angular resolution, and hence the ability to detect more detail in a radio map, can be improved to $\lambda/2D$, but seldom better, unless there is reliable *a priori* information about the source. Improvement beyond this limit usually leads to misleading results. In theory, one can obtain extraordinarily high resolution because the brightness distribution is an analytic function and, by the process of analytic continuation, small angular details might be derived. This process, sometimes called superresolution, is of only limited usefulness because the first and higher derivatives of the observed brightness must be derived with great accuracy, a process limited by the inevitable presence of noise and by limited knowledge of the antenna beam shape (see also Section 10.9.1).

### 8.9.1 Beam Smoothing and Convolution

We now quantify the effect of scanning any brightness distribution $T_B(\theta, \phi)$ with a telescope beam. Figure 8.30 is a cut through a normalized beam pattern $G_n(\theta, \phi)$ pointing at a specific position $(\theta, \phi)$ on the sky. In Section 8.3.3, Eqn 8.24 provide the integral relationship between $T_B(\theta, \phi)$, $G_n(\theta, \phi)$, and the measured antenna temperature $T_A(\theta, \phi)$ for this particular pointing direction; $T_A(\theta, \phi)$ is evidently a weighted mean of $T_B(\theta, \phi)$ over the solid angle of the beam.

Scanning implies moving the beam to a series of new positions, offset from the first, and performing the summation again at each offset. The reader should recognize this process as akin to correlation (Appendix 1), heuristically described in Section 7.2 (on autocorrelation spectrometers). Two functions (different in the case of cross-correlation) are placed over

---

[17] The two factors of 1.15 are coincidental.

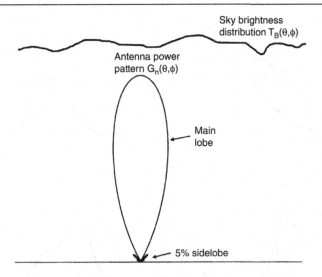

Figure 8.30. A cut through the antenna beam pattern $G(\theta, \phi)$ at constant $\phi$ with the power plotted linearly with respect to the on-axis peak.

each other, their product is calculated point by point[18] and the results summed to produce a single value at each offset. If the beam is symmetrical then the outcome of scanning across multiple offsets is identical to cross-correlating the beam pattern and the pattern of sky brightness temperature. In the more general case, where the beam is asymmetrical, there is an important difference. This is illustrated in the schematic diagram (Figure 8.31) in which the sky brightness temperature distribution and the beam pattern are simplified to functions of one angle $\theta$ only; the variable angular offset which describes the action of scanning is $\theta_o$.

Figure 8.31 illustrates why convolution implies reversing the direction of the beam pattern. The receiver output as a function of scan angle is the convolution of the angular structures of the beam and the source (in this case a point). The right-hand side of the beam meets the source first, so the beam pattern traced out in the receiver power output is automatically flipped around. This is a characteristic of the convolution process. The outcome of the scanning process is a third function, $T_A(\theta_o)$, the observed, smoothed, antenna temperature:

$$T_A(\theta_o) = \frac{1}{\Omega_A} \int_{2\pi} T_B(\theta) G_n(\theta_o - \theta) d\theta, \tag{8.32}$$

where $\Omega_A$ represents the integration of the beam over the single angular coordinate. Note that, compared with the equation describing cross-correlation, the beam function is 'flipped' around before the integration is carried out. Figure 8.31 shows pictorially why this comes about.

---

[18] It is simpler to envisage the process when continuous functions are sampled at discrete points. In the formal integral the separation between the samples tends to zero.

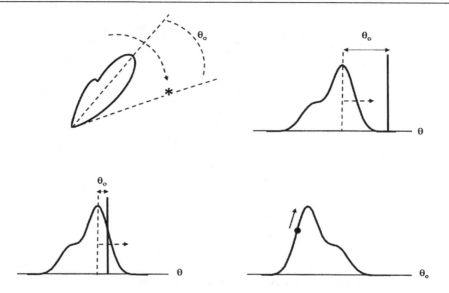

Figure 8.31. Convolution by a telescope beam. (upper left) A view in polar coordinates of a one-dimensional beam with a single highly asymmetric sidelobe, captured at the starting point of a scan across a point source represented by the star. (upper right) The same situation in Cartesian coordinates, with the source represented by the solid vertical line. (lower left) Part of the beam has now moved over the source. (lower right) The receiver output as a function of scan angle is the convolution of the angular structures of the beam and the source (in this case a point).

For both angular dimensions the process of convolution is written as

$$T_A(\theta_o, \phi_o) = \frac{1}{\Omega_A} \int_{4\pi} T_B(\theta, \phi) G_n(\theta_o - \theta, \phi_o - \phi) d\Omega \quad . \tag{8.33}$$

This is the aptly named *smoothing function* of the sky brightness and the antenna beam. We can write Eqn. 8.33 succinctly as

$$T_A(\theta, \phi) = T_B(\theta, \phi) * G_n(\theta, \phi), \tag{8.34}$$

where $*$ stands for convolution.

Figure 8.32 is an illustration of beam smoothing in the sky plane in one dimension, showing a series of four point sources of equal strength scanned with a Gaussian-shaped antenna beam. The left-hand two sources are well separated (by $3.5\theta_{1/2}$) whereas the right-hand two sources (separated by $0.85\theta_{1/2}$) are blended and their dual nature has been lost. Depending on the illumination taper, $\theta_{1/2}$ is $\sim \lambda/D$ ; thus if $D = 500\lambda$ then detail is lost on scales smaller than 1/500 radians (0.115 degrees). In general, only details larger than the FWHM of the beam, i.e. $\sim \lambda/D$, can be clearly discerned in any map of the sky.

The sky brightness temperature distribution can conveniently be envisaged as a fine grid (spaced at several times smaller than the beam) of point sources of different strengths. Equivalently, in the Fourier domain, the sky can be represented by a spectrum of angular

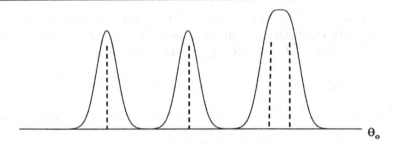

Figure 8.32. Beam smoothing. A series of four point sources of equal strength have been scanned with a Gaussian-shaped antenna beam. The left-hand two sources are well separated whereas the right-hand two sources are blended and their dual nature has been lost.

frequency components with units of cycles per radian.[19] The loss of detail in scanning the sky can be expressed in terms of these Fourier components. From the convolution theorem (Appendix 1), the spectrum of the convolved function is the product of the spectra of the two input functions. Thus we can write for the sky spectrum equivalent to Eqn 8.34:

weighted spectrum = true spectrum × transfer function,

all expressed in cycles per radian. The equivalent to smoothing by convolution in the sky plane is, therefore, a down-weighting of the high frequency Fourier components of the true sky brightness temperature distribution by a 'transfer function' which is the Fourier transform of the antenna power beam. In Section 8.3 we saw that the latter is the autocorrelation of the aperture distribution. Thus the spectrum of the smoothed sky distribution is 'band limited', i.e. cut off beyond a maximum number of cycles per radian.

The low-pass angular filtering effect of the antenna beam can be understood with reference to Figure 8.20, which is a simplification to one dimension of a uniformly illuminated square antenna of dimension $u = D/\lambda$ wavelengths. The triangular autocorrelation function (bottom left) acts as the angular frequency transfer function, i.e. as a low-pass angular spectral filter. Angular frequency components of the sky brightness distribution lower than $u$ cycles per radian are passed by the antenna but with reduced magnitude. Angular frequency components of the sky higher than $u$ cycles per radian are completely cut off. On the bottom right of Figure 8.20 is the corresponding $sinc^2$ power beam with which the sky brightness temperature distribution is convolved. In practice the aperture is likely to be circular and the illumination will be tapered towards the edge (see Section 8.8.1) and the beam will have lower sidelobes than indicated in Table 8.1.

This argument applies generally, including to interferometer arrays. Only those Fourier components of sky brightness which correspond to non-zero values in the filtered angular frequency spectrum are recorded in the output of a telescope scan across the sky.[20] To link in with the discussion on interferometry in Chapters 9 and 10, one should note that direction

[19] Akin to cycles per second for the frequency spectrum of a time series. A component exhibiting five oscillations over an angle of 0.01 radians has an angular frequency of 500 cycles per radian.
[20] For the single-dish case we have described the beam and the sky in terms of angular coordinates $\theta, \phi$ referred to the pointing direction.

cosines $l, m$ can also be used, where $l = \sin\theta$ and $m = \sin\phi$. The telescope is pointed in direction $l_0, m_0$ and if the angles are small, the projection factor $1 - l_0^2 - m_0^2$ can be neglected in the expression for the solid angle. The equivalent to Eqn 8.33 is

$$T_A(l, m) = \frac{1}{\Omega_A} \int \int T_B(l, m) G(l - l_0, m - m_0) dl \, dm, \tag{8.35}$$

and the Fourier transform of $T_A(l, m)$ is the product of the Fourier transforms of $T_B(l, m)$ and $G(l, m)$.

### 8.9.2 Sampling in Angle

In order to make a fully sampled brightness temperature map of an area of sky with a single telescope one needs to decide an appropriate separation between the series of telescope pointings. This is simply another application of the sampling theorem (Appendix 3), except that in this case the separation between the points is in angle rather than time.

We have seen that an antenna of dimension $u = D/\lambda$ is only sensitive to Fourier components of the sky brightness temperature distribution out to $u$ cycles per radian. Since these are the highest frequencies, requiring to be sampled at the Nyquist rate, points should be taken at a rate $> 2u$ per radian; the corresponding angle $\Delta\theta < 1/2u$ radians. For example, if $D = 500\lambda$, $\Delta\theta < 1/1000$ radians or $< 0.0573°$; i.e. the rate should be more than two points per FWHM of the main beam. In practice samples are usually taken at a rate of about three points per FWHM.

The classic map of the radio sky at 408 MHz ($\lambda = 0.74$ m; see Figure 14.3 and the Supplementary Material) provides a real-life example. The data were taken using three separate telescopes: the Jodrell Bank 76 m, the Effelsberg 100 m, and the Parkes 64 m. The resolution of the map was set by the smallest telescope, the maps made with the larger telescopes being convolved down. The published map resolution is $0.85° = 1.28/u = 1.28\lambda/D$ for the 64 m telescope, implying significant tapering of its illumination pattern. The map data points are separated in angle by $0.25°$ (3.4 points across the beam; 16 points per square degree) and hence the map of the 41 259 square degrees of the sky is made up from $\sim 660\,000$ separate points.

### 8.9.3 Effects of Sidelobes

Both the near and far sidelobes can have serious effects on radio astronomical observations; one practical consideration is that the far sidelobes may be sensitive to terrestrial sources of radio interference which would not otherwise be a nuisance. Interference from satellites can also enter via the far sidelobes, and this will be an increasingly serious problem in the future. Sidelobes will also pick up radiation from the ground around the telescope and this contribution to the system noise will vary as a function of azimuth and elevation.

Sidelobes are a particular issue when dealing with the extended emission regions in the Milky Way, in particular from the hydrogen line or synchrotron radiation. A large field (hundreds or thousands of square degrees) may have to be mapped, including regions of both high and low surface brightness. Observations of a complex region where the surface brightness is low within the primary beam may then be contaminated by the sidelobe pickup of more intense galactic radiation (or the ground) outside the primary beam. Complete mapping of the surroundings may require a deconvolution of the observations, using the techniques to be described in Chapter 10, in order to recover the true surface brightness.

Sections of the radio map of the Milky Way at 408 MHz before and after correction for the effect of sidelobes are shown in the Supplementary Material. The features marked are artefacts arising from the difficulty of making absolute calibration measurements on different days under different observing conditions. The minimization of sidelobes was, therefore, an important factor in choosing the offset feed design for the 100 m GBT, with its freedom from aperture blockage.

### 8.9.4 Pointing Accuracy

Whatever the observations being undertaken it is vital to know the position of the telescope's beam on the sky at all times. How accurately does one need to know this position? An often used rule of thumb is to require that pointing errors $\Delta\theta$ be no worse than about than one-tenth of the beam's FWHM ($\theta_{1/2}$). To see why, we need to assume a beam shape. Any antenna beam shape will be approximately parabolic near the peak and so, with respect to a position offset $\Delta\theta$ in one angular coordinate, its normalized gain can be written as

$$G(\Delta\theta) = 1 - \frac{1}{2}\left(\frac{2\Delta\theta}{\theta_{1/2}}\right)^2. \tag{8.36}$$

Substituting $\Delta\theta = 0.1\theta_{1/2}$ gives $G(\Delta\theta) = 0.98$, i.e. a 2% loss in signal. Alternatively the beam can be approximated by a Gaussian, in which case

$$G(\Delta\theta) = \exp\left[-4\ln 2 \left(\frac{\Delta\theta}{\theta_{1/2}}\right)^2\right] \tag{8.37}$$

and, for $\Delta\theta = 0.1\theta_{1/2}$ one obtains $G(\Delta\theta) = 0.972$, i.e. a 2.8% loss in signal. If the pointing errors have the same magnitude in both angular coordinates then the resultant error is $\sqrt{2}$ times larger and for a symmetrical beam the loss in signal will be 4% for the parabolic profile and 5.4% for the Gaussian profile.

To maintain the gain within a few per cent of the maximum value during the observations, the *combined* pointing error should, therefore, not exceed about one-tenth of the half-power beamwidth. And the observer should remember that, from the peak, the gain falls off quadratically; thus, for an error $\Delta\theta = 0.2\theta_{1/2}$ in one coordinate, the loss in signal is 8% (parabolic profile) or 10.5% (Gaussian profile). Such errors are usually deemed to be unacceptable.

Establishing an acceptable[21] pointing model for a large paraboloid can be an exacting and time consuming process. A geometric model of the telescope structure and its location

---

[21] Such that the highest operating frequency is set by factors other than pointing.

on the geoid is the starting point. The *a priori* knowledge of the telescope will, however, not be perfect – for example the axes will not be aligned perfectly, the encoders may have offsets, and the structure will certainly deform under gravity. At low elevations, refraction in the troposphere must be taken into account; sources appear at a higher elevation than they would in vacuo, effectively lowering the radio horizon.[22] An extensive series of positional offset measurements, for a network of well-known strong sources, must be undertaken and used to refine the model as a function of position in the sky.

### *8.9.5 Source Confusion*

When a single radio telescope is being used to detect discrete sources then it may well, if the receiver is very sensitive, hit a practical survey flux density limit before the limit set by receiver noise and telescope performance. This is due to 'confusion', when the surface density of detectable sources is suffiently high that a significant number fall, by chance, closer together than the beam FWHM and hence cannot easily be distinguished as separate objects. For a given telescope the beam is larger at low frequencies and hence confusion sets in more quickly; it is often a significant consideration below a few GHz and, for the biggest telescopes, below $\sim 10$ GHz. A useful rule of thumb is that blends of individually recognizable sources (detected at $\geq 5\sigma$) become a statistically significant issue for the catalogue when they are, on average, less than five times the FWHM apart, corresponding to a surface density of one source per $\sim 25$ beam areas.[23]

Confusion is a more subtle issue than than this. In addition to the effects of receiver noise, there are low level fluctuations in a radio map arising from the superposition of objects in the population of faint sources below the survey limit. Condon (2007) gives the following empirical equation for the rms confusion 'noise' $\sigma_c$:

$$\sigma_c \approx 0.2\nu^{-0.7}(\theta_{1/2})^2, \tag{8.38}$$

where $\sigma_c$ is measured in mJy beam$^{-1}$, the frequency $\nu$ in GHz and the beamwidth in arcmin. The standard advice is that sources fainter than $5\sigma_c$ may be affected by confusion and cannot be relied upon in the absence of any additional evidence (see also Section 10.12.1).

### *8.9.6 Source Positions*

For surveys of discrete sources a basic issue is the accuracy of the celestial positions assigned to catalogued objects. Assuming that the pointing model is accurate to the level

---

[22] More details about telescope pointing are given in the book by Baars (2007). Note also that the structure will deform due to differential thermal expansion as a function of time of day and cloud cover. Temperature measurements, taken at a relatively few well-chosen points on the structure, can be used as inputs to a correction algorithm.

[23] If the distribution of sources is purely random then Poisson statistics apply. The probability of finding $n$ sources within unit area (the beam area) is $P(n) = m^n e^{-m}/n!$, where $m$ is the mean probability. For a surface density of 25 sources per beam area, $m = 0.05$; hence the probability of finding $n = 2$ sources in one beam area is $\sim 0.08\%$.

proposed in Section 8.9.4 then the position of a source detected with a high signal-to-noise ratio ($>$10) can be determined simply by scanning the telescope beam across it in orthogonal directions, noting the angular positions of the maxima. Such observations are in fact part of the process of determining the pointing model of the telescope in the first place, using sources whose positions are already known from high-resolution interferometric measurements.[24]

If, however, the source begins to approach the noise level set by a combination of receiver and confusion noise then the positional accuracy will be limited by random fluctuations. In one angular coordinate the rms uncertainty $\delta\theta$ in the derived position of the source is approximately given by (Condon 1997)

$$\delta\theta \approx \frac{\theta_{1/2}}{\text{SNR}}, \tag{8.39}$$

where SNR is the peak signal-to-noise ratio. Thus, for a five-sigma detection the rms positional uncertainty is about $0.2\theta_{1/2}$ in one coordinate and $0.28\theta_{1/2}$ overall. An important caveat must, however, be noted. Equation 8.39 applies only for sources which are very much smaller than the beam. In practice radio sources have a wide range of angular sizes and some, apparently compact, sources will in fact have structure approaching the size of the beam. Equation 8.39 would still apply if such sources were symmetric, but real radio sources associated with active galaxies (Chapter 16) are a complex combination of cores, asymmetric jets, and lobes while star-forming galaxies exhibit a wide range of complex morphologies. Caution must therefore be exercised when one is seeking to use beam centroids to identify radio source counterparts in other wavebands.

## 8.10 State-of-the-Art Radio Telescopes

The astronomical requirements for single-dish radio observations have dictated a thrust towards ever higher frequencies; this places demands on maintaining the surface profile and the precise pointing of the dish. At the same time, the ever-pressing need for greater sensitivity requires the largest possible aperture; this places demands on structural integrity in the face of wind and gravity loads. These requirements interact, and experience shows that the cost of a fully steerable parabolic dish scales with its diameter $D$ and with its upper frequency limit $\nu_{max}$ roughly as $D^{2.6}\nu_{max}^{0.8}$. Designers therefore have to strike compromises driven by the science programmes for which the dish is built. The books by Baars (2007) and by Baars and Kärcher (2017) describe both the criteria involved in parabolic dish design and many approaches to meeting them. The dish structure is, however, never the whole story and continuous innovations in feed and receiver technology can maintain a decades-old telescope at the research forefront; a good example is the 13-beam L-band system on the 1960s-vintage Parkes 64 m telescope (see Section 8.7). We now briefly describe a few

---

[24] In practice offsets are often determined using a fitting algorithm based on the measured power at five points: one on the nominal position, straddled by two in azimuth and two in elevation.

leading single-dish telescopes, primarily used for radio astronomy, and draw attention to other notable examples; pictures can be found in the Supplementary Material.

### 8.10.1  FAST and Arecibo

Cost per unit area can be reduced by accepting a restriction on the sky coverage and the upper frequency limit; this was the route taken by the 300 m Arecibo telescope in Puerto Rico and the Five Hundred Metre Aperture Spherical Telescope (FAST) in Guizhou Province, China. Both telescopes have perforated sheet-aluminium surfaces and operate at frequencies up to several GHz. FAST has supplanted Arecibo as the world's largest single-aperture radio telescope. It follows the bold concept pioneered in the 1960s at Arecibo in having a reflecting surface built in a deep natural bowl in local karst topography. With their focal structures suspended on cables above stationary reflectors both telescopes perforce have limited sky coverage. With a collecting area equivalent to a 200 m conventional paraboloid the Arecibo telescope can cover a zenith angle range of $\pm20°$ while FAST offers coverage of $\pm40°$. The reflecting surface of Arecibo has a fixed spherical profile and, to correct for aberrations, its focal structure (weighing 1000 tonnes!) has secondary and tertiary mirrors. The surface of FAST is active and can be pulled by over 2000 steel cables into a parabolic profile over a diameter of 300 m. As the source moves across the sky the parabolic area moves under computer control. Since no correction optics are needed, the prime-focus structure of FAST is much lighter than that of Arecibo and the optics allow for a 19-beam L-band horn array compared with the seven-beam equivalent at Arecibo.

### 8.10.2  Large Steerable Dishes with Active Surfaces

The Green Bank Telescope (GBT) in West Virginia, USA, commissioned in 2001, is the largest of a trio of large fully steerable parabolic dishes principally aimed at decimetre and centimetre wavelength observations. The first of these was the 76 m dish at Jodrell Bank, UK, commissioned in 1957. The next was the 100 m dish at Effelsberg, Germany, commissioned in 1972. The Effelsberg dish pioneered the use of the principle of homologous design, which allows the reflecting surface to retain a paraboloidal shape despite the distorting effects of gravity. The GBT also uses this principle and, with its surface panels adjustable by over 2000 actuators under computer control, an overall surface accuracy of 240 µm rms has been achieved; this allows useful observations down to 3 mm wavelength to be made when the atmospheric opacity is low. The GBT was the first large radio telescope to use an offset feed support carrying the secondary reflector; the advantage is an unblocked aperture and lower sidelobe levels. The large receiver laboratory can accommodate a large suite of low-noise receivers. The location at an existing National Radio Astronomy Observatory site benefits from a radio-quiet zone a hundred kilometres square.

The 64 m Sardinia Radio Telescope (SRT) in Italy was commissioned in 2013. It has an active surface controlled by over 1000 actuators; the specification of 150 μm rms enables operation to 3 mm wavelength with good efficiency. Designed for flexible operation for a variety of uses, including space communications, the SRT's innovative optics permit a wide range of receivers to be operated at several focal stations. The 65 m Tian Ma telescope in Shanghai, China, was commissioned in 2012, in part to support China's Lunar Exploration project. It has Cassegrain optics and active surface control giving an overall accuracy of 300 μm rms, allowing good performance up to 50 GHz.

### 8.10.3 Millimetre and Sub-Millimetre Wave Dishes

The Franco-German Institut de Radio Astronomie Millimetrique (IRAM) operates a 30 m radio telescope, designed for millimetre-wave observations, on the Pico Veleta (altitude 285 m) near Granada, Spain. Commissioned in the mid-1980s the telescope is a homologous design, with an overall rms surface accuracy of 70 μm; it is equipped with a suite of receivers that cover the frequency range 80–280 GHz, intended primarily for spectroscopic studies of the interstellar medium. It has a Cassegrain configuration, with the receiver cabin located behind the primary surface at the focal plane. To overcome the atmospheric emission from water vapour, it has a nodding secondary that allows comparison of the observing field with a reference field nearby.

The 15 m James Clerk Maxwell Telescope (JCMT), originally conceived and operated by a UK, Dutch, and Canadian consortium, began operations in 1987. It is now operated by the East Asian Observatory. The telescope is situated near the summit of Mauna Kea in Hawaii at an altitude of 4092 m; it has surface accuracy 24 μm rms, enabling it to operate at sub-mm wavelengths. Receivers are located at both Cassegrain and Nasmyth foci; the SCUBA2 receiver, operated at one of the Nasmyth platforms, is a 5120-element bolometric array operating at both 450 and 850 μm wavelengths for large-area surveys.

The APEX (Atacama Pathfinder Experiment) telescope is operated by a consortium of the Max Planck Institut für Radioastronomie (MPIfR) at 50%, the Onsala Space Observatory (OSO) at 23%, and the European Southern Observatory (ESO) at 27%. It is situated on the Llano Chajnantor, Chile, close to the ALMA array at an altitude of 5105 m. The antenna is a modified ALMA prototype dish with diameter 12 m and rms surface accuracy 17 μm, enabling it to operate at frequencies above 1000 GHz. The range of instruments includes several SIS-based receivers for spectroscopy and the LABOCA (345 GHz) and SABOCA (805 GHz) bolometer array receivers for broadband continuum surveys.

## 8.11 Further Reading

*Supplementary Material* at www.cambridge.org/ira4.
Baars, J. W. M. 2007. *The Paraboloidal Reflector Antenna in Radio Astronomy and Communication*, Springer.
Baars, J. W. M., and Kärcher, H. J. 2017. *Radio Telescope Reflectors*. Springer.

Balanis, C. A. 2005. *Antenna Theory and Design*, 3rd edn. Wiley.

Christiansen, W. N., and Högbom, J. A. 1985. *Radiotelescopes*, 2nd edn. Cambridge University Press.

Clarricoats, P. J. B., and Olver, A. D. *Corrugated Horns for Microwave Antennas*. IEE Electromagnetic Waves Series, vol. 18.

Maddalena, R. (NRAO). 2016. *Observing Techniques with Single Dish Radio Telescopes*. www.gb.nrao.edu/~rmaddale/Education/ObservingTechniques.pdf.

Max Planck Institut für Radioastronomie. 2010. *European Single Dish School in the Era of Arrays*, https://www.mpifr-bonn.mpg.de/ESSEA2010.

Rumsey, V. H. 1966. *Frequency Independent Antennas*. Academic Press.

Stanimirovic, S., Altschuler, D., Goldsmith, P., and Salter, C. 2002. *NAIC–NRAO School on Single-dish Radio Astronomy: Techniques and Applications*, Astronomical Society of the Pacific Conference Series, vol. 278.

# The Basics of Interferometry

The combination of wind and gravity has always militated against building very large single dishes. Since radio wavelengths are hundreds to millions of times longer than optical wavelengths all single-aperture radio telescopes are, therefore, hindered by diffraction effects and their angular resolution is limited to the range arcmins to tens of arcmins.[1] By contrast the resolution of ground-based optical telescopes is $\leq 1$ arcsec whilst that of the diffraction-limited Hubble Space Telescope is $\sim 50$ milliarcsec. The application and development of radio interferometry, building on the rapidly developing arts of electronics and signal processing, overcame this handicap. In their 1947 studies of the Sun, McCready, Pawsey, and Payne-Scott already recognized that an interferometer's response to an extended source amounted to determining a particular value of the Fourier transform of the source brightness distribution. This insight was broadly recognized in the radio astronomy community, and informed much of the work in Sydney, Cambridge, and Manchester in the following decades.

As the art of interferometry progressed, interferometers were used with multiple spacings to measure samples of the Fourier transforms of extended sources; these samples could then be inverted to give satisfactory maps of the brightness distributions if the spacings were carefully chosen.[2] Fourier concepts, reviewed in this chapter, became the natural language for discussing the brightness distribution across sources. Finally, Ryle at Cambridge formulated Earth-rotation synthesis, in which the rotation of the Earth is used to vary the orientation and effective length of interferometer baselines, yielding an extensive sampling of the Fourier transform of the sources each day. The *aperture-synthesis arrays* developed in the 1970s and 1980s, such as the Westerbork Synthesis Radio Telescope (WSRT) in the Netherlands, the Australia Telescope Compact Array (ATCA), the Very Large Array (VLA) in the USA, the MERLIN array in the UK, and international VLBI arrays all use this principle. Many other large synthesis arrays have subsequently been developed around the world including the Low Frequency Array (LOFAR), centred in the Netherlands, the Giant Metre Wave Radio Telescope (GMRT) in India at metre wavelengths, and the Atacama Large Millimetre Array (ALMA) in Chile at mm and sub-mm wavelengths. At metre and centimetre wavelengths the Square Kilometre Array in Australia and South Africa will become the dominant instrument in the 2020s and beyond.

---

[1] The latter being comparable to the angular diameter of the Sun and the Moon.
[2] The angular resolution of an array with maximum spacing $d_{km}$ is approximately $2\lambda_{cm}/d_{km}$ arcsec.

Interferometer arrays combine the data from many small antennas to produce results which are, in part, equivalent to those from a large single antenna. In single-antenna work, however, we do not have to be concerned with the phase of the incoming waves with respect to the reception system; all our measurements are in terms of intensity, in which knowledge of the incoming phase is lost. In interferometry this is no longer the case. While the absolute signal phase has no significance, interferometry intimately depends on determining the *relative* phase of signals at different reception points. One therefore has to be careful to maintain the timing and the frequency standard at each antenna and to define and establish the geometry of the system both on the sky and on the Earth. The more complicated data path from collection to output (often an image) is inevitably harder to grasp than it is for a single antenna.

This chapter therefore focusses on the basic principles of an interferometer, starting with the simplest two-element system and using concepts which would have been familiar to Michelson in his classic optical work on stellar diameters over 100 years ago. We make clear the similarities and the differences between adding interferometers (see Section 8.2) and multiplying (correlation) interferometers and, concentrating on the latter, introduce the concept of the complex fringe visibility as a sample of the Fourier transform of the source brightness distribution associated with a particular baseline vector. We discuss the effect of a finite receiver bandwidth and show how measurements of the visibility amplitude and phase yield information on a source's structure and its celestial position. The power of interferometric delay and phase measurements is particularly evident in radio astrometry and geodesy; the relative positions of the elements of an interferometer array thousands of kilometres across may be determined with an accuracy of a few millimetres and the positions of compact sources can be determined to tens of microarcseconds. While the basic theory of interferometers can be developed from elementary considerations we stress that correlation interferometers actually measure the spatial and temporal coherence of the incoming radiation field; we will outline these ideas in simple terms. Spatial coherence and the response to source structure is captured by the Van Cittert–Zernike theorem (see Section 9.6). Temporal coherence, captured in the Wiener–Khinchin theorem (see Sections 5.1 and 7.2 and Appendix 1), reveals itself in the effect of the finite receiver bandwidth which restricts the field of view.

## 9.1 The Basic Two-Element Interferometer

An early example of an interferometric observation of radio sources, shown in Figure 9.1, illustrates some principal features of interferometry. The signals received by two fixed antennas, also shown in the diagram, are added, amplified by a radio receiver, and detected, to give the total power as a function of time. There is a slowly varying level of background noise, partly generated by the radio receiver itself, and partly coming from the Galaxy. As the Earth rotates, first the extragalactic source Cygnus A and then the supernova remnant Cassiopeia A pass through the antenna patterns. The recording shows the signal from each source, modulated by a quasi-sinusoidal oscillation. This arises from the alternately constructive and destructive interference of the signals from the two antennas, with the

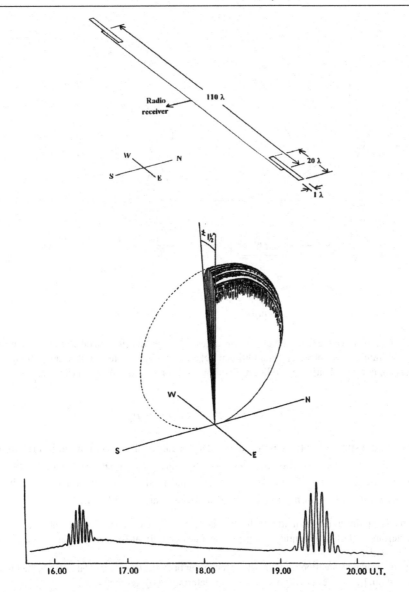

Figure 9.1. The radio sources Cyg A and Cas A recorded with an early interferometer system (Ryle *et al.* 1950). The antenna elements were dipole arrays 20λ by 1λ, giving a fan beam sliced by the interference pattern.

average source power and the envelope of the modulation both tracing the antenna gain pattern. The characteristic quasi-sinusoidal oscillation is called the fringe pattern.[3]

---

[3] The term 'fringe' comes from early optical observations of diffraction at a shadow edge; it was then extended to the optical two-slit interference pattern seen by Young over 200 years ago, and is now applied to interference patterns in practically every kind of interferometer.

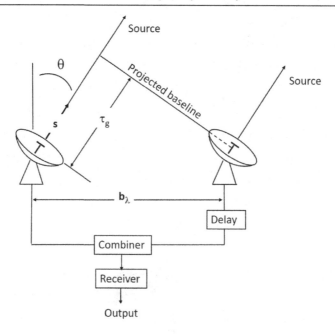

Figure 9.2. Geometry of the two-element interferometer. The geometric path delay $\tau_g$ is compensated by the delay circuit in the receiver; an additional delay $\tau_i$ can be inserted by this delay circuit. The two signals are combined either by addition (Section 9.1.2) or by multiplication (Section 9.1.3).

### 9.1.1 One-Dimensional Geometry

Moving from the specific to the generic, the basic elements of a two-element interferometer are shown in Figure 9.2. For this introductory discussion we make a series of simplifying assumptions which will be lifted in later sections (and further in Chapter 10) as we discuss increasingly realistic situations. The initial assumptions are as follows:

– the source is a point in the far field which generates plane waves at a single frequency $\nu$;
– the interferometer has two identical antennas, operates entirely at frequency $\nu$, and is sensitive to only one polarization;
– there are no propagation effects either before reception or in the transfer of data within the interferometer system, hence the relative delay is purely determined by geometry.

The baseline vector $\mathbf{b}$ (metres) or $\mathbf{b}_\lambda$ (wavelengths) connects the phase centres of the two antennas; if these are identical, any convenient reference point such as the antenna vertex will do;[4] one of the antennas is designated the reference. The radio source S under observation is in the direction given by a unit vector $\mathbf{s}$. The projection of $\mathbf{b}$ perpendicular to $\mathbf{s}$ is the *projected baseline*, length $b \cos \theta$. The projection along $\mathbf{s}$ is the geometrical path

---

[4] If the antennas are different then care must be taken to understand the antenna geometry, which governs how each chosen reference point moves in the direction of the unit vector $\mathbf{s}$ as the Earth rotates – see also Section 9.8.2.

length difference between the signal arriving first at the reference antenna and then at the other; it can be expressed in several different ways depending on the context. The physical path length difference

$$\Delta l_g = \mathbf{b} \cdot \mathbf{s} = b \sin \theta \quad \text{(metres)} \tag{9.1}$$

is useful if the delay is to be compensated with RF cable.[5] It is equivalent to a time delay

$$\tau_g = \Delta l_g/c = b \sin \theta/c \quad \text{(seconds)}, \tag{9.2}$$

which is relevant when we discuss signal coherence. The path difference can also be expressed as the number of cycles of the incoming signal,

$$\Delta l_g/\lambda = b \sin \theta/\lambda = b_\lambda \sin \theta \quad \text{(RF cycles)}; \tag{9.3}$$

this is the number of fringe spacings per radian and hence is relevant to the angular resolution. Finally, the relative phase between the two signals is

$$\phi = 2\pi \nu \tau_g = 2\pi b_\lambda \sin \theta \quad \text{(radians)}, \tag{9.4}$$

which is appropriate when discussing interferometric position measurements. All the above are, of course, time dependent as the source moves across the sky and $\theta$ changes.

### 9.1.2 The Adding Interferometer

The simplest situation, used in the observation of Figure 9.1, is an adding interferometer, in which signals from two identical antennas, diameter $D$, separated by spacing $d$, are fed into a single receiver. Figure 9.3 shows the resultant polar diagram traced out by a solitary point source; it is an interference pattern with fringe spacing $\lambda/d$, contained in the polar diffraction diagram of a single antenna, width $\sim \lambda/D$. The total noise power is recorded, so the signal from the point source is recorded on top of the receiver noise. In the receiver the two signals, in the form of voltages, are added, square-law detected and time-averaged. Let the two signal voltages be $V_A$ and $V_B$ and the independent noise voltage generated by the receiver be $V_{rec}$. Working in terms of angular frequency $\omega = 2\pi \nu$, the receiver output $R_{A+B}$ is given by

$$R_{A+B}(\tau_g) = \langle [V_A \cos \omega t + V_B \cos \omega(t - \tau_g) + V_{rec}]^2 \rangle, \tag{9.5}$$

and expanding the right-hand side yields

$$\langle [V_A^2 \cos^2 \omega t + V_B^2 \cos^2 \omega(t - \tau_g) + 2V_A V_B \cos \omega t \cos \omega(t - \tau_g) + V_{rec}^2] \rangle. \tag{9.6}$$

Here we have ignored the cross-terms between the signals and the receiver noise since they have no long-term phase relationships and hence average out to zero. The third term can be written[6] as $\langle [\cos \omega \tau_g + \cos(2\omega t - \omega \tau_g)] \rangle$, the second part of which oscillates at the RF

---

[5] Taking the velocity factor in the cable into account.
[6] Using the identity $\cos A \cos B = [\cos(A - B) + \cos(A + B)]/2$.

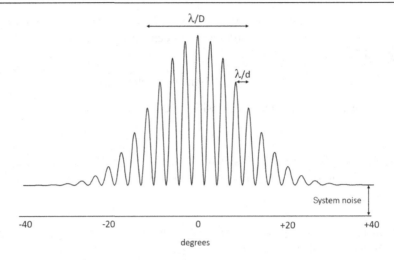

Figure 9.3. The polar diagram of a two-element adding interferometer. In this example the antenna diameters were 2.4λ and the interferometer spacing was 20λ.

frequency and hence does not build up with averaging. Since the time average of $\cos^2 \omega t$ is $1/2$, we are left with

$$R_{A+B}(\tau_g) = \tfrac{1}{2}\left[(V_A^2 + V_B^2) + V_{rec}^2\right] + \langle V_A V_B \cos \omega \tau_g \rangle. \tag{9.7}$$

The first, bracketed, term represents the approximately constant system noise power, whilst the time average of the other, interference, term varies slowly and cosinusoidally as $\tau_g$ changes. The result can be seen in Figure 9.1, which shows the response of a first-generation radio interferometer to the two strongest sources in the radio sky. Typically sources are much weaker and the receiver noise power dominates, a situation which emphasizes the undesirable property of an adding system. The stability of its output is affected by receiver gain fluctuations (see Section 6.8) and these fluctuations in the output may be comparable with, or larger than, the interference term.

For some astronomical applications (e.g. pulsars and other time-variable sources), the gain fluctuations in the total-power terms may not pose a serious problem. However, for the study of weak sources requiring integrations over long periods, the signals are always combined by cross-correlation (see Section 9.1.3), which removes the total-power terms.

Adding interferometers are the basis of phased arrays, introduced in Chapter 8, and discussed further in Chapter 10 where they are used directly to form instantaneous beams. The beam pattern of a uniformly spaced phased array feeding a single receiver (Section 8.2) can be understood by adding more signal voltages to Eqn 9.5.

### 9.1.3 The Correlation Interferometer

The cross-correlation $R_{xy}(\tau)$ of two amplitudes $x(t)$ and $y(t)$ is defined as the time-averaged product of the two amplitudes, with one delayed by time $\tau$:

$$R_{xy}(\tau) \equiv \langle x(t) y(t - \tau) \rangle. \tag{9.8}$$

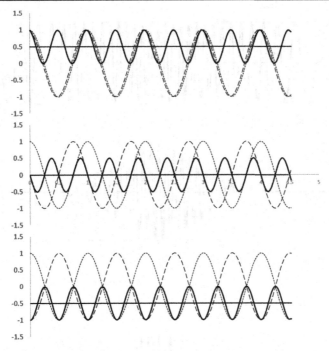

Figure 9.4. Correlation of voltage inputs (after R. Perley).

This has the dimensions of power and so it can be called the *cross-power product*. The correlation of the two elementary signal voltages is, therefore,

$$R_{AB}(\tau_g) = \langle V_A \cos \omega t \, V_B \cos \omega(t - \tau_g) \rangle, \tag{9.9}$$

which, as above, reduces to a fringe term:

$$R_{AB}(\tau_g) = \langle \tfrac{1}{2} V_A V_B \cos \omega \tau_g \rangle. \tag{9.10}$$

The simplification, compared with the adding interferometer, arises from the fact that only $V_A$ and $V_B$, which come from the same source, bear a definite phase relationship to each other. All other cross-terms have no persisting phase relationships and, oscillating at RF frequencies, do not add up over the integration period. Since the total-power terms do not correlate, a correlation interferometer is insensitive to receiver gain variations.

The reception pattern which results can be understood with the aid of Figure 9.4, which illustrates Eqn 9.10. The dotted and broken lines show two cosinusoidal signals, one phase shifted with respect to the other. The solid lines represent their multiplication and the associated time averages. As the source direction s changes, the relative phase $\omega \tau_g$ between the voltages changes. At the top there is almost zero phase shift, in the middle there is a 90° phase shift, while at the bottom the waves are in antiphase. As the relative phase continues to increase the average value moves back to its original value. The fringe amplitude therefore oscillates, with a shift of one cycle of RF phase corresponding to one fringe cycle.

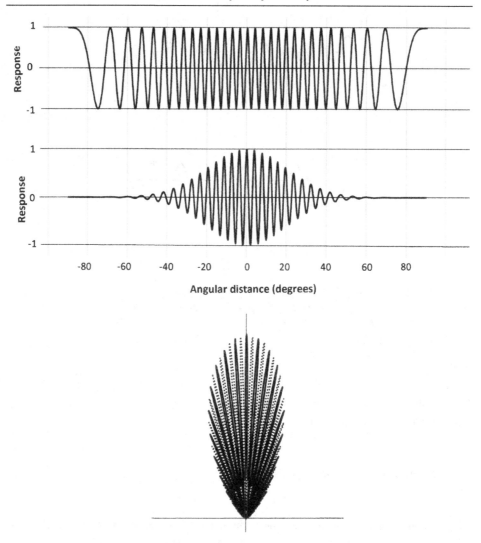

Figure 9.5. The power reception pattern (normalized to unity) of a correlation interferometer with (upper) fixed isotropic or (lower) low-gain antenna observing a monochromatic source as it traverses the beam pattern. Angles are measured from the perpendicular bisector to the baseline. The plot is at the bottom, shown in polar coordinates with the negative responses dotted.

Figure 9.5 illustrates the power reception pattern for an East–West baseline, 15 wavelengths long, lying along the Earth's equator and observing a point source on the celestial equator. Over the course of the day the source moves parallel to the baseline from horizon to horizon and is directly overhead at meridian transit, at which time it lies perpendicular to the baseline, hence $\theta = 0°$. The top diagram shows the unmodulated fringe pattern which would be obtained if the antennas had isotropic gain patterns. The following points should be noted:

- the fringes oscillate about a zero mean (no total power term);
- there are 15 maxima on each side of the centre line, i.e. from the zenith, at $0°$, to the horizon ($\pm 90°$);
- around baseline transit the small-angle approximation $\sin\theta \approx \theta$ is valid and the pattern is closely sinusoidal; the angular separation between successive positive or negative lobes is $\lambda/b = 1/15$ rad, i.e. $\sim 4°$, corresponding to $b_\lambda = 15$ fringe cycles per radian;
- away from baseline transit the projected baseline $b\cos\theta$ becomes foreshortened and the fringe spacing increases; the variation due to the $\sin\theta$ term in the delay remains clear;
- close to any direction $\theta$, whilst the fringe spacing is larger than at transit, the fringes are equally spaced cosinusoids.

For fixed antennas, pointing vertically, the source passes through the envelope of their combined primary beam in a similar fashion to Figures 9.1 and 9.3. For clarity, the beam in Figure 9.5 has been drawn with a broad beam (FWHM $\sim 45°$), as is appropriate for low-gain antennas; the resultant fringe pattern is shown in both Cartesian and polar coordinates. In the latter the negative lobes are dotted.

For the system described, the rotation of the Earth at a rate $\Omega_E \sim 7.3 \times 10^{-5}$ rad s$^{-1}$ moves the source through the fringes at a maximum *fringe frequency* $b_\lambda \Omega_E$ Hz.[7] For a $15\lambda$ baseline this is only $\sim 1$ mHz but for a VLBI baseline of $b_\lambda = 10^8$ the peak fringe frequency would be 73 kHz. To sample these fringes properly the correlator would have to read out independent samples at twice this rate (the Nyquist sampling rate, see Appendix 3). In practice this is not necessary, as the fringe frequency can be slowed down to near zero by the continuous adjustment of a time delay $\tau_i$ inserted in the transmission line from one of the antennas. As we will see in Section 9.2, this is particularly important for a wide-bandwidth signal, in which the fringe frequency would otherwise vary over the band.

### 9.1.4 Steps Towards Practicality

In the analysis so far we have greatly simplified the interferometer system. We now need to extend the analysis to real-life interferometers. The detailed analysis of practical systems was given in the classic text by Thompson, Moran, and Swenson (2017, TMS) and we are largely guided by their formalism in this chapter and the next.

The first step is to recognize that antennas have specific performance characteristics, represented by the direction-dependent effective antenna area $A(\mathbf{s})$. In forming the cross-correlation, the product of the two radio frequency voltages, representing the power received from a point source, will be proportional to the effective area of the antenna pair, $A(\mathbf{s}) = \sqrt{A_1(\mathbf{s})A_2(\mathbf{s})}$, and to the specific flux density of the point source $S_\nu$. The cross-correlation in Eqn 9.10 can then be written as

$$R_{AB}(\tau_g) = A(\mathbf{s})S_\nu \cos\omega\tau_g = A(\mathbf{s})S_\nu \cos(\omega\mathbf{b} \cdot \mathbf{s}/c), \qquad (9.11)$$

---

[7] The minimum fringe spacing is $1/b_\lambda$ rad, hence the time to pass through one fringe is $1/b_\lambda\Omega_E$ seconds; the fringe frequency is $b_\lambda\Omega_E$ Hz. If the source is not on the celestial equator, the rate is $b_\lambda\Omega_E\cos\delta$ where $\delta$ is its declination.

or alternatively

$$R_{AB}(\mathbf{s}) = A(\mathbf{s})S_\nu \cos(2\pi \mathbf{b}_\lambda \cdot \mathbf{s}), \tag{9.12}$$

and, in the small-angle approximation,

$$R_{AB}(\theta) \approx A(\mathbf{s})S_\nu \cos(2\pi b_\lambda \theta). \tag{9.13}$$

The cross-correlation can be seen to be simply the power received from the point source by an antenna of area $A(\mathbf{s})$ multiplied by the fringe pattern.

The next step is to recognize that the region of sky under study will have a finite angular size and to consider what effects this has on the cross-correlation. Consider first a point source whose true position is close to, but not exactly on, the chosen reference position $s_0$ defined by the condition $\tau_g = \tau_i$. The reference direction $s_0$ is called the *phase tracking centre*. Since $\tau_i$ compensates for the geometrical time delay, it can be called the equalizing or compensating time delay. The direction to the source, with respect to the phase tracking centre, can be written

$$\mathbf{s} = \mathbf{s}_0 + \boldsymbol{\sigma}, \tag{9.14}$$

where $\boldsymbol{\sigma}$ is a small vector normal to $s_0$. (It must be normal, since both $\mathbf{s}$ and $s_0$ are unit vectors; see Figure 9.18). Since the geometric delay associated with $s_0$ is exactly compensated by the instrumental time delay, only the small differential associated with $\boldsymbol{\sigma}$ affects the time delay in most of the analysis that follows. This differential time delay introduces limitations on the field of view when the interferometer has a finite bandwidth. We develop this formalism in Section 9.2.

In practice we observe not only point sources but also extended sources within the field of view. The basic analysis for this situation is presented in Section 9.4, to prepare for the more extensive treatment in Chapter 10.

### 9.1.5 The Frequency Domain Approach

There is an alternative, completely equivalent, way of analysing the response of an interferometer. Instead of considering the cross-correlation between voltages in the time domain the process can be described in terms of the radio frequency domain; the two form a Fourier pair. The analysis in the frequency domain leads naturally to the complex representation that is commonly used in practice. The Fourier transform of a cross-correlation $R_{xy}(\tau)$ is, by the convolution theorem, the product of the transform of $x(t)$ and the complex conjugate of the transform of $y(t)$,

$$S_{xy}(\nu) \equiv X(\nu)Y^*(\nu), \tag{9.15}$$

and is known as the *cross-spectrum power density* (see Appendix 1). This can be used to describe the interferometer output, as an extension to the description of the total power output of a single antenna. The Fourier transform of a monochromatic signal is a delta function at the frequency $\nu$, and the Fourier transform of the time-delayed signal $y(t - \tau_g)$

has a phase shift of $2\pi\nu\tau_g$, so it follows that the cross-spectral product described above becomes

$$S_{AB}(\nu) = A(\mathbf{s})S \exp(i2\pi\nu\tau_g) = A(\mathbf{s})S \exp(i2\pi\mathbf{b}_\lambda \cdot \mathbf{s}). \tag{9.16}$$

This is clearly the complex equivalent of Eqn 9.13. The interferometric calculations can therefore be carried out in either the frequency or the time domain, depending on which representation is more convenient.

## 9.2 Finite Bandwidths and Fields of View

Thus far we have been considering monochromatic signals. Most radio sources, however, emit over a wide range of frequencies. The interferometer usually has a small fractional bandwidth, and the radio spectrum of a continuum source across the band is then effectively flat, simplifying the analysis in this section. All receiving systems have a finite bandwidth $\Delta\nu$; the radio noise is therefore quasi-Gaussian, so the signal at one frequency is uncorrelated with the signals at adjacent frequencies. A wider bandwidth improves sensitivity; however, the combination of the responses at the different frequencies across the band restricts the angular range over which fringes appear.

This is a chromatic aberration effect equivalent to the 'white light fringe' in optics. The fringe spacing is inversely proportional to the frequency, but, at the position where the delay is perfectly compensated, constructive interference will occur at all frequencies within the observing bandwidth (the central 'white light fringe'). If the source is displaced from this direction by a small amount $\Delta\theta$, the phase of the fringe pattern will change by a different amount at different frequencies across the band. If the displacement is large enough, the signals at one end of the band may be interfering destructively when the signals at the other end of the band are interfering constructively. As a result, the net fringe amplitude will be reduced and, if the displacement is large enough, the fringe will disappear almost completely. The resultant effect is illustrated in Figure 9.6. Here we see a fringe pattern that

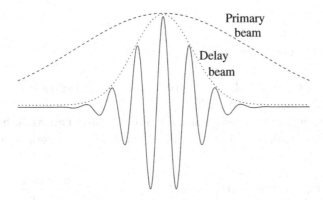

Figure 9.6. The delay beam effect in an interferometer. The width of the delay beam depends on the bandwidth of the receiver.

would, for a small bandwidth, fill the primary telescope beam pattern; instead the fringes are confined to a narrower beam whose width is determined by the bandwidth; the narrowed interferometric response is therefore known as the *delay beam*. The centre of the delay beam is kept at the peak of the primary beam by continuously adjusting the instrumental delay $\tau_i$ to track the source.

The delay beam can best be understood in terms of the *coherence time* $\tau_{coh} \sim 1/\Delta\nu$ of the incoming noise signals, introduced in Chapter 5. To obtain a high degree of correlation between the signals in the two arms of the interferometer they must be synchronized to well within $\tau_{coh}$, placing a limit on the differential delay error $\Delta\tau_g \ll 1/\Delta\nu$. Differentiating Eqn 9.2 with respect to $\theta$ to obtain the change in delay for the displacement angle $\Delta\theta$ gives

$$\Delta\tau_g = \frac{b\cos\theta\,\Delta\theta}{c}. \tag{9.17}$$

Again, for simplicity taking $\theta = 0°$, the condition for high coherence is $b\Delta\theta/c \ll 1/\Delta\nu$, which can be more memorably be written as

$$\Delta\theta \ll \left(\frac{\nu}{\Delta\nu}\right)\left(\frac{\lambda}{b}\right), \tag{9.18}$$

where $\nu/\Delta\nu$ is the inverse of the fractional bandwidth and $\lambda/b$ is the fringe spacing. Thus, for a $\sim 10\%$ fractional receiver bandwidth, satisfactory correlation occurs only within an angle about $\pm 10$ fringe spacings across; this is inadequate for general use. The answer is to split the band into many narrow channels which are correlated independently.

A more detailed calculation of the delay beam effects, using the flat-spectrum approximation, starts by calculating the interferometer response to a point source when the receiver has a square bandpass $G(\nu)$, which is unity over a bandwidth $\Delta\nu$ centred at $\nu_0$ and zero elsewhere. Since the radio noise is quasi-Gaussian, the signal at a given frequency is independent of the signals at other frequencies, and the single-frequency response given in Eqn 9.16 can be summed over the bandpass. Again for simplicity, the source direction is assumed to be nearly normal to the interferometer baseline. The response becomes the *cross-product power* $P_{AB}$, given by

$$P_{AB} = \int_{-\infty}^{\infty} S_{xy}(\nu, s)G(\nu)d\nu, \tag{9.19}$$

so that $P_{AB}$ is a function of the time delay $\tau_g$:

$$P_{AB}(\tau_g) = \int_{\nu_0-\Delta\nu/2}^{\nu_0+\Delta\nu/2} A(\nu, s)S(\nu)\exp(-i2\pi\nu\tau_g)\,d\nu. \tag{9.20}$$

Therefore, assuming that the effective area and flux are constant across the bandpass (usually a good assumption), the fringe pattern is modulated by the Fourier transform of the band shape:

$$P_{AB}(\nu_0, \tau_g) = A(\nu_0, s)S(\nu_0)\Delta\nu\exp(-i2\pi\nu_0\tau_g)\frac{\sin\pi\Delta\nu\tau_g}{\pi\Delta\nu\tau_g}. \tag{9.21}$$

Since the source flux is approximately constant over the band, but does vary with the frequency $\nu_0$, it is convenient to divide by $\Delta\nu$ to give the power density (the power per unit bandwidth). The final term can then be recognized as the sinc function, and Eqn 9.21 becomes the *cross-spectrum power density*:

$$S_{AB}(\nu_0, \tau_g) = A(\nu_0, s)S(\nu_0) \exp(-2\pi i \nu_0 \tau_g) \operatorname{sinc}(\Delta\nu\tau_g). \tag{9.22}$$

If the compensating time delay $\tau_i$ is not identical to the geometric delay $\tau_g$, Eqn 9.22 takes the form

$$S_{AB}(s) = A(\nu_0, s)S(\nu_0) \operatorname{sinc}[\Delta\nu(\tau_g - \tau_i)] \exp[-2\pi i \nu_0 (\tau_g - \tau_i)]. \tag{9.23}$$

This expression shows that the fringe oscillations, $\exp(-2\pi i \nu_0 \tau_g)$, are severely reduced in amplitude by the delay beam term, $\operatorname{sinc}[\Delta\nu(\tau_g - \tau_i)]$, when the delay is of order $1/\Delta\nu$ or larger. The instrumental time delay is therefore chosen to be close to the geometric time delay for a given antenna pointing direction. However, as can be imagined from an examination of Figure 9.6, a source can be well inside the primary beams of the antennas but sufficiently off-axis to be outside the delay beam, and it will then produce only weak fringes.

Practical receiving systems will not have perfectly square bandpass shapes, but similar delay beam effects are still present. As noted above, the shape of the delay beam is effectively given by the Fourier transform of the gain function $G(\nu)$. The square bandpass approximation is adequate for many practical purposes but for highly precise work corrections for bandpass shapes are required (see Chapters 10 and 11).

The compensation for the geometric delay removes the relative phase $\phi$ (Eqn 9.4) for one specific direction. But, since the fringe spacing varies with direction, at an angle $\theta + \Delta\theta$ there is an extra contribution $\Delta\phi$, where

$$\Delta\phi = 2\pi\nu\Delta\tau_g = \frac{2\pi b \cos\theta}{\lambda}\Delta\theta \equiv \frac{2\pi\nu b \cos\theta}{c}\Delta\theta. \tag{9.24}$$

In Chapter 10 we discuss imaging with multi-element interferometers; at a specific frequency, high fidelity imaging is confined to a region in which the phase difference $\Delta\phi$ across the field is small, for example much less than a radian. Imaging centred on other points is possible if a different delay is introduced, and the imaging process is repeated for each part of the map. The problems of wide-field imaging are discussed further in Chapter 11.

## 9.3 The Basis of Position Measurements

Measurements of interferometer delay and phase can be exploited to determine the geometry of the baseline vector and the celestial position of a radio source with exquisite precision. Here we outline the basic ideas regarding interferometric position measurement under the assumption that the baseline geometry is perfectly known.

If the delay compensation is precise then one can, for the purposes of this discussion, treat the centre of the band $\nu_0$ as the monochromatic frequency. The change $\Delta\phi$ in the relative phase for a positional shift $\Delta\theta$ transverse to the fringe pattern is given by Eqn 9.24. When the source direction s is perpendicular to the baseline vector **b** (i.e. at $\theta = 0°$) it can readily

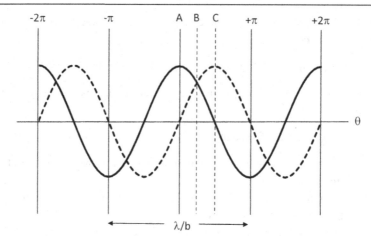

Figure 9.7. Cosine and sine responses for a point source close to transit at three positions, A, B, and C.

be seen that a sideways shift by one fringe cycle, i.e. $\Delta\theta = \lambda/b$, corresponds to a relative phase shift of one RF cycle, $\Delta\phi = 2\pi$, thereby moving the correlated interferometer output through one cycle. Equivalently Eqn 9.24 shows that a change in the time delay $\tau_g - \tau_i$ by one RF cycle, i.e. by one wavelength of light travel time, changes the position of the fringe pattern in the sky by one fringe cycle. While most easily seen around $\theta - 0°$ the $1:1$ correspondence of cycles between transverse shifts (fringes) and longitudinal shifts (time delay and RF phase) applies to the behaviour of the interferometer in the region of sky close to any angle.[8]

To extract the requisite phase information from a given baseline the signals are correlated twice, in parallel, both with the nominal compensating delay $\tau_i$ and also with $\tau_i + \lambda/4c$. The extra quarter wavelength of time delay shifts the second set of fringes transversely by a quarter cycle with respect to the other. Together they constitute the so-called 'cosine' and 'sine' fringes, and they also play a vital role in interferometric imaging as introduced in the next section and treated in detail in Chapter 10.

Figure 9.7 illustrates the essential principle of position determination for the same East–West equatorial configuration as described at the start of Section 9.2. A point source close to transit is at three different positions, A, B, C, relative to the cosine (solid line) and sine (broken line) fringes. The spacing of the fringes is set by the baseline length in wavelengths, while their exact position on the sky is determined by the setting of the compensating delay with respect to the geometric delay, $\tau_g - \tau_i$. The source can have any displacement relative to the cosine and sine fringe patterns but, for the chosen (simplest) geometry, only shifts perpendicular to the direction of the fringes (i.e. in Right Ascension) matter; shifts in the orthogonal direction (in Declination) are parallel to the fringes and hence do not affect the phase.

---

[8] This is easily seen by substituting $\Delta\phi = 2\pi$ in Eqn 9.24, yielding $\Delta\theta - \lambda/b \cos\theta$; since $b \cos\theta$ is the projected baseline, the shift is equal to the local fringe spacing.

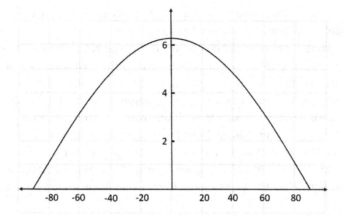

Figure 9.8. Differential phase in radians, for an interferometer with an East–West equatorial baseline tracking a source on the celestial equator from horizon to horizon ($\pm 90°$).

The amplitudes of the outputs from the two correlators are $R_{\cos}$ and $R_{\sin}$ and the arbitrary displacement of the source with respect to the two fringe patterns changes the magnitude of the correlations. The ratio defines the phase difference $\Delta\phi$ with respect to the fringe pattern via

$$\Delta\phi = \tan^{-1}(R_{\sin}/R_{\cos}) \quad \text{radians.} \tag{9.25}$$

Thus, for example:

- at position A, $\Delta\phi = 0$, the source is exactly where it was expected *a priori* and hence lies on the peak of a cosine fringe and at the zero of a sine fringe;
- at position B, $\Delta\phi = \pi/4$, the source is one-eight of a fringe spacing away from the expected position;
- at position C, $\Delta\phi = \pi/2$, the source is one-quarter of a fringe spacing away from the expected position.

This simplified analysis assumes that the initial position of the source is known to within a fringe period. Usually this is not the case, and a single phase measurement will be ambiguous since the same phase can be obtained for a transverse shift of an integer number of fringes ($n \times 2\pi$). The solution is to track the source across the sky. The projected baseline and hence the fringe spacing varies; therefore $\Delta\phi$, which is expressed in terms of fringe periods, varies continuously.

This is illustrated in Figure 9.8, which shows the variation of $\Delta\phi$ for a point source on the celestial equator (Declination $0°$) tracked from horizon to horizon with an East–West equatorial baseline of length $10\,000\lambda$. The actual position of the source is one minimum fringe spacing (0.0001 radians or 20.6 arcsec) in Right Ascension from its expected position. The peak phase error of $2\pi$ or $360°$ occurs at meridian transit; it decreases before and after, where the baseline is foreshortened and the fringe spacing is larger. In the general case of a source not on the celestial equator and a non-EW baseline, the plot of $\Delta\phi$ against time is part of a cosinusoid, which is complete if the source is circumpolar. A fit to this cosinusoid enables the source position to be calculated.

For strong sources a variation in $\Delta\phi$ of $< 10°$ might well be detectable; this corresponds to less than one- thirty-sixth of the minimum fringe spacing and hence, for point sources, astrometric measurements can be carried out on angular scales much smaller than the formal resolving power. Note that, since $\Delta\phi$ is expressed in terms of fringe spacing, longer baselines are more sensitive to source position errors; the fringe period $\lambda/b$ decreases and hence $\Delta\phi$ becomes proportionately larger. To give an extreme example, on a VLBI baseline of 100 M$\lambda$ a variation in $\Delta\phi$ of $10°$ corresponds to a position difference of $\sim 60$ microarcsec. It is therefore not surprising that VLBI measurements of quasars are the primary standards for the International Celestial Reference Frame for astronomical positions.

There is a major caveat to all this discussion, however. We have assumed that the baseline vector **b** is known precisely. If it is not then errors in **b** will appear in the plot of phase variations and will corrupt the determination of the astrometric position. In practice, therefore, both factors need to be taken into account in any solution for the baseline–source position geometry. In the early years of radio interferometry the short baselines employed could be measured with traditional surveying techniques and the positions of a suite of the strongest sources determined directly. Since then progress has been step by step, using sources to determine baselines and vice versa. We will return to these points in the discussion of VLBI in Chapter 11, where we summarize the current state of the art in astrometry and geodesy.

## 9.4 Dealing with Finite Source Sizes

Michelson's stellar interferometer, an adding system, was used to measure the angular diameters of stars by observing the diminution of the visibility of interference fringes as the two mirrors of his interferometer were moved further apart. He recognized that there is a Fourier transform relation between the fall of fringe visibility with interferometer spacing and the size and brightness distribution across the stellar source, but he was only able to observe the intensity of the fringes (via the eye or photography) rather than both their amplitude and phase. In this section we examine the response of a two-element radio interferometer to a source of finite size and develop the Fourier relationship involving both the amplitude and the phase of the response. At this stage we assume that the source is sufficiently compact that the small-angle approximation applies.

### 9.4.1 The Situation in One Dimension

It is conceptually easiest to start with the situation in one dimension, building on the concepts already established in this chapter. Figure 9.9 shows a cut through an asymmetrical source brightness distribution $B(\theta)$ and a set of uniformly spaced cosine and sine fringes. Note that at this stage we are setting aside the modulation of the fringe pattern arising from

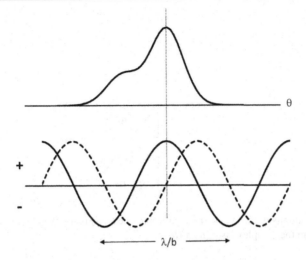

Figure 9.9. Finding a single Fourier component of a brightness distribution. The output of an interferometer is the product of the source brightness distribution and the fringe pattern (sine or cosine).

the antenna primary beam and the delay beam, which is acceptable if the source is compact. The dotted vertical line indicates the position of a fiducial point in the source, here chosen to be the brightest point. The maximum of the cosine fringe pattern is arbitrarily chosen to lie at the same point. In practice this would mean that the source–baseline geometry is perfectly known and the compensating delay has been correctly set for that specific sky position.

The action of a correlation interferometer is to multiply the brightness distribution by one of the two fringe patterns; this forms a complex gain pattern which is equivalent to a summation of the responses to a closely spaced set of point sources, as in the analysis in Section 9.1.4. The correlated output of the cosine channel for a baseline $b_\lambda$ is

$$R_{\cos}(b_\lambda) \propto \int_{\text{source}} B(\theta) \cos(2\pi b_\lambda \theta)\, d\theta, \qquad (9.26)$$

where $b_\lambda = b/\lambda$ is the number of fringe cycles per radian; thus $b_\lambda \theta$ corresponds to a fraction of a fringe period. Similarly, for the sine channel,

$$R_{\sin}(b_\lambda) \propto \int_{\text{source}} B(\theta) \sin(2\pi b_\lambda \theta)\, d\theta. \qquad (9.27)$$

These equations are evidently related to the Fourier cosine and sine transforms of $B(\theta)$ respectively. The integration of the cosine pattern over the source picks out the even component of the brightness distribution; the sine pattern picks out the odd component. Both channels are needed since in general $B(\theta)$ is asymmetrical. It is also evident that as the fringe period becomes smaller with respect to the source the response from the positive-going and negative-going parts of the fringes will increasingly tend to cancel each other out; we say that the source becomes resolved.

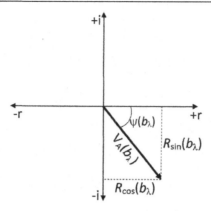

Figure 9.10. The outputs of the cosine (real axis) and sine (imaginary axis) correlators on a baseline $\mathbf{b}_\lambda$ combine to form the complex visibility $V(b)$.

For a given baseline $\mathbf{b}_\lambda$ the correlator outputs, each of which is a real number, can be combined into a single complex Fourier component defined by $V(b_\lambda) = R_{\cos}(b_\lambda) - iR_{\sin}(b_\lambda)$, the choice of sign being arbitrary but conventional. This is simply an explicit case of the algebra of complex numbers and can be illustrated by an Argand diagram (Figure 9.10).

This definition of $V(b_\lambda)$ enables us to write a single equation,

$$V(b_\lambda) \propto \int_{\text{source}} B(\theta)e^{-2\pi i b_\lambda \theta}\, d\theta, \tag{9.28}$$

rather than two, as in Eqns 9.26 and 9.27 above (see Appendix 1). By an extension of Michelson's terminology, $V(b_\lambda)$ is called the *complex fringe visibility* $V(b_\lambda) \equiv V_A(b_\lambda)e^{-i\psi(b_\lambda)}$. Its magnitude is [9]

$$V_A(b_\lambda) = [R_{\cos}(b_\lambda)^2 + R_{\sin}(b_\lambda)^2]^{1/2}, \tag{9.29}$$

and its phase is

$$\psi(b_\lambda) = \tan^{-1}\frac{R_{\sin}(b_\lambda)}{R_{\cos}(b_\lambda)}. \tag{9.30}$$

The *visibility phase* is the shift of the Fourier component $V(b)_\lambda$ with respect to the assumed position of the source, as discussed in Section 9.3. It is measured in terms of the fringe spacing corresponding to that Fourier component: one fringe period $= 2\pi$ radians. The component $V(b_\lambda)$ is necessarily complex since Fourier components with different spatial shifts are needed to describe an arbitrary brightness distribution $B(\theta)$, which in general will be asymmetrical.

Figure 9.11 gives an illustration of these arguments. The target source C has a rectangular positive (real) brightness distribution which is spatially offset from the *a priori* assumed sky position. This asymmetric source can be constructed by adding together two other

---

[9] Note that this is the *amplitude* of a Fourier component describing an *intensity* distribution.

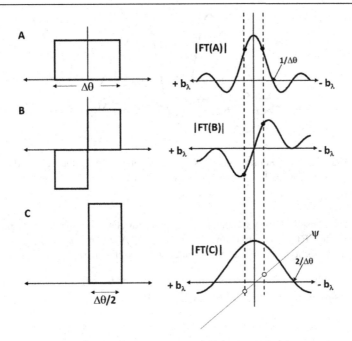

Figure 9.11. Symmetric (A) and antisymmetric (B) functions and the amplitudes of their Fourier transforms and the asymmetric function (C) formed from A + B, with the amplitude and phase of its Fourier transform.

real functions: a half-height, double-width rectangular source A centred on the assumed position (and hence symmetric) and an antisymmetric double-width source B also centred on the origin but with one half negative and the other positive. The function B is called the doublet pulse and is the convolution of a rectangle with positive- and negative-going delta functions. The output of the cosine correlator $R_{cos}(b_\lambda)$ responds only to the perfectly symmetric function A and, for any given baseline, provides a point on the amplitude of its Fourier transform, $|FT(A)|$, which is a sinc function. The output of the sine correlator $R_{sin}(b_\lambda)$ responds only to the perfectly antisymmetric function B and, from the convolution theorem, $|FT(B)|$ is antisymmetric, being the multiplication of a sinc function and a sine wave. The Fourier amplitude of C is $[|FT(A)|^2 + |FT(B)|^2]^{1/2}$, which is another sinc function. The asymmetry of C is captured by its Fourier phase $\psi(b_\lambda) = \tan^{-1}[|FT(B)|/|FT(A)|]$, which is a linearly increasing function of baseline length. This latter characteristic (the shift theorem in Fourier formalism) is simply a result of the spatial shift of the source from the origin, which corresponds to an increasing fraction of the fringe spacing as the baseline length increases (see Section 10.8 and Appendix 1). Notice the following.

– For a baseline $b_{\lambda,i}$ the output of the cosine channel $R_{cos}(b_{\lambda,i})$ is a particular point (indicated by a solid circle) on $|FT(A)|$. If the parity of the baseline is reversed, on taking the other antenna as the reference, $b_{\lambda,i}$ becomes $-b_{\lambda,i}$ but $R_{cos}(-b_{\lambda,i})$ does not change since since $|FT(A)|$ is symmetrical.

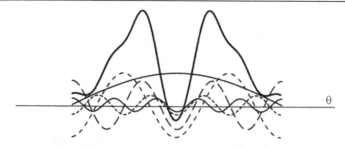

Figure 9.12. The essence of Fourier synthesis in one dimension. A double source (bold solid line) constructed from five complex Fourier components.

– On reversing the baseline parity the output of the sine correlator has the same magnitude but the opposite sign since |FT(B)| is antisymmetric, i.e $R_{\sin}(-b_{\lambda,i}) = -R_{\sin}(b_{\lambda,i})$.
– The sign of the visibility phase reverses on changing the baseline parity.

This overall behaviour is characteristic of the Fourier transforms of real functions and we will discuss its implications further in Chapter 10.

### 9.4.2 The Essence of Fourier Synthesis

Interferometer arrays provide the means to synthesize high resolution images of radio sources, and the description of how this is achieved forms the basis of most of the remainder of this chapter and of Chapter 10. But first it is useful to see how a simple image can be made from a limited number of Fourier components.

Figure 9.12 shows a one-dimensional example. Five complex Fourier components with different angular (or spatial) frequencies corresponding to five different baselines are added together to synthesize an image of a double source. The shorter baselines provide the lower-angular-frequency components, the longer baselines the higher ones. The amplitude of the Fourier component has an obvious interpretation whilst its phase encodes the shift of that component relative to an assumed position (see the previous section); note that here the components have different phase shifts. Although the amplitudes and relative phases in Figure 9.12 are perfectly known, the limited number of components results in an imperfect image with unwanted fluctuations and a negative region. Since the sky brightness is always positive, the negative region in the middle is non-physical. The fluctuations arise because not all the Fourier components required for a high fidelity reconstruction have been collected. Negative regions arise due to the particular inability of a correlation interferometer to measure the constant offset-zero baseline Fourier component;[10] the average brightness (a constant level) within the primary beam is therefore unavailable. We will touch on this latter point in the next section and return to consider all these points in detail in Chapter 10.

---

[10] Two telescopes cannot be in the same place at the same time.

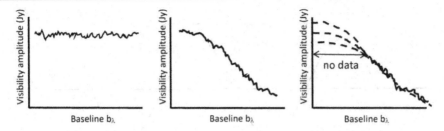

Figure 9.13. Simulated visibility amplitudes: (left) an unresolved source; (centre) a resolved source with good coverage of short baselines; (right) a resolved source which was not observed on short baselines.

### 9.4.3 Simple Sources and their Visibility Amplitudes

During the early years of interferometry the visibility phase was difficult if not impossible to measure. This was due to a combination of imperfect knowledge of the source–baseline geometry, delay drifts in the equipment, and variable propagation effects in the ionosphere and atmosphere. Radio astronomers often had to rely on interpreting the visibility amplitudes, which are largely unaffected by these issues, by fitting the Fourier amplitudes of simple analytic functions to their data. To this day VLBI observations at short mm wavelengths do not preserve phase information and so need to make some use of model fitting. We briefly outline this process in one dimension as another practical step towards understanding interferometric imaging, as developed in the rest of this chapter and in Chapter 10.

Figure 9.13 illustrates some initial points; it shows simulated visibility amplitudes as a function of baseline length (in wavelengths) for three simple sources. On the left the amplitude does not change, within the noise, as the baseline length increases and hence the source is unresolved; its angular diameter must be $\ll 1/b_{\lambda,\mathrm{max}}$ radians. In the centre the amplitude falls and the source is resolved; its angular diameter must be comparable to $1/b_{\lambda,\mathrm{max}}$. In addition, sufficient short baselines are available that a good estimate of the 'zero-baseline' flux density can be obtained by extrapolation. On the right the source is also resolved but the coverage of short baselines is not sufficient for a reliable extrapolation to establish the zero-baseline flux density.

Moving on to specific functions, the simplest is a rectangle of extent $\Delta\theta$ whose Fourier transform is a sinc function, $\sin(\pi b_\lambda \Delta\theta)/(\pi b_\lambda \Delta\theta)$. As illustrated in Figure 9.14, the first zero occurs when exactly one fringe cycle fits within the extent of the source, i.e. when $\lambda/b = \Delta\theta$, corresponding to a spatial frequency of $b_\lambda$ cycles per radian.

A physically more realistic function is a uniform disk, which is appropriate for modelling the Sun at centimetric wavelengths and shorter. The Fourier transform is $2J_1(\pi b_\lambda \Delta\theta)/(\pi b_\lambda \Delta\theta)$ and the first zero falls at $1.22/\Delta\theta$ wavelengths. As schematized in Figure 9.15 the visibility amplitude of the Sun (dotted lines) can be successively sampled (solid blocks) with a single baseline whose length is varied between observations. At frequencies $\geq 30$ GHz the Sun's angular diameter is essentially the same as at visible wavelengths, i.e. $\Delta\theta \approx 0.5° \equiv 0.0087$ radians, and so the first minimum will occur at

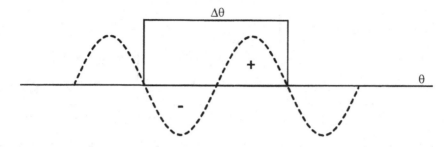

Figure 9.14. Zero visibility amplitude of a rectangular source, when one fringe cycle exactly fits the width of the source.

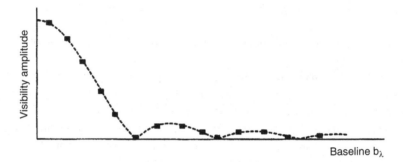

Figure 9.15. The one-dimensional visibility amplitude function of a uniform disk source.

$b \approx 140\lambda$. Note that in the complex transform the second and fourth maxima are phase reversed and hence can be drawn with negative amplitudes. When there is no knowledge of the visibility phase only positive amplitudes can be recognized.

The most useful single-source model is a Gaussian. Figure 9.16(left) shows a one-dimensional cut through a Gaussian with two size parameters marked, the FWHM $= \theta_{1/2}$ and the $1/e$ radius $r_e$ at which the amplitude has fallen to 0.37 of its maximum value; note that $r_e = \theta_{1/2}/2(\ln 2)^{1/2} \approx 0.60\theta_{1/2}$.

Uniquely, the Fourier transform of a Gaussian function is another Gaussian and, for a symmetric (circular) source, the visibility amplitude has the simple form

$$V_A(b_\lambda) = S_0 e^{-(\pi b_\lambda r_e)^2}, \qquad (9.31)$$

where $S_0$ is the zero-baseline flux density. Figure 9.16(right) shows the visibility amplitude as a function of baseline. Notice that, because a Gaussian does not have a distinct edge, the visibility amplitude decreases monotonically as the baseline increases; this contrasts with the visibility function of a source with a defined edge, such as a uniform disk, which exhibits well-defined minima. Table 9.1 shows how $V_A(b_\lambda)/S_0$ falls as $\theta_{1/2}$ increases with respect to the fringe spacing $\lambda/b = 1/b_\lambda$. A Gaussian source is 10% resolved when

Table 9.1. *Resolution of a*
*Gaussian source*

| $V_A(b_\lambda)/S_0$ | $\theta_{1/2}/b_\lambda$ |
|---|---|
| 0.9 | 0.17 |
| 0.5 | 0.44 |
| 0.1 | 0.80 |
| 0.029 | 1.0 |

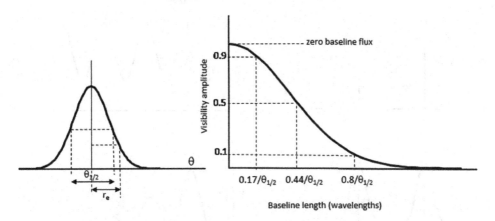

Figure 9.16. The one-dimensional visibility amplitude function of a cut through a Gaussian source (shown on the left). On the right, the visibility amplitude as a function of baseline.

$\theta_{1/2}$ is 17% of the fringe spacing and 90% resolved when $\theta_{1/2}$ is 80% of the fringe spacing; when $\theta_{1/2}$ is equal to the fringe spacing a Gaussian source is almost maximally resolved.

Double sources are usually modelled as pairs of points or Gaussians. A few representative cases are shown in Figure 9.17. The visibility amplitude function of equal point double sources has the form $[1 + \cos(2\pi b_\lambda \theta)]$ and the first minimum occurs when the fringe spacing is equal to half the separation of the double source,[11] i.e. when $b_\lambda = 1/(2 \Delta \theta)$; at this point the contributions from the two components cancel. The components add when the fringe spacing is equal to the separation of the double source and the pattern then repeats for integer multiples of these values. Unequal point doubles show less variation in visibility amplitude, with the maximum equal to the sum of the components and the minimum equal to their difference. Double Gaussians can be understood with the aid of the convolution theorem. The source is the convolution of two points with a single Gaussian; thus its Fourier transform is the product of the transform of the double and that of a Gaussian (see the Supplementary Material).

---

[11] Note the factor 2 difference from the rectangular source.

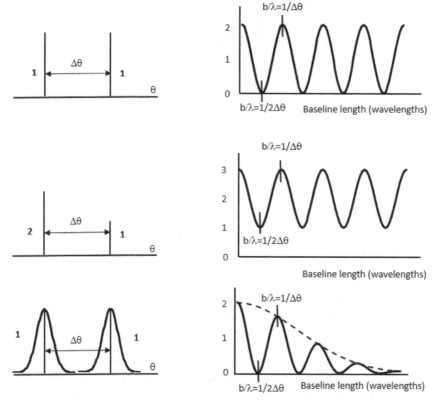

Figure 9.17. (left) Simple double sources and (right) their visibility amplitude functions.

Whatever the model, and two-dimensional variations on the above can easily be constructed, the process is the same. By varying the parameters of the model the difference between the predicted visibility amplitude and the measured values is minimized to give the best fit.

## 9.5 Interferometry in Two Dimensions

Having outlined the basic principles in one dimension we now move on to describe the first practical situation in which a two-element interferometer observes a patch of sky with a finite area. We proceed in a stepwise fashion as follows:

(i)   the generalized response is derived in coordinate-free form;
(ii)  coordinate systems are chosen for the baseline and for the sky directions;
(iii) some highly specific limitations are imposed on the geometry and an exact two-dimensional Fourier transform relationship between the complex visibility and the source brightness distribution is derived.

In Chapter 10 we will relax the geometric limitations and examine the circumstances under which the two-dimensional Fourier transform relationship remains an adequate approximation. We will also consider the additional complications arising for antenna arrays with non-coplanar baselines observing wide fields of view.

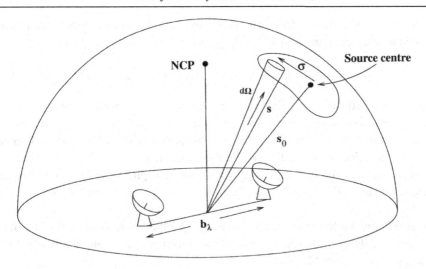

Figure 9.18. The contribution of a small receiving element in the direction **s**, solid angle d$\Omega$, to the response of an interferometer; NCP is the North Celestial Pole.

### 9.5.1 Coordinate-Free Description

To establish a coordinate-free description we might start from the response to a single point source (Eqn 9.16) and add the responses[12] from many point sources each with a different flux density and direction and hence with a different phase term $\exp(-i2\pi \mathbf{b}_\lambda \cdot \mathbf{s})$. However, for an extended source it is more logical to describe the situation in terms of a continuous distribution of specific sky brightness, $B_\nu(\mathbf{s})$. The geometry is illustrated in Figure 9.18, which shows the radiation received from a small element of an extended source subtending a solid angle d$\Omega$ in the direction **s**.

A direction $\mathbf{s}_0$ (invariably the direction of maximum antenna gain) is chosen as the phase tracking centre and, following the definition in Eqn 9.14, the vector from the phase tracking centre to the source element is $\boldsymbol{\sigma}$. The cross-spectrum power density produced at a frequency $\nu_0$ by this element of the source will have the same form as Eqn 9.23, but with the source flux $S(\nu_0)$ replaced by the radiating source element $B_{\nu_0}(\mathbf{s}_0 + \boldsymbol{\sigma})$; to maintain generality, the delay beam effects are explicitly included. The cross-spectrum power density is then

$$S_{xy}(\nu_0, \mathbf{s}_0 + \boldsymbol{\sigma}) = A(\mathbf{s}_0 + \boldsymbol{\sigma}) \, \text{sinc}(\Delta \nu \, \tau_g) B_{\nu_0}(\mathbf{s}_0 + \boldsymbol{\sigma}) \, \exp[-i2\pi \nu_0 (\tau_g - \tau_i)] \, d\Omega. \quad (9.32)$$

For the purposes of this discussion we will seek the correlator output in units of relative power density rather than its absolute value (invariably the total flux scale is calibrated by reference to a standard point source, and this process automatically takes into account the antenna areas and the gains of the receivers). The *relative* antenna area $A(\boldsymbol{\sigma})$, incorporating the delay beam term, has value unity in the direction $\mathbf{s}_0$ and is therefore used to

---

[12] Because the radiation from celestial sources is incoherent random noise, the intensities add and the system may be said to be linear in intensity; this immediately suggests that Fourier theory may be applicable.

express Eqn 9.32 in a simpler form. The total correlator output is the cross-spectral density integrated over the entire radio sky:

$$S_{xy}(\nu_0, \boldsymbol{\sigma}) = \int_{4\pi} \mathcal{A}(\boldsymbol{\sigma})B_{\nu_0}(\boldsymbol{\sigma}) \exp[-i2\pi\nu_0(\tau_g - \tau_i)]\,d\Omega. \tag{9.33}$$

In practice $\mathcal{A}(\boldsymbol{\sigma})$ effectively restricts the integration to a much smaller solid angle.

At this point, several approximations are in order: the overall size of the patch of sky will be assumed to be small compared with the response pattern of the delay beam, and hence its effects can be neglected. A corollary is that the centre frequency of the receiver bandpass, $\nu_0$, can be taken as the defining frequency. In many applications the single-frequency approximation is good enough provided that the width $\Delta\theta$ of the patch is consistent with Eqn 9.18.

For simplicity, we let $\nu$ (without the subscript) be the centre frequency and write the geometric time delay explicitly in terms of the direction vectors and the baseline $\mathbf{b}_\lambda$ in wavelengths, as in Eqn 9.12; this yields

$$S_{xy}(\mathbf{s_0}) = \int \mathcal{A}(\boldsymbol{\sigma})B_\nu(\boldsymbol{\sigma}) \exp\{-i2\pi[\mathbf{b}_\lambda \cdot (\mathbf{s_0} + \boldsymbol{\sigma}) - \nu\tau_i]\}\,d\Omega. \tag{9.34}$$

The geometric phase term $\exp\{-i2\pi[\mathbf{b}_\lambda \cdot \mathbf{s_0}]\}$ relates to the phase tracking centre and is compensated by the instrumental delay term $\exp(i2\pi\nu\tau_i)$. Equation 9.34 therefore simplifies to

$$V(\mathbf{b}_\lambda) = \int \mathcal{A}(\boldsymbol{\sigma})B_\nu(\boldsymbol{\sigma}) \exp(-i2\pi\mathbf{b}_\lambda \cdot \boldsymbol{\sigma})\,d\Omega, \tag{9.35}$$

and thereby becomes a fuller definition of the *complex visibility* $V(\mathbf{b}_\lambda)$, introduced in Section 9.4.1. Recalling the definition of flux density in Eqn 5.12, it can be seen that the complex visibility has units of flux density and is often referred to as the *correlated flux density*. As it is the intensity collected via a fringe pattern, the complex visibility is the complement of the effective flux density received by a single antenna.

Equation 9.35 can be regarded as the fundamental equation for a practical interferometer.[13] The principal observables are the amplitude and phase of $V(\mathbf{b}_\lambda)$, which are obtained from the quadrature correlations for that value of $\mathbf{b}_\lambda$, the phase being referred to the tracking centre $\mathbf{s_0}$. While Eqn 9.35 is related to the Fourier transform of the specific brightness distribution $B_\nu(\boldsymbol{\sigma})$, further restrictions on the geometry must be imposed in order to arrive at the actual two-dimensional transform relationship.

### 9.5.2 The u, v Plane and the 2D Fourier Transform

We now need to write Eqn 9.35 in terms of practical coordinate systems both for the baseline vector $\mathbf{b}_\lambda$ and for the sky offset vector $\boldsymbol{\sigma}$. The three orthogonal components of $\mathbf{b}_\lambda$ are conventionally expressed in a right-handed rectilinear coordinate system as $u, v, w$, as illustrated in Figure 9.19, all coordinate distances being expressed in wavelengths.

---

[13] Some authors take the relative gain term $\mathcal{A}(\boldsymbol{\sigma})$ outside the integral and define the complex visibility purely in terms of an integral over $B_\nu(\boldsymbol{\sigma})$.

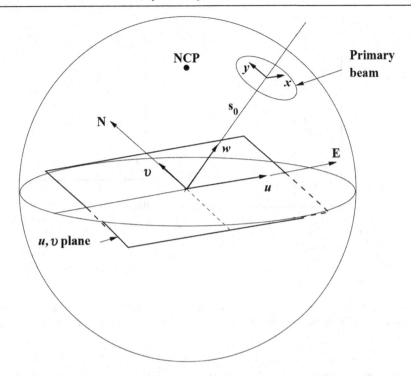

Figure 9.19. The geometrical relationship between an interferometer, a celestial source, and the $u, v$ plane, seen in relation to the celestial sphere. The components of the offset $\sigma$ with $x$ parallel to $u$ are $x, y$; NCP is the North Celestial Pole.

Figure 9.19 also shows the source plane (part of the celestial sphere), with coordinates that can be expressed as celestial angular coordinates $x, y$ in the small-angle approximation (see later in this section).

The analysis is simplified if the unit vector in the delay tracking direction $s_0$ defines the direction of $w$ while, in the perpendicular plane, $u$ and $v$ are in the projected easterly (Right Ascension) and northerly (Declination) directions. This plane, perpendicular to the source direction, is widely known as the $u, v$ plane. It is this plane in which the Fourier transform of the source brightness distribution is measured. The $w$ axis always points towards the phase tracking centre, defined by $s_0$, as it moves throughout the observation; $w$ is therefore the geometric delay expressed in wavelengths $w = c\tau_g/\lambda$.

The $u, v$ plane is best understood from the point of view from the source, looking down at the Earth. From this viewpoint one can naturally see that we need the projection of the baseline onto this orthogonal plane. The projected baseline appears as a line which can be decomposed into a component parallel to the equator at the point nearest the source (the $u$ component) and a component along the line between this point and the North Pole (the $v$ component.). These projections change as the Earth rotates. We will come back to these issues in Chapter 10 but here we restrict our analysis to a single instant of time.

In order to deal with the wide fields of view to be considered in Chapter 10, it is necessary to describe the direction of the offset vector $\sigma$ over the entire celestial sphere. For this

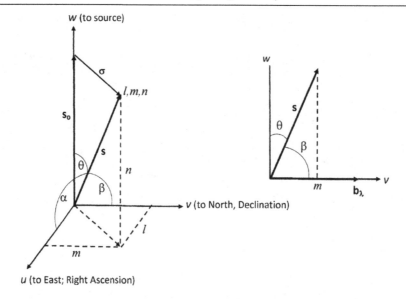

Figure 9.20. (left) The direction cosines $l, m, n$ of a unit vector $\mathbf{s}$ with respect to the $u, v, w$ coordinate system used to specify the baseline vector (not shown). (right) A specific 1D case, where the baseline vector is shown.

the *direction cosine* formalism of three-dimensional (3D) geometry is appropriate.[14] The direction cosines $l, m, n$ of a 3D unit vector are the projections $l = \cos\alpha$, $m = \cos\beta$, $n = \cos\gamma$ onto a set of coordinate axes. For a unit vector, $n$ is not independent since $\cos^2\alpha + \cos^2\beta + \cos^2\gamma = l^2 + m^2 + n^2 = 1$; hence

$$n = \sqrt{1 - l^2 - m^2}. \tag{9.36}$$

The coordinate axes for $l, m, n$ are chosen to be aligned with $u, v, w$ and the geometry is then as shown on the left-hand side of Figure 9.20; note that the $l, m, n$ coordinates refer to the spatial domain while the $u, v, w$ coordinates refer to the spatial frequency domain. On the right-hand side of Figure 9.20 a particular 1D situation is shown in which $\mathbf{s}$ lies in the $v, w$ plane and the baseline $\mathbf{b}_\lambda$ lies along $v$. In this simpler situation the geometrical delay expressed in wavelengths is $\mathbf{b}_\lambda \cdot \mathbf{s} = v \cos\beta = vm$; in three dimensions the delay is given by

$$\mathbf{b}_\lambda \cdot \mathbf{s} = ul + vm + wn. \tag{9.37}$$

This direct link with the delay is the essential reason why direction cosines are used for interferometer formalism.[15] Since $\sigma = \mathbf{s} - \mathbf{s}_0$ and $\mathbf{s}_0$ is a unit vector, by definition we have

$$\sigma = l, m, \sqrt{1 - l^2 - m^2} - 1, \tag{9.38}$$

---

[14] In 3D geometry it is the direction cosines which naturally occur rather than the direction angles themselves. For analysing the response of a single dish it is more convenient, and sufficient, to use the direction angles $(\theta, \phi)$ defined locally with respect to the beam maximum.

[15] Note that $m$ can be defined as $\sin\theta$, as in our introductory discussion in the first part of this chapter; when the angles are small, $m = \sin\theta = \theta$.

and the argument in the exponential term in Eqn 9.35 becomes

$$2\pi \mathbf{b}_\lambda \cdot \boldsymbol{\sigma} = 2\pi[ul + vm + (\sqrt{1 - l^2 - m^2} - 1)w]. \tag{9.39}$$

We will return to the general case and its ramifications in Chapter 10. However, our understanding of the interferometer response is aided by considering first a particular situation in which the baseline vector is restricted to lie in the $u, v$ plane and hence $w = 0$. This corresponds to the situation in Figures 9.18 and 9.19 when the source is instantaneously perpendicular to the baseline and hence the relative delay for the reference direction $s_0$ is zero. Equation 9.39 then becomes

$$2\pi \mathbf{b}_\lambda \cdot \boldsymbol{\sigma} = 2\pi(ul + vm). \tag{9.40}$$

As illustrated in one dimension in Figure 9.21, the $l, m$ plane containing the offset vector $\boldsymbol{\sigma}$ is tangent to the celestial sphere in the direction of $s_0$ (the phase centre at $l = 0$, $m = 0$). By definition the $l, m$ and $u, v$ planes are parallel. In the 1D case illustrated the point P, in the direction of $\mathbf{s}$, is determined by the direction cosine $\cos \beta = \sin \theta = m$; in the 2D case it becomes $l, m$ and the corresponding delay is $ul + vm$ wavelengths. The element of solid angle $d\Omega$ (shown greatly exaggerated in Figure 9.21) in the direction $\mathbf{s}$ projects onto an element of area $dldm$ in the $l, m$ plane:

$$d\Omega = \frac{dldm}{\cos \gamma} = \frac{dldm}{\sqrt{1 - l^2 - m^2}}, \tag{9.41}$$

where $\theta$ is the angle between $s_0$ and $\mathbf{s}$ in Figure 9.21. Equation 9.35 therefore becomes

$$V(u, v) = \frac{1}{\sqrt{1 - l^2 - m^2}} \int \int A(l, m) B_\nu(l, m) e^{-i2\pi(ul+vm)} dldm, \tag{9.42}$$

and now $V(u, v)$ is an exact 2D Fourier transform of a modified source brightness:

$$V(u, v) \overset{\text{FT}}{\longleftrightarrow} \frac{A(l, m) B_\nu(l, m)}{\sqrt{1 - l^2 - m^2}}. \tag{9.43}$$

Given the restriction $w = 0$, this expression is valid for all directions in the sky. Notice again that, because of the direct link with phase delay, directions in the brightness distribution and the relative antenna area are parametrized in terms of the cosine (or sine) of the angles, *not* in terms of the direction angles themselves. The direction cosines automatically account for the change in fringe spacing with the angle from $s_0$, as discussed in Section 9.1.3. When expressed in terms of celestial coordinates, i.e. the Right Ascension and Declination of the pointing centre $(\alpha_0, \delta_0)$ and of a point in the field $(\alpha, \delta)$, the direction cosines can be written (e.g. Myers *et al.* 2003) as $l = \cos \delta \sin(\alpha - \alpha_0), m = \sin \delta \cos \delta_0 - \cos \delta \sin \delta_0 \cos(\alpha - \alpha_0)$. The direction cosines $l, m$ can then be interpreted as Cartesian coordinates in the tangent plane in Figure 9.21. In practice this geometry is taken care of in the data reduction software.

In many cases of interest, however, the required field of view is sufficiently compact that several other simplifications are valid which make the above Fourier transform relationship easier to appreciate and to use.

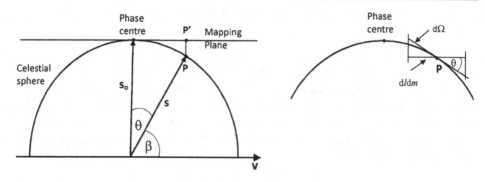

Figure 9.21. (left) A 1D projection of the $l, m$ mapping plane which is tangent to the celestial sphere in the direction of $s_0$. (right) The element of solid angle $d\Omega$ in the direction $s$ (represented in two dimensions by the line tangential to the celestial sphere at P) projects into an element of area $dl dm$ in the mapping plane $l, m$.

– The small-angle approximation applies and so in Eqn 9.42 the direction cosines can actually be replaced by the angular offset coordinates $x$ and $y$ (measured from $s_0$) as illustrated in Figure 9.19, where $x$ is parallel to $u$.
– The term $\sqrt{1 - l^2 - m^2} = \cos \gamma$ tends to unity as $\gamma$ becomes close to zero.
– The antenna beam size is large compared with the field of view and, if the radio telescopes are pointed well, the variation of gain with angle will be negligible. In this case the relative gain $A(l, m)$ can be set to unity.

With these simplifications, which commonly apply, Eqn 9.42 becomes

$$V(u, v) \approx \int B(x, y) \exp[-i2\pi (ux + vy)] \, dx dy \tag{9.44}$$

or

$$V(u, v) \xleftrightarrow{\text{FT}} B(x, y). \tag{9.45}$$

Therefore, a single interferometer observation, in which both the amplitude and the phase of the complex visibility are measured, evaluates the Fourier transform of the source brightness distribution for a particular value of the spatial frequency, that of the baseline vector, projected onto the $u, v$ plane. This is illustrated in Figure 9.22. At this point we have arrived at a situation which is the 2D equivalent of Figure 9.12.

A single interferometer observation provides only one complex number, but an assemblage of observations will sample the *visibility function* $V(u, v)$, which can then be inverted to yield the source brightness distribution $B(x, y)$. This was illustrated in one dimension in Figure 9.12. The ways in which this can be done in practice are explored further in Chapter 10.

The small-angle approximation may be good enough for many observations of individual sources carried out with the large aperture-synthesis arrays in use today. However, survey observations are increasingly requiring wide fields and every observer should be aware of the approximations that lead to the simple form given in Eqns 9.44 and 9.45. We will consider some of the relevant issues in Chapters 10 and 11.

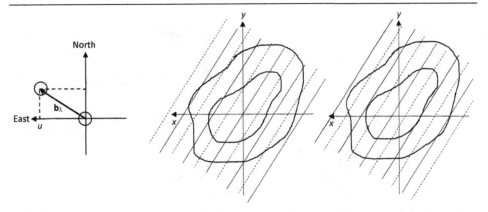

Figure 9.22. (Left) The projected components $u$, $v$ of a baseline $\mathbf{b}_\lambda$; (centre) cosine fringes with spacing $1/|\mathbf{b}_\lambda|$ radians in position angle $\tan^{-1} u/v$ across the face of a source; (right) the corresponding sine fringes.

## 9.6 Coherence

Thus far we have largely described the behaviour of interferometers in terms of the monochromatic approximation, although the effect of broad receiver bandwidths was analysed in Section 9.2. In this section we recognize explicitly that an interferometer correlates two noise signals and that its action is quantitatively to measure the degree of similarity between them. Since we know that the self-noise generated by one receiver bears no relationship to that from the other, an interferometer can be seen to be an instrument whose function is to measure the degree of coherence (or equivalently correlation) between the noise-like signals *entering* each antenna. The travelling wavefronts striking the antennas are fluctuating rapidly in time but the time-averaged correlations between receiver voltages, generated from different parts of the radiation field, yield statistically stable results. For our present purpose we shall ignore the statistical correlations between the receiver-generated noises (see Section 9.8.1).

Figure 9.23 illustrates the concept of temporal coherence. Wavefronts from a single source arrive at two different points $x_1$ and $x_2$, corresponding to antennas A and B, where they generate voltages $V(x_1, t)$ and $V(x_2, t)$. Note that in this schematic illustration the noise envelope is drawn as if it were already band-limited on arrival at the antenna whereas the band-limited voltages are actually defined in the receivers. The averaged autocorrelation in time *at either antenna* $\langle V(x_1,t)V^*(x_1,t+\tau)\rangle$ or $\langle V(x_2,t)V^*(x_2,t+\tau)\rangle$ reveals the statistical similarity of the band-limited noise to a delayed copy of itself and measures the degree of temporal (or longitudinal) coherence. The Fourier transform of this coherence function is the power spectrum of the radiation field; this is the Wiener–Khinchin theorem (see Appendix 1). This thinking also links up with the concept of the coherence time $\tau_{\mathrm{coh}}$ and the coherence length $c\tau_{\mathrm{coh}}$ illustrated in the diagram. The intensity of the radiation field $I(x, t) = \langle V(x, t)V^*(x, t)\rangle$ is obtained when $\tau = 0$; it does not vary with position as long as the source is in the far field and there are no differential propagation effects. The *degree of*

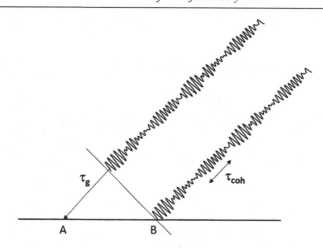

Figure 9.23. Temporal coherence.

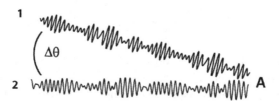

Figure 9.24. Spatial coherence I. Incoherent noise signals from sources separated in angle.

*temporal coherence* at a particular point in space,

$$\gamma(\tau) = \frac{\langle V(t)V^*(t+\tau)\rangle}{I(t)} \tag{9.46}$$

characterizes the nature of the signal. A monochromatic signal gives $\gamma = 1$ for all $\tau$; pure random noise corresponds to $\gamma = 0$ except at $\tau = 0$. Manmade signals are partially coherent while astronomical sources of spectral lines, particularly masers whose linewidth can be as small as a few hundred Hz, can also be termed partially coherent.

   An interferometer compares the noise signals at two points across a wavefront and hence is sensitive to their spatial (or lateral) correlation. Figure 9.24 illustrates the concept of spatial coherence. Incoherent noise signals from distant sources 1 and 2 separated in angle by $\Delta\theta$ (here greatly exaggerated) arrive at antenna A, where they combine to produce a resultant voltage $V(x_1, t)$. Note that, as in Figure 9.24, the noise envelope is drawn as if it were already band-limited on arrival at the antenna. At a point B spatially separated from A the signals from sources 1 and 2 will have a different relative phase shift and so the resultant voltage $V(x_2, t)$ will be different from $V(x_1, t)$. Despite this, $V(x_1, t)$ and $V(x_2, t)$ will retain a degree of similarity; they are different complex additions of the same inputs and so will be partially correlated even though the sources themselves are incoherent. If we then allow

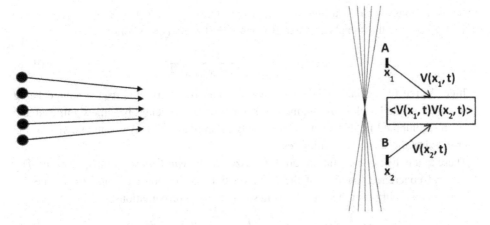

Figure 9.25. Spatial coherence II. The wavefronts from several incoherent sources (solid circles) arriving at an interferometer.

the comparison of the voltages generated at A and B (see Figure 9.23) to extend over time $\tau$ (as above), we can define a general *mutual coherence function*

$$\Gamma(x_1, x_2, \tau) = \langle V(x_1, t)V^*(x_2, t + \tau)\rangle. \tag{9.47}$$

It is important to note that spatial coherence does not depend on the absolute time $t$ at which the data are taken (as long as the source is non-variable) nor does it depend on the absolute locations of the antennas; as long as the source is in the far field, only their separation $|x_1 - x_2|$ matters.

In the interferometer the geometrical delay $\tau_g$ must be compensated in order that the same part of an incoming wavefront is compared at A and B. But the compensation can be correct only for one direction (say source 1); for source 2, in a slightly different direction, the delay compensation will be imperfect and hence the mutual coherence will fall. This reduction in temporal coherence gives rise to the 'delay beam' effect, as analysed in Section 9.2. If in Eqn 9.46 we set $\tau = 0$ then we are assuming that sources 1 and 2 are well within the delay beam, in which case Eqn 9.46 becomes the *spatial coherence function*.

Figure 9.25 complements Figure 9.24. A set of incoherent point sources (any extended source can be regarded as a sum of incoherent point sources) lies in the far field of the baseline AB of length $|b|$. Each source gives rise to plane waves striking the baseline at a different angle. As noted above, if the sources are well within the delay beam then we can ignore effects due to temporal (or longitudinal) incoherence. The plane wavefronts have no phase relationships laterally but each contributes to the combined electric field (a vector sum) seen at antenna A, at location $x_1$, and antenna B, at location $x_2$, so there will be some degree of correlation, i.e. $\langle V(x_1, t)V^*(x_2, t)\rangle$, even though they are spatially separated. When $|b|$ is small, i.e. when A and B are close to the crossing point of the waves in Figure 9.25, the differential delays are small, the combined fields at A and B are very similar, and hence their phases exhibit a high degree of correlation. The *degree of spatial coherence* $\gamma(x_1, x_2)$ between the signals at A and B when $\tau = 0$ is defined as follows:

$$\gamma(x_1, x_2) = \frac{\Gamma(x_1, x_2)}{[I(x_1, t)I(x_2, t)]^{1/2}}. \tag{9.48}$$

A high degree of spatial (or lateral) coherence can be interpreted as the source being unresolved on that baseline. As $|b|$ increases, the differential delays become significant and hence the combined fields become increasingly dissimilar. The loss in lateral coherence is interpreted as the source becoming resolved.

These qualitative considerations are formalized in the *van Cittert–Zernike theorem*. The 2D lateral coherence function of the radiation field is the Fourier transform of the 2D brightness distribution of the source; using small angle approximations,

$$\langle V(x_1, t)V^*(x_2, t)\rangle = \int\int B(x, y)e^{-i2\pi(ux+vy)}dxdy, \tag{9.49}$$

where $u = (x_1 - x_2)/\lambda$ and $v = (y_1 - y_2)/\lambda$. Equation 9.49 is in the same form as Eqn 9.44; thus the visibility function is another name for the spatial correlation function.

This section has given an elementary consideration of coherence in the context of interferometry. For an in-depth treatment the reader is referred to Thompson, Moran, and Swenson (2017, TMS).

## 9.7 Propagation Effects

As incoming radiation propagates through the Earth's atmosphere it encounters time-dependent variations in optical depth and path length delay, which are different for each antenna. The variations in absorption, with associated emission of atmospheric noise, reduce the visibility amplitude and decrease the interferometric signal-to-noise ratio. At frequencies of 15 GHz and below the tropospheric-induced amplitude and excess noise effects are usually relatively small (a few per cent) even at sea level; at mm wavelengths this is no longer the case and is part of the reason why telescopes are sited at high altitudes (see the discussion in Sections 4.4 and 9.7.1). The ionosphere absorbs at low frequencies but again the effects are small, being typically less than 1 per cent at 100 MHz.

The more problematic effect arises from the spatial variations in refractive index over each antenna. Figure 4.2 showed a schematic snapshot of a complex phase screen exhibiting spatial fluctuations in refractive index. These variations impose delays on an incoming plane wavefront, which emerges as randomly irregular. In the regime of scintillation (as discussed in Chapter 4) the delay variations are large and lead to amplitude modulations because of interference effects. In most radio astronomical situations the phase variations arising in the ionosphere and troposphere are much weaker and do not give rise to amplitude effects; they do however affect the visibility phase.

Figure 9.26 illustrates a small part of a weak atmospheric phase screen covering a single interferometer. The solid lines represent radiation from a source passing directly overhead and perpendicular to the baseline. The grey cloud schematically represents the different path

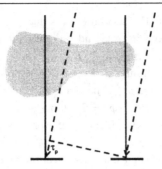

Figure 9.26. Effect of phase delay variations.

lengths over each antenna, variations which give rise to differential propagation delays $\tau_p$ that affect the visibility phase. As discussed in Section 9.3, on a single baseline the effect is to make the source position appear different to its true position (as shown) or equivalently to rotate the fringe pattern on the sky. These effects are highly frequency dependent and cause the fringe pattern to fluctuate. At frequencies below a few GHz fluctuations in the ionospheric plasma are dominant. Above a few GHz, variations in the poorly mixed water vapour in the troposphere play the principal role. Unless calibrated by some means, the interferometer is said to be phase unstable and cannot make good positional measurements or contribute to making good images.

The discussion so far has referred to the line of sight to one target source. Phase calibration can be effected by taking the difference between the complex visibilities on the target and a nearby compact source whose precise position is known *a priori* (see Section 9.8.2). This only works if the differential phase delay between target and calibrator is sufficiently small (much less than a radian) that the relevant part of the wavefront is still effectively plane. The angle over which this holds true defines the *isoplanatic patch*.

### 9.7.1 Troposphere

Above $\sim 2$ GHz, wind-blown variations in the unmixed water vapour in the troposphere are the dominant source of phase fluctuations; this is despite the fact that only a few per cent of air molecules are water. At typical radio frequencies the tropospheric path length delay is frequency independent and hence non-dispersive. The additional path length at the zenith is

$$\Delta L_{\text{trop}} \approx 0.228P + 6.3\text{PWV} \quad \text{cm}, \tag{9.50}$$

where PWV is the precipitable water vapour. The first and larger term is due to the well-mixed dry air; it is solely pressure dependent and hence is slowly varying; with the pressure $P$ measured in millibars, $\Delta L_{\text{trop,dry}} \approx 230 \pm 5$ cm at sea level, falling statistically with altitude $h$ (km) as $e^{-h/8.5}$. Accurate pressure measurements enable $\Delta L_{\text{trop,dry}}$ to be modelled at the sub-mm level. The second term is due to the poorly mixed water vapour and can be highly variable both in time and line of sight. At temperate latitudes at sea level the PWV is typically $\sim$ 2 cm; thus $\Delta L_{\text{trop,wet}} \approx$ 13 cm with a maximum in summer (PWV $\sim$ 4 cm) and a minimum in winter (PWV $\sim$ 1 cm); the decrease with altitude is, statistically, by a factor $e^{-h/2}$. The sea-level zenith path delay is $\Delta\tau_{\text{trop}} \sim$ 8 ns and the corresponding phase delay $\Delta\phi = 2\pi\nu\Delta\tau_{\text{trop}} = 2\pi\nu\Delta L_{\text{trop}}/c$ is $\sim$ 100 radians at $\nu = 2$ GHz. The non-dispersive nature of the phase delay means that it is directly proportional to $\nu$.

The topic of tropospheric turbulence and its effect on interferometers is treated in detail in TMS, Chapter 13. The tropospheric phase fluctuations are due to 'clouds' (albeit invisible in clear-air conditions) of water vapour at a characteristic altitude of $\sim$ 1 km; the 'clouds' have a wide range of sizes and are wind-blown over the interferometer at typical speeds of $\sim$ 10 m s$^{-1}$. The resultant phase variations show statistically quasi-stable characteristics as a function of the separation of the antennas (i.e. the baseline length $|b|$) at which they are sampled. They can be described by means of a so-called *phase structure function*, which involves an ensemble average:

$$D_\phi(b) = \langle [\Phi(x) - \Phi(x - b)]^2 \rangle, \tag{9.51}$$

where $\Phi(x)$ is the phase at point $x$ and $\Phi(x - b)$ is the phase at the point $x - b$. The rms deviation of the interferometer phase is

$$\sigma_\phi(b) = \sqrt{D_\phi(b)}. \tag{9.52}$$

For a specific atmospheric model the fluctuation spectrum can be understood from considerations of Kolmogorov turbulence. In practice the atmosphere does not conform to an ideal model and conditions vary but, typically, out to baselines of $\sim$ 5 km, $\sigma_\phi(b)$ increases as $b^\alpha$ with $\alpha$ in the range $\sim 0.5 \pm 0.2$. Beyond $\sim$ 5 km $\sigma_\phi(b)$ 'saturates' and longer baselines suffer no worse phase fluctuations. This is the so-called 'outer scale' of turbulence, which is, not surprisingly, the size of large (liquid) clouds.

The mean height of the phase screen is low ($\sim$ 1 km) and at centimetric wavelengths the isoplanatic patch may be up to ten degrees across and the typical timescales for significant phase changes can be tens of minutes or longer. Since the isoplanatic patch is invariably much larger than the field of view set by the primary beam of the antenna, only one atmospheric phase term applies to all sources within the field of view. However, even on short km-scale baselines, this variable phase still needs to be calibrated to maximize the dynamic range in the image (Section 10.7). At mm wavelengths the effects of the troposphere are much greater and in Section 11.3 we outline the more involved calibration techniques required.

### 9.7.2 Ionosphere

Below $\sim$ 2 GHz, temporal and spatial variations in the electron density of the dispersive ionospheric plasma (see Chapter 4) play the principal role. At a frequency $\nu$ well above the plasma frequency the additional path delay in traversing a vertical path through the ionosphere is

$$\Delta L_{iono} = -40.3\left(\frac{TEC}{1\ TECU}\right)\left(\frac{\nu}{1\ GHz}\right)^{-2}\ cm, \tag{9.53}$$

where *TEC* is the total electron content, i.e. the integrated electron column density in TEC units, with 1 TECU $= 10^{16}$ m$^{-2}$. The *TEC* is highly variable with time of day, season, and level of solar activity;[16] it may be a few TECU at night and $\sim 50$ TECU in the day but can be up to an order of magnitude higher when the Sun is highly active. At 2 GHz, with $TEC = (5-50)$ TECU, $\Delta L_{iono} \approx (50-500)$ cm, which illustrates the similarity with the tropospheric zenith delay at this frequency.

The dispersive nature of the ionospheric path length delay means that the phase delay is inversely proportional to frequency. It is given by (Mevius *et al.* 2016)

$$\Delta\phi = -8.45\left(\frac{TEC}{1\ TECU}\right)\left(\frac{\nu}{1\ GHz}\right)^{-1}\ rad. \tag{9.54}$$

For interferometry, what matters is the difference in phase between nearby lines of sight and this presents a challenge for low frequency work. For example, at 150 MHz with $\Delta\phi = 10^3$ rad, a path length difference of only 0.1 per cent is equivalent to 1 radian.

Like the troposphere, the ionosphere exhibits a power-law turbulence spectrum as a function of antenna separation but now on spatial scales up to hundreds of km (e.g. Mevius *et al.* 2016) and highly variable. In addition, travelling ionospheric disturbances (TIDs) produce time-varying fluctuations. Medium-scale TIDs can show variations of a few per cent in TEC over wavelengths of hundreds of km and velocities from 100 to 700 km s$^{-1}$; large-scale TIDs exhibit variations out to thousands of kilometres.

This evidence indicates that interferometry at metre wavelengths is significantly more challenging than at centimetre wavelengths. For a given observing frequency, the size of the isoplanatic patch is a complicated combination of the spatial scales involved: the height of the phase screen, the correlation lengths of the turbulence, and the baseline length as well as the state of the ionospheric 'weather'. The reader should consult Lonsdale (2005) and Kassim *et al.* (2007) for further explanations and instructive illustrations of the different regimes. In practice, on baselines out to $\sim 10$ km the phase variations may be coherent across the primary beam. On longer baselines the isoplanatic patch becomes smaller than the beam and direction-dependent calibration schemes are required for wide-field imaging. On LOFAR baselines of hundreds of km, at $\nu = 150$ MHz the isoplanatic patch is typically 1 degree across and the phase varies by 1 radian on timescales of minutes (Jackson *et al.* 2016). The challenges of low frequency imaging are further addressed in Section 11.6.

---

[16] See TMS, Chapter 13, for more details.

## 9.8 Practical Considerations

### 9.8.1 Point-Source Sensitivity

In Chapter 6 we derived Eqn 6.12, which gives the point-source sensitivity of a single antenna equipped with a total-power radiometer. The result is repeated here:

$$\Delta S_{\text{rms}} = \frac{2kT_{\text{sys}}}{A_{\text{eff}}(\Delta \nu \, \tau)^{1/2}} \quad \text{Jy}. \tag{9.55}$$

The extension to an interferometer[17] relies on the fact that the noise power from a square-law detector (see Section 5.3) is the same as that from a correlator fed with two identical voltages. This is schematized in the upper diagram in Figure 9.27.

In the lower diagram in Figure 9.27, an elementary interferometer consisting of two identical antennas is observing a point source. The antennas are connected to a correlator and each is also equipped with an identical radiometer with a square-law detector. Since the source-generated voltage inputs to the correlator will have equal amplitudes, the correlated output power and the source-generated contribution to the power from each detector must all be equal. The effective collecting area $A_{\text{eff}}$ of the two-element interferometer is, therefore, equal to the effective collecting area of one of the individual elements.

Now consider the limit where the antenna temperature contributed by the point source is much smaller than the receiver noise temperature $T_{\text{sys}}$ (the usual situation) and hence can

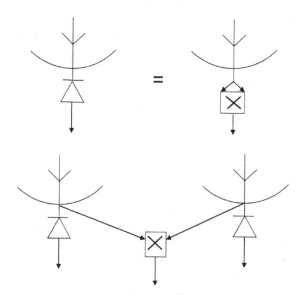

Figure 9.27. Interferometer noise power.

---

[17] This derivation is based on the treatment given in the book Condon and Ransom (2016), *Essential Radio Astronomy*, Chapter 3.

be ignored. The receiver noise voltages from the two elements are, therefore, completely uncorrelated and hence add incoherently; on the other hand the inputs of the receiver noise voltages to the square-law detectors are identical and hence add coherently. The correlator output noise power is, therefore, $\sqrt{2}$ *lower* than the noise power from the individual square-law detectors. Since $A_{eff}$ is the same, the point-source sensitivity of a two-element interferometer is $\sqrt{2}$ *better* than the sensitivity of a *single* antenna. Equation 9.55 therefore becomes

$$\Delta S = \frac{2kT_{sys}}{A_{eff}(2\Delta\nu\,\tau)^{1/2}}. \tag{9.56}$$

In the real world, a two-element interferometer will seldom conform to this idealized version. Frequently, dissimilar antennas, with different effective areas and receiver noise temperatures, will form the pair. In this case $A_{eff}$ becomes $(A_{eff,1}A_{eff,2})^{1/2}$ and $T_{sys}$ becomes $(T_{sys,1}T_{sys,2})^{1/2}$. Using the definition of system equivalent flux density (SEFD) in Section 6.4.3, a practical expression for the rms flux-density sensitivity per polarization for a single baseline can then be written as

$$\Delta S = \left[\frac{SEFD_1\,SEFD_2}{2\Delta\nu\tau}\right]^{1/2}. \tag{9.57}$$

With Eqn 9.57 as the basis, two comments should be made.

- For a combination of two polarizations, $\Delta S$ is lower by a factor $\sqrt{2}$.
- For a digital system (see Appendix 3) the losses in the correlation process will increase the ideal noise level by a factor $1/0.64 = 1.56$ for single-bit (two-level) and $1/0.88 = 1.14$ for two-bit (four-level) quantization.

The rms sensitivity of an adding (total-power) interferometer is $\sqrt{2}$ higher and hence equal to the sum of the two antennas, because its peak response includes, in effect, an autocorrelation term for each antenna. This does not mean, however, that a total-power interferometer would be better, for two reasons. First, the total-power interferometer has practical disadvantages, as noted in Section 9.1.2, and second, for synthesis imaging it is not the instantaneous total power that is desired but the fringe visibility. No technique gives a better signal-to-noise ratio than cross-correlation in determining the fringe visibility.

### 9.8.2 Amplitude and Phase Calibration: Basic Ideas

The output numbers from the correlation system depend on many things. The signal amplitude depends primarily on the intensity of the source, its angular size compared with the fringe spacing, and the effective areas of the antennas; the rms noise level depends (as above) on the characteristics of the receiving systems. Both also depend directly on the electronic gain through the system. It is possible, therefore, to establish the absolute visibility amplitude on a flux density scale from a comparison of the signal level with the correlated noise level, with the latter determined from a detailed knowledge of the observing system. This has been the practice in VLBI since its inception (see Section 11.4). However, the simplest, most practical, route to calibrating the visibility amplitude scale is to observe a compact source of known flux density. This flux calibrator is ideally a point source but slightly resolved sources with simple structures can also be used.

Having established the flux scale, a series of measurements of the amplitude and phase of the target source will contain fluctuating errors, which must be corrected to yield the complex visibility as a function of time. Within our present small-sky-patch assumption, the absolute reduction in amplitude off-axis due to the primary and delay beams can be discounted and at centimetre wavelengths the main contributions to amplitude errors, which are typically at the few per cent level, will be a combination of atmospheric absorption, imperfect knowledge of the antennas' effective areas as a function of elevation, and errors in the antenna pointing model. The principal errors which affect the scientific interpretation of the data arise in the measurements of visibility phase due to:

(i) imperfect knowledge of the interferometer geometry;
(ii) variations in the atmospheric path delay and through the receiver systems (see Section 9.7);
(iii) errors in transferring frequency and time between the antennas (see the next section).

An important initial step is to define the baseline vector correctly; this must connect the phase centres of the, often dissimilar, antennas. For a parabolic dish the phase centre of the feed is that point from which spherical waves appear to emanate or converge. In practice, a reference point, not necessarily the phase centre, is used to define a reference baseline. The offset of the phase centre from the reference point is calculated as a function of the pointing direction, and the physical baseline is calculated in the software. For more details consult Section 4.6 of TMS.

The basic step in the calibration of the complex fringe visibility is achieved by including in the observations one or more nearby compact sources with precisely known positions, i.e. with uncertainty much smaller than the fringe spacing. At long wavelengths there may be several such sources within the primary beam of the antennas and the delay beam of the interferometer. At centimetric wavelengths it is, however, usually necessary to calibrate the phase by observing a known source outside the primary beam, but within the isoplanatic patch. With parabolic dishes this involves mechanically swinging all the antennas of the array to the calibration source and resetting the delays. The loss of observing time on the required source is well compensated by the improvement in calibration and stability. Beam switching every few minutes is often needed to keep track of atmospheric delays and instrumental phase shifts. Calibration sources must be chosen with care if the interferometer has a high angular resolution (as in VLBI) and it may be necessary to correct for the angular structure of each calibration source in order to provide phase corrections of the required accuracy (see Section 11.4).

At ionospherically dominated metre wavelengths the situation is much more compli-cated. Not only is the isoplanatic patch small and the phase variations more rapid but, in contrast with the tropospherically dominated regime, the isoplanatic patch is often sig-nificantly smaller than the field of view of the low-gain antennas which constitute the array. This makes the problem of phase calibration much more challenging, and complex

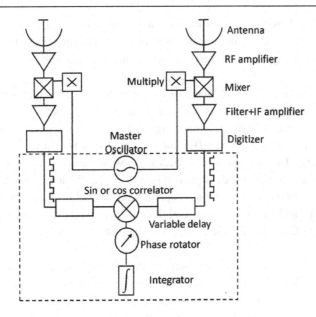

Figure 9.28. Outline of a practical interferometer system; the broken-line box indicates the central station. Many details have been omitted for clarity.

algorithms have had to be designed to correct for the spatially dependent phase variations. We will return to the vital topic of phase calibration for synthesis arrays in Chapters 10 and 11.

### 9.8.3 Outline of a Practical Signal Path

In an interferometer the antenna–receiver combination captures incoming electric fields and turns them into voltages, which must then be brought together and compared. A schematic outline of a typical signal path is shown in Figure 9.28. It must be noted, however, that not only are many details omitted but also interferometer architectures are many and varied. Figure 9.28 and the discussion in this section should, therefore, only be taken as a general guide.

Except for the shortest millimetre wavelengths, the signal is first amplified at the observing frequency by a low-noise amplifier chain at each antenna. These antennas may be separated by large distances and it is usually impractical to transmit the amplified RF signal directly to the correlator. Invariably a heterodyne receiver (as described in Section 6.6) is used to translate the signals to a lower, intermediate, frequency, at which it is easier to filter, digitize, and then transmit. The RF signals are therefore each mixed with, ideally identical, local oscillator signals followed by IF filters and additional amplification. In many cases, frequency conversion will be made more than once and the final conversion may result in a frequency band extending from (nearly) zero frequency to the upper limit of the original bandpass; this is known as conversion to baseband.[18] As pointed out in Section 6.6 the combination of mixing and subsequent filtering preserves the relative amplitudes and phases of signal components (i.e. their internal coherence) within the receiver bandpass.

Interferometry demands that the local oscillator is coherent, i.e. that it has exactly the same frequency at the two antennas. There are many ways to achieve this but typically a lower frequency (e.g. 5 MHz) reference signal derived from a master oscillator (often a hydrogen-maser frequency standard) at the central station is sent to each receiver over a transmission line. This is usually an optical fibre, with the reference signal carried as a modulation of laser light. The individual local oscillators at the antennas are phase-locked to an appropriate multiple of the reference signal. Since the delay over the fibre is subject to changes in its physical environment, its path length is continuously monitored by sending the signal back to the central station. The (variable) one-way path delay is then half the round trip delay.

The downconverted IF signals are sent to the central station. If the baselines are sufficiently short then the signals can be sent via coaxial cable or waveguide, but in most installations today they are sent via optical fibre. The signals are, most commonly but not invariably, digitized at baseband frequency before transmission. If digitized at the antenna, the digital clock rate is derived from the local oscillator signal and the transmitted digital data are time-stamped to ensure correlation of the correct bits. If the transmission is in analogue form, the IF signal is digitized at the central station where the master oscillator is situated. The varying geometric delay $\tau_g$ is compensated by the difference $\tau_i$ in the settings of the digital delays inserted in the signal lines before they reach the correlators. The time-dependent changes in the path delay of the frequency reference signal, and hence the changes in the local oscillator at each antenna, are compensated before correlation. The exact means by which this is achieved depends on the installation but, in some systems, the compensation is carried out by adjustments to the geometrical delay model. The digitized signals are then cross-multiplied in independent sine and cosine correlators.

There is one further nuance to be considered, which arises from the use of heterodyne receiver architecture. The fact that the compensating delay $\tau_i$ is effected at the IF frequency rather than at the RF frequency (as has been assumed in all our discussion up to this point) means that an additional, rapidly variable, phase term appears in the correlator output; this must be compensated before the cross-multiplication products can be integrated. The output phase of an interferometer operated entirely at RF (see Eqn 9.4) is, after delay compensation,

$$\phi_{RF} = \omega_{RF}(\tau_g - \tau_i), \tag{9.58}$$

which vanishes if $\tau_g = \tau_i$. The delay compensation rotates the fringes in such a way that they stay in a fixed position relative to the source. However, the actual phase achieved at IF is

---

[18] If there are many conversions, there is effectively only a single local oscillator frequency, since the heterodyne operation translates only the initial frequency band.

$$\phi_{IF} = \omega_{RF}\tau_g - \omega_{IF}\tau_i, \tag{9.59}$$

which is non-zero for $\tau_g = \tau_i$. The phase difference is

$$\phi_{RF} - \phi_{IF} = (\omega_{IF} - \omega_{RF})\tau_i = -\omega_{LO}\tau_i \tag{9.60}$$

for the particular choice of frequency conversion $\omega_{IF} = \omega_{RF} - \omega_{LO}$. Thus, as the Earth rotates and the geometrical delay changes, this internally generated phase difference gives rise to a fringe rate (see Section 9.1.3) as if the interferometer were operating at the LO rather than the RF frequency. For this specific choice of frequency conversion the fringes rotate in the opposite sense. A compensating phase rotation is applied after the sine and cosine correlators. The orthogonal data channels are then integrated to produce the two components of the complex visibility $V(\mathbf{s_0})$.

Before concluding it must be emphasized once more that there is a host of missing details in, and many architectural variations on, the above sketch. It should be noted that:

- invariably the bandpass is split, with digital filters, into multiple frequency channels which are correlated separately;
- both hands of polarization are usually observed;
- some interferometers (at mm wavelengths) use both sidebands after mixing.

Further explanation is beyond the scope of this book.

## 9.9  Further Reading

*Supplementary Material* at www.cambridge.org/ira4.

Cotton, W. D. 2011. Ionospheric Effects and Imaging and Calibration of VLA Data. EVLA Memo 118.

Taylor, G. B., Carilli, C. L., and Perley, R. A. (eds.) 1999. *Synthesis Imaging in Radio Astronomy* II. Astronomical Society of the Pacific Conference Series, vol. 180. (Also known as the 'White Book'.)

Thompson, A. R., Moran, J. M., and Swenson, G. W. Jr. 2017. *Interferometry and Synthesis in Radio Astronomy*, 3rd edn. Wiley.

# 10

# Aperture Synthesis

Interferometry has been the vital tool for obtaining much higher angular resolution than provided by single apertures. From the modest beginnings in the 1950s and 1960s, large and complex interferometer systems have been built to map the distribution of brightness across the sky. Some concentrate on imaging small-diameter radio sources, and their angular resolution can greatly exceed the resolving power of the largest optical telescopes on the ground or in space. At the other extreme, low-resolution arrays with many baselines can image faint distributed emission. The essential link between interferometer observations and the brightness distribution of a source is the Fourier transform, as the analysis in Chapter 9 demonstrated. An array of radio telescopes, their outputs separately combined pairwise to form all possible interferometric combinations, is called an *aperture-synthesis array*. In simple terms, the source is being observed with a collection of two-element interferometers, each with a cosine-wave and sine-wave gain pattern rippling across the source, with spatial frequency and orientation determined by the particular baselines, as illustrated in Figure 9.21. For an array with $N$ elements there are $N(N-1)/2$ interferometer pairs; $N-1$ since a telescope cannot form a baseline with itself and a factor 2 fewer since baseline AB is the same as BA apart from a phase reversal.

This chapter addresses the issues involved in producing a total intensity image of a broadband continuum radio source from a connected-element array operating at centimetric wavelengths. In Chapter 11 we address the additional challenges posed by:

– spectroscopy and polarimetry;
– wide fields of view;
– observations at metre and millimetre wavelengths;
– independent-element networks operating across continents.

We begin with a brief historical overview and some pen portraits of leading centimetric arrays.

## 10.1 Interferometer Arrays

The concept of aperture synthesis grew gradually in a series of transit interferometer experiments at the Cavendish Laboratory in Cambridge, UK, at the Radiophysics Laboratory of Australia's CSIRO in Sydney, and at the University of Sydney. At the Cavendish

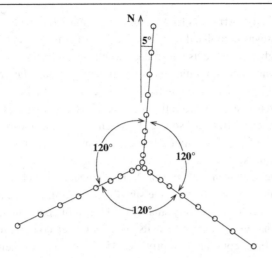

Figure 10.1. The 27 antennas of the VLA. Each arm of the Y is 20 km long.

Laboratory, Ryle and Neville (1962) showed that an individual baseline could provide an extended coverage of Fourier components by the continuation of observations while the Earth rotates on its axis; this changes the orientation and effective length of the baseline, which follows part of an ellipse in the $u, v$ plane (see Section 10.6). This technique immediately became known as Earth-rotation synthesis, and their work led to the building of the three-element Cambridge One-Mile Telescope, the first Earth-rotation synthesis radio telescope (Ryle 1962). As the technique became widely adopted the name was shortened, since the rotation of the Earth was understood; an aperture-synthesis array or aperture-synthesis telescope is now the common description. The technique is heavily dependent on the availability of computer power; the first systems essentially grew alongside the early development of digital computers.

To simplify the image analysis, all the early arrays used East–West baselines. The Cambridge One-Mile Telescope had two fixed and a single movable dish, and the $u, v$ plane was filled in further by moving this single element on separate days. It was followed at Cambridge by the eight-element Five Kilometre Telescope (Ryle 1972); compared with the One-Mile it produced superior, higher-resolution, images. At about the same time the Westerbork Synthesis Radio Telescope in the Netherlands (Baars *et al.* 1973) was commissioned. It remains in full-time operation, in largely its original configuration, and with it we begin a brief summary of current synthesis arrays.

The Westerbork Synthesis Radio Telescope (WSRT) originally had ten fixed and two movable dishes; subsequently two more movable dishes were added to speed up the data gathering. The array is currently (2018) being transformed into an efficient 21 cm L-band survey facility by replacing the front-ends in 12 telescopes by phased array feeds (PAFs, see Section 8.7), which will greatly increase its field of view.

The Jansky Very Large Array (JVLA), of the US National Radio Observatory, is situated in New Mexico and was completed in 1980. The VLA (as it was first called) combined a big jump in the number of elements with a departure from an East–West configuration. The array, in its most compact configuration, is shown in the Supplementary Material. It consists of $N = 27$ identical 25 m dishes; the individual dishes, connected by an optical-fibre data network, are arranged in a Y-configuration (Figure 10.1) to provide excellent coverage of the $u, v$ plane (see Section 10.6) for sources between the North Celestial Pole and declination $-30°$. The arms of the Y are approximately 21 km long, giving a maximum baseline length of over 36 km. All the dishes can be moved on rail tracks between fixed stations, allowing the spacings between the antennas to be changed. At each station there is a connection to an optical fibre that conducts the signals to the central correlator. There are four basic configurations, each successively smaller by a factor of approximately three, although hybrid configurations are often used. In its most compact configuration the maximum spacing is $\sim 1$ km. Each configuration provides 351 independent baselines. Thus a single brief time average can provide enough data for a usable map if the source is sufficiently simple; this is known as the 'snapshot' mode. The array can also be subdivided into smaller subarrays for simultaneous observations in separate programmes that do not need the full array. Alternatively, the entire array can be co-phased to serve as a single telescope having the geometrical collecting area of a 130 metre aperture; in this mode, it has been used to gather telemetry data from spacecraft during planetary encounters. An extensive system upgrade, completed in 2012, provides essentially complete coverage of the radio spectrum from 1 GHz to 50 GHz in eight different bands, with additional low frequency bands at $\sim 70$ and $\sim 350$ MHz.

*eMERLIN*   The Multi-Element Radio-Linked Interferometer Network, begun in the 1970s and completed in 1990, is a more extended array; it has six fixed telescopes (five 25 m and one 32 m in diameter) distributed across a large part of central England, with a maximum baseline of over 200 km (Figure 10.2).[1] Originally connected by radio links, the array is now connected by optical fibres to the central site at Jodrell Bank, Cheshire. For a significant fraction of observations, the 76 m Lovell Telescope (equivalent in area to nine 25 m dishes) is included in the array. The JVLA and eMERLIN are complementary, with eMERLIN having the longer baselines and therefore the higher angular resolution at a given frequency, and the JVLA having greater collecting area and denser $u, v$ coverage. Data from the JVLA and eMERLIN can be combined to gain the advantages of both arrays.

*ATCA*   The Australia Telescope Compact Array near Narrabri, New South Wales, was opened in 1988. Up to 2017 it was the leading centimetric array in the southern hemisphere. It consists of six 22 m dishes (of which five are movable), distributed along an 6 km East–West baseline.

---

[1] A 25 m telescope at Goonhilly (shown in Figure 10.2) will be added to the network in 2020, doubling the baseline and improving $u, v$ coverage for low declination sources.

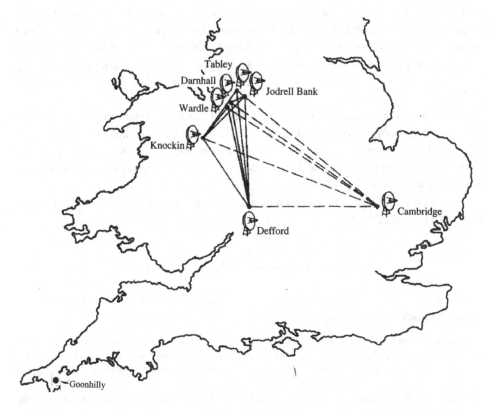

Figure 10.2. The Multi-Element Radio-Linked Interferometer Network eMERLIN.

*ASKAP* The Australia Telescope SKA Pathfinder is one of two new southern-hemisphere arrays nearing completion at the time of writing (2018), both of which are precursors of the SKA. ASKAP has thirty-six 12 m dishes in fixed locations with baselines up to 6 km. It is primarily aimed at observations around 1 GHz (including redshifted atomic hydrogen) and, along with the WSRT, is pioneering the use of phased array feeds to enhance greatly the survey speed of the array.

*MeerKAT* This was originally the Karoo Array Telescope, in South Africa, and is the other southern-hemisphere array nearing completion at the time of writing. It will consist of sixty-four 13.5 m dishes in fixed locations with maximum baselines up to 8 km. When completed it will be the largest and most sensitive radio telescope in the southern hemisphere until the SKA1-MID is completed on the same site in the mid-2020s.

*GMRT* The Giant Metre-Wave Radio Telescope near Pune, India, which began operating in 1995, is designed to operate at longer wavelengths (frequencies from 50 MHz to 1.4 GHz). However, it resembles centimetric arrays in many respects and hence is included here rather than in Section 11.7. The array consists of thirty 45 m dishes in fixed locations; there is a central 1 km 'core' containing 12 dishes; the other 18 are in a Y-shaped configuration providing baselines up to 25 km. At these wavelengths a large collecting area could be constructed by using a novel dish design with a lightweight reflector surface.

*ATA* The Allen Telescope Array in Hat Creek, California, which began operations in 2007, was designed principally for the Search for Extraterrestrial Intelligence (SETI). It has forty-two 6 m offset parabolic dishes with baselines to 300 m. The design includes many technical innovations to allow the array to carry out wide-area surveys over a large range of frequencies ($\sim$ 1 to $\sim$ 10 GHz).

Other arrays operating at metre and mm/sub-mm wavelengths are described in Sections 11.6 and 11.3 respectively.

## 10.2 Recapitulation on the Visibility Function

Before moving on to the formalities of aperture synthesis it is useful to summarize the basic properties of the visibility function developed in Chapter 9.

(i) An interferometer acts as a spatial filter and, for a restricted range of angles from the phase tracking centre, a baseline $u, v$ picks out a single complex Fourier component; this is the visibility $V(u, v)$ of the source brightness distribution $B(l, m)$. The visibility is perforce Hermitian, i.e. $V(u, v) = V^*(-u, -v)$ since $B(l, m)$ is real.

(ii) The *visibility amplitude* is the magnitude of the Fourier component. The visibility phase describes the sideways shift (expressed as a fraction of the spatial period and with respect to a fiducial source position) which maximizes the overlap between the Fourier component and $B(l, m)$.

(iii) Measurement of the *visibility phase* requires a precise knowledge of the geometry of the baseline (with the potential to be affected by differential propagation delays in the atmosphere and the interferometer system itself), so as to be able to maintain the position of the fringes precisely with respect to a specific position on the sky.

(iv) If, as the baseline length increases, the visibility amplitude systematically decreases, the source is said to be resolved.

(v) A correlation interferometer can never measure the true brightness of an extended source whose size is comparable with, or larger than, the primary beam of the antennas. It cannot form very short or zero baselines and hence 'resolves out' such extended structure. It can, however, measure the brightness or the total flux density of a compact source.

(vi) The 2D complex visibility function is equivalent to the 2D spatial coherence function of the incoming radiation; this is the van Cittert–Zernike theorem. Only the separation between the telescopes in a given baseline matters; their absolute position is immaterial.

(vii) Combining many different samples of the complex visibility enables one to reconstruct an estimate of the brightness distribution of an extended source; the points at small $u, v$ distances record large-scale structure, and the fine-scale structure is measured at large $u, v$ distances.

## 10.3 The Data from an Array

We now move on to generalize the analysis in Chapter 9 for an array of $N$ antennas. Each pair of array elements $i, j$ is combined as one of the $N(N-1)/2$ independent interferometers and, during an observation, the cosine and sine correlators evaluate the cross-power products $R_{cos,ij}$ and $R_{sin,ij}$ from the two voltage amplitudes, as in Eqns 9.26 and 9.27, and perform the average over a pre-set integration time. When Earth-rotation synthesis is used to obtain a more complete set of visibilities in the $u, v$ plane, each set of $N(N-1)/2$ visibilities has a time tag specifying a particular value of $(u_k, v_k)$ for the $k$th time interval of integration. The array elements, either mechanically steerable paraboloids or electrically steerable sections of a phased array, follow the radio source as the Earth rotates. The delay tracking system also follows the source, continuously adjusting the instrumental time delays $\tau_i$ to compensate for the changing geometric time delays $\tau_g$; the latter are calculated with respect to a plane wave from the direction of the phase tracking centre $s_0$ and with respect to a single fiducial point in the array.

As shown in Figure 9.28 the instrumental delay system operates on the data streams from the individual array elements on an antenna-by-antenna basis; this is an important point, which will recur when we discuss calibration in Section 10.7. With such an architecture the time averages of the cross-power products form estimates of the fringe visibility $V_{ij}$ for all the interferometer pairs in the array.

The set of fringe visibilities $V_{ij}(u_k, v_k)$ taken over the observation period is a discrete matrix of complex quantities in the $u, v$ plane, and the calculation of a brightness map from this set of visibilities is carried out by a discrete Fourier transform. It is, nevertheless, conceptually convenient to continue to view the process in its integral (i.e. continuous) form for the next steps in the analysis.

## 10.4 Conditions for a 2D Fourier Transform

In Section 9.5.2 we developed Eqn 9.39 for the differential delay appropriate to a point lying $\sigma$ away from the phase tracking centre $s_0$. In the rest of that section we simplified the situation to the special case where $s_0$ is perpendicular to the baseline and hence the $w$-term, which is the delay for the direction $s_0$ expressed in wavelengths, is exactly zero. Let us now allow $s_0$ to be at an arbitrary angle to the baseline and thus include the full phase term in Eqn 9.42:

$$V(u, v, w) = \frac{1}{\sqrt{1 - l^2 - m^2}} \iint A(l, m) B_\nu(l, m) e^{-i2\pi(ul, vm, (\sqrt{1 - l^2 - m^2} - 1)w)} dl\, dm. \quad (10.1)$$

It is already clear that this equation, unlike Eqn 9.42, cannot be a modified Fourier transform since the visibility is now a function of three variables (the three dimensions of the baseline) while the brightness distribution is a function of only two (the sky is in the far field and hence there is no depth information). This equation is not easy to solve (see Section 11.5) so we want to know the circumstances under which we can retain the simplicity of the 2D Fourier transform. This is equivalent to requiring that, as illustrated in Figure 9.21, we can use the flat mapping plane tangential to $s_0$ (and hence to $l = 0, m = 0$) rather than the celestial sphere. The distortions arising from the term $(\sqrt{1 - l^2 - m^2} - 1)w$ must, therefore, become negligible and hence $|l|$ and $|m|$ must be small – but how small? Taking the first-order approximation, we have

$$(\sqrt{1 - l^2 - m^2} - 1)w \approx (l^2 + m^2)w/2; \quad (10.2)$$

thus the effect of the $w$-term on the phase is $\pi(l^2 + m^2)w$. If the overall field of view suitable for 2D imaging with insignificant distortion is $\theta_{2D}$ then $l^2 + m^2 = (\theta_{2D}/2)^2$ and we can write the phase error as $\pi w(\theta_{2D}/2)^2$. Taking the worst-case scenario in order to provide an upper limit to $\theta_{2D}$, we have $w_{max} = b_{max}/\lambda$, which occurs when the source is low in the sky and lies along the direction of the longest baseline in the array. Since the synthesized beam diameter $\theta_{synth} \approx \lambda/b_{max}$ radians the phase error is then approximately $\pi \theta_{2-D}^2/4\theta_{synth}$. The usual limit adopted for this error is 0.1 radians (which of course will be less on the shorter baselines) and hence we arrive at a conservative limit for the overall 2D field of view in radians:

$$\theta_{2D} < \frac{0.4\sqrt{\theta_{synth}}}{\pi} \sim \frac{\sqrt{\theta_{synth}}}{3}. \tag{10.3}$$

Two representative examples help to bring this limit to life. If $\theta_{synth} = 1$ arcsec (roughly the maximum resolution of the JVLA at 1.4 GHz) then $\theta_{2D} < 150$ arcsec; this is about ten times smaller than the primary field of view of its 25 m telescopes. If $\theta_{synth} = 0.15$ arcsec (the resolution of eMERLIN at L-band) then $\theta_{2D} < 60$ arcsec, which is smaller than the size of the Hubble Deep Field ($\sim 200$ arcsec). While the majority of images of discrete sources can be, and so far have been, made under the restrictive 2D field condition, increasingly radio astronomers want to image the entire primary field of view. The problem is especially challenging at long wavelengths where the array elements may be dipole-like with a large primary field of view (see Section 11.6). In Section 11.5 we will discuss the more complex interferometer geometry involved and outline some algorithmic solutions which have been developed to allow wide-field imaging.

When describing the many steps involved in making an image from visibility data it helps to stick to the simplest 2D imaging case. Thus, to reiterate, for a restricted field of view $[\mathcal{A}(l, m)/\sqrt{1 - l^2 - m^2}] \rightarrow 1$, and we can write the Fourier pair as

$$V(u, v) = \iint B(l, m)e^{-2\pi i(ul+vm)}dldm,$$
$$B(l, m) = \iint V(u, v)e^{2\pi i(ul+vm)}dudv. \tag{10.4}$$

When the angular distances $(x, y)$ from the reference direction permit the small-angle approximation, the direction cosines $l, m$ can be written as angular offsets $x, y$, so that Eqns 10.4 become

$$V(u, v) = \iint B(x, y)e^{-2\pi i(ux+vy)}dxdy,$$
$$B(x, y) = \iint V(u, v)e^{2\pi i(ux+vy)}dudv. \tag{10.5}$$

The next practical step in the imaging process is the calibration of the data, but we defer this discussion to Section 10.7 in order to continue our development of the synthesis formalism. Let us for the moment assume that the data are perfectly calibrated. The set of $N(N-1)/2$ visibilities $V_{ij}(u_k, v_k)$ then form a sampled approximation to the Fourier transform of the source brightness distribution. The sampling of the Fourier plane by any practical array is necessarily incomplete, with the missing spacings corresponding to unknown components in the Fourier transform plane; as a result the initial synthesized image inevitably suffers some distortion (see Section 10.6.2).

## 10.5 The Spatial Frequency Transfer Function, or $u, v$ Coverage

The image distortions depend in detail on how the $u, v$ plane is sampled. Several different names, which are all essentially synonymous, have been attached to the set of samples, viz. the *spatial frequency transfer function;* the *spectral sensitivity function* or *diagram;* the $u, v$ *coverage.*

We start by expressing the sampling as a *spatial frequency transfer function*, which we call $S(u, v)$ to represent its origin. In Section 8.3 we showed that the Fourier transform of the power pattern of an antenna is the autocorrelation of the current excitation pattern across the aperture. The same holds true for interferometers: $S(u, v)$ is the autocorrelation of the excitation pattern which is now spread across multiple antennas.

The spatial frequency transfer function $S(u, v)$ plays an equivalent role to the temporal frequency transfer function $H(\nu)$ introduced in the discussion of filters in Section 6.6. The function $H(\nu)$ operates in the frequency domain, and its Fourier transform gives the response of the filter in the time domain to a voltage impulse $\delta(t)$. By the same token, if a 2D point source $\delta(x, y)$ is observed by an interferometer array, the image in the spatial domain appears as a spread pattern which is the 2D Fourier transform of $S(u, v)$.

It is worthwhile expanding on this point. The Fourier transform of $S(u, v)$ with unity amplitude and zero phase at each sampled point gives the response to a point source. This pattern is generally known as the synthesized beam or 'dirty beam', which we call $D(l, m)$; it is the map that would be generated by a computation of the Fourier transform of the visibilities from observing a point source such as an unresolved quasar. Note that this is not the same as a telescope diffraction pattern, since it is derived from *complex amplitudes*, while the diffraction pattern describes *power*. As a simple example, consider the case of visibilities sampled within a circle of radius $\rho$ (measured in wavelengths), densely enough that the Fourier integral is a good approximation. The resulting synthesized beam, using the small-angle approximation $r$ (since the function is circularly symmetric) instead of the direction cosine, is

$$b_0(r) = J_1(2\pi \rho r)/(\pi \rho r). \tag{10.6}$$

This has a central maximum but its sidelobes are both positive and negative. The power pattern is the square of this function, which can be immediately recognized as the Airy function (Figure 8.21), the diffraction pattern for a uniform circular aperture. The synthesized beam plays the role of a diffraction pattern, but with several distinctions: it is

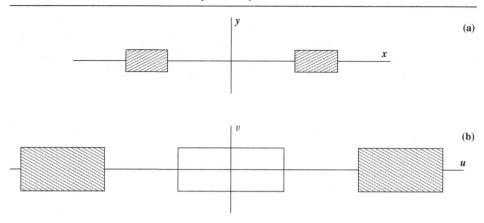

Figure 10.3. The coverage, or 'support', in the $u, v$ plane for a two-element interferometer: (a) the interferometer antennas in the $x, y$ plane; (b) the support in the $u, v$ plane.

a calculated function residing in the computer, and it must have both positive and negative values (since the total power terms are neglected, its integral over the sky must be zero). The term 'dirty beam' derives from the arguably unpleasant-looking character of its sidelobe pattern. This is made worse by the presence of 'holes' in the $u, v$ coverage.

It is therefore important to determine the region of the $u, v$ plane over which $S(u, v)$ is non-zero; this is the *support domain*. This concept was introduced by Bracewell (1962), who referred to $S(u, v)$ as the *spatial sensitivity diagram*. In the example of Figure 10.3 a simple interferometer has two equal and uniformly filled rectangular elements. The *support* $S(u, v)$ in the $u, v$ plane, defined by the 2D cross-correlation of the excitation $E(\xi, \eta)$ of the two antennas, is shown on the same scale, where both $\xi, \eta$ and $u, v$ are measured in wavelengths. The open area in the centre of Figure 10.3(b) is the support domain for an adding interferometer system, which is sensitive to the total power as well as to the interference signal between the spaced telescopes. This central area corresponds to the range of Fourier components available to each single telescope aperture and detects the total received power. An example of early observations with such a total-power two-element interferometer was shown in Figure 9.1, where the recording shows the sinusoidal fringes of each radio source superimposed on the total-power signal; the extent of the fringes is limited by the reception pattern of the antenna, defined by the Fourier components in the central part of $S(u, v)$. Aperture-synthesis arrays, however, are invariably of the correlation type, which eliminates the central part since they are not sensitive to the total power.

The discussion so far has assumed that all parts of $S(u, v)$ are treated equally. However, the amplitude of any part of $S(u, v)$ may be varied during the process of synthesizing the final map, by applying weighting factors to the individually recorded visibilities. We will discuss the effects of various weighting functions in Section 10.8.1.

## 10.6 Filling the *u, v* Plane

The brief summaries in Sections 10.1 and 10.2 lead us to consider the ways in which more independent visibilities can be collected, allowing better images of the sky brightness distribution to be constructed. The first two approaches discussed below are relatively cheap to implement; the second two are capital intensive, and the final one simply takes advantage of existing investment.

(i) The rotation of the Earth is used to change the projected baselines throughout an extended observation. This technique was developed in the early 1960s for the original Cambridge synthesis work and for the long-baseline observations at Jodrell Bank (Rowson 1963). We will discuss the essentials of the method in the next section.

(ii) Multifrequency synthesis (MFS): By changing the frequency of observation the length of the baseline vector, measured in wavelengths, can be altered. Thus, at least in a limited fashion (the baseline is not rotated), one can 'move' physically fixed antennas. Coupled with Earth rotation, MFS over frequency ranges of tens of per cent allows one to fill in substantial gaps in single-frequency *u, v* coverage. As will be discussed in Section 10.13, algorithms have been developed to solve for spectral variations across the face of the brightness distribution under study.

(iii) More antennas can be added to the array, which increases the number of baselines $N(N-1)/2$. From the beginning, aperture-synthesis array designers have had to seek a balance between an 'ideal' placement of the antennas (and indeed to decide what 'ideal' means, given the science goals) and various practical constraints. There is no single solution for all circumstances.

(iv) The antennas can be moved around into different configurations. This approach, which is applicable if the source does not vary during the period of the observations, was pioneered in the original Cambridge synthesis observations and adopted at the WSRT. Different configurations are also used routinely on the JVLA, the ATCA, and the mm-wave arrays (see Chapter 11) ALMA, the NOEMA Interferometer, and the Smithsonian Millimetre Array.

(v) Data can be combined from different arrays, for example the JVLA and eMERLIN, as mentioned in Section 10.1.

### 10.6.1 The *u, v* Coverage and Earth Rotation

Having laid the groundwork we now describe how, in practice, $S(u, v)$ is built up over the course of an observation making use of the rotation of the Earth. We provide a basic introduction sufficient to enable the interested reader to tackle the more sophisticated treatments in TMS, Chapter 4, and elsewhere, in particular for astrometric and geodetic measurements with VLBI (see Chapter 11).

In Chapter 9 we described the geometrical relationship between the tangent plane of the sky, defined by the coordinates $l, m$ and the $u, v$ plane, but we did not explicitly locate the interferometer elements on the surface of the rotating Earth. A baseline in 3D space is defined by the differences between two sets of antenna coordinates in an appropriate reference frame. As a vector (although it is not strictly a vector since its direction is not defined modulo 180°) the physical baseline does not depend on the choice of frame, so the latter comes down to a matter of convenience. There are several options but the standard has become a right-handed Cartesian frame, the X axis of which lies in the Earth's equatorial plane with the positive direction pointing at the Greenwich meridian; in celestial

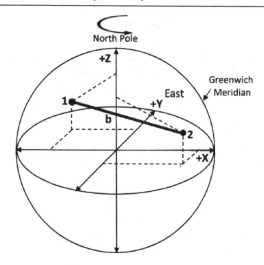

Figure 10.4. A physical baseline in an Earth-fixed frame.

coordinates the positive X axis therefore points towards the direction $H = 0^h$, $\delta = 0°$.[2] The Y axis also lies in the equatorial plane, with the positive direction pointing East and thus towards the direction $H = -6^h$, $\delta = 0°$. The positive Z axis points along the Earth's spin axis north towards the celestial pole at $\delta = 90°$. This is illustrated in Figure 10.4.

Antenna positions on the Earth's surface are described by the projections $P_X, P_Y, P_Z$ onto the X, Y, Z axes; these components are always quoted in metres. The physical baseline vector between telescopes 1 and 2 has components whose magnitudes are

$$b_X = P_{X,1} - P_{X,2}, \quad b_Y = P_{Y,1} - P_{Y,2}, \quad b_Z = P_{Z,1} - P_{Z,2} \quad \text{m.} \quad (10.7)$$

These components can then be converted into wavelengths to make the link with the $u, v, w$ plane and, following TMS, we label these $X_\lambda, Y_\lambda, Z_\lambda$. Recall from Chapter 9 that the $u, v$ plane is that on which the baseline vector is projected as seen from the direction of the target source; it is natural to take it as orthogonal to this direction. The $w$ axis then points towards the target source, and $w$ is the interferometer delay $\tau_g$ expressed in wavelengths.

In Figure 10.4 the rotating Earth is shown at one instant of time looking from the direction of a source at $\delta = +53°$ and well after transit at the Greenwich meridian. As the rotation proceeds the changing components of the projected baseline are expressed in the $u, v, w$ coordinate frame. For a source at celestial coordinates $H, \delta$ the coordinate transformation (see TMS Chapter 4 for more details) between the components in the $u, v, w$ frame and the baseline components $X_\lambda, Y_\lambda, Z_\lambda$ is

$$\begin{bmatrix} u \\ v \\ w \end{bmatrix} = \begin{bmatrix} \sin H & \cos H & 0 \\ -\sin\delta\cos H & \sin\delta\sin H & \cos\delta \\ \cos\delta\cos H & -\cos\delta\sin H & \sin\delta \end{bmatrix} \begin{bmatrix} X_\lambda \\ Y_\lambda \\ Z_\lambda \end{bmatrix}. \quad (10.8)$$

---

[2] With this fixed definition for the direction of +X the hour angle $H$ is defined relative to the Greenwich meridian; this formulation is applicable to any array but is particularly appropriate for VLBI arrays. The alternative is to define X with respect to the local meridian, in which case $H$ is the local hour angle.

We identify $u$ as the E–W component of the projected baseline and $v$ as the N–S component. Viewed from the source the $u, v, w$ components vary during the course of the day; Eqn 10.8 represents this, with $H$ increasing from $0^h$ to $24^h$ (0 to $2\pi$ radians). (It is unlikely that the reader will ever have to use Eqn 10.8 since interferometric analysis packages will do the job behind the scenes.) In order to simplify the analysis in Chapter 9, we considered the special case where the source lies symmetrically on the midline between two antennas so there is no relative delay and $w = 0$. In Section 10.4 we extended the argument to the case where $w \neq 0$ but could still be ignored; in Section 11.5 we explore the wide-field situation where $w$ must be taken fully into account.

Before considering some examples of the effect of diurnal rotation on the baseline's projection in the $u, v$ plane we need to reiterate an important feature of interferometers: for every $u, v$ point we can automatically add a second one. This feature arises because either telescope can act as the reference, and the phase of the complex visibility with one parity is just the negative of the other; they are Hermitian conjugates, $V_{ij}(u_k, v_k) = V_{ij}^*(-u_k, -v_k)$, exhibiting reflection symmetry through the origin. Heuristically, as pointed out by Christiansen and Hogböm (1985), swapping the two antennas around swaps the additional $\pi/2$ radians phase shift in the sine correlator channel over to the other side; this is mathematically equivalent to changing the sign of the channel which provides the imaginary part of $V_{ij}(u_k, v_k)$. The pairs of visibility points are, therefore, not independent, no new information is implied and the resolution is not doubled. Both points are, however, required.[3] The visibility function is the Fourier transform of the sky brightness distribution; the latter is a real function, and to recover a real function from the inverse transformation of the complex visibility measurements demands that they conform to the Hermitian property $V_{ij}(u_k, v_k) = V_{ij}^*(-u_k, -v_k)$. In effect we are using our prior knowledge of the properties of the sky to complete the $u, v$ coverage. The use of prior knowledge of the sky will recur over the course of this chapter. To illustrate these points consider Figure 10.5, which shows a compact array of three antennas envisaged to lie around the North Pole, at the origin of coordinates. Their positions have components (in metres from the origin, not marked) only in the plane of the equator.

The $u, v$ plane is viewed from a source directly overhead, i.e. from the direction of the North Celestial Pole ($\delta = 90°$); the scale in wavelengths is not marked since it depends on the observing frequency. As described above, when the order of the telescopes is reversed the visibility on a baseline becomes the complex conjugate of the original $V(u, v)$; a second point $V^*(-u, -v)$ reflected through the origin, can therefore be added. Thus, at a particular instant there are $3 \times 2$ $u, v$ points and as the Earth rotates they will trace out three concentric circles in the $u, v$ plane, repeating every $12^h$. This simplest geometry illustrates the general point that Earth rotation causes the projected baseline to move around the $u, v$ plane.

In Section 10.5 we introduced the concept of a spatial frequency transfer function for an interferometer as an extension of the single-dish case, i.e. as the autocorrelation of the aperture distribution. When the extent of an array is large compared with the size of the elements it can be approximated as a collection of point antennas, and its spatial frequency

---

[3] These arguments are a complement to the discussion in Section 9.4.1 and Figure 9.12.

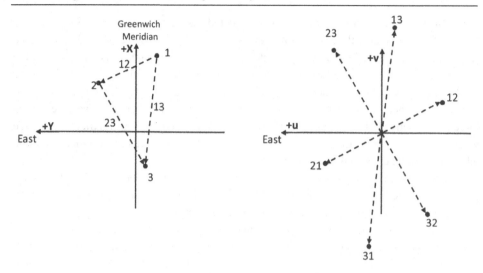

Figure 10.5. Three-antenna array illustrating the Hermitian property: (left) the antennas positioned around the North Pole (at the centre of the axes); (right) the $u, v$ plane viewed from $\delta = 90°$.

support will also be a discrete array of points. In the two-element interferometer in Figure 10.3, if the elements were represented by a pair of delta functions, the autocorrelation would be three delta functions, two at plus and minus the spacing and a central total-power spike having twice the height. The simple $u, v$ coverage in Figure 10.5 can, therefore, also be understood as the autocorrelation of the antenna positions minus the central peak at zero shift, which is absent in multiplying interferometers. Some other elementary examples are shown in Figure 10.6.

A purely E–W baseline, at whatever latitude, produces tracks which are centred on the origin; for sources away from the poles the circular tracks become elliptical, with an axial ratio depending on the declination of the source. For a source on the celestial equator, $\delta = 0°$, the ellipse degenerates to a single line and hence such interferometers provide no N–S resolution at all. A pure N–S baseline also produces an elliptical track but with its centre displaced from the origin along the $v$ axis.

In general a baseline will have components in the X, Y (equatorial) plane and along the Z (polar) axis. The major and minor semi-axes of the resulting $u, v$ ellipse are $(X_\lambda + Y_\lambda)^{1/2}$ and $(X_\lambda + Y_\lambda)^{1/2} \cos \delta$ respectively and the offset from the origin in the $v$ direction is $Z_\lambda \cos \delta$. While the tracks, on the right in Figure 10.6, are symmetrical with respect to the $v$ axis, this is not a general feature. Depending on the antenna locations in latitude and longitude different parts of the $u, v$ track appear, or are eliminated, as the source rises or sets at the antenna locations; the available track may therefore be asymmetrical. Finally, since only the length and orientation of the baseline matter (see the discussion on spatial coherence in Section 9.6), one is free to place one end of any baseline 'vector' in the centre of the axes of the $u, v$ plane.

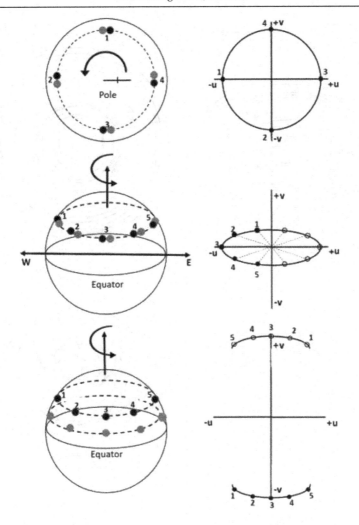

Figure 10.6. Elementary baselines (formed by the grey and black pairs of circles) and their $u, v$ tracks. (top) An E–W baseline viewed from $\delta = 90°$. (middle) An E–W baseline viewed from an intermediate declination; the source is not visible throughout the day but the Hermitian visibilities provide the second track (open circles). (bottom) An N–S baseline viewed from an intermediate declination.

These various characteristics are most easily seen together in the $u, v$ coverage from an array with a relatively small number of antennas such as eMERLIN. The eMERLIN array has six elements, which are spread over distances up to $\sim 200$ km. Such a sparsely filled network can produce maps only by Earth-rotation synthesis over an extended period. The 15 baselines trace out the $u, v$ coverage, or equivalently the spectral sensitivity function, shown in Figure 10.7. As noted above, since the physical extent of the antenna elements is very small compared with the baseline lengths, the broadening of the $u, v$ tracks cannot be

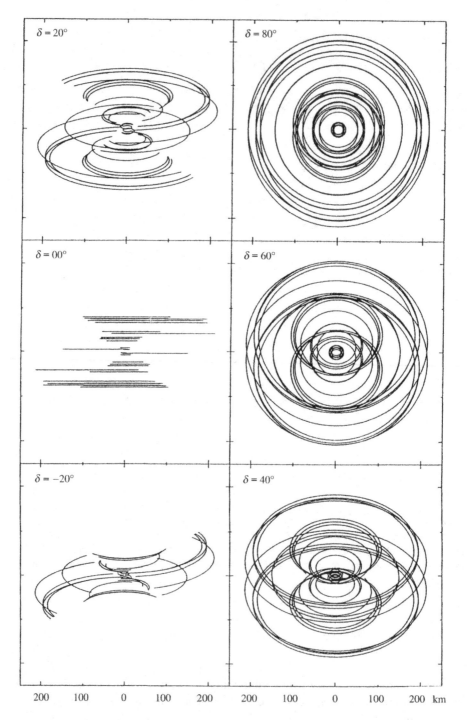

Figure 10.7. Results of the MERLIN interferometer array using six telescopes: the single-frequency $u, v$ coverage for complete tracking observations at six declinations.

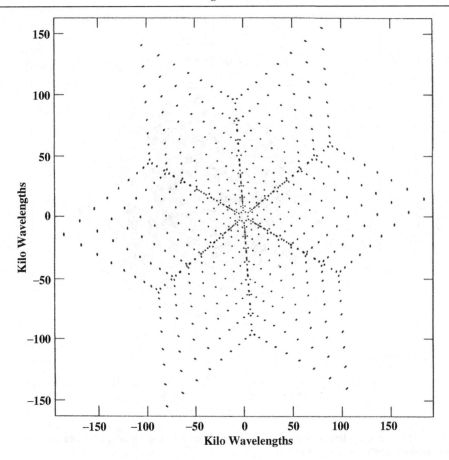

Figure 10.8. The instantaneous coverage, or 'support', in the $u, v$ plane, of the VLA for an observation at the zenith.

represented in this figure. The broadening of the tracks is, however, utilized in the imaging technique of 'mosaicing', which we discuss in Section 11.5.4.

Figure 10.8 shows the instantaneous or 'snapshot' $u, v$ coverage of the VLA for a source at the local zenith (in this case at $\delta = 30°$). It can be thought of as a more extensive version of Figure 10.5. With $N = 27$ elements the VLA has 351 separate baselines and hence, as a result of Hermitian symmetry, there are 702 discrete points in the snapshot $u, v$ coverage. As noted earlier, this coverage may be sufficient for a useful map to be made without the need for Earth rotation.

The advantage that Earth rotation gives in covering the $u, v$ plane is seen in Figure 10.9, which plots the coverage for an eight-hour VLA observation at declination 30°; comparison with Figure 10.8 shows a much more completely filled $u, v$ plane, but still with a hole at the centre due to the physical limitations on small telescope separations. Even with Earth rotation the $u, v$ coverage is necessarily much more sparse in very long baseline interferometry (VLBI). Here the antennas may be thousands of kilometres apart and one

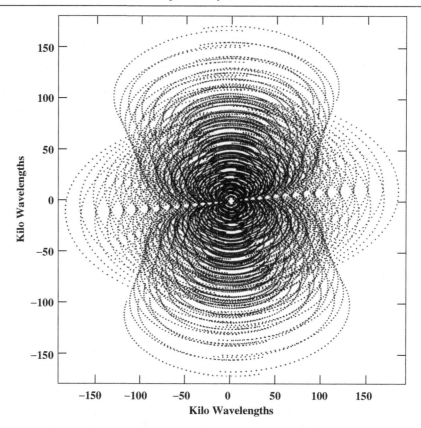

Figure 10.9. The $u, v$ plane coverage for an eight-hour tracking observation with the VLA, for observations at declination $\delta = 30°$.

antenna may even be on a spacecraft tens to hundreds of thousands of kilometres above the Earth's surface.

### 10.6.2 The Effect of Incomplete $u, v$ Coverage

Each baseline produces one measurement, $V_{ij}(u_k, v_k)$, per integration so the effect of the discrete coverage of the $u, v$ plane is to sample the complete, continuous, visibility function $V(u, v)$; in other words sampling is equivalent to multiplication by a spatial frequency transfer function $S(u, v)$. Hence we can write

$$V(u, v) \times S(u, v) = V(u, v)_{\text{obs}}, \tag{10.9}$$

that is, (FT of sky) $\times$ (transfer function) = (observed visibility).

Using the convolution theorem we can therefore also write, for the Fourier partners of the above,

$$B(l, m) \ast D(l, m) = B^d(l, m), \tag{10.10}$$

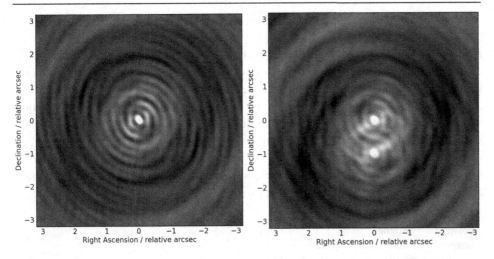

Figure 10.10. (left) The dirty beam for an eight-hour MERLIN track at $\delta = 60°$; (right) the dirty map of a double source made with this $u, v$ coverage. The corresponding CLEAN image is shown in Figure 10.14(b). (Courtesy of Adam Avison.)

that is, (true sky) $*$ (dirty beam) = (dirty map).

The dirty map is, therefore, the convolution of the true sky distribution with the 'dirty beam' or point spread function.

One can see that the appearance of the dirty map is determined by the characteristics of the spatial frequency transfer function and, in particular, by the unmeasured Fourier components (see also Figure 9.12), which correspond to missing information. The form of the sidelobes will depend on the geometry of the unfilled area – in fact this form will have a configuration proportional to the Fourier transform of the unfilled area. To allow a quantitative discussion we refer to the VLA snapshot $u, v$ coverage in Figure 10.8.

– Information from spatial frequencies $> u_{max}$ cycles per radian (i.e. $> b_{\lambda,max}$ wavelengths) is not collected; this limits the angular resolution in the reconstructed synthesized map to $\sim 1/b_{\lambda,max}$ radians. In Figure 10.8 the hexagonal pattern falls within a circle of radius 200 k$\lambda$ and hence the synthesized beam diameter will be $\sim 1/(2 \times 10^5)$ radians or $\sim 1$ arcsec.

– Information is not collected in the gaps. If the synthesized map of the sky has extent $l, m$ then, in order to obey the Fourier sampling theorem (see Appendix 3), in the $u, v$ plane it must be sampled at least as frequently as $\Delta u = 1/l$ and $\Delta v = 1/m$ to avoid aliassing. In Figure 10.8 the gaps $(\Delta u, \Delta v)$ are typically 15 k$\lambda$ to 20 k$\lambda$ in extent and hence one can expect dominant distortions and fluctuations in the map on angular scales $6.7 \times 10^{-5}$ to $5 \times 10^{-5}$ radians ($\sim 14$ to $\sim 10$ arcsec).

– At the centre, there is a hole in the coverage since the array has a minimum spacing $(u_{min}, v_{min})$; the consequence is that, for any extended source, structures larger than the angular dimension $(\sim 1/u_{min}, \sim 1/v_{min})$ will go undetected (see Section 10.11 for further discussion on this point).

Figure 10.10 shows the resultant dirty beam and dirty map for a sparse transfer function (MERLIN at $\delta = 60°$; see Figure 10.7). As a direct result of the missing information, the dirty map is not only incomplete but has non-physical negative regions.

## 10.7 Calibrating the Data

At the end of Section 10.4 we set aside the issue of calibration in order to focus on the essential formalities of 2D imaging. We now turn to the practical steps by which a data set is prepared for translation into an image. The essential steps are the removal of bad data ('flagging') and calibration. We defer discussion of flagging to Section 10.14. Calibration means turning the numerical outputs from the cosine and sine correlators into complex visibilities on a known scale and correcting for their time and frequency dependence.

The process requires an appreciation of the wide range of effects which can arise during the signal's passage through the atmosphere and the instrument. The list can become long and increasingly subtle, particularly in the case of:

– imaging with high dynamic range;
– polarization imaging;
– imaging with wide fields of view;
– astrometry, geodesy, and imaging with VLBI.

We will encounter these effects in Chapter 11. Here we seek only to establish those basic features of interferometric calibration that are sufficient for imaging limited fields of view with adequate dynamic range. For simplicity, assume that the positions of the antennas are known, to a small fraction of a wavelength, and that the tracking centre $s_0$ is well within the delay beam.

The magnitude of the signal voltage input to the correlator from one of the antennas in a baseline is a product of the source flux density (modified by absorption in the atmosphere), the antenna voltage gain $\mathcal{A}(l, m)$, and the overall electronic voltage gain. The phase of this voltage is affected by variations in the path delay through the atmosphere and in the electronic path through to the correlator. We can describe these time-variable effects *for each antenna i* by means of a complex antenna gain $g_i(t) = a_i(t)e^{i\phi_i(t)}$. The observed visibility on a baseline $i, j$ can then be written in terms of the true visibility:

$$V_{ij}^{\text{obs}}(t) = G_{ij}(t)V_{ij}^{\text{true}}(t) + \epsilon_{ij}(t), \tag{10.11}$$

where

$$G_{ij}(t) = g_i(t)g_j^*(t) = a_i(t)a_j(t)e^{i[\phi_i(t) - \phi_j(t)]} \tag{10.12}$$

and $\epsilon_{ij}(t)$ is the thermal noise term (given by Eqn 9.57), which is distinct *for each baseline* and hence is additive in Eqn 10.11 (there are other terms, which we will encounter in Section 10.14). Given a source model (in this context most simply a point) the $V_{ij}^{\text{obs}}(t)$ data can be factorized into telescope-related complex gains $g_i(t)$ (see also Section 10.10.2).

With this in mind the calibration process can be broken down into a series of steps:

– set the flux density scale;
– correct for frequency dependence (bandpass calibration);
– correct for time dependence.

We consider the first two together and then separate the third into amplitude and phase corrections.

### 10.7.1 Flux and Bandpass Calibration

The visibility amplitude has units of flux density (see Section 9.5) and hence the observed amplitudes, which at this stage are simply a set of numbers, must be rescaled into numbers representing correlated janskys. As first outlined in Section 9.8.2, this is most often achieved by observing a strong compact source of known flux density and transferring the $a_i$ gain solutions to the data on the target source. The flux calibrator is chosen from one of a small group of sources with accurately known flux densities that vary only slightly (a few per cent) and slowly with time. Flux calibrators (Figure 10.11) need not be perfect point sources; for example in the northern hemisphere 3C48, 3C138, 3C147, and 3C286 are commonly used as standards although they are partially resolved by the JVLA and eMERLIN, depending on the baseline length and frequency. The resolution effects can be corrected using source models. Ultimately the flux standards rely on measurements of the strongest (but larger) sources such as Virgo A, Cygnus A, etc. The most intense source at low frequencies, Cassiopeia A, decreases in flux by about 1% per year. For many years the well-documented measurements of Baars *et al.* (1977) provided the ultimate flux standards, but a new set of flux standards for centimetric interferometry has been established, via reference to the planet Mars, by Perley and Butler (2013).

The instrumental response across the reception band will not be constant with frequency and this dependence must be corrected. The correlated data are, in practice, in the form of many discrete narrowband (typically 1 MHz) frequency channels. On each baseline the time-dependent gains $G_{ij}(t)$ in Eqns 10.11 and 10.12 can be generalized by including a frequency-dependent term $B_{ij}(\nu)$ which changes very slowly, thus:

$$G_{ij}^{obs}(\nu, t) = G_{ij}(t)B_{ij}(\nu). \tag{10.13}$$

Data from a short integration on a strong continuum source (which could be the same source as that used for the flux calibration) can be factorized into frequency-dependent complex gains for each channel for each telescope; together these represent the 'bandpass' of each receiver chain. The spectrum of the bandpass calibrator must be featureless over the band of interest but, as long as it is well known, any spectral curvature can be accounted for in the software. The telescope-based solutions which equalize the gains across each bandpass are then applied to the data on the target source. Bandpass calibration is a routine step in the analysis of continuum data but the quality of this calibration plays a critical role in spectroscopy (see Section 11.1), since frequency-dependent errors will appear as spurious features in the spectra.

### 10.7.2 Amplitude Calibration

There will be temporal variations in the individual antenna amplitude gains $a_i(t)$ arising from a combination of changes in:

– atmospheric absorption (see Section 9.7);
– the electronic path gains;

– antenna pointing, due to a combination of imperfectly modelled gravitational deformation at different elevations, wind loading, differential heating of the structure especially at dawn and dusk, and atmospheric refraction at low elevations.

At centimetre and longer wavelengths these amplitude effects are generally relatively small and vary only slowly in time. Repeated checks on the variations are made by switching between the target source and a nearby compact source, with the angular separation and the timescale of switching set by the requirements of phase calibration (see below). The antenna-based calibration solutions are then interpolated in software for the times when the target source is being observed. By this means the visibility amplitudes can be externally calibrated to a few per cent or better. At millimetre wavelengths, however, atmospheric attenuation is more severe and additional procedures are required (see Section 11.3). For VLBI, because of the problem of identifying suitable compact calibration sources, a completely different primary strategy based on the rms noise level is adopted (see Section 11.4).

### 10.7.3 Phase Calibration

Phase calibration is a more challenging task than amplitude calibration since the visibility phase, i.e. the relative positioning of the Fourier components, has a much greater effect on the synthesized image than their amplitudes. Oppenheimer and Lim (1981) were among the first to demonstrate the effect of swapping over the amplitudes and phases of the Fourier transforms of two distinct images; the appearance of each inverse-transformed 'hybrid' image is dominated by the phase. There are many other examples on the web, often based on human faces. At all wavelengths the phase variations are often severe enough to preclude useful imaging without external calibration.

Under our idealized restriction that there should be no varying geometric terms in the path delay, temporal variations in the path delays to each antenna occur due to a combination of changes in:

– the troposphere and the ionosphere (see Section 9.7);
– the electronic signal paths to the correlator.

In addition, interferometry requires that the frequency standard (the effective local oscillator) at each antenna should be identical (see Section 9.8.3). Hence phase errors can also occur due to:

– imperfect corrections in the the real-time frequency transfer system;
– unmonitored drifts in the independent atomic frequency standards used in VLBI.

An extensive treatment of these issues is given in TMS, Chapters 7 and 9.

Combining all these factors into the relative phase stability of a baseline sets an arbitrarily defined interferometer 'coherence time' (to be distinguished from the RF coherence time $\tau_{coh} \sim 1/\Delta\nu_{RF}$), during which the position of the fringe pattern is stable to a fraction

Table 10.1. *Representative coherence times at metre and centimetre wavelengths in 'good' and 'bad' propagation conditions*

| Frequency | Good | Bad |
|---|---|---|
| 70 MHz | 1 min | 5 s |
| 150 MHz | 3 min | 30 s |
| 500 MHz | 10 min | 1 min |
| 1.5 GHz | >30 min | 5 min |
| 5 GHz | >30 min | 5 min |
| 22 GHz | 5 min | 30 s |

of a fringe period and the complex visibility samples can be added coherently without significant loss. A relative phase error $\Delta\phi$ will reduce the amplitude by $\cos(\Delta\phi)$ and hence, for a probability distribution of phase errors $p(\Delta\phi)$, the reduction in amplitude will be

$$\int_{-\infty}^{\infty} p(\Delta\phi)\,d(\Delta\phi)\cos(\Delta\theta). \tag{10.14}$$

If the phase fluctuations have a Gaussian probability distribution with rms deviation $\sigma$ from a zero mean,

$$p(\Delta\phi)\,d(\Delta\phi) = \frac{1}{\sigma\sqrt{2\pi}}e^{-(\Delta\phi^2/2\sigma^2)}, \tag{10.15}$$

integration then yields a reduction in the amplitude to $e^{-\sigma^2/2}$ of its true value (to first order, $e^{-\sigma^2/2} \approx 1 - \sigma^2/2$). To limit the loss due to phase fluctuations to $< 1\%$ requires $\sigma < 0.14$ rad, i.e. an rms of $< 8°$.

Usually the atmospheric variations dominate and the relative phase variations may not be Gaussian; also they depend on the separation of the antennas (see Section 9.7). Empirically the values listed in Table 10.1 are reasonable approximations to the coherence times on many interferometers operating at metre and centimetre wavelengths. But at periods of high solar activity the coherence times at 1.5 GHz and below can be markedly less than those in Table 10.1.

How frequently one needs to switch to a nearby calibration source and how far away that source can be depends on the coherence time and the size of the isoplanatic patch, as outlined in Section 9.7. However, in the frequency range 1.5–15 GHz and for instruments ranging from the VLA-A array, through eMERLIN, and on to the international VLBI arrays (baselines of tens of km to thousands of km), observing a calibrator up to 5 degrees from the target source once per 10 minutes usually produces results acceptable for an initial map. Faster switching and smaller target–calibrator distances inevitably produce improved results; in contrast, on shorter baselines the requirements on temporal and spatial separations can be relaxed somewhat. The derived phases on a calibrator $\phi_i$ are, with corresponding amplitudes $a_i$, interpolated to provide antenna-based corrections to the data on the target source.

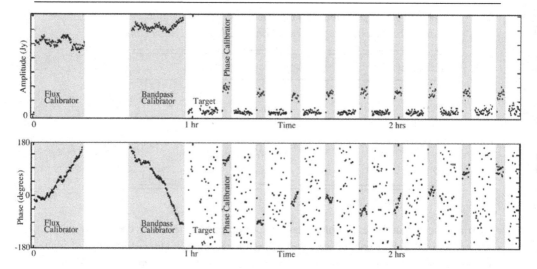

Figure 10.11. External calibration of visibility data on a single baseline: (top) amplitude; (bottom) phase. In the first hour a flux calibration source has been observed, followed by a bandpass calibration source, each for ~ 20 minutes; in some cases this can be the same source. In the remainder of the plot the interferometer switches every ten minutes from the target source to the phase calibration source. (Courtesy of Nick Wrigley.)

At the longer wavelengths calibration of the ionosphere-dominated phase fluctuations (see Section 9.7.2 ) becomes increasingly challenging, and we defer discussion to Section 11.6. At millimetre wavelengths the phase calibration problem is also harder than at centimetric wavelengths. We briefly discuss the calibration of mm-wave arrays in Section 11.3.

## 10.8 Producing the Initial Map

### *10.8.1 Gridding and Weighting the Data*

Synthesis imaging is facilitated not only by working within a field of view suitable for a 2D Fourier transform but also by the use of the fast Fourier transform (FFT) algorithm, whose speed makes the whole process viable. If data are equispaced in one Fourier domain and the number of values in the set is an integer power of 2, the FFT carries out the transformation into the other Fourier domain enormously faster than the direct Fourier transform (DFT). For a data set with $n$ values the speed advantage is $n/\log_2 n$; thus, for $n = 1024$ the factor is 102.4 while for a $1024 \times 1024$ square matrix of values it is 52428.8.

Not all this speed advantage can be realized, however, since the observed visibility data, which are irregularly located on the $u, v$ plane, first have to be turned into values on a grid with uniform spacing. The maximum extent of the visibility data is in general different in the $u$ and $v$ directions but, for simplicity, let us consider the case where the grid is square with equal spacing wavelength increments in both the $u$ and $v$ directions. There is a variety of ways to grid the data, as first described by Thompson *et al.* (1974), but the now-standard

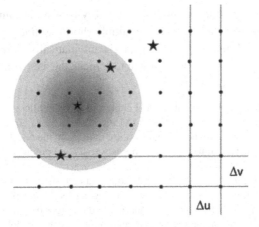

Figure 10.12. Gridding. An observed visibility sample (the star symbol at the centre of the shaded circle) is distributed over nearby points on a uniform grid, using a convolution function indicated by the graded circular patch; the process is repeated for all samples.

method is to convolve each observed data point with a function $G(u, v)$ which distributes its flux amongst the surrounding grid points. The process is illustrated in Figure 10.12 and, starting from the relationships developed in Section 10.6.2, can be written as:

$$V(u, v)_{\text{grid}} = V(u, v)_{\text{obs}} \times S(u, v) * G(u, v). \tag{10.16}$$

One should always be aware that, when regularly spaced visibility data are Fourier transformed into the image plane, multiple copies of the image field regularly spaced at $1/\Delta u$ are automatically produced (see Appendix 1, Figure A1.1(c)). Furthermore, if the synthesized image size is not large enough to contain all the significant sources of emission in and around the target area, the far-off-centre emission will be aliassed and appear at the wrong position within the synthesized field. This is directly analogous to aliassing in the time domain (see Appendix 3). If the synthesized image is chosen to be too small, $\Delta u$ is too large and does not sample the observed data finely enough.

To eliminate aliassing one would need to convolve with a 2D sinc function whose Fourier transform is rectangular and goes to zero exactly at the edge of the chosen image field. It turns out that to achieve this sharp cutoff the sinc function would have to be evaluated at every element in the $u, v$ grid for every observed visibility point; the resulting computational load would then nullify the advantage of the FFT! The mathematical details of practical gridding functions are beyond our scope, but special functions have been constructed which minimize potential aliassing while not requiring too large a computational load. Nevertheless, when dealing with data from most modern arrays the gridding step dominates the Fourier transform step.[4]

At this point in the image construction process one can introduce additional multiplicative weightings $W(u, v)$ of the sampled visibility data, whose effect is to modify $S(u, v)$

---

[4] Further convolving functions can be applied at gridding time, to account for direction-dependent phase and amplitude effects; see Sections 11.5 and 11.6.

and hence the shape of the dirty beam. This flexibility is desirable since synthesis arrays often produce many more data points from multiple short baselines close to the origin of the $u, v$ plane. As a result the synthesized beam is automatically weighted towards lower resolution, which may or may not be desirable to achieve the astronomical goal. There are three standard responses to this situation.

– *Natural weighting*    Accept the concentration of data from short baselines and a larger synthesized beam diameter. However, in occupied cells in the $u, v$ data grid, weight the contributing points according to the inverse square of their signal-to-noise ratio (as in the weighted mean in basic error propagation). This scheme maximizes point source sensitivity and produces the lowest rms noise in the image.
– *Uniform or density weighting*    Assign a weight to a grid cell inversely proportional to the local density of visibility points and hence make the sampling across the $u, v$ plane more uniform. The effect is to de-emphasize the short baselines and allow the longer baselines to have more effect. This scheme provides higher angular resolution and minimizes the sidelobes at the expense of a higher noise level in the image.
– *Robust weighting*    This allows the user to select an adjustable compromise between extremes of natural and uniform weighting; this scheme is also called optimum or Briggs weighting.

The difference between these choices depends on the $u, v$ coverage but may result in changes to the synthesized beam diameter and to the rms noise level in the image by factors up to 2.

### 10.8.2  Dimensions in the Image and the Visibility Planes

The pixel size $\Delta x$ in the image is set by the requirement to Nyquist-sample the synthesized beam. For simplicity, just consider the largest extent of the $u, v$ coverage, say $u_{max}$. The minimum (half-power) size of the synthesized beam in one direction is therefore $\sim 1/u_{max}$, depending in detail on the weighting function $W(u, v)$. Satisfying the Nyquist sampling theorem (see Appendix 3) requires $\Delta x < 0.5/u_{max}$, and commonly a pixel size three times smaller than the smallest beam dimension is chosen, as is the case for sampling single telescope beams (see Section 8.9.2). There is little or no advantage in choosing finer-scale gridding.

One then chooses the number of pixels $N$ to be such that the image size (invariably square) is large enough to include all the emission from the target region above the noise level. The choice of $N$ is a trade-off between practicality (smaller images are quicker to process) and conservatism about one's knowledge of the target region and its surroundings. To take full advantage of the FFT, $N$ is invariably chosen as a power of 2 and typically will be 1024 or 2048 for the present restricted 2D imaging case.

The choice of dimensions in the image plane automatically has implications for the $u, v$ plane. Thus the pixel spacing implies 'predictions' of the visibility function beyond $(0.5/u_{max})^{-1} = 2u_{max}$ cycles radian$^{-1}$. We will return to this point in the next section. The image size $N\Delta x$ implies that the grid spacing in the visibility plane $\Delta u$ is given by $1/N\Delta x$ and, as noted previously, aliasing will occur if the image is too small or, equivalently, the $u, v$ gridding is too coarse to capture the variations in the visibility function.

## 10.9 Non-Linear Deconvolution

As we described in Section 10.6.2, the effect of missing data in the $u, v$ plane is to produce distortions in the synthesized image. The maximum baseline sets an upper limit to the angular resolution; the shortest baseline sets a lower limit to the angular scale of low-brightness emission which can be imaged; the sharp transition to no data beyond the maximum baseline and the 'holes' between baselines produce sidelobes in the synthesized beam. Some of these sidelobes must be negative since the total power in the field is not measured. As a result the typical 'dirty map' from a synthesis array is only of limited use since the sidelobes from bright compact regions may obscure potentially important weaker structure.[5]

The image obtained from a set of visibility data with zero assigned to unmeasured regions was originally called the principal solution since it is only one of the images consistent with the data. The actual visibility function could have arbitrary values at the points where no measurements are made but their contribution to the synthesized image, after convolution with the dirty beam, must vanish identically. The visibility data are therefore consistent with an infinite set of 'invisible distributions' (e.g. Bracewell and Roberts 1954). Clearly it is the act of zeroing the visibility function in the unmeasured regions which leads to the unphysical and undesirable features of the dirty map. Whilst it is the easiest route to an image, we know full well that $V(u, v)$ does not plunge down to zero as soon as we stop measuring it! The challenge in the 1970s was, therefore, to develop a method which provides plausible visibility values where none were measured and thereby to make the synthesized image behave more sensibly. This challenge was met by Jan Högbom with his development of the CLEAN method (Högbom 1974). CLEAN seeks to undo the effect of convolution of the true sky brightness distribution by the dirty beam, or point spread function (psf), and hence is dubbed a 'deconvolution' algorithm.

An aside is pertinent here. In the general arena of digital image processing there is a range of deconvolution methods which improve the appearance of the image, typically by increasing the resolution and/or by removing a blurring effect. A standard approach is to divide the image transform by the Fourier transform of the psf, whose multiplicative effect downweights the high spatial frequencies; division therefore returns them to their true values. All such *linear* deconvolution algorithms depend on reweighting the measured Fourier components which, for a fully filled aperture, are continuous out to the diffraction limit. In Wiener filtering a signal-to-noise ratio term dependent on the spatial frequency is included in the reweighting function to prevent excessive noise amplification. This addition of *a priori* information improves the reliability of the reconstruction.

---

[5] While this is always true, the first and second generation E–W arrays were arranged to sample the $u, v$ plane with uniform spacings in order to concentrate the sidelobes at well-defined positions across a synthesized image; these were termed 'grating' or 'ring' sidelobes for obvious reasons. This arrangement makes the dirty image relatively easy to interpret, without deconvolution, if the grating lobes fall outside the emission region of interest. However, the downside of this array configuration is the degeneracy of repeated interferometer spacings, which limits the coverage of the $u, v$ plane for a given number of antennas. As an example the WSRT has 14 antennas, hence 91 baselines, but fewer than 40 different spacings for any one configuration. See the example in the Supplementary Material.

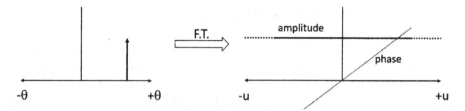

Figure 10.13. Fourier transform of an off-centre delta function.

Linear deconvolution methods are not applicable to synthesis imaging; for such imaging the aperture is not filled and the process of division in the unmeasured regions of the $u, v$ plane would be undefined. The inverse problem, from visibilities to image, is therefore said to be ill-posed: the visibility data are not, of themselves, sufficient to determine the sky brightness distribution unambiguously. The sidelobes are simply the image plane representation of our imperfect knowledge of the visibility function. The solution must lie in supplying information, over and above the measured data, based on *a priori* knowledge of, or assumptions about, the properties of the target region. Note that we have already supplied *a priori* information in the form of the Hermitian property of the visibility plane, which follows because the sky brightness is real and positive.

### 10.9.1 The CLEAN Algorithm

The underlying philosophy of CLEAN can be simply described. If a dirty map $B^{\mathrm{d}}(x, y)$ is dominated by a response which closely resembles the dirty beam $D(x, y)$ then one need not agonize about invisible distributions but immediately jump to the conclusion that there is a single point source convolved with $D(x, y)$.[6] To represent the sky, a delta function, with flux density determined from the height of the central peak in $B^{\mathrm{d}}(x, y)$, is placed in a blank image. Note that the act of placing a single point source in the image automatically makes a prediction about the behaviour of the visibility function between the measured values. The amplitude is the same over the entire $u, v$ plane and the phase, $e^{-i2\pi(ul+vm)}$, systematically increases as the distance from the origin increases; this is simply the Fourier shift theorem, illustrated in one dimension in Figures 9.11 and 10.13 (see also Appendix 1). In this sense CLEAN is a non-linear deconvolution algorithm, in that it generates new values for the visibility function where none previously existed. These are new *spatial* frequencies and there is a direct analogy with the action of a non-linear mixer, which generates new *temporal* frequencies (harmonics and mixture products) from the original input signals.

The philosophy of recognizing point sources can be developed to include extended distributions; thus a significant correlation between $B^{\mathrm{d}}(x, y)$ and $D(x, y)$ suggests a contribution from a point source at the position of maximum correlation. At the position of the peak in the image one therefore subtracts a $D(x, y)$ pattern from the entire map centred at this pixel.

---

[6] 'If it looks like a duck, if it walks like a duck, if it quacks like a duck – then it is a duck!'

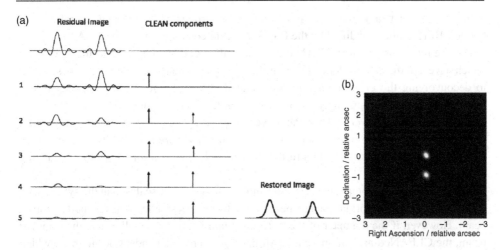

Figure 10.14. (a) A 1D schematic illustration of the CLEAN process, showing the original (noise-free) image progressively reduced, the CLEAN components, and the restored image. (b) The CLEAN image made from the dirty beam and dirty map in Figure 10.10.

However, as this pixel will have contributions from the sidelobes of all the other point sources in the image, one should only subtract a scaled-down version, $\gamma D(x, y)$, where $\gamma < 1$ (typically $\sim 0.1$) is the 'loop gain'. This process is illustrated in one dimension at the top of Figure 10.14(a). The process of subtracting 'CLEAN components' is repeated sequentially[7] until the residual sidelobes in the image are reduced to below the thermal-noise level (Section 10.12).

At the end of the process, if all has gone well, we are left with a noise-like residual image plus a 2D distribution of point sources. The latter distribution is not only hard to interpret but is noisy on the pixel-to-pixel scale ($< 0.5/u_{max}$), which, by definition, is smaller than the synthesized beam diameter. The reason is that the basic CLEAN process has no memory between iterations and enforces no correlation between neighbouring pixels. There can be, therefore, spurious fine-scale detail in the component distribution, which corresponds to incorrect predictions about the visibility function, i.e. extrapolations beyond $u_{max}$. To control this effect one simply smooths the component map with a 'restoring beam', usually a Gaussian whose FWHM is the same as that of the central peak in $D(x, y)$. Note that the effect of smoothing by the conventional CLEAN beam is to weight down heavily the information from the longest baselines,[8] and thus the Fourier transform of the restored CLEAN map does not fit the observed visibility data. The last step in producing a realistic reconstruction of the sky is to add back the residual noise image. This final image has the

[7] The mathematical justification for CLEAN came only after its widespread adoption. The visibility function of a point source is a pair of real and imaginary sine waves (the alternative to the description in terms of amplitude and phase used in Figure 10.13). One may therefore appreciate that Schwarz (1978) was able to show that CLEAN is equivalent to the least squares fitting of sinusoids to the measured visibility data. The sum of these sinusoids 'predicts' the visibility function over the $u, v$ plane, implicitly interpolating in the gaps and extrapolating beyond the maximum baseline.

[8] See Wilkinson (1989b) and the discussion in Section 9.4.3.

units janskys per beam area, which is related to the true sky brightness temperature (see Section 10.12), albeit modified by the Gaussian beam convolution in the CLEANed area.

Despite these caveats the CLEAN process usually works well and the resulting images, which have a uniform well-defined resolution, are easy to interpret visually. It is best suited to deconvolving the effect of $D(l, m)$ on isolated compact emission features but it still works well, perhaps surprisingly so, on more complex sources. An intervention which can help the process to converge when the $u, v$ coverage is sparser than desirable is to restrict the regions in the image from which subtractions of $D(l, m)$ are allowed. These are termed CLEAN 'windows' and represent a further way to add *a priori* information to the imaging process.

The set of CLEAN components contains information on finer scales than the synthesized beam diameter. This information may be explored by using a smaller than standard restoring beam. However, if the diameter of the restoring beam is $\alpha$ times smaller than the standard beam, the CLEANed area of an image will have $\alpha^2$ more independent beam areas, which could be more than the visibility data can reliably constrain. *Superresolution* is therefore of limited usefulness for imaging any but the simplest sources; it should be used with extreme care (see also Section 8.9).

CLEAN images can have significant artefacts. For example, as first recognized in early VLA snapshot surveys, flux densities of compact sources can be systematically underestimated. It is also possible to 'over CLEAN' the noise in an image if too lax a stopping criterion is used. Finally, since CLEAN represents the sky as a set of independent point sources, another problem, not unexpectedly, occurs in representing regions of smoothly varying brightness. Here CLEAN tends to produce spotty or corrugated artefacts, which arise from the fact that when $\gamma D(x, y)$ is subtracted from a smooth area it leaves an inverted copy of itself imprinted on the residual image. The next CLEAN iteration will, therefore, tend to find a peak at positions around the first negative sidelobes, which are now the most positive features. This can induce spurious regularities whose underlying cause is imperfect implicit interpolations into the $u, v$ gaps (see Schwab 1984 and Wilkinson 1989b). As a corollary, CLEAN is also profligate in seeking to describe a large smooth area of image brightness with huge numbers of point sources when, at another extreme, the area might be described with an elliptical Gaussian requiring only six parameters: peak flux density; sky position $x, y$; diameters $\Delta x, \Delta y$; position angle of major axis. Since its introduction in the 1970s a series of successful modifications to the basic process have been devised. These include: the Cotton–Schwab algorithm (Schwab 1984), which greatly speeds up the subtraction process and allows multiple fields to be handled simultaneously; multi-scale CLEAN (Cornwell 2008), which models the sky brightness by a summation of components of emission having different size scales; WSCLEAN (Offringa *et al.* 2014), which handles better the imaging requirements of wide fields and low frequency arrays (see Chapter 11). Given the 'industrial' scale of its usage, in various forms and circumstances and over many decades, a large amount of empirical knowledge has accrued about the CLEAN algorithm and the characteristics of the images it produces. These are best learned by handling real data, with experienced advice on tap.

## 10.10 Correcting the Visibility Data

### *10.10.1 Closure Quantities*

Accurate visibility phase information is vital for reconstructing images from interferometer data. The Fourier components need to be correctly aligned if the image is not to contain spurious features, whether or not deconvolution is required. In the 1960s and early 1970s it was possible to calibrate the visibility phase with arrays a few km in size, and ground-breaking radio images with resolutions of 5–10 arcsec were produced. At the same time VLBI, with its potential of milliarcsec imagery, was beginning to reach a level of technical maturity, and intra- and intercontinental arrays began to be coordinated. But a combination of imperfect knowledge of telescope and source positions, frequency drifts between the independent atomic clocks at each site, and uncorrelated atmospheric propagation delays meant that these VLBI systems were phase-unstable, thereby rendering 'classical' synthesis imaging impossible.

Progress was made by recognizing that there is always some visibility phase information available in an array, a fact which had been acknowledged in the early 1950s at Jodrell Bank (Jennison 1958) but forgotten by the radio astronomy community until it was rediscovered by Rogers *et al.* (1974) in the context of VLBI. This is the so-called *closure phase*.

This important idea is best appreciated by considering Figure 10.15, which shows three telescopes constituting a triangle of baselines. There is an atmospheric inhomogeneity over telescope 1 corresponding to a propagation delay of $\tau_{\mathrm{prop}}$ and hence a phase delay of $2\pi\nu\tau_{\mathrm{prop}}$. Over telescopes 2 and 3 there is no propagation delay. The phase assigned to the spatial frequency component on a baseline $ij$, the 'observed' phase $\Phi_{ij}$, is the true visibility phase $\psi_{ij}$ plus the difference between the propagation phase shifts for the two telescopes. Thus, proceeding clockwise around the triangle and taking care with the ordering, on baseline 12 we would measure

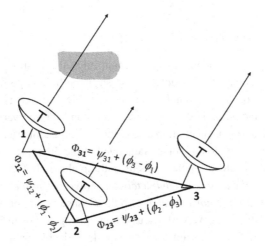

Figure 10.15. Illustration of closure phase.

$$\Phi_{12} = \psi_{12} + \phi_1 - \phi_2 = \psi_{12} + \phi_1, \tag{10.17}$$

and similarly on baseline 31

$$\Phi_{31} = \psi_{31} + \phi_3 - \phi_1 = \psi_{31} - \phi_1. \tag{10.18}$$

On baseline 23 there are no propagation delays, hence we have simply

$$\Phi_{23} = \psi_{23}. \tag{10.19}$$

This is illustrated schematically in Figure 10.15. The single propagation delay affects two of the baselines but not the third, and if one forms the arithmetic sum of the observed phases then the unwanted phase shift $2\pi \nu \tau_{\mathrm{prop}}$ cancels out exactly, leaving a good observable, the closure phase $C_{ijk}$. Thus, in this example,

$$C_{123} = \Phi_{12} + \Phi_{23} + \Phi_{31} = \psi_{12} + \psi_{23} + \psi_{31}. \tag{10.20}$$

Since the visibility function is Hermitian, $\psi_{ij} = -\psi_{ji}$ and Eqn 10.20 can also be written as

$$C_{123} = \Phi_{12} + \Phi_{23} + \Phi_{31} = \psi_{12} + \psi_{23} - \psi_{13}. \tag{10.21}$$

In general there will be propagation phase shifts associated with each telescope but nevertheless, on forming $C_{123}$, their effects cancel identically. Thus *the closure phase is a linear combination of visibility phases containing information only about the brightness distribution of the source.* The cancellation of telescope-related phase shifts can be impressive; see Figure 10.16.

The closure phase emerges naturally from taking ratios of Eqns 10.11 and 10.12 describing the observed visibilities and the time-variable complex gain 'miscalibrations' for each telescope in a baseline pair. Thus, around a triangle of baselines (ignoring the noise terms and taking the time dependence as implicit) we obtain

$$\frac{V_{12}^{\mathrm{obs}} V_{13}^{\mathrm{obs}}}{V_{23}^{\mathrm{obs}}} = \exp[i(\psi_{12} + \psi_{23} - \psi_{13} + \phi_1 - \phi_2 + \phi_2 - \phi_3 - \phi_1 + \phi_3)]$$

$$= \exp[i(\psi_{12} + \psi_{23} - \psi_{13})], \tag{10.22}$$

and the argument of the exponential can be recognized as the closure phase in Eqn 10.21.

The vital criterion is that the unwanted phase shifts be 'telescope factorizable', i.e. not related to the correlation of the signals from the two telescopes as is $\psi_{ij}$. This is true for propagation effects (including system-related effects) and for any variations in frequency standards at each telescope. Cancellation also extends to all geometrical information (baseline vectors and source positions) which is encoded in the difference in the time delays to each telescope. Thus the closure phase tells us nothing about absolute source or telescope positions. Note that *if the source is completely unresolved them the closure phase should be identically zero within the noise level* and so the closure phases provide a definitive test of the purity of the data paths. We will discuss 'non-closing' errors in Section 10.14.4.

A 'closure amplitude', unaffected by telescope-factorizable gain variations, was first introduced by Twiss *et al.* (1960) (although Smith 1952 used ratios of visibility amplitudes from three telescopes and looked ahead to something akin to the closure amplitude). Like

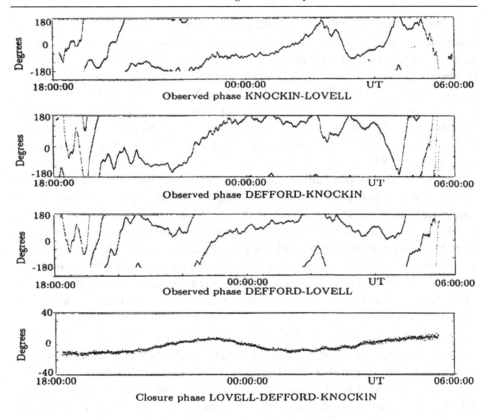

Figure 10.16. Observed phases on three MERLIN baselines and the corresponding closure phase.

the closure phase, it was forgotten until rediscovered, by Readhead *et al.* (1980). The closure amplitude can be formed from a ratio of four observed visibility amplitudes:

$$\frac{V_{12}^{obs} V_{34}^{obs}}{V_{13}^{obs} V_{24}^{obs}} = \frac{A_{12}A_{34}a_1a_2a_3a_4}{A_{13}A_{24}a_1a_2a_3a_4} = \frac{A_{12}A_{34}}{A_{13}A_{24}}. \qquad (10.23)$$

In an $N$-antenna array there are many possible ways of forming closure phases but only $(N-1)(N-2)/2$ are independent. This can be understood simply. There are $N(N-1)/2$ baselines but they are corrupted by only $N-1$ telescope-based phase errors since the absolute phase at the reference telescope has no significance and can be set to zero. The number of independent closure phases is therefore $N(N-1)/2-(N-1) = (N-1)(N-2)/2$. There are $N$ telescope-related amplitude errors and so the number of independent closure amplitudes is $N(N-1)/2 - N = N(N-3)/2$.

### 10.10.2 Self-Calibration

The challenge of making images when the interferometer phase cannot be measured was first addressed by Baldwin and Warner (1976, 1978). They introduced the idea of 'phase-less' aperture synthesis. This idea was related to some in X-ray crystallography, which is not surprising since both disciplines require the target to be reconstructed by the manipulation of imperfect samples in the Fourier plane. Baldwin and Warner coined the term 'hybrid mapping' for imaging with a combination of calculated phases and observed amplitudes. The reintroduction of the closure phase idea moved hybrid mapping forward. In particular, Readhead and Wilkinson (1978; see also Wilkinson 1989a) combined the $(N-1)(N-2)/2$ independent $C_{ijk}$s with $N-1$ calculated phases to produce estimates of the $N(N-1)/2$ visibility phases. Hybrid mapping was an important stage in the development of modern synthesis imaging, and its development can be traced in the papers by Ekers (1983), Pearson and Readhead (1984), and Wilkinson (1989a).

The direct use of *baseline-related* closure phases, while useful for small-$N$ arrays, rapidly becomes unwieldy as $N$ grows. For example, with $N = 27$, as in the JVLA, there are 325 independent closure phases to deal with. In the late 1970s, therefore, radio astronomers at the VLA reached the conclusion that array calibration should be antenna-based (see e.g. Ekers 1983), and Schwab (1980) developed an antenna-based correction algorithm aimed at VLA data. At around the same time, and independently, Cornwell and Wilkinson (1981) developed a similar antenna-based method whose first target was correcting data from the MERLIN array.[9] Antenna-based correction became known as 'self-calibration', for reasons which will become clear.

Our understanding starts from Eqns 10.11 and 10.12. The true visibility $V_{ij}^{\text{true}}(t) = A_{ij}(t)e^{i\psi_{ij}(t)}$ corresponds to one of the desired spatial frequency components, and the task is to solve for the independent amplitude and phase errors and hence recover $V_{ij}^{\text{true}}(t)$. It should be clear from the discussion above that, if one makes changes only to the $a_i$s and the $\psi_i$s, the visibility data from which the image is constructed must, perforce, be consistent with the closure quantities. Thus the closure and self-calibration approaches are essentially equivalent. The antenna-based approach allows much simpler data handling and more flexible use of *a priori* knowledge of the observed data.

Schwab (1980) solved for the complex gains appropriate to a particular integration period by a non-linear minimization of the weighted sum of squares $S$ of the moduli of the error vectors, between the observed data $V_{ij}^{\text{obs}}$ and the predicted data $g_i g_j^* V_{ij}^{\text{true}}$, determined from a 'trial map' (see below). Formally the minimization can be written

$$S = \sum_{i j i < j} w_{ij} |V_{ij}^{\text{obs}} - g_i g_j^* V_{ij}^{\text{true}}|^2, \tag{10.24}$$

where the summation is over all baselines in the array. A vector diagram applicable to the visibility phases on one baseline is shown in Figure 10.17. Weights $w_{ij}$ equal to the inverse of the variance of the data, $1/\sigma^2$, can be chosen to utilize the signal-to-noise information.[10] The visibility data are then corrected by applying the inverse of the gain solutions to the data, i.e. for the $i$th telescope, $[g_i(t)]^{-1} = [a_i(t)]^{-1}e^{-i\phi_i(t)}$.

As described well by Pearson and Readhead (1984), the optimum phase solutions are determined iteratively, moving between the visibility and image planes, in a procedure

---

[9]  At that time the array was called the Multi-Telescope Radio-Linked Interferometer (MTRLI).

[10]  Cornwell and Wilkinson (1981) gives a heuristic description allowing one to visualize how an antenna-based algorithm determines the telescope-based phase errors.

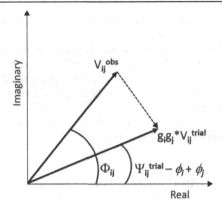

Figure 10.17. Vectors showing the predicted data and the observed data for a particular integration period on one baseline. The error vector (broken line) is invariably dominated by the differences between the phases.

pioneered by Readhead and Wilkinson (1978). The observed amplitudes (determined by external calibration) and estimated visibility phases are used to make a distorted dirty map, which is then CLEANed to produce a 'trial map' (the point sources from CLEAN). At any stage the trial map is only an approximation to the final one and so *all* the estimated visibility phases will contain errors. If the phase errors are significant then spurious brightness components, both positive and negative, will appear across the image. The trial map can therefore be regarded as the true map plus an error map. CLEAN supplies *a priori* image-plane constraints which drive the process to convergence as the iterations proceed. It removes the predictable sidelobes due to the incomplete $u, v$ coverage and, by enforcing positivity and compactness, it also steadily filters out the error map. The whole process is repeated until the CLEAN map stops changing.

The term self-calibration came about because the process solves for errors in the visibility data whilst at the same time using these data to form an image of the sky. This can be achieved only if the visibility data are partially redundant. In loose terms, if there are more than enough data to image the desired region of the sky in the absence of errors, the additional information can be used to solve for the errors. This picture makes it clear that imaging complex sources with sparse arrays and self-calibration is doubly difficult (see e.g. Wilkinson 1989a). Before concluding this section a few additional points must be made.

– The timescale for which the gain solutions are calculated (Eqn 10.24) is usually different from the primary integration time and can be chosen by the astronomer making the image. The primary integration time is set by the requirement not to 'time-smear' the data (see Section 10.14) while the 'solution interval' is the timescale over which the gains do not change significantly. At the end of the imaging iterations, the gain solutions should be smoothly varying to within the noise; 'glitches' are often associated with manmade RFI which has not been excised in the initial data-editing stage.
– We have implicitly assumed that the signal-to-noise ratio on each baseline is high. In practice this may not be the case. However, as long as the signal-to-noise ratios, on the timescale of the solution interval (see below), are at least 3 : 1, self-calibration is likely to provide an improvement in the

image. If this is not the case then great care must be taken since spurious point sources can be created in images made from noisy data.[11]

– We have presumed that there is only one complex gain error per telescope per solution interval, i.e. that the whole of the target field is in the same isoplanatic patch. This is usually the case at centimetre and millimetre wavelengths. For wide-field imaging at metre wavelengths there may be many isoplanatic patches within the field of view of the primary beam. A brief discussion of this situation is given in Section 11.6.

– Self-calibration has changed little since the mid-1980s (e.g. Cornwell 1986), albeit the arrays for which it is used are becoming increasingly complex.[12]

## 10.11 Missing Short Spacings

We have alluded several times to the facts that correlation interferometers do not measure total power and that short baselines are often not available; the implications were briefly discussed in Section 10.6.2. The observer must always be aware that by missing low-spatial-frequency Fourier components the array is insensitive to broad, smoothly varying, brightness regions and that this can have an impact on the scientific interpretation. Wilner and Welch (1994) analysed the effect quantitatively for models of low brightness emission and $u, v$ coverage, well-sampled out to $u_{max}$ but with no coverage inside $u_{min}$. For a Gaussian distribution with $\theta_{1/2} = 1/u_{min}$ (see Section 9.4.3) only 3% of the central brightness is recovered. This is a quantitative confirmation of the rule of thumb that structures larger than $\sim 1/u_{min}$ (or $\sim 1/v_{min}$) can go undetected. The classic manifestation of the lack of short spacings is that the source appears to sit in a negative-going bowl. An example is shown in Figure 10.18.

The lack of short spacings is most pressing when the primary beam $\theta_{PB}$ is comparable to or smaller than the target region. Ideally the $u, v$ plane should be sampled more frequently than $1/2\theta_{PB}$, but this is impossible owing to the physical limitation on small antennna separations. There are several potential solutions:

– scan the target regions with a single dish whose diameter is larger than the shortest baseline;
– add data from a separate array of smaller antennas;
– combine mosaics of the target region.

These situations occur most frequently in mm- and submm-wave synthesis imaging (Section 11.3). Mosaicing is discussed in Section 11.5.4.

## 10.12 Flux Density and Brightness Sensitivity

The rms noise level is measured, using a routine in the array data-reduction package, from the pixel statistics in an emission-free, sidelobe-free, area of the synthesized image. It may

---

[11] In situations where the signal-to-noise ratio in a coherence time is low, for example in the optical and millimetre and sub-mm bands, it can be advantageous to use the averaged-up closure quantities rather than the visibilities themselves; see e.g. Cornwell (1987); Wilkinson and Woodall (1991), Chael *et al.* (2018).

[12] In massively redundant arrays, such as HERA, novel applications of the closure relation are now emerging (Carilli *et al.* 2018).

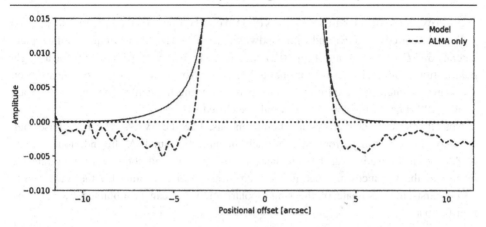

Figure 10.18. The effect of missing short spacings. A simulated model brightness distribution 'observed' with a single ALMA configuration shows the negative bowl. (Courtesy of Adam Avison.)

be possible to find such a region before the deconvolution step has been carried out. If this is not possible then care should be taken not to have CLEANed the image deeply into the noise. The convolution of the resulting low-level CLEAN components with a Gaussian restoring beam will alter the noise statistics.

We can derive expressions for the noise level of an array starting from Section 9.8.1, which gave the $1\sigma$ point source sensitivity of a two-element single polarization interferometer whose elements have identical performance. We now explicitly include a term $1/\eta_{corr}$ dependent on the digitization level at the correlator.[13] An array with $N$ antennas provides $N(N-1)/2$ independent two-element interferometers and very often the antennas and their receivers have essentially identical performance. The single-polarization $1\sigma$ point source sensitivity of such an array is therefore

$$\Delta S = \left(\frac{1}{\eta_{corr}}\right) \frac{2kT_{sys}}{A_{eff}[N(N-1)\Delta\nu\,\tau]^{1/2}}. \tag{10.25}$$

When $N$ is large, $[N(N-1)]^{1/2}$ tends to $N$ and hence Eqn 10.25 becomes

$$\Delta S = \left(\frac{1}{\eta_{corr}}\right) \frac{2kT_{sys}}{NA_{eff}(\Delta\nu\,\tau)^{1/2}}. \tag{10.26}$$

Allowing for the correlator efficiency, the single-polarization point-source sensitivity of an interferometer array with a large number of antennas of identical performance therefore approaches that of a single antenna with an area equal to the sum of the areas of its constituents. Following the argument in Section 9.8.1, an equivalent expression is

$$\Delta S = \left(\frac{1}{\eta_{corr}}\right) \frac{SEFD}{N(\Delta\nu\,\tau)^{1/2}}. \tag{10.27}$$

[13] $\eta_{corr} = 0.64$ for the single-bit (two-level) case or 0.88 for the two-bit (four-level) case. With multi-bit digitization, and hence more levels, the signal-to-noise penalty falls but there are many subtleties in the derivation of the actual value; see TMS, Chapter 8.

As an example consider a $\tau = 3 \times 10^4$ s (8.3 hr) observation with the JVLA ($\eta_{\text{corr}} = 0.93$) at C-band, centred on 6 GHz and with bandwidth $\Delta\nu = 2$ GHz. At this frequency the typical antenna $SEFD \approx 300$ Jy, and Eqn 10.27 therefore yields $\Delta S \approx 1.5$ microJy for a single polarization and $\approx 1.1$ microJy for dual polarization. Note that 8.3 hr is the time spent on the source of interest; including the time spent observing the calibration source, the total 'wall clock' time for the observation would be close to 10 hr.

Not all interferometer arrays are homogeneous, like the JVLA. The rms noise then depends on the actual performance of each antenna. For small $N$ the individual values of $\Delta S$ for each baseline can be calculated from Eqn 9.56 and combined (by adding the inverse of the variances and taking the reciprocal) to give a value for the array. However, a first-order estimate of the single-polarization $\Delta S$ can be obtained by rewriting Eqn 10.26 as

$$\Delta S \approx \left(\frac{1}{\eta_{\text{corr}}}\right) \frac{2k\overline{T_{\text{sys}}}}{A_{\text{eff,total}}(\Delta\nu\,\tau)^{1/2}} \text{ Jy},\tag{10.28}$$

where $\overline{T_{\text{sys}}}$ is a representative average system temperature of the set of receivers and $A_{\text{eff,total}}$ is the combined effective collecting area of the telescopes. In practice a variety of factors ensure that $\Delta S$ is always higher than predicted:

- the process of self-calibration, which feeds back noisy estimates of the complex gain errors. Cornwell (1981) analysed the effect, and for a point source and a good signal-to-noise ratio on each baseline the noise level is increased by $\sqrt{(N-1)/(N-3)}$, corresponding to 1.29 for $N = 6$ (e.g. eMERLIN) and to 1.04 for $N = 27$ (JVLA).[14]
- the weighting scheme applied at the time of gridding the data (Section 10.8.1). The equations above refer to natural weighting, where all the visibility points are treated equally. The other types of weighting will increase $\Delta S$ by tens of per cent (robust) and possibly up to a factor 2 (uniform).
- low-level residual errors in the data due to non-closing effects (see Section 10.14.4) and unrecognized RFI.

Source confusion (Section 10.12.1) can also be a factor at low resolution, and in general a noise level in the image within a factor $\sqrt{2}$ of the theoretical value is often thought acceptable.

For point sources the equations for $\Delta S$ can be interpreted simply in units of flux density, and they apply regardless of the sparsity of the array or the solid angle of the synthesized beam $\Omega_{\text{synth}}$ (the 'beam area'); the signal-to-noise ratio for a source of a given flux density will be the same. But when the emission is extended, the beam area matters since we are now measuring the source's brightness, equivalent to flux density per beam area.[15] The brightness temperature sensitivity of an array is therefore inversely proportional to its angular resolution squared. A few examples are illuminating.

---

[14] A plausibility argument for the noise increase is that there are a reduced number of 'good observables' with which to work. Thus instead of $N(N-1)$ good data points (an amplitude and phase on each baseline) one has $2N$ fewer, to give a total of $N(N-3)$. The noise level is therefore increased by a factor $\sqrt{(N-1)/(N-3)}$.

[15] If the beam can be approximated by a symmetrical Gaussian then $\Omega_{\text{synth}} = 1.13\theta_{1/2}^2$, where $\theta_{1/2}$ is the FWHM of the Gaussian.

We start with a compact source of flux density $S$. Using Eqn 5.13 we can obtain a lower limit to its brightness temperature:

$$T_B > \frac{S\lambda^2}{2k\Omega_{synth}}. \tag{10.29}$$

By definition a compact source's solid angle is unknown and so we must use the solid angle of the synthesized beam as an upper limit. The equality in Eqn 5.13 therefore becomes a limit in Eqn 10.29; the brightness temperature of the source must be at least as high as $T_B$.

The small size of synthesized beams from large interferometer arrays means that the detectable brightness temperatures are much higher than for single-dish observations. For example, a 10 mJy point source observed at $\lambda = 0.2$ m with a single dish with beam diameter $\theta_{1/2} = 10$ arcmin ($2.9 \times 10^{-3}$ radians) must have a brightness temperature greater than $\sim 15$ mK. In contrast, a medium-sized array such as the JVLA might produce a synthesized beam $\theta_{1/2} = 1$ arcsec at the same wavelength, in which case the brightness temperature limit ($1\sigma$) increases to 5400 K. In a VLBI array the resolution is in the milliarcsec range, at $\lambda = 0.2$ m, and the point-source brightness temperature limit typically exceeds $10^8$ K. The most intense compact radio sources in the centres of quasars have brightness temperatures exceeding $10^{12}$ K (see Section 16.3.10).

At the other extreme, compact arrays, often consisting of identical antennas and receivers, may have baselines comparable with the diameter of large single dishes. Arrays like this are appropriate for measuring the very low brightness temperatures of sources on the scale of arcminutes. An example target is the faint Sunyaev–Zel'dovich (S–Z) decrement associated with the hot gas in clusters of galaxies (see Chapter 17).

In this case an estimate of the brightness temperature sensitivity can be obtained by rewriting Eqn 10.25 in terms of the rms brightness temperature $\Delta T_B$ associated with an rms flux density $\Delta S$ (Eqn 5.13). The result is

$$\Delta T_B = \left(\frac{1}{\eta_{corr}}\right) \frac{T_{sys}\lambda^2}{NA_{eff}(\Delta\nu\,\tau)^{1/2}\Omega_{synth}}. \tag{10.30}$$

Now, for the array, $\Omega_{synth} = \alpha(\lambda/D)^2$ where $D$ is its overall dimension and $\alpha$ is a numerical factor which depends on the particular weighting $W(u, v)$ applied to the spatial frequency transfer function. Then, if the individual antennas have circular apertures of diameter $d$ and aperture efficiency $\eta$, we have $A_{eff} = \eta\pi d^2/4$. Substitution of these two expressions into Eqn 10.30 yields

$$\Delta T_B = \left(\frac{1}{\eta_{corr}}\right) \frac{4T_{sys}D^2}{N\alpha\eta\pi d^2(\Delta\nu\,\tau)^{1/2}}. \tag{10.31}$$

The numerical factor $4/\alpha\eta\pi$ is not identical to unity but is sufficiently close that a useful approximate form of Eqn 10.31 is

$$\Delta T_B \approx \left[\frac{T_{sys}}{(\Delta\nu\,\tau)^{1/2}}\right]\left(\frac{D^2}{Nd^2}\right), \tag{10.32}$$

which is the radiometer equation for a single dish multiplied by the reciprocal of the array-filling factor, i.e. the ratio of the summed areas of the antennas and the area of the synthesized aperture. Note that we have ignored the factor $1/\eta_{\text{corr}}$ since the the single-dish digital-data acquisition system will also contribute a similar factor. This result is a complement to the beam dilution factor for single antenna beams, which was discussed in Chapter 5.

### *10.12.1 Source Confusion*

In Section 8.10.3 we introduced the issue of source confusion in the context of single-aperture telescopes and noted that, in addition to receiver noise, there are low-level fluctuations arising from the superposition of faint sources. Filled-aperture observations have the merit of detecting sources regardless of their angular size but this is not true for interferometer observations. In view of the difference between the resolution of single apertures and that of most synthesis images, Eqn 8.38 is not appropriate. In synthesis images the rms fluctuations due to confusion can be estimated from the relation derived by Condon *et al.* (2012):

$$\sigma_{\text{con}} \approx 1.2 \ \mu\text{Jy} \left( \frac{\nu}{3.02 \ \text{GHz}} \right)^{-0.7} \left( \frac{\theta}{8 \ \text{arcsec}} \right)^{10/3}. \tag{10.33}$$

However, the reader should note that under the general heading 'confusion' there are other, systematic, effects to take into account. The effect of data imperfections is discussed in Section 10.15.2 whilst the effects of strong sources within the primary beam or in its sidelobes are introduced in Sections 11.5.3 (wide fields), 16.9.1 and 16.9.3 (source counts), and 11.6.1 (low frequency imaging).

## 10.13 Multifrequency Synthesis

As mentioned in Section 10.6 the limited coverage of the $u, v$ plane in sparse interferometer arrays can be improved with multifrequency synthesis (MFS). Since the spatial frequency to which a physical baseline is sensitive depends directly on the observing frequency, many spatial frequencies can be captured by simultaneously observing over a range (at least tens of a per cent) of narrower frequency bands and treating them all separately in the imaging process. At any instant, therefore, a single physical baseline can produce a range of points in the $u, v$ plane which lie along a line pointing to the origin. Combined with Earth's rotation the effect is to produce a much better filled $u, v$ plane than is available with a single frequency band. Figure 10.19 shows the effect for the eMERLIN array. From the resultant visibility data one can construct an image at a central reference frequency, which has much reduced reconstruction errors for sources extending over many synthesized beam areas.

The compensating challenge is to take account of the variations in the source brightness distribution as a function of frequency. Effectively one has to create images of the total intensity and of the spectral index – although the latter may not need to be very accurate. In some cases the spectral errors arise from simply ignoring the spectral index variations,

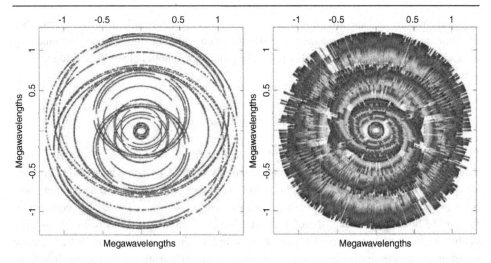

Figure 10.19. Multifrequency synthesis. (left) An example of narrowband MERLIN $u$, $v$ coverage: the gaps are associated with calibrations. (right) Multifrequency coverage, with the high frequencies in black and the lower frequencies in grey. (Courtesy of Nick Wrigley.)

which may be below the thermal noise level. When this is not the case (typically in high-dynamic-range imaging) the spectral errors must be recognized and removed. To do this, Conway *et al.* (1990) developed an extension to CLEAN which they termed 'double deconvolution'; other algorithms have since been developed, by Sault and Wieringa (1994) and Rau and Cornwell (2011). Explaining how these algorithms work is, however, beyond our present scope.

Multifrequency synthesis is only applicable to sources with a continuum spectrum, and the necessary broadband frequency coverage inevitably involves greater susceptibility to RFI. An additional challenge for high-dynamic-range wide-field imaging is that the frequency dependence of the primary beam of the antenna elements must be taken into account (see Section 11.6.2). Despite these caveats MFS is now a routine tool for synthesis imaging, and much experience has been accrued about its use.

## 10.14  Data Limitations

### 10.14.1  Flagging

For radio maps of the highest quality, detailed scrutiny of the visibility data is essential. Removing the errors, often called 'flagging', is the first task at the start of the imaging process. The principal sources of bad data are: telescope(s) off source; multifarious equipment malfunctions; and severe RFI. Some of these may be identified and tagged by the real-time software. Flagging the data can be tedious and time consuming, and so a first pass is carried out with automatic error recognition software. Other software can then display the data in a variety of ways (by telescope, by baseline, by spectral channel, etc.) to enable the astronomer to identify where further manual flagging is necessary. Each array will have the required software and specific on-line guidelines. If, as is increasingly the case, the number of elements in the array is large then the astronomer can often be quite ruthless in removing data and yet can retain enough for the task. The philosophy should always be 'using no data is better than using bad data'.

### 10.14.2 Time Averaging

In Section 9.1 we introduced the idea of the natural fringe rate, which falls to zero for a source at the pointing centre if the geometrical delay is perfectly compensated. For sources away from the pointing centre the delay is not perfectly compensated and hence there will be a residual fringe rate which increases with distance from the centre. The requirement for a wide field of view can, therefore, come into conflict with the desire to average the visibilities[16] for as long as possible in order to reduce the size of the data set and hence speed up the processing; over-averaging results in 'time smearing'. It should be clear from the discussion in Section 10.8.2 that time averaging the data over cell sizes comparable with $1/N_{pix}\Delta x$ will smear the data over that $u, v$ scale which captures the variations in the visibility function associated with the edge of the field.

The book TMS, Chapter 6, discusses the effect in some detail but a few basic points can be made here. Since the averaging is along the $u, v$ tracks, the sky plane distortions take the form of broadly *azimuthal* stretching (most obvious for compact sources), which becomes worse towards the edge of the field. The effect is, however, complicated (see Bridle and Schwab 1989). It depends on the source position and baseline direction and there is no simple analytic formulation; instead the effect must be calculated numerically. The averaging time limit becomes more stringent as the fringe period gets smaller, i.e. for long baselines and high frequencies. The on-line 'user guide' for each array will give numerical tables of acceptable averaging times for different combinations of maximum baseline and observing frequency. In practice, averaging times range from $\sim 0.1$ s to $\sim 10$ s but in general the lesson is to average up the data as little as possible.

### 10.14.3 Frequency Averaging

In Section 9.2 we discussed the effect of a finite observing bandwidth and how the delay beam limits the field of view. The generic effect of averaging over frequency is to smear the visibility data in a radial direction, and the resultant error in the image is also a *radial* smearing (hence compact sources are stretched into ellipse-like shapes), which gets worse towards the edge of the field. A rough rule of thumb for the field of view (Eqn 9.18) is the minimum fringe spacing divided by the fractional bandwidth. Note that *the averaging effect is dependent only on the baseline length and the channel bandwidth* and is independent of

---

[16] Note that the analysis software coherently averages the orthogonal (cosine and sine) components of the visibility to ensure that the noise statistics are correctly preserved. Averaging the amplitudes and phases separately does not do this.

the central observing frequency. We can see this by substituting for $\nu = c/\lambda$ in Eqn 9.18, yielding $\Delta\theta \ll c/(b\Delta\nu)$.

Again, TMS, Chapter 6, discusses the effect in some detail but it is obvious that to maximize the field of view the bandwidth $\Delta\nu$ needs to be small. This requirement has consequences for the number of frequency channels which are correlated independently. For continuum observations typical channel bandwidths in current digital correlators are $\sim 1$ MHz (give or take a factor of order unity); thus to cover a band of, say, 2 GHz the correlator must be designed to handle 2000 or more channels. Since it needs to produce sine and cosine outputs from each of four polarization combinations, the actual number of channels will be eight times greater. Correlator configurations are flexible and, as for time averaging, the on-line user guide for each array will have numerical tables of fields of view for different baselines and correlator configurations.

### 10.14.4 Non-Closing Effects

Our assumption so far has been that all the phenomena which affect the observed complex visibilities can be recognized and corrected on a telescope-by-telescope basis. There are, however, low-level errors which are baseline-related and so do not close; they are not corrected by self-calibration algorithms. Some of these causes of error are as follows.[17]

– *Time averaging*  As well as smearing in the $u, v$ plane (Section 10.14.2), averaging the data over periods which do not capture rapid amplitude and phase variations, often due to tropospheric effects, will lead to non-closing errors.
– *Frequency averaging over non-matched bandpasses*  A toy model indicates the potential for non-closing amplitude errors. An identical four-telescope array is observing a point source whose contribution to the correlation is therefore the same on all baselines; hence the closure amplitude (Eqn 10.23) should be unity. Now let the bandpass associated with each telescope be slightly different, i.e. (1) $\Delta\nu$, (2) $\Delta\nu + \delta\nu$, (3) $\Delta\nu + 2\delta\nu$, and (4) $\Delta\nu + 3\delta\nu$. The cross-correlated powers will be proportional to the products, and one can see that the closure amplitude in Eqn 10.23,

$$\frac{V_{12}^{obs} V_{34}^{obs}}{V_{13}^{obs} V_{24}^{obs}} \propto \frac{[(\Delta\nu)^2 + \delta\Delta\nu][(\Delta\nu)^2 + 5\delta\Delta\nu]}{[(\Delta\nu)^2 + 2\delta\Delta\nu][(\Delta\nu)^2 + 4\delta\Delta\nu]} \neq 1, \tag{10.34}$$

deviates from unity.[18] Non-closing phase errors can arise from non-matched variations in the phase characteristics across the band. However, as noted above, most correlators now operate in a mode where the RF band, perhaps 0.5–2 GHz in extent, is split into a thousand or more channels. By calibrating and correlating these well-defined narrow channels independently, the errors due to frequency averaging are greatly reduced.
– *Coherent RFI, usually picked up via the sidelobes*  Since the RFI emanates from a specific location it will have different amplitudes and phases in different antennas and the cross-correlations will not close. The effects are likely to fluctuate in time.

---

[17] Here we do not consider errors in the digital correlator system. Some additional effects come into play for wide-field imaging and these will be discussed in Section 11.5.

[18] We have made the simplifying assumption that $\delta\nu \ll \Delta\nu$.

– *Polarization impurities* (Rogers 1983) *can give rise to closure errors*    Massi *et al.* (1997) analysed
   the effect for the EVN and showed that closure phase errors up to 0.5° could arise from polarization
   impurities of order 5%. Receiver polarization purities are now better than this. Non-closing errors
   from this effect vary with the parallactic angle.
– *The Sun*    The Sun is a very strong radio source ($> 10^6$ Jy at centimetric wavelengths) and can be
   seen through the far sidelobes of an antenna. However, its angular size (0.5°) means that 'the Sun
   in the sidelobes' is mainly a problem for low-resolution arrays.

### 10.15  Image Quality

#### *10.15.1  Signal-to-Noise Limits*

In the idealized case, where there are no residual errors in the data, the image is said to be
*noise limited*. The rms thermal noise level $\sigma = \Delta S$, as in Section 10.12, then determines
the minimum detectable brightness within a given field of view. A common requirement
in noise-limited images is to count the number and flux density of compact sources; the
challenge is to decide on the weakest reliable detections. Take as an example an image of
$1000 \times 1000$ resolution elements. Gaussian statistics indicate that a positive beam-sized
feature, whose peak is $3\sigma$ above the noise, has about one chance in a thousand of being
due to a noise fluctuation. Such a feature would, therefore, be a likely detection *if one
is looking at a given position*. Searching for random sources in the field is, however, an
entirely different matter; in a million resolution elements there will up to a thousand $3\sigma$
random fluctuations and there will even be a few $5\sigma$ fluctuations. Searches for sources
must, therefore, choose a significance level that is appropriate to the *a priori* probability of a
chance fluctuation. In practice the noise is never perfectly Gaussian and as a result the image
invariably has more fluctuations than theory would suggest.[19] As a matter of common
practice, therefore, unless the statistical properties of the data set and the *a priori* hypotheses
have been carefully defined and understood, astronomers adhere to the following rules of
thumb: $10\sigma$ detections are seldom false, $5\sigma$ detections are not necessarily convincing, and
$3\sigma$ detections should be viewed very skeptically.

Finally, if the image combines a low noise level with low angular resolution, there
will be many faint unresolved sources in the field around or below the noise level. This
constitutes 'hidden' information, which can be exploited in source count work using the
statistical $P(D)$ technique to be outlined in Section 16.9.4. In imaging applications these
low-level sources will not be recognized by the deconvolution algorithm and their sidelobes
will therefore contribute to the noise level. The beam-sized responses to stronger sources
may also begin to overlap. This is the issue of confusion introduced in Section 10.12.1.
With current arrays confusion can be a problem with beam sizes of tens of arcseconds to
arcminutes. In the future, with the SKA, confusion will become an issue at resolutions of
arcseconds.

---

[19] Noise is measured at the same $u, v$ locations as visibility data; therefore image noise will have the same
   spatial characteristics as real sources. On occasions when the noise level of the image is low and the $u, v$
   coverage has significant gaps, the data reduction process can 'invent' faint sources.

## *10.15.2 Dynamic Range*

The *dynamic range* is the ratio of the peak brightness in the image (usually an unresolved feature) to the rms fluctuations in a blank area. If the image is noise limited then the dynamic range is set solely by the strength of the brightest source within the field of view and the noise level. Scientifically important images may, therefore, have low dynamic ranges simply because they do not contain any bright sources. If there is a bright source in the field then the requirement for high dynamic range, and hence the ability to see faint sources in the presence of a bright source, is low residual errors in the visibility data.

Despite careful flagging, calibration, self-calibration, and deconvolution, the corrected visibilities will have residual errors in both amplitude and phase, in particular due to the non-closing errors. We can thus regard each corrected visibility on baseline $ij$ as the true visibility *multiplied* by a residual complex gain factor $g^r_{ij}$, to which is *added* a non-closing error term $NC_{ij}$ and the noise term $N_{ij}$:

$$V^{\text{obs}}_{ij} = g^r_{ij} V^{\text{true}}_{ij} + NC_{ij} + N_{ij}. \tag{10.35}$$

The image that results from taking the Fourier transform of these data can be expressed as

$$B(l, m) = B^{\text{true}}(l, m) + B^{\text{sp}}(l, m) + N(l, m), \tag{10.36}$$

where $B^{\text{true}}(l, m)$ is the actual brightness distribution and $B^{\text{sp}}(l, m)$ is the sum of the spurious responses that result from imperfect correction of amplitude and phase errors and imperfect interpolation into the gaps in the $u, v$ plane;[20] they both sit on top of the noise background $N(l, m)$. When there is a strong source in the field, the largest contributions $B^{\text{sp}}(l, m)$ come from data errors. If these contributions dominate the noise background then they determine the effective signal-to-noise ratio in the image; in this situation the image is said to be *dynamic range limited*.

The manner in which the visibility errors contribute to $B^{\text{sp}}(l, m)$ depends both on their size and on their distribution in the $u, v$ plane. To diagnose the problem(s) limiting the dynamic range the radio astronomer needs to develop a pictorial understanding of how patterns and symmetries in the $u, v$ plane Fourier transform into the image plane and vice versa (see also TMS, Chapter 10). In brief:

- A narrow feature in the $u, v$ data will become a broad feature in the image. A ring-like feature in the $u, v$ plane (errors along a baseline track) becomes a ring-like feature in the image.
- Amplitude errors give rise to symmetric (even) features in the image, whilst phase errors produce antisymmetric (odd) features. The discussion in Section 9.4.1 should make this clear.
- Multiplicative errors transform into an error pattern in $B^{\text{sp}}(l, m)$ which convolves the image; the same pattern therefore appears to be 'attached' to all discrete image features and will be more apparent around bright ones.
- Additive errors produce features in $B^{\text{sp}}(l, m)$ which are unrelated to the astronomical structures.

---

[20] This complements the discussion at the end of Section 10.10.2.

Two other useful points to note are:

– A phase error $\delta\phi$ on a data point is equivalent to a relative amplitude error $\delta A = \sin\delta\phi \approx \delta\phi$ for small angles; thus if $\delta\phi = 1°$, $\delta A = 0.017$.
– A few 'bad' $u, v$ samples, missed in the flagging stage, may well not be seen in the image since their effect is heavily diluted by the many millions of samples which constitute a typical Earth rotation synthesis data set. On the other hand, low-level persistent errors add up.

If the residual errors are quasi-randomly distributed, as might result from imperfect tropospheric calibration using a nearby reference source (Section 10.7), then the spurious image features will not be coherent over many synthesized beams. At centimetric wavelengths, dynamic ranges of up to $\sim 1000 : 1$ can be achieved with external calibration; self-calibration can then largely eliminate these telescope-related errors and increase the dynamic range by one or two orders of magnitude.

The self-calibrated image may still not be thermal noise limited, because of the baseline-related errors. Even small errors, often in the form of constant offsets, can add up to produce visible effects. For example relative errors $\delta A$ could produce coherent spurious features at the level of $\sim \delta A / \sqrt{N_{base}}$, where $N_{base}$ is the number of baselines. Thus, purely for illustration, if $\delta A = \pm 0.005$ (0.5%) on all baselines and $N_{base} = 351$ (as in the JVLA) the dynamic range could be limited to $\sim 4000 : 1$. Clearly, to achieve dynamic ranges one or two orders of magnitude higher,[21] all the sources of baseline-related errors must be very carefully controlled. If the highest dynamic range is required, baseline-related corrections can be determined from observations of (perfect) point sources in the knowledge that the closure phases should end up as zero and the closure amplitudes as unity.

For an extensive discussion of the issues involved in synthesis imaging with high dynamic range see Braun (2013) and the references in the Supplementary Material.

### 10.15.3 Fidelity

By fidelity is meant the trustworthiness of the recovered brightness distribution, in other words, the difference between the synthesized brightness distribution and the true distribution. The latter is, of course, unknown. Over and above the effect of data errors, the on-source contribution to $B^{sp}(l, m)$ is a complicated combination of the target's brightness distribution, the gaps in the $u, v$ coverage, and the success of the deconvolution algorithm in interpolating across them (see Section 10.9.1). The 'fidelity errors' will, therefore, be significantly greater than would be expected from the dynamic range. There is no analytical formula relating fidelity to the quality of coverage of the $u, v$ plane and the deconvolution method. It should be noted, however, that for many astronomical requirements it may not matter that the brightness of a particular feature in the source is uncertain to a few per cent. But, to detect more or less subtle temporal variations, especially motion, one must develop confidence in the reliability of specific (usually compact) features in the reconstructed images. Realistic simulations can provide added confidence, if required. Many 'blind test'

---

[21] As is routinely achieved by the JVLA; the SKA will then push the requirement by a further one or two orders of magnitude.

simulations have demonstrated levels of fidelity that are sufficient for a wide range of scientific results from synthesis imaging to be accepted by the astronomical community.

In conclusion we stress that recognizing and reducing errors in the data, and in the images constructed from them, is at the heart of successful aperture synthesis. This section has given just a very brief introduction to this important subject.

## 10.16 Further Reading

*Supplementary Material* at www.cambridge.org/ira4.

Christiansen, W. N., and Högbom, J. A. 1985. *Radiotelescopes*, 2nd edn. Cambridge University Press.

Taylor, G. B., Carilli, C. L., and Perley, R. A. (eds.) 1999. *Synthesis Imaging in Radio Astronomy II*. Astronomical Society of the Pacific Conference Series, vol. 180. (Also known as the 'White Book'.) Since the publication of the White Book. a wealth of continuously up dated information on all aspects of aperture synthesis has become available from the rolling series of pedagogic workshops given at different times under the aegis of: the US National Radio Astronomy Observatory; the European Radio Interferometry School; the Australia Telescope National Facility. Reports and presentations from these workshops are freely available on the web.

(TMS) Thompson, A. R., Moran, J. M., and Swenson, G. W. 2017. *Interferometry and Synthesis in Radio Astronomy*, 3rd edn. Springer.

*The Pynterferometer: A Synthesis Array Simulator* The link between the quality of a synthesis image and the filling of the $u, v$ plane can be explored with the *Pynterferometer* package (Avison and George 2013).[22] This package allows the user to vary the number of antennas and their configuration and to examine the Fourier filtering effect on a user-supplied 'true' image. The software can be freely downloaded from www.jb.man.ac.uk/pynterferometer/.

---

[22] Avison, S., and George, S. J. 2013. *European J. Phys.* **34**, 7. arXiv:1211.0228.pdf.

# 11

# Further Interferometric Techniques

In Chapters 9 and 10 we explained the basis of connected-element interferometry and aperture-synthesis. These two chapters introduced the heart, but are far from the whole, of modern interferometry and aperture synthesis. In this chapter we briefly cover the main extensions to the basic ideas. They are:

(i) spectral line imaging (Section 11.1), where extra calibration of the system is required and large data volumes are encountered;

(ii) polarization imaging (Section 11.2) where working with Stokes parameters requires more cross-correlations and system calibration steps;

(iii) imaging at mm and sub-mm wavelengths (Section 11.3), where calibrating the turbulent troposphere provides a much greater challenge than at centimetre wavelengths;

(iv) very long baseline interferometry (VLBI; Section 11.4), where the telescopes operate independently and precision geodesy and astrometry become powerful science imperatives in addition to imaging;

(v) wide-field imaging (Section 11.5), which requires a more complete version of the formalism, the implementation of which places greater demands on computing power;

(vi) imaging at metre wavelengths (Section 11.6), where the combination of ultra-wide fields and the turbulent ionosphere provides the greatest aperture-synthesis challenge of all.

Each section could warrant a scientific monograph in its own right. Our treatment is intended only to introduce the reader to the essence of the new formalism and technical demands. More detailed information can be found in the references, in the reports and presentations available from international synthesis workshops, and in the guides for observations and data analysis for specific arrays.

## 11.1 Spectral Line Imaging

Observing in 'spectral line mode', where the overall band is divided into multiple, contiguous, frequency channels is now the norm in synthesis imaging. Not only does this allow the astronomer to study the astrophysical sources of spectral lines, but it also allows continuum imaging with wide fields of view not limited by the overall bandwidth (see Section 9.2). On a practical level narrow channels also allow the observer to remove narrowband RFI from the data before embarking on the rest of the analysis. The analysis shares many common features with broadband imaging; it does, however, involve additional complexities and requires more steps, which we now outline.

### 11.1.1 Observational Choices

New choices must be made before the observations are undertaken. In brief these are: (i) making sure that the correct observing frequency has been calculated; (ii) selecting the overall frequency band to be covered and the channel width for the required velocity resolution; (iii) employing the right combination of sensitivity and angular resolution to achieve an appropriate surface brightness in the image. We now outline the rationale for each choice.

(i) *Observing frequency* The Doppler shifts due to the array's motion through space must be corrected for a fixed frame of reference. The heliocentric frame includes contributions from the spin of the Earth ($0.5 \, \text{km s}^{-1}$) and its $30 \, \text{km s}^{-1}$ orbital motion around the Sun. If corrections for the Moon and the planets are made, the heliocentric frame becomes the 'barycentric' frame. The 'Local Standard of Rest' (LSR) is based on the average motion of stars in the solar neighbourhood (see Section 14.2). The 'Galactocentric' frame takes into account the rotation of the Milky Way. Observations of galactic objects are usually corrected to the LSR while extragalactic observations are usually referred to the heliocentric frame.[1] The shifts are sky-position dependent, and software is available to make the requisite calculations. For connected element arrays a fixed Doppler correction can be applied for the observations; further corrections can be made off-line if required.

(ii) *Channel width and the overall frequency band* As explained in Section 3.6, the width of a line is a measure of the radial velocity dispersion $\delta v$ of the gas within the reception beam; the observable frequency width is then $\delta \nu = \nu \delta v / c$. To sample a line adequately the channel width must be several times narrower than the width of the line. Velocity dispersion is the fundamental physical parameter and linewidths range from $< 1 \, \text{km s}^{-1}$ for galactic masers with low internal motions to hundreds of $\text{km s}^{-1}$ for extragalactic megamasers. Line-emitting regions often coexist with, or lie in front of, sources of continuum emission. The observer therefore selects a band which is wide enough to include a range of line-free channels so that both types of emission can be imaged. Modern correlators are sophisticated combinations of specialized digital electronics and software. The data can be split into sub-bands (also called 'spectral windows' or 'IFs' in VLBI) with each divided into many spectral channels. The correlator can also be reconfigured to 'zoom in' on one of the sub-bands and make it smaller, with concomitantly narrower channels. More than one sub-band can be used to observe well-separated lines at the same time, for example H I at $1.4 \, \text{GHz}$ and the OH maser lines in the range $1.6–1.7 \, \text{GHz}$.

(iii) *Sensitivity and angular resolution* The combination offered by the array must provide sufficient brightness sensitivity (see Section 10.12). As an illustration, H I (with brightness temperatures 50–200 K) can only be imaged with resolution $\geq 5$ arcsec with JVLA even though it has a geometrical collecting area of $13\,000 \, \text{m}^2$. It was the ambition to image H I with resolutions ten times higher which first spawned the idea of the Square Kilometre Array (Wilkinson 1991). At the other extreme, non-thermal masers are accessible to study with milliarcsecond resolution using VLBI. Note that in calculating the sensitivity it is the channel width $\delta \nu$, not the overall bandwidth $\Delta \nu$, which is appropriate.

---

[1] Relative velocities of nearby galaxies are quoted in $\text{km s}^{-1}$: for more distant galaxies the redshift $z$ is used: the line is observed at $\nu_0 / (1 + z)$.

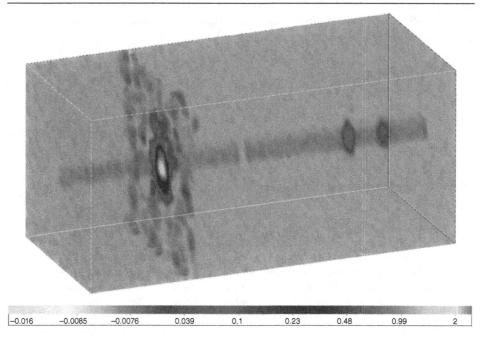

| −0.016 | −0.0085 | −0.0076 | 0.039 | 0.1 | 0.23 | 0.48 | 0.99 | 2 |

Figure 11.1. ALMA data cube for the galaxy NGC3256 covering a field 2 × 2 arcmin; the long axis is frequency, from 115 to 111.5 GHz. The grey scale represents intensity, and the continuous dark line along the axis is emision from warm dust in the centre of the galaxy, with a central gap between the observing sub-bands; the enhanced emission is from different spectral lines, each occupying a few tens of frequency channels. (Courtesy of Anita Richards.)

## *11.1.2 Data Analysis Issues*

The quality of the bandpass calibration (Section 10.7.1) is critical since frequency-dependent errors will appear as spurious features in the spectra. The calibrator should be observed for long enough that the signal-to-noise ratios of the bandpass channels are significantly higher than those for the target source.

If the line source is associated with continuum emission then the two should be dealt with separately; software allows the subtraction of the continuum (line-free) channels from the line-containing channels. The continuum channels are used to make a broadband image in the conventional way, as described in Chapter 10. The line data are analysed on a channel-by-channel basis to produce a series of independent 'channel maps'. In the case where the line data have too low a signal-to-noise ratio for self-calibration, the complex gain solutions from the continuum data can be applied to the line data.

Multiple channel maps can often best be interpreted when combined (stacked together) to form a 3D data cube (two sky coordinates and a radial velocity); an example is given in Figure 11.1. These cubes can can contain large amounts of data; for example, one from ALMA might consist of millions of spatial pixels and thousands of velocity channels. With the SKA the load will be even higher. A new challenge is therefore to write software to enable the astronomer to get to grips, scientifically, with the torrent of data which spectro-scopic imaging can produce. A simple first step is to rotate a data cube around; this can give an immediate impression of the shape of the line-emitting region and the motions of the gas within it. However, this is just the start and a whole arena of data visualization and associated computer vision techniques have been, and are being, developed to help the astronomer comprehend his/her data (e.g. van der Hulst *et al.* 2017).

## 11.2 Polarization Imaging

Thus far we have dealt with imaging only the total intensity of the incoming radiation, but synchrotron-emitting radio sources often exhibit weak (a few per cent) linear polarization while cosmic masers and pulsars can show higher degrees of both linear and circular polarization. The extensions to the theory of total-intensity imaging are quite complicated but they are well understood. In this section we focus on the main points appropriate for imaging linear polarization close to the centre of the field of view. In so doing we have been principally guided by pedagogic presentations by R. Perley (NRAO).

### 11.2.1 Basic Formalism

In Section 7.6 we described how orthogonal polarizations can be separated by means of carefully designed feed structures. In Section 7.5 we noted that to observe linear polariza-tion it is preferable (although not mandatory) to employ circularly polarized feeds. Since this is the dominant choice for connected-element arrays and VLBI we will proceed on this basis.[2]

In order to collect all the polarization information from an interferometer, four correla-tions between the two orthogonal states are required. Assuming initially that each antenna and receiver chain produces voltages associated with only $R$ and $L$ components of the incident field, the correlations for a single baseline are (excluding the complex-voltage gain factors for simplicity):

$$
\begin{aligned}
\langle R_i R_j^* \rangle &= (I_{ij}^v + V_{ij}^v)/2, \\
\langle L_i L_j^* \rangle &= (I_{ij}^v - V_{ij}^v)/2, \\
\langle R_i L_j^* \rangle &= (Q_{ij}^v + iU_{ij}^v)/2, \\
\langle L_i R_j^* \rangle &= (Q_{ij}^v - iU_{ij}^v)/2.
\end{aligned}
\tag{11.1}
$$

These equations follow directly from the definitions of the Stokes parameters in Section 7.5. The quantities $I_{ij}^v, Q_{ij}^v, U_{ij}^v, V_{ij}^v$ are the so-called Stokes visibilities, the Fourier transforms of the images in the different Stokes parameters; with this terminology the

---

[2] Note that the WSRT, LOFAR, the ATCA, and ALMA use linear feeds; there is a completely equivalent formalism for this situation.

visibilities in Chapters 9 and 10 are $I_{ij}^v$. Rearranging the terms in Equations 11.1 and then simplifying the notation yields

$$
\begin{aligned}
I_{ij}^v &= \langle R_i R_j^* \rangle + \langle L_i L_j^* \rangle \equiv RR + LL, \\
Q_{ij}^v &= \langle R_i L_j^* \rangle + \langle L_i R_j^* \rangle \equiv RL + LR, \\
U_{ij}^v &= i(\langle L_i R_j^* \rangle - \langle R_i L_j^* \rangle) \equiv i(LR - RL), \\
V_{ij}^v &= \langle R_i R_j^* \rangle - \langle L_i L_j^* \rangle \equiv RR - LL.
\end{aligned}
\tag{11.2}
$$

The usual nomenclature refers to $RR$ and $LL$ as the correlations of the parallel hands, and $LR$ and $RL$ as the correlations of the cross-hands. In the general case, where all Stokes parameters are significant, the correlations of both parallel hands are required to measure the total intensity. In Chapters 9 and 10, however, we implicitly assumed the correlation of only a single polarization state, either $RR$ or $LL$; this is acceptable since, for most astrophysical sources, $V \approx 0$. In practice both parallel hands are correlated and $I$ is formed from the average.

On the basis of Eqns 11.2, polarization imaging is, in principle, straightforward. The four visibility data sets are calibrated and then imaged separately, as described in Chapter 10. Note that: (i) resolution elements in $Q$ and $U$ images can be negative and hence the imaging process cannot use positivity as a constraint; (ii) the $I$, $Q$, and $U$ images can be combined to produce an image in linearly polarized intensity or in fractional linear polarization with an associated orientation (modulo $180°$) of the $\mathcal{E}$-vector for each resolution element (see Section 2.2); (iii) for many sources the $V$ image will be noise-like, which provides a check on the quality of the calibration.

Unfortunately things are not so simple in practice since, compared with intensity-only imaging, two extra calibration steps are required: first, because the antennas and the system electronics do not deliver perfectly separated $R$ and $L$ voltages to the correlators; second, because measurements involving only $R$ and $L$ do not directly determine the position angles of the electric vectors on the sky. In practice, making the requisite calibration measurements and applying corrections appropriate for linear polarization imaging around the field centre has become a well-established procedure. We will outline here the practical steps usually adopted in synthesis polarimetry. A more formal approach using the *measurement equation* is outlined in Appendix 4.

### 11.2.2 Calibration for Linear Polarization

The first steps are the same as for intensity imaging. The bandpasses and amplitudes and phases for the two parallel hands $RR$ and $LL$ are determined independently, by a combination of external phase referencing and self-calibration. The phase corrections derived from one of the parallel hands can be used to correct the cross-hand data. To maintain a constant phase difference between the data from the parallel hands their phases are both established relative to the same reference antenna. However, the phase difference between the $R$ and $L$ voltages from this antenna are not known *a priori*; hence the position angle of the $\mathcal{E}$-vector on the sky is also arbitrary until calibrated (see below).

The next step is to measure and remove the effects of imperfect polarization purity received from the antennas. In the present context the dominant imperfection arises from the fact that practical feed and receiver systems respond to elliptical rather than perfect circular polarization and each hand will be corrupted by leakage from the other hand. Furthermore, the algebra greatly simplifies if, as is the norm, the deviations from polarization purity are small and higher-order terms can be neglected. The responses per antenna can then be written quite simply in terms of pure circulars with small, complex, 'D-terms' which capture the cross-talk between them (Conway and Kronberg 1969).

With no additional feed rotations (see below) we can write, for a given antenna,

$$R^{\text{obs}} = R + D_R L,$$
$$L^{\text{obs}} = L + D_L R,$$

(11.3)

where again the complex voltage gains have been omitted for simplicity. The corrupting D-terms are accessible from the cross-correlations and can be measured in two complementary ways. The simplest method is to observe a short scan on a strong compact *unpolarized* ($\ll 1\%$) source, examples of which are usually listed in an Observer's Guide for each array. In this specific case the spurious signals in the cross-channels can be written as

$$L_i^{\text{obs}} R_j^{\text{obs}*} = (D_{L_i} + D_{R_j}^*) I_{ij},$$
$$R_i^{\text{obs}} L_j^{\text{obs}*} = (D_{R_i} + D_{L_j}^*) I_{ij},$$

(11.4)

which shows that a fraction of the total intensity appears as a spurious cross-hand correlation. The polarization leakages from $I$ are typically of a similar size (1%–5%) to the cross-hand correlations on a weakly polarized target source; this emphasizes the need to determine the polarization leakage level to a few tenths of a per cent. The great advantage of observing a calibration source with no intrinsic polarization is that the cross-hand signals are purely instrumental and therefore provide a direct measurement of the polarization leakages. The second method involves observing a strong compact *polarized* source, as shown in Appendix 4 for a single-beam polarimeter. However, since compact sources are often variable, one cannot rely on knowing the polarized component; this has to be solved for as part of the calibration process. Conway and Kronberg (1969) showed how to do this.

For an alt-az mounted telescope, a direction defined at the feed rotates with respect to the sky over the course of the day (this is called 'field rotation' in optical astronomy). The rotation is measured by the changing parallactic angle $\Psi_P$, which is the angle between a line drawn to the zenith through the source (equivalently the local vertical drawn on the feed) and a line drawn from the source to the North Pole (Figure 11.2). If the antenna is at latitude $\phi$ then, for a source at hour angle $H$ and declination $\delta$,

$$\Psi_P = \tan^{-1}\left(\frac{\sin H \cos\phi}{\sin\phi \cos\delta - \cos\phi \sin\delta \cos H}\right).$$

(11.5)

The linearly polarized intensity $P$ of the calibration source is defined as $Q + iU = P e^{i2\chi}$, where $\chi = (1/2)\tan^{-1}(U/Q)$ is the orientation of the $\mathcal{E}$-vector (Section 7.5). These relations are defined relative to North, so when the source is observed away from the

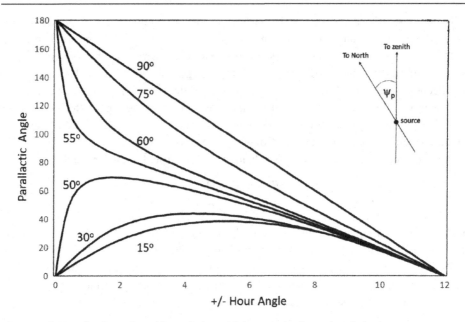

Figure 11.2. Parallactic angle and its variation with hour angle, for a given latitude.

meridian the apparent angle of $P$ is rotated by $2\Psi_P$. There is no effect on the parallel hands.[3] For the cross-hands we can write

$$\langle R_i L_j^* \rangle = (Q_{ij}^v + iU_{ij}^v)e^{i2\Psi_P}/2,$$
$$\langle L_i R_j^* \rangle = (Q_{ij}^v - iU_{ij}^v)e^{-i2\Psi_P}/2. \tag{11.6}$$

Simplifying the terminology as in Eqns 11.2:

$$Q_{ij}^v = RLe^{i2\Psi_P} + LRe^{-i2\Psi_P},$$
$$U_{ij}^v = i(LRe^{-i2\Psi_P} - RLe^{i2\Psi_P}). \tag{11.7}$$

One can now see how alt-az mounts allow the instrumental leakage to be separated from the source polarization if data are taken over a wide range of parallactic angle. Figure 11.3 makes the point graphically. Since the instrumental term is not related to the parallactic angle it remains constant, while the source-related term rotates. This separation works provided that the calibration source is at a declination where $\Psi_P$ varies significantly during the observations. At low declinations the range of $\Psi_P$ is small, while at high declinations the range is large; near the local zenith $\Psi_P$ can change too rapidly for accurate observations.

---

[3] The effect of the parallactic angle on the parallel-hand correlations on a given baseline is to rotate the phases by the difference of the parallactic angles of the two antennas. For the cross-hands the phases are rotated by the sum of the angles. Note that for antennas at different latitudes (as in VLBI arrays), $\Psi_P$ will be different and hence the phases on a baseline will be rotated by different amounts. Parallactic angle corrections are deterministic and are applied at an early stage of the analysis.

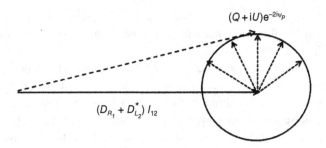

Figure 11.3. Schematic, showing the constant instrumental cross-polarization term and a rotating term $(Q + iU)e^{-2i\Psi_P}$ associated with the polarized calibrator (after R. Perley).

Finally one has to establish the orientation of the $\mathcal{E}$-vector of linear polarization; this is encoded in the $R$–$L$ phase difference at the reference antenna. This phase difference can be derived from a short observation of the cross-hand correlations for a strong linearly polarized source whose intrinsic position angle is well known.[4] An $R$–$L$ phase difference has the effect of rotating the position angle of $P = Q + iU$ (Eqn 7.27), and from Eqns 11.6 one can see that for a point source at the field centre the $R$–$L$ phase difference is equal to twice the apparent angle of the $\mathcal{E}$-vector. In practice, since the observation will be made off the meridian, the apparent angle will include the parallactic angle.

Low frequency ($\leq 2\,\text{GHz}$), wide-band, and wide-field observations all provide further calibration challenges. Birefringence in the ionosphere can produce time-variable rotation of the plane of linear polarization. The cross-hand leakage terms will be frequency dependent since, for example, the polarizers behind the feeds will not deliver exact $90°$ rotation across wide fractional bandwidths. The instrumental polarization across the voltage beam of a parabolic dish is non-zero away from the symmetry axes, and if the feed is offset from the antenna vertex (as for example in the JVLA) the beam polarization effects are magnified.

## 11.3 Aperture Synthesis at Millimetre Wavelengths

The mm to sub-mm waveband, from $\sim 30\,\text{GHz}$ to $\sim 1000\,\text{GHz}$, provides access to unique information about the cool ($T < 100\,\text{K}$) Universe, from dust and molecules in cold clouds and star- and planet-forming regions in the Milky Way to dusty star-forming galaxies at high redshift. Star-forming regions exhibit a forest of emission lines from the $\sim 200$ molecular species detected in the Milky Way.[5] More than 60 species have been detected in external galaxies and the brightest provide redshifts when the visible band is dust-obscured. Continuum radiation can be seen from dusty galaxies at high redshifts owing to the sharply rising spectrum in the Rayleigh–Jeans regime. As mentioned in Section 2.3, for optically

---

[4] In the northern hemisphere the standard calibrator is 3C286, whose intrinsic position angle at centimetric wavelengths is $(33 \pm 1)°$; other position angle calibrators will be found in the Observers Guide for a given array.
[5] www.astro.uni-koeln.de/cdms/molecules.

thin dust $I_\nu \propto \nu^{3-4}$ and in a given observing band the dimming effect of distance, evident in the total bolometric flux, is compensated as more intense dust emission is shifted into the band (see Figure 16.2).

While the basic approach shares many similarities to centimetric work, the challenges of interferometry are compounded at wavelengths one to two orders of magnitude shorter. Radio telescopes with the required surface accuracy are necessarily small (typically 10–15 m). The noise contribution of the troposphere is higher, owing to the opacity of water vapour and oxygen, and so there is a lower limit to the achievable system temperature. For a given bandwidth, therefore, the sensitivity per baseline is lower than at centimetric wavelengths. Furthermore, the troposphere is often highly variable, which adds greatly to the challenge of calibrating the visibilities. Finally, the structure of molecular-line emitting objects is often so complex that a series of observations, with different pointings and/or with reconfigured arrays, is needed to obtain sufficient coverage of the $u, v$ plane.

### 11.3.1 Phase and Amplitude Calibration

Rapid changes in the visibility phase, if not monitored and corrected, can cause loss of coherence (Section 10.7.3) and fluctuations in the apparent position of the source (Section 9.7). These effects are worse on longer baselines and, if wrongly averaged out, can make an unresolved source appear to be resolved in the radio equivalent of optical 'bad seeing'.

An outline of the characteristics of the troposphere of relevance to radio interferometry was given in Section 9.7.1. If both are measured in the same units the excess zenith path length due to water vapour $L_w \approx 6.3$ PWV; thus, for example, $L_w \approx 6.3$ mm for PWV = 1 mm. Ideally the fluctuations in path length should be followed to $\sim \lambda/20$ or better; hence at $\lambda = 1$ mm ($\nu = 300$ GHz) the path changes should be tracked to $\sim 0.05$ mm. Even on high mountain sites the PWV can change by more than this on the timescale of minutes or less, thus large phase corrections are needed at frequent intervals.

The necessary calibration can be done conventionally by fast switching to an external reference source and/or by measuring the PWV over each telescope using total-power water vapour radiometers (WVRs) working in either the 22 GHz or 183 GHz absorption lines. The clear-sky brightness temperature $T_B^{\text{atm}}$ on the line is a measure of the opacity and hence of the PWV. The 183 GHz WVRs on the ALMA telescopes (e.g. Nikolic *et al.* 2013) measure $T_B^{\text{atm}}$ every second and convert the results to predictions of phase for later corrections to the observed complex visibilities. Nikolic *et al.* reported that in humid (PWV = 2.2 mm) unstable daytime conditions the ALMA WVR system reduced the fluctuations from $\sim 1$ mm to $\sim 0.16$ mm on a baseline of 650 m. Water vapour radiometers therefore make observations possible over a wide range of tropospheric conditions and can also identify times when the PWV is low enough to permit observations at shorter wavelengths. External phase referencing (offsets of $< 3°$ and rapid switching cycles of 20 s) combined with WVR measurements have been used with significant success for phase calibration on longer (2.7 km) ALMA baselines (Matsushita *et al.* 2017). There is, however, a fundamental problem with WVR-based phase correction; the presence of liquid water in clouds breaks the relationship between $T_B^{atm}$ and inferred path delay since water drops radiate but have little effect on the overall refractive index. The accurate routine phase calibration of mm-wave arrays on baselines of many km remains a challenging work in progress (e.g. Matsushita *et al.* 2017).

As noted in Section 4.4, absorption in the troposphere has a doubly deleterious effect on the signal-to-noise ratio provided by a radio receiver; not only is the incoming signal attenuated but also the noise temperature of the receiver is increased by the concomitant emission. This is particularly apparent in the sub-mm band. As explained in Chapter 6, the characteristics of a multi-stage receiver can be summarized by a power gain $G$ and a noise temperature $T_{rec}$, the latter being dominated by the characteristics of the first stage. If the atmosphere is at temperature $T_{atm}$ and has optical depth $\tau$ then the system noise temperature including the atmospheric contribution is

$$T_{sys} = T_{rec} + (1 - e^{-\tau})T_{atm}. \qquad (11.8)$$

Taking account of the signal attenuation, the signal-to-noise ratio is

$$\frac{e^{-\tau}T_{source}}{T_{rec} + (1 - e^{-\tau})T_{atm}} \equiv \frac{T_{source}}{e^{\tau}[T_{rec} + (1 - e^{-\tau})T_{atm}]}. \qquad (11.9)$$

In other words the effective $T_{sys}$, referred to the top of the atmosphere, is the ground-based value multiplied by the inverse of the attenuation loss. An example makes the effect clear. At the ALMA Chajnatnor site at 5000 m altitude a typical PWV value of 3 mm corresponds to a zenith $\tau = 0.15$ at 230 GHz. For $T_{atm} = 270$ K the atmospheric emission will be 38 K; thus if $T_{rec} = 75$ K then the effective $T_{sys} = 1.16(38 + 75) = 131$ K. If the source is at an elevation of 30° the optical depth doubles and hence $T_{sys} = 1.35(70 + 75) = 196$ K. At higher frequencies optical depths are greater and the dominant effect of the atmospheric opacity is even more apparent. Since the optical depth can fluctuate quickly the system must be calibrated frequently. At ALMA the system-noise calibration at each antenna is carried out, every 10 to 20 minutes, by means of two absorbent loads at different physical temperatures placed in front of the horn feeds.

To check the absolute performance of the array, compact thermally emitting solar system objects (Chapter 12) are used, whose sizes, physical temperatures, and hence flux densities are known. Mars, Uranus, and Neptune are standard targets; minor planets and some of the moons of Jupiter and Saturn are also used. However, on the longer ALMA baselines, these objects are resolved, and so many bright flat-spectrum AGNs are used as secondary calibrators; their secular variability is constantly monitored (Bonato *et al.* 2018).

### 11.3.2 System Requirements and Current Arrays

The engineering requirements on mm-wave antennas are stringent. First the Ruze formula (Section 8.3.3) indicates that a surface accuracy of tens of microns is necessary. Second, since the primary beam (FWHM) of a 12 m antenna at 350 GHz ($\lambda = 0.86$ mm) is $\sim 18$ arcsec, the antenna needs to point to within $\sim 2$ arcsec (see Section 8.10.2). This is extremely challenging in the face of variations in wind, differential solar heating, and the effects of gravity on the structure as it moves around. Since 2 arcsec pointing is not good enough for work at the highest frequencies, differential pointing corrections can be carried out with respect to well-defined point sources. Third, the antenna must be able to move very quickly across angles of a few degrees to allow for successful amplitude and phase calibration. Millimetre-wave antennas therefore need to combine great robustness with high precision and hence are more expensive per unit area than antennas at longer wavelengths.

There are many other challenging requirements. Baselines need to be established with sub-mm accuracy. Suites of superconducting receivers operating at physical temperatures of 4 K must be maintained. The local oscillator distribution system must be very stable to avoid coherence losses. The correlator must cover broad bands with many narrow channels to detect spectral lines and to provide the sensitivity for calibration using continuum sources.

At the time of writing the leading mm-wave arrays in operation are:

*NOEMA* The Northern Extended Millimetre Array of the Institut de Radio Astronomie Millimétrique is located on the Plateau de Bure in the French Alps (altitude 2550 m). When completed in 2019 it will have twelve 15 m antennas which can be moved on E–W and N–S tracks to provide baselines up to 760 m. Four suites of receivers cover atmospheric windows from 3 mm to 0.8 mm (72–373 GHz).

*SMA* The Smithsonian Submm-Wave Array on Mauna Kea, Hawaii (altitude 4080 m), has eight 6 m antennas which can be configured to provide baselines up to 783 m. The receivers cover the wavelength range 1.7 to 0.7 mm (180–418 GHz) in three bands.

*ALMA* The Atacama Large Millimetre Array, a collaboration between the European Organisation for Astronomical Research in the Southern Hemisphere (ESO), the US National Science Foundation (NSF), and the National Institutes of Natural Sciences (NINS) of Japan in cooperation with the Republic of Chile, is located on the Llano de Chajnantor in Chile (altitude 5000 m; frequency range 31–950 GHz). ALMA is the largest array working in this wavelength regime. Its main array comprises fifty 12 m antennas, which can be moved into different configurations, from compact (baselines to 150 m) to extended (baselines to 16 km). The Atacama Compact Array (ACA) is a subset of four 12 m antennas and twelve 7 m antennas which are close in separation in order to improve ALMA's ability to study objects with a large angular size.

A generic issue confronting all these arrays is that many interesting emission regions are extended and have relatively low surface brightness. In Section 10.12 we showed that the surface brightness sensitivity of a partly filled array is approximately given by the radiometer equation for a single dish multiplied by the reciprocal of the array-filling factor; the latter is maximized if the dishes are close packed. If, however, the emission is extended on scales comparable to, or larger than, the antenna primary beam (which is small because of the short wavelengths), the 'missing short spacings' issue (see Section 10.11) becomes relevant. Additional steps must then be taken to recover the appropriate Fourier components. ALMA has the ACA for this purpose; another interferometric approach is mosaicing (Section 11.5.4). However, if some necessary low-spatial-frequency components are still missing, the interferometer data must be complemented with data from a single dish, as outlined at the end of Section 11.5.4.

## 11.4 Very Long Baseline Interferometry (VLBI)

The interferometer shown in Figure 9.28 is a 'connected-element' system, which operates in real time with all the signals travelling along dedicated transmission lines or optical fibres. But, as the telescope separations are increased in the quest for higher resolution, it becomes increasingly hard to maintain a real-time connection. The latter is, however, not vital. The stability of atomic frequency standards and the wide bandwidth capabilities of modern digital recording permit the telescopes and their receiving systems to be operated independently while still maintaining long-term coherence; this allows arbitrarily long baselines, including baselines to Earth-orbiting telescopes, to be explored. The correlation is carried out, perhaps weeks later, at a central facility when all the recorded data from telescopes in a VLBI array have been collected together. Note, however, that the challenge of real-time operation is being met by some arrays on a regular basis, using data sent over public fibre networks.

### 11.4.1 VLBI versus Connected-Element Interferometry

The principles of the VLBI technique are discussed in TMS, Chapter 9, and updates on the current methodologies can be found in the proceedings of international interferometer schools and workshops. Here we provide a starting point for further study. At the outset it is important to stress that, for many astrophysical targets and at centimetric wavelengths, modern-day VLBI analysis is very similar to connected-element interferometry and the discussion in Chapter 10 is entirely relevant. The distinctive features of VLBI can be summarized as follows.

(i) *Frequency standards* To maintain synchronization to well within a coherence time across the array and over the course of the observation, each observing station has its own highly stable frequency standard (sometimes called the local clock) from which the various local oscillators and digital timing signals used in the receiver are generated. These clocks suffer from both systematic and random drifts. The stability on short timescales ($\sim 100$ s) is set by a crystal oscillator but on longer timescales an atomic frequency standard, usually a hydrogen maser, takes over. Hydrogen masers typically exhibit a frequency stability[6] $\sim 3 \times 10^{-15}$ over periods of 1000–10 000 s; for illustration, the associated phase error, $2\pi (3 \times 10^{-15}) \nu \tau_{int}$ radians, corresponds to $\sim 5°$ for $\nu = 5$ GHz and an integration time $\tau_{int} = 1000$ s.

(ii) *Timing* For initial data synchronization at the correlator, each observing station must determine the Universal Coordinated Time (UTC; see Appendix 2) independently. Determining UTC to $\sim 1$ μs was a major challenge in the early days of VLBI and there are heroic stories of atomic clocks being flown around the world to achieve clock synchronization in the 1960s.[7] GPS signals

---

[6] Usually expressed in terms of the Allan variance, which is normalized to be independent of frequency; see TMS, Chapter 9, for more details.

[7] The development of VLBI is described in Clark (2003); a brief summary is given in TMS, Chapter 9.

now allow the routine determination of UTC to tens of nanoseconds anywhere on Earth, but note that this is still equivalent to an antenna location error of several metres.

(iii) *External effects* Over and above the drifts in the independent frequency standards and receiver chains there is a range of external effects that must be taken into account to establish the *a priori* delay phase model in the correlator. The reason is that the fringe spacings are tiny and so the computer model to 'steer' the fringes over time must be ultra-precise. In short, the geometric effects due to errors in the geocentric positions of the antennas (including variations due to relative motions of the Earth platform), the sky positions of the source(s) under observation (including relativistic effects), and the Earth's orientation (diurnal rotation, precession and nutation of the spin axis, motion of the poles) must all be known to significantly less than a fringe period for the highest-precision geodetic and astrometric work (see Section 11.4.6). In addition to the purely geometric terms, fluctuations in the tropospheric and ionospheric propagation delays, which have little or no correlation across the array and so contribute with full effect, can affect the delays and phases very significantly. In modern-day VLBI these are the dominant sources of delay and phase errors.

(iv) *Calibration sources* There may be no true point-like calibration sources available to use for flux-scale and visibility amplitude calibration and for external phase referencing. For flux-scale and amplitude calibration the observed correlation coefficients can be converted to a flux density scale using measurements of the system performance at each antenna (Section 11.4.3); for external phase referencing the (typically simple) angular structure of the calibrator can be modelled and the change in its centroid as a function of baseline taken into account.

(vi) *Brightness limitation* The high angular resolution of VLBI means that only sources with high brightness temperatures can be detected (see Section 10.12). Thermal emitters are currently off-limit.

### 11.4.2 Outline of a VLBI System

The general layout of a VLBI system based on recorded data is illustrated in Figure 11.4. After heterodyning to one or more intermediate frequencies for amplification and manipulation, the signal is converted down to 'baseband', because the digital recording capacity is limited. The local oscillators for each heterodyne stage are synthesized from a reference signal generated from the master oscillator.

A digital format unit samples the baseband signal at the Nyquist rate (twice the bandwidth) with the amplitude usually quantized with one or two bits (two or four levels); TMS, Chapter 8, explains the trade-offs involved in this choice. The timing reference for the digitization is provided by pulses generated from the master oscillator. The bit sequence is then digitally split into a number of user-selectable sub-bands. As a current example the baseband data for each hand of polarization might be split into eight sub-bands, each 64 MHz wide. The correlator centres provide on-line software which allows the astronomer to assess the various trade-offs involved in these choices. The data are also split into time blocks. Each block has a 'time tag' that allows the correlator electronics to reconstruct the original time series; within each block the regularly sampled bits themselves provide the time base. Finally, the data are recorded on high capacity hard disk drives.

The correlation of the digitized data from the array is then carried out at the central correlator facility. Because the stations have independent times with respect to UTC, there

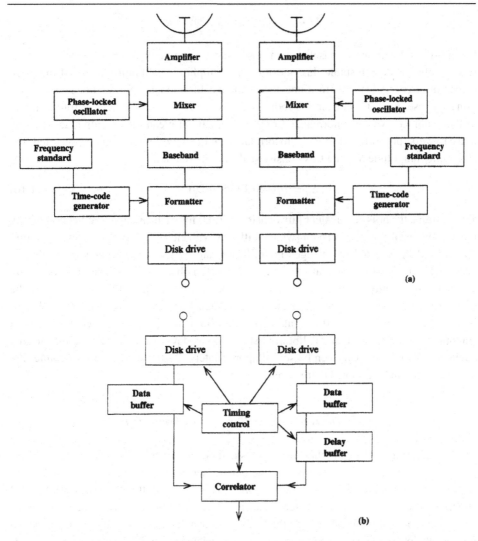

Figure 11.4.  An outline of a VLBI system, showing (a) the two independent receiver stations, (b) the correlator system.

are initially unknown delay offsets between the data streams, and the correlations must be evaluated for a range of station delay values so that the maxima can be established (see Section 11.4.4). The principles and practice of digital correlation are beyond the level of this introductory text but TMS, Chapter 8, provides an excellent overarching discussion; the interested reader is also referred to Romney (1999) and Deller *et al.* (2011).

### 11.4.3 Amplitude Calibration

For most combinations of baseline and observing frequency there are no *a priori* point-source calibrators with stable flux densities. The amplitude calibration in VLBI therefore takes a different route to that in connected-element calibration, as follows. For each integration on a baseline $i,j$ the correlator produces a cross-correlation coefficient $\rho_{ij}$ (proportional to the visibility) which is normalized by the geometrical mean of the individual receiver noise powers (proportional to the autocorrelations). Each unitless $\rho_{ij}$ can be converted into a visibility amplitude $S_{ij}$ on a flux density scale via the relation

$$S_{ij} = b\rho_{ij}(\text{SEFD}_i\text{SEFD}_j)^{1/2}. \tag{11.10}$$

To determine the noise equivalent flux density (SEFD) for each antenna (see Section 6.4.3), measurements of the antenna gain (which will vary with elevation) and system temperature (which will be affected by changes in spill-over and by the sky temperature within the primary beam) must be taken at each station during an observing run. The proportionality constant $b$ takes into account correlator-specific scaling factors (see TMS, Chapter 9). The derived amplitudes can be checked using measurements on a compact source. Even though it may be resolved, the fall-off in peak amplitude with increasing baseline length should be smooth. An incorrect antenna calibration shows up as deviations on the baselines to that antenna, whose SEFD can then be scaled appropriately. By these means the absolute flux density scale can be established to $< 5\%$.

### 11.4.4 Delay and Phase Corrections; Fringe Fitting

The independent clocks and other frequency standards within a VLBI array all have different, slowly variable, offsets which must be determined from the data and then corrected. This particular feature of VLBI data analysis is called fringe fitting and deserves special attention. The response of an interferometer is the vector sum of the individual frequency components within the bandpass; as the basis for understanding the fringe-fitting process, we trace the phase of a single component as it is translated through the system. Our analysis is based on that of Moran (1989).

The received signal phases are $e^{i2\pi\nu t}$ at antenna $i$ and $e^{i2\pi\nu(t-\tau_g)}$ at antenna $j$, where $\tau_g$ is the geometrical delay. The phases of the two local oscillators can be written $2\pi\nu_{lo}t + \psi_i$ and $2\pi\nu_{lo}t + \psi_j$, where $\psi_i$ and $\psi_j$ are terms associated with slowly varying systematic and random frequency offsets of the independent frequency standards at the two stations. After mixing and filtering to translate the signal to baseband we have

$$\phi_i = 2\pi(\nu - \nu_{lo})t - \psi_i, \tag{11.11}$$

$$\phi_j = 2\pi(\nu - \nu_{lo})t - 2\pi\nu\tau_g - \psi_j. \tag{11.12}$$

After including the effect of the station clock errors $\tau_i$ and $\tau_j$ with respect to UTC the phases of the recorded signals become

$$\phi_i = 2\pi(\nu - \nu_{lo})(t - \tau_i) - \psi_i, \tag{11.13}$$

$$\phi_j = 2\pi(\nu - \nu_{lo})(t - \tau_j) - 2\pi\nu\tau_g - \psi_j. \tag{11.14}$$

The geometric delay is compensated by advancing the $j$ data stream by the correlator's model of the delay $\tau_g'$.[8] Noting that the adjustment takes place at the baseband frequency $(\nu - \nu_{lo})$, we have

$$\phi_j = 2\pi(\nu - \nu_{lo})(t - \tau_j + \tau_g') - 2\pi\nu\tau_g - \psi_j. \tag{11.15}$$

The interferometer phase is obtained by subtracting Eqn 11.15 from Eqn 11.13 to yield

$$\phi_{ij} = 2\pi(\nu - \nu_{lo})(\tau_j - \tau_i) + (\psi_j - \psi_i) + 2\pi\nu(\tau_g - \tau_g') + 2\pi\nu_{lo}\tau_g'. \tag{11.16}$$

As explained in Section 9.8.3 the last term in Eqn 11.16 arises because the delay is not implemented at RF; it is removed by multiplying the output data stream by $e^{-i2\pi\nu_{lo}\tau_g'}$. After this phase rotation the final interferometer output phase is

$$\phi_{ij}' = 2\pi(\nu - \nu_{lo})\Delta\tau_c + \Delta\psi_{ji} + 2\pi\nu\Delta\tau_g, \tag{11.17}$$

where the instrumental error terms are $\Delta\tau_c = \tau_j - \tau_i$ (the differential clock error) and $\Delta\psi_{ji} = \psi_j - \psi_i$ (the differential LO phase error); $\Delta\tau_g = \tau_g - \tau_g'$ is the difference between the true geometric delay and the correlator model delay and therefore contains the information on the source–baseline geometry (see Section 11.4.6).

In the ideal case $\Delta\tau_c$, $\Delta\psi_{ji}$, and $\Delta\tau_g$ are all zero and the peak of the fringe pattern sits exactly on the fiducial sky position, i.e. the fringes have been 'stopped'. In reality perfection is never achieved and the time-dependent station clock and frequency offsets need to be determined and removed from the correlated data. Before outlining this process we make a few initial points.

- As $\nu$ changes across the band (or sub-band) the clock error $\Delta\tau_c$ (the first term in Eqn 11.17) causes a phase slope $\partial\phi_{ij}'/\partial\nu$ across the band. This will reduce the resultant of the vector summation unless the phase change $2\pi\Delta\nu\Delta\tau_c$ is much less than 1 cycle of phase, i.e. unless $\Delta_c \ll 1/\Delta\nu$ (the coherence time).
- Systematic differences between LO frequencies produce instrumental fringe rates and hence linearly time-dependent phase changes $\Delta\psi_{ji}$. A frequency difference $\delta\nu$ results in a phase change $\Delta\psi_{ji} \sim 2\pi\delta\nu t_{corr}$ over a correlator integration time $t_{corr}$. This will tend to wash out the fringes unless $\delta\nu \ll 1/t_{corr}$.
- Differential drifts between the station clocks by $\delta(\Delta\tau_c)$ over a time $\delta t$ also produce time-dependent phase changes and thus frequency drifts. Starting from the first term in Eqn 11.17 it is easy to show that

$$\frac{\delta(\Delta\tau_c)}{\delta t} \equiv \frac{\delta\nu}{\nu}. \tag{11.18}$$

---

[8] The compensating delay was called $\tau_i$ in Chapter 9.

With this background in mind the task is to estimate the interferometer phases, delays, and fringe rates for the $N$ sub-bands recorded for each polarization. In the early days of VLBI this fringe-fitting process was done baseline by baseline but, as in the discussion in Section 10.10.1, most contributory factors are antenna dependent. Schwab and Cotton (1983) developed an extension of their self-calibration technique (Section 10.10.2) which solves for the first derivatives of the phase in frequency (the delay errors) and time (the fringe rates) on an antenna-by-antenna basis. The process is called *global fringe fitting*.

Its essence can best be understood by moving to a simplified description of the interferometer phase, which ignores the specific origin of the terms. We can write

$$\Delta\phi_{ij} = \phi + \frac{\partial\phi}{\partial\nu}\Delta\nu + \frac{\partial\phi}{\partial t}\Delta t, \tag{11.19}$$

where $\phi$ is the phase error corresponding to a particular time and frequency. Expressing Eqn 11.19 in antenna-based terms:

$$\Delta\phi_{ij} = (\phi_i - \phi_j) + \left(\frac{\partial\phi_i}{\partial\nu} - \frac{\partial\phi_j}{\partial\nu}\right)\Delta\nu + \left(\frac{\partial\phi_i}{\partial t} - \frac{\partial\phi_j}{\partial t}\right)\Delta t. \tag{11.20}$$

To gain access to the antenna errors, one of the antennas in the array is defined as the reference and its phase $\phi_r$, delay $\partial\phi_r/\partial\nu$, and fringe rate $\partial\phi_r/\partial t$ are set to zero. On baselines to the reference antenna and over a series of 'solution intervals' each comparable to the *a priori* coherence time, the fringe rates and delays for the other antennas are determined from the correlated data. The output of the algorithm is a table of delay, phase, and fringe rates for each antenna relative to the reference antenna, which can be used to correct the observed data. The reader should note that this outline is highly simplified. In practice data from all baselines in the array, weighted by their signal-to-noise ratios, are used to determine the best estimate of the antenna errors; this is the reason for the word 'global' in the name of this process. For further details the interested reader should consult Walker (1989) and Cotton (1995). As the *a priori* knowledge of VLBI systems has improved, fringe fitting is now a routine step.

### 11.4.5 Basic VLBI Analysis for Imaging

Despite the additional calibration steps involved in VLBI[9] the accrued experience of the technique and the knowledge of array parameters means that it is straightforward for an astronomer to make high-resolution images. The main steps are:

– remove bad points from the data ('flagging');
– calibrate the visibility amplitudes using SEFD information;
– correct for the (channel dependent) instrumental delays and LO offsets with respect to a reference antenna. Data from a short observation of a strong compact source whose position is well known are used with fringe fitting to align the delays and phases between the baseband channels and remove the instrumental fringe rates;

---

[9] For an overview of VLBI calibration see Diamond (1995).

– starting from these estimates, fringe-fit the target source data (if it gives a good signal-to-noise ratio) over the observing run. If the target source is weak then a nearby compact reference source is used to establish the phases, delays, and rates, which are then interpolated and applied to the target source data. After fringe fitting the phases should be flat in each sub-band and the sub-bands aligned;

– correct for amplitude and phase variations across the sub-bands (the bandpass correction, Section 10.7.1);

– proceed as for connected-element imaging, as in Chapter 10; the next step would be self-calibration.

### 11.4.6 Geodesy and Astrometry

Some of the most important contributions of VLBI have come from non-imaging applications in the form of ultra-precise measurements of celestial source positions, measurements of antenna positions and their change with time, and measurements of variations in the rate and direction of the Earth's rotation. The foundation of this success was, and is, the establishment of the positions of a sky-wide net of radio sources at cosmological distances which constitutes an inertial reference frame. Systematic work to establish this frame began in the 1970s. The latest incarnation is the second generation of the International Celestial Reference Frame (ICRF2). The equivalent in geodesy is the International Terrestrial Reference System (ITRS).

### 11.4.7 Methods

In Section 9.3 we described the principle behind source-position measurements using the interferometric phase. However, fluctuations in atmospheric propagation delays, during the time it takes the antennas to move from one source to another, destroy the phase connection between sources separated by one or more radians. Connection amounts to knowing the exact number of interference fringes between the sources; unmodelled fluctuations may insert $2\pi n$ ambiguities between the measured phases. This can be resolved by using a band of frequencies, observing the group delay $\tau = (1/2\pi)\partial\phi/\partial\nu$ rather than the phase delay $\tau = (1/2\pi)\phi/\nu$ as the primary observable. The group delay offers an accuracy $\propto 1/\Delta\nu$, set by the effective bandwidth $\Delta\nu$; while this is less than the fringe width, the measurements can be tied together across the sky without ambiguity.

The instantaneous geometric time delay is

$$\tau_g = [b_X \cos\delta \cos H(t) - b_Y \cos\delta \sin H(t) + b_Z \sin\delta]/c \qquad (11.21)$$

where $b_X, b_Y, b_Z$ are the geocentric components of the baseline (see Section 10.6.1); $\alpha, \delta$ are the source's Right Ascension and declination, and $H(t) = ST - \alpha$ is the hour angle at which the observation is taken.[10] Note that the rotation of the Earth causes the relative separation between the antennas to change during the light travel time from one to the other; this is called the *retarded baseline* effect (Cohen and Shaffer 1971) and corrections

---

[10] Strictly, the ST is the Greenwich Apparent Sidereal Time or GAST; see Appendix 2.

are made relative to a single reference antenna. The components of the baseline vector can be determined from a series of measurements of $\tau_g$ for point sources at known positions. Conversely, if the baseline vector is known, delay measurements enable the source positions to be determined.

Using the generic version of the time delay $\tau_g = \mathbf{b} \cdot \mathbf{s}/c$ (Section 9.1.1) rather than the explicitly parametrized Eqn 11.21 allows one to see quickly that a delay error $\Delta \tau$ will translate into a source position error $\Delta \mathbf{s} \sim c\Delta \tau/\mathbf{b}$. Delay errors arise from many effects (see below), which must be understood and accounted for. In particular an incorrect value for an antenna position, thus a baseline error $\Delta \mathbf{b}$, results in $\Delta \tau_g \sim \Delta \mathbf{b}/c$. On these basic principles the success of modern VLBI astrometry and geodesy has been built. However, to reach the present state of the art many subtle technical and conceptual hurdles have had to be overcome.

The main contributions to the observed delay are given by

$$\tau_{obs} = \tau_{geom} + \tau_{instrument} + \tau_{tropos} + \tau_{ionos} + \tau_{rel} + \tau_{structure} + \epsilon_{noise}. \qquad (11.22)$$

The terms have obvious interpretations apart from $\tau_{rel}$, which includes special and general relativistic effects, and $\tau_{structure}$, which accounts for fact that the centroid of the source changes with the resolution. The science is mainly in $\tau_{geom}$. Special relativistic effects include diurnal aberration, which shifts the apparent source position periodically by $\sim 20$ arcsec over the course of a year. The general relativistic effect of the gravitational field of the Sun affects source positions over the whole sky; at the solar limb the bending is 1.75 arcsec and it is still $\sim 4$ milliarcsec in a direction 90 degrees away from the Sun. For the highest precision work the effects of the Moon and the planets are also taken into account. After calibration the instrumental terms are not the dominant terms in the delay error budget.[11]

The breakthrough in position measurements using group delay came from the development of the *bandwidth synthesis* technique (Rogers 1970). The residual error in the correlator delay model can be determined from phase slope as a function of frequency, which is provided by the fringe-fitting process. Clearly the wider the range of frequencies involved, the more accurately can that slope be determined. The idea of bandwidth synthesis is, therefore, simultaneously to record many widely spaced frequency bands, at the expense of reducing their widths. This is a good trade-off since, for bright sources, the thermal noise fluctuations $\epsilon_{noise}$ in Eqn 11.22 are below the level of the systematic errors. The current approach is to record up to eight narrower sub-bands within the X-band range 8.2–8.95 GHz and to fit the slope of phase versus frequency using fringe fits from all eight bands covering a frequency range, and hence an effective bandwidth, $\Delta\nu = 720$ MHz. This is shown schematically in Figure 11.5. With achievable signal-to-noise ratios $\gg 10 : 1$, multi-band group delays can be determined to 10 ps or 3 mm of path length.

---

[11] In positional VLBI the phase calibration technique, which involves injecting master-oscillator-derived frequencies into the signal path at each station, plays a vital part in the determination of instrumental delays; see TMS, Chapter 9, for more details.

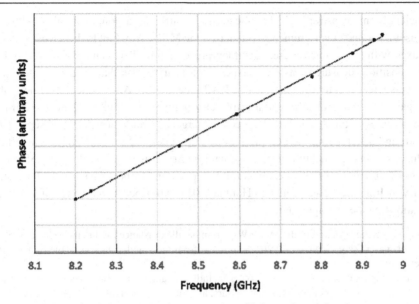

Figure 11.5. Schematic of phase versus frequency; the grid shows the sub-bands.

The main perturbations to the measured X-band delays are the total ionospheric and tropospheric propagation paths and fluctuations therein. The path delays due to the troposphere and the ionosphere were discussed in Section 9.7. To separate them, advantage is taken of the fact that the tropospheric delay is non-dispersive whilst the ionospheric group delay is proportional to $1/\nu^2$. Simultaneously with the X-band observations, data are also taken in multiple sub-bands in a second frequency range (2.22–2.35 GHz; the S-band) and the multi-band group delay determined at this lower frequency. The ionospheric contribution to the delay can then be calculated from

$$\delta\tau_{\text{iono}} = \frac{\nu_S^2}{\nu_X^2 - \nu_S^2}(\tau_X - \tau_S). \tag{11.23}$$

The slowly varying tropospheric dry path can be calculated from meteorological measurements; the wet path is estimated from atmospheric models and the variations are derived from the data taken on many sources during a geodetic observing run (see e.g. Schuh and Böhm 2013).

### 11.4.8 Astrometry

Several decades of observations by international teams have led to the production of the ICRF2 catalogue (Fey *et al.* 2015) which contains 3414 extragalactic radio sources, with median positional accuracies of 100–175 microarcsec; 295 of them are defining sources with positional accuracy $\sim$ 40 microarcsec. An improved catalogue, ICRF3, with 4536 sources became effective in 2019. The *absolute* positional accuracy of ICRF2 is, however, insufficient for many astrophysical applications, in particular trigonometrical parallaxes. To make useful distance measurements well across the Milky Way (out to distances $\sim 10\,\text{kpc}$), parallaxes with $\sim 10$ microarcsec accuracy are required. This can be achieved only by *relative* position measurements with respect to a nearby (less than a few degrees) phase reference source, ideally drawn from the ICRF2 catalogue. An excellent overview of the techniques and results of differential astrometry is given by Reid and Honma (2014; see Further Reading). Rapid switching between the target and reference sources[12] separated by $\Delta\theta$ (rad) and differencing the delays gives a relative delay $\Delta\tau = \Delta\theta\Delta\mathbf{b}/c$. The effect is to reduce the absolute delay errors by a factor $1/\Delta\theta$; thus if $\Delta\theta = 2° = 0.035$ rad, the reduction factor is $\sim 29$. This enables relative positions to be derived with the required accuracy of 10 microarcsec. Reid and Honma (2014) described a wide range of astrometric applications. Some examples are:

– constraints on the structure of the Milky Way from parallax distances and radial velocity measurements of masers (e.g. Sanna *et al.* 2017) and also from the angular motion of the Sun about the galactic centre;
– parallax distances to a variety of stellar types including young stellar objects and X-ray binaries;
– parallax distances and proper motions of pulsars, allowing their birth places to be estimated;
– positions, combined with radial velocities and accelerations, of megamaser spots in the nucleus of NGC4258 enable the distance, $7.6 \pm 0.23$ Mpc, to the galaxy to be measured. With observations of galaxies over greater distances the Megamaser Cosmology Project aims to determine the Hubble constant to $\sim 3\%$;
– the gravitational light bending near the Sun is consistent with the predictions of general relativity to one part in $10^4$;
– tracking the position of the Cassini probe during its infall through the atmosphere of Titan enabled the vertical profile of the windspeed to be measured.

## *11.4.9 Geodesy*

Very long baseline interferometry is carried out on a rotating reference frame with antennas fixed to the surface of the Earth and linked to the International Terrestrial Reference System;[13] the measurements are made with respect to the inertial frame ICRF2. The observed geometric delay is affected by many small perturbations and hence contains a wealth of information about the behaviour of the Earth in space. A classic reference is Sovers, Fanselow, and Jacobs (1998); for a more recent review and new results consult Schuh and Böhm (2013) and Schuh and Behrend (2012). A short summary gives a flavour of the depth of the discipline.

The surface of the Earth is not stable – there is a variety of deformations, including tectonic plate motions (10–100 mm per year, lateral), postglacial rebound (several mm per year, vertical; since the Ice Ages Scandinavia has uplifted by 300 m); solid Earth tides due to the Moon and Sun ($\sim 300$ mm over 12 hr, mainly vertical); local tidal loading ($\sim 20$ mm over 12 hr, mainly vertical); local atmospheric loading ($\sim 20$ mm over a year,

---

[12] The VERA array antennnas (Section 11.4.11) have dual beam systems and can observe both sources at the same time.
[13] www.iers.org.

mainly vertical). The VLBI method is also sensitive to the instantaneous orientation of the Earth in space. It responds to the variation of the rotational period (UT1) with respect to international atomic time (UTC) and to changes in the spin axis due to precession ($\sim 30$ arcsec per year), nutation ($\sim 20$ arcsec in 18.6 years) and the periodic ($\sim 500$ milliarcsec in 433 days) and wandering components of polar motion. These spin-axis related terms constitute the Earth orientation parameters (EOPs), which link the terrestrial to the celestial reference frames. The EOPs are a unique contribution of VLBI.

Two results illustrate the power of the technique (for references and figures see Schuh and Böhm 2013). The rates of continental drift were first established with VLBI. For example, measurements over 30 years show that Europe and North America (the Wettzell–Westford baseline) are separating at a rate of $16.91 \pm 0.03$ mm yr$^{-1}$. The difference between the time (UT1) associated with the fluctuating spin rate of Earth and the atomic-standard derived UTC (Appendix 2) can be measured to 3–5 µs (the average over a day). The Earth is currently slowing by an average of 2 ms per day with respect to atomic time, with respect to the definition of the UTC second. There is also a range of fascinating higher-order perturbations which can be studied.

### *11.4.10 Space VLBI*

A VLBI station need not be on Earth; a radio telescope in space, with the VLBI electronics of Figure 11.4, can be an interferometer element (see Schilizzi *et al.* 1984 and Hirabayashi *et al.* 2000). The stable frequency standard can be in space, or it can be at a ground station, with the reference frequency relayed to the satellite, where a secondary oscillator is locked to the reference. Its signal is then returned to the ground, where a comparison with the original standard allows corrections to be made. The astronomical signal is relayed to the ground by a separate radio link and recorded as in a standard VLBI system. The orbit is changing with time, which complicates the reduction process but also allows data to be taken from a wide range of baselines with only a small set of ground telescopes.

The first space VLBI experiment used a NASA TDRSS satellite as the orbiting element (Levy *et al.* 1986); its results are described in Linfield *et al.* (1989). The first dedicated VLBI spacecraft, VSOP/HALCA, was launched by Japan's Institute for Space and Astronautical Sciences in 1997 and operated until late 2005. The reflector had a diameter of $\sim 8$ m and the orbit took it from 12 000 to 27 000 km above the Earth's surface, providing resolutions about three times higher at a given wavelength than Earth-based arrays. The HALCA system and the first results are collected in Hirabayashi *et al.* (2000); later results are summarized in Hirabayashi (2005). The second dedicated space VLBI project, RadioAstron, is led by the Astro Space Center of the Lebedev Physical Institute in Moscow, Russia. The spacecraft was launched in July 2011 and ceased operation in January 2019. The reflector is 10 m in diameter and the orbit takes it out to 350 000 km from the Earth, thus providing resolutions $> 30$ times those available in Earth-based arrays (see also Section 16.3.10). For the latest information the reader should consult the RadioAstron website.[14]

---

[14] www.asc.rssi.ru/radioastron/.

## 11.4.11  VLBI Arrays

There is a range of VLBI arrays, some of which focus largely on astronomy, some exclusively on astrometry and geodesy, and some which do both.

*VLBA*   The Very Long Baseline Array of the USA is a dedicated facility consisting of ten 25 m identical telescopes. Their locations extend at northern latitudes from New Hampshire to the state of Washington, and at southern latitudes from St Croix in the Virgin Islands to the island of Hawaii. The central correlator is located at the NRAO Science Operations Center, Socorro, New Mexico.

*EVN*   The European VLB Network is a cooperative arrangement of radio telescopes in the UK, The Netherlands, Germany, Italy, Poland, Russia, Ukraine, China, and Japan. The central correlation facility is the Joint Institute for VLBI in Europe (JIVE) at Dwingeloo, The Netherlands. Periodically the EVN and the VLBA cooperate to form a worldwide network. At the time of writing the EVN operates in real-time fibre-connected mode for $\sim 25\%$ of the total time allocated for EVN operation; this percentage will grow with time.

*East Asian VLBI Networks*   The Japanese VLBI Network (JVN) and the Japanese dedicated astrometric array (VERA) with its four dual-beam antennas work independently. VERA will also work in cooperation with the three-element Korean VLBI Network (KVN; radio.kasi.re.kr/kvn/) to form the kaVA.

*Australian VLB Array*   consists of the Australia Telescope Compact Array and single dishes at Parkes, Tidbinbilla, Hobart Ceduna, and Perth. It is the only VLBI array in the southern hemisphere.

*IVS*   The International VLBI Service for Geodesy and Astrometry coordinates global VLBI resources for positional VLBI. At various times this includes 45 antennas sponsored by 40 organizations located in 20 countries. The IVS Coordinating Center is located at Goddard Space Flight Center in Greenbelt, MD. The next-generation coordinated facility VLBI2010 is planned to have more small fast-slewing antennas and a much enhanced multiband receiving system.

*GMVA*   The Global Millimeter Wave VLBI Array is a cooperative arrangement of many mm-wave telescopes under the auspices of five different international organizations. They come together about twice per year to make coordinated network observations.

*EHT*   The Event Horizon Telescope is another international coordinated network of independent telescopes; it includes the phased-up ALMA. The purpose is to make ultra-high-resolution observations at 1.3 mm wavelength, particular targets being the massive black holes at the centre of the Milky Way and the elliptical galaxy M87.

## 11.5  Wide-Field Imaging

In Section 9.2 we began the discussion of wide fields of view in elementary terms. In Section 9.5.2 we developed the general interferometer formalism leading to the full imaging Eqn 10.1. In Section 10.4 we discussed the restrictive field-of-view conditions under which the imaging equation becomes a more tractable 2D Fourier transform relation. These 2D fields of view are often smaller than the primary beam of the antenna elements, and many observing programmes, ranging from deep surveys of discrete sources to the study of complex emission regions in the Milky Way, now require the ability to image within the entire primary beam and beyond. If the phases from off-centre directions within the beam are not corrected then a 2D image becomes progressively distorted.

We describe off-centre phase effects and ways to mitigate them in Section 11.5.1. Wide-field imaging within the primary beam is also affected by: bandwidth smearing (considered in Sections 9.2 and 10.14.3); time-averaged smearing (considered in Section 10.14.2), and direction-dependent and time-dependent variations in the primary beam response (to be outlined in Section 11.5.2). In addition there will always be discrete sources outside the primary beam but within its sidelobes, sources which are strong enough that their sidelobes, visible in the synthesized image, need to be subtracted from or *peeled* out of the data, as will be outlined in Section 11.5.3. Finally, if the emission region to be imaged is larger than the primary beam then a new approach, mosaicing, is required; this is outlined in Section 11.5.4. The breakdown of atmospheric isoplanicity in low frequency imaging is deferred to Section 11.6.

### 11.5.1 The w-term and its Effects

Equation 10.1 is the general 3D imaging equation for a single interferometer. The phase factor involves $w$, the baseline component towards the source and the celestial sphere described by $\sqrt{1 - l^2 - m^2}$. For 2D imaging, points away from the phase tracking centre are effectively projected onto the $w = 0$ tangent plane (see Section 9.5 and Figure 9.21); larger offsets shift a source further away from its true direction. As the Earth rotates, $w$ changes and so do all the position shifts; in the 2D approximation this can be ignored.

The w-term can be brought to life with two simplified antenna configurations, as shown in Figure 11.6; for clarity the baseline between them has no N–S component. On the left-hand side a source at the zenith (broken lines) is transiting the baseline, so normally $w$ would be zero; however, one antenna is displaced towards the source by $w$ wavelengths, giving a new contribution $2\pi w$ to the on-axis phase delay. For the off-axis direction the phase delay is easily shown to be $2\pi(u \sin\theta + w \cos\theta)$ and hence, with respect to the on-axis phase delay, $\Delta\phi = 2\pi[u \sin\theta + w(\cos\theta - 1)]$. In terms of direction cosines this can be written as $2\pi[ul + w(\sqrt{1 - l^2} - 1)]$, which demonstrates the link with the full-phase Eqn 10.1. The right-hand side of Figure 11.6 shows the antennas on the same level but with the source well past baseline transit. The w-term for the phase tracking centre ($l = 0, m = 0$) is compensated at the correlator, but in other directions the component of the baseline towards the source $w'$ is different and so its phase is not correctly compensated (see also Section 9.2).

Cornwell and Perley (1992) laid out the fundamental issues and described a method for constructing wide-field images based on the general 3D equation. This turned out to be too computationally expensive to be of practical use. Another method is to stitch together many 2D images ('facets') to cover the wide field in a technique called 'polyhedral imaging'

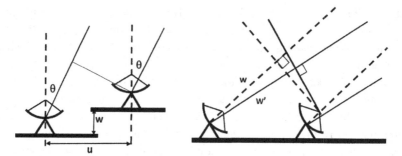

Figure 11.6. The geometrical origin of the $w$-term.

(Perley 1999; TMS, Chapter 11). This approach, which has difficulties for sources which extend over facet boundaries, has been superseded by others that deal with the whole field of view. The first such method was $w$-projection (Cornwell *et al.* 2008), which we now outline.

The general visibility equation can be rewritten as

$$V(u, v, w) = \int \int \frac{\mathcal{A}(l, m) B_\nu(l, m)}{\sqrt{1 - l^2 - m^2}} e^{-i2\pi(ul+vm)} G(l, m, w) \, dl \, dm, \qquad (11.24)$$

where

$$G(l, m, w) = e^{-i2\pi(\sqrt{1-l^2-m^2}-1)w}. \qquad (11.25)$$

This is just the 2D imaging equation multiplied, inside the Fourier integral, by the differential phase factor captured by Eqn 11.25. In other words, from any point in the field of view the incident waves falling on the array are planar (far-field, Fraunhofer, diffraction) but correcting to a single phase tracking centre involves summing over spherical waves, as in near-field, Fresnel, diffraction. The formalism of $w$-projection closely resembles Fresnel diffraction.

Heuristically, for each visibility sample $V(u, v, w)$, the challenge is to disentangle the differently phased contributions (Eqn 11.25) from each $l, m, n$ point (pixel) within the field of view. If we knew the flux at each pixel, that is if we knew $B_\nu(l, m)$, then the *forward* projection to a particular $V(u, v, w)$ would be straightforward. But we do not know $B_\nu(l, m)$ and we also want to project *back* to the $w = 0$ plane starting from each $V(u, v, w)$ point. A general back projection is not possible. However, a good approximation is to associate the observed amplitude with different phases appropriate for each pixel in the field of view. This must be done for each discrete $u, v$ point in the data set since $u, v, w$ change continuously. Clustered around each point in the $u, v, w = 0$ plane, therefore, there will be other complex visibility points, each at a slightly different $u, v$ location appropriate for a direction within the field of view. The situation can also be described as a convolution since, applying the convolution theorem in two dimensions, one can write

$$V(u, v, w) = V(u, v, w = 0) * \mathcal{G}(u, v, w), \tag{11.26}$$

where $\mathcal{G}(u, v, w)$ is the Fourier transform of $G(l, m, w)$ with respect to $u, v$; Cornwell *et al.* (2008) provides the full mathematical details. Happily, the time-consuming convolution step need only be done once, at the same time as gridding (see Section 10.8.1). Thereafter 2D imaging proceeds largely as normal.

Wide-field techniques are advancing all the time, and in Section 11.6 we mention the most recent algorithm, WSCLEAN, devised for very wide fields of view from dipole-type arrays. Suffice it to say that astronomers now experience little or no difference between 2D imaging and wide-field imaging, apart from the speed of analysis.

### 11.5.2 Effects of the Primary Beam

Thus far in our discussion of synthesis imaging we have ignored all effects of the primary beam (PB, i.e. $A(l, m)$) on the data. This is often acceptable when working within the restricted areas appropriate for 2D imaging. However, synthesis data are now commonly taken over wide frequency bands and the astronomer often requires images with a wide field of view and a high dynamic range. In this environment the data analysis has *a fortiori* to contend with direction-dependent and time-dependent changes of the PB, which are not corrected by the standard process of bandpass calibration. The PB effects involve some mixture of the following.

(i) *Intrinsic asymmetries in the beam pattern*  These arise from a combination of (sometimes off-axis) feed illumination and blockage from the focus support structure. For all but polar-mounted antennas the parallactic angle changes with time and points offset from the beam centre will rotate with respect to the beam pattern; this causes the complex gains to be direction and time dependent. This behaviour is inconsistent with the basic assumption of self-calibration (that the complex gain errors are direction independent) and limits the achievable dynamic range.

(ii) *Time-variable pointing errors*  These result in amplitude and phase modulations which become larger for sources nearest to the edge of the PB. This behaviour is also inconsistent with basic self-calibration.

(iii) *Position-dependent polarization response*  As mentioned in Section 11.2.2 the polarization response inevitably varies across the voltage beam of a parabolic dish. It is obviously necessary to understand the pattern for obtaining linear polarization images across a wide field. For total-intensity imaging, cross-polar leakage from sources in the outer reaches of the PB can limit the achieved dynamic range.

(iv) *Dissimilar antennas in the array*  On any one baseline the voltage pattern response $A(l, m)$ is the geometric mean of the responses of two antennas. The available software packages can take the different baseline responses into account if supplied with accurate beam models for each antenna.

(v) *Frequency dependence*  In wide-band observations the beam size is inversely proportional to frequency and, hence, away from the beam centre sources will appear increasingly fainter at higher frequencies, causing an apparent spectral steepening unless the PB's frequency dependence is taken into account.

The 'take home' point of this section is that accurate 2D beam models, ideally out into the near-in sidelobes, have become vital for extracting the optimum performance from modern synthesis arrays. The narrowband effects were discussed by Bhatnagar *et al.* (2008) while Bhatnagar *et al.* (2013) extended the analysis to broadband observations. As is the case for the corrections of phase errors due to the *w*-term, visibility amplitude corrections for known direction-dependent gain terms can be applied at gridding time.[15] The gain-correction algorithms, which go under the collective title of A-projections, are part of the suite of software now available to users of all major instruments.

### 11.5.3 Peeling

The dirty beam responses of strong sources located in the outer edges of the PB or in its sidelobes can give rise to ripples within the target field. Sources with flux densities in the range 10–100 mJy can pose problems for deep imaging at the μJy level, and there are up to ten such sources per square degree at L-band. The systematic removal of these far-out confusing sources is known as *peeling*. However, because of imperfect knowledge of the behaviour of the PB, one cannot in general deconvolve their responses without taking additional direction-dependent correction steps.

After correcting the visibility data with respect to an external calibration source one makes a 2D facet image in the direction of the brightest confusing source (whose position will be well known). One or more self-calibration cycles will then correct the amplitudes and phases for this specific direction and the confusing source can then be subtracted from the visibility data. Since the self-calibration corrections are only valid for one specific direction, they must be deleted from the data before moving on to the next brightest confusing source and the process repeated. With this one-by-one approach the sidelobes from any number of confusing sources can be removed from the data. Before making the final wide-field image, all the self-calibration corrections in the peeling steps should be deleted.

### 11.5.4 Mosaicing

Radio astronomers wish to combine the resolution of an interferometer array with a single telescope's ability to map extended regions with high surface brightness sensitivity. Imaging larger areas requires *mosaicing*, that is combining visibility data from a pattern of discrete pointings across the face of the target region. The technique is most often used with mm or sub-mm arrays owing to the small size of their primary beams and the extended nature of many galactic targets. It is also used with centimetric and decimetric arrays for imaging extended regions and for surveys of discrete sources.

The first such method involved constructing images from each pointing separately and then combining them, on a pixel-by-pixel basis, into a wide-field image. A weighted addition of the overlapping images can give a close-to-uniform sensitivity except near

---

[15] This is another case where prior knowledge is used in the image construction process.

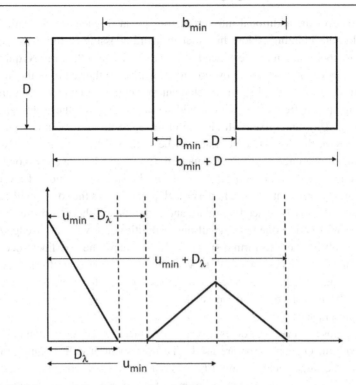

Figure 11.7. Interferometer dimensions and spatial frequencies.

the edge of the field. The VLA NVSS (Condon *et al.* 1998) and FIRST (Becker *et al.* 1995) discrete-source surveys used this method. For extended structures one can, however, do better (e.g. Cornwell 1988; Holdaway 1999) by using 'hidden' spatial frequency information.

As we discussed in Section 10.5, the $u, v$ coverage is the autocorrelation of the excitation pattern spread across the antennas in an array. Thus each $u, v$ point is convolved with the autocorrelation of the antennas' aperture distribution (assume for simplicity that they are identical). When the antennas are relatively close-packed this can help to fill in $u, v$ gaps, and a better image is the result. As an illustration consider the elementary interferometer shown in Figure 11.7, which is similar to Figure 10.3. At the top two closely spaced square antennas, of side $D$, separated by the minimum baseline in the array, $b_{min}$, are pointing at a source directly overhead. The spatial frequency sensitivity is shown underneath. At a wavelength $\lambda$ the interferometer is sensitive from $(b_{min} - D)/\lambda$ to $(b_{min} + D)/\lambda$ or, equivalently, $u_{min} - D_\lambda$ to $u_{min} + D_\lambda$. Thus each *interferometer* is sensitive to a range of spatial frequencies but, in practice, only their average is produced by the correlator (see also the discussion in Section 11.5.1). Each *antenna* is sensitive to the lower spatial frequencies, from zero (the mean brightness level) to $D/\lambda$ or $D_\lambda$. This information can be recovered if the autocorrelation data from each antenna are also collected.

One can recover the additional interferometric spatial frequency information from the pattern of discrete pointings. Note first that to avoid aliassing in the constructed wide-field image the pointings must be spaced in angle $\Delta\theta$ by less than the Nyquist criterion $\lambda/2D$, as for single-dish imaging. Now consider the pixel in the centre of the first pointing and hence at the peak of the PB. In the elementary 1D interferometer in Figure 11.7 the corresponding spatial frequency point would be at the peak of the right-hand triangle, centred at $u_{min}$. The same pixel will also be observed at the next pointing along but, for this new direction, the projected baseline will be slightly different and the pixel will fall part of the way down one side of the PB pattern. In Figure 11.7 the corresponding spatial frequency point will therefore fall part way down the side of the side of the right-hand triangle, giving it a different weight. In general all points across the source will be observed several times with closely spaced $u, v$ data and different weights. To make the optimum use of all the information one re-grids all the visibilities and, with knowledge of the PB, 'jointly deconvolves' them to form a single widefield map. The non-linear deconvolution scheme automatically makes use of all the visibility data contributing to each image pixel, effectively linking together the data sets. This approach was introduced by Cornwell (1988) and has since been evolved by him and collaborators. The details of these algorithms are beyond our present scope.

Planning a mosaic survey involves a series of choices. A common pointing pattern is a uniform grid but other patterns are used. The time spent at each pointing depends on the array and the target field; it could be 12 hr or more for a complete track to obtain the highest available sensitivity and to optimize the $u, v$ coverage. The properties of the antenna primary beams need to be well understood. There is no 'one size fits all' recipe, and mosaicing with wide bandwidths is currently a research topic. The section on mosaicing in the Guide to VLA Observing provides a range of practical advice.[16]

It must be stressed that mosaicing cannot supply spatial frequency information in the central $u, v$ 'hole' if the autocorrelation data are not measured or, alternatively, in the gap between the telescope diameter $D_\lambda$ and $u_{min} - D_\lambda$ if they are. In the latter case the mosaiced source will again sit in a negative bowl, albeit shallower than before. As previously noted in Section 10.11, these defects can be remedied with $u, v$ information obtained from Fourier transforming a single-dish map of the region. The dish diameter should be large enough that, in the spatial frequency coverage, $D_\lambda$ stretches out to $u_{min}$ and preferably beyond.

## 11.6 Low Frequency Imaging

Radio astronomy began and developed through the 1950s and into the 1960s mostly at frequencies below a few hundred MHz, i.e. at metre wavelengths. The reasons were simple: the receivers were easier to build and antennas with reasonable gain could be constructed out of wires. Advances in technology then allowed the scientific interest to migrate into the centimetre and millimetre bands but now, after decades of relative neglect, there is a renewed interest in observations at metre wavelengths, increasingly fuelled by the abil-

---

[16] https://science.nrao.edu/facilities/vla/docs/manuals/obsguide/modes/mosaicking.

ity to construct large arrays of low-gain elements (Section 11.6.2) and the computing power to deal with wide fields. Low frequency continuum work naturally focusses on non-thermal emitters with steep radio spectra. The wide fields now available from phased arrays are well suited for conducting large population studies of active galaxies (Section 16.8), while metre-wave images of 'old' electrons in individual sources provide diagnostics of the transfer of energy into the intracluster gas (Section 16.3.6). A compelling spectroscopic opportunity involves the study of redshifted neutral hydrogen from the early phases of the Universe (Chapter 17).

Existing dish arrays, in particular the GMRT and the JVLA, are making new contributions to low frequency astronomy, but state-of-the-art procedures now involve synthesizing images from electronically driven phased arrays, requiring the development of new algorithms. The challenges are great and the landscape of techniques is ever evolving. In the next section we content ourselves with highlighting the principal issues.

### *11.6.1 The Challenges*

The new challenge of imaging in the low frequency regime is the emergence of significantly greater direction-dependent effects (DDEs) than those discussed in Sections 11.5.1 and 11.5.2. The main issues are outlined below.

*Geometric phase effects* For an array with a maximum baseline $b_{max}$ and antennas of diameter $D_\lambda$ the ratio of $\theta_{2-D} \sim \sqrt{\lambda/b_{max}}/3$ (Eqn 10.3) and the size of the PB, $\sim\lambda/D_\lambda$, decreases as $\lambda^{1/2}$. Thus, for a given array, uncorrected 2D imaging becomes increasingly less appropriate at lower frequencies.[17] In Section 11.5.1 we outlined the ideas behind the $w$-projection algorithm, which was the first to allow a single 2D field to be constructed over areas much larger than $\theta_{2-D}$. Phased (aperture) arrays constructed from low-gain dipole-like elements arrays have much larger PBs (typically tens of degrees or more) than dish-based arrays and this automatically exacerbates the 3D to 2D imaging challenge. In addition $w$-projection works best at small zenith angles, where $w$ is small; large fields of view automatically span a wide range of zenith angles and in this environment $w$-projection can become slow. This has led to various evolutions of the method. The wide-field imaging algorithm WSCLEAN (Offringa *et al.* 2014) offers significant speed advantages over $w$-projection whilst placing a greater demand on computer resources.

*Primary beam effects* In phased-aperture arrays, allowance is made for beam forming and beam pointing by means of complex weights applied to the signals from the low-gain elements together with delay and phase gradients across all or parts of the array (see Section 8.2). In practice many antenna elements are often combined into 'tiles' or on larger scales into complete 'stations', and one or more beams are formed from the combinations. When tracking a given field with any electronically formed beam the projected area of the group of antenna elements contributing to that beam will change with zenith angle;

---

[17] The JVLA A-array (maximum baseline 36 km) uncorrected 2D imaging is marginal for $\lambda = 6$ cm (5 GHz) and definitely requires $w$-correction for imaging at $\lambda 20$ cm and $\lambda 90$ cm. At $\lambda = 90$ cm uncorrected 2D imaging is marginal even for the JVLA D-array (maximum baseline 0.9 km).

the overall shape, and hence also the effective gain, of that beam will therefore change with time. These PB variations are much larger than those for dish-based arrays and the properties of the PB will be frequency dependent. All these effects must be incorporated into the imaging algorithms to extract maximum performance. As noted in Section 11.5.2 the direction-dependent effects of the PB can be represented by convolutions applied during the visibility gridding step (A-projections).

*Ionospheric effects*   We outlined the general properties of the fluctuating ionosphere in Section 9.7.2 and alluded briefly to its impact on low frequency interferometry. Ionospheric fluctuations pose formidable phase-correction problems. Local phase gradients in the ionosphere produce shifts in source positions which, on baselines longer than $\sim 5$ km, can be different at different points across the field. The shifts are also functions of time, giving rise to the phenomenon of source motion, which is strikingly apparent in time-lapse videos of wide-field images. Self-calibration can be carried out using bright sources within the wide field itself but the breakdown of isoplanicity poses problems since the complex gain is not constant across the field. Techniques are being developed to deal with this problem; one approach is to peel out the sources one by one, starting with the strongest (see Section 11.5.3). Such 'faceted self-calibration' can work well but at the expense of large amounts of computer time. Full-field LOFAR maps are now routinely made with the Dutch baselines. On long baselines ($\gg 50$ km) with LOFAR (see Section 11.6.2) imaging is (currently) concentrated on individual objects using conventional calibration techniques, although identifying suitably bright compact calibration sources within $\sim 1°$ of the target source can be a problem. Calibrating low frequency interferometer data remains a research topic.

*Intense sources*   Emission from the Sun and the strongest sources, e.g. Cas A and Cyg A in the north and Cen A in the south, can corrupt the measured visibilities, and it is advisable to avoid including them in the PB; even so their flux can also leak in via sidelobes well outside the PB. If the response covers the whole sky, as in LOFAR-low (see Section 11.6.2), intense sources will always affect the data. The solution is to peel them away, as described in Section 11.5.3. Whilst the response down the side of the PB, or in the sidelobes, may be imperfectly known, the process works well enough to be a routine stage in the data analysis.

*Surface density of discrete sources*   Extragalactic radio sources predominantly have steep spectra and hence their effect is greater than at higher frequencies, so that at low resolution the blend of sources may give rise to confusion noise (see Sections 8.9.5 and 10.12.1). At higher resolutions and greater signal-to-noise ratios the sky model for deconvolution must include many individual discrete sources.

*Radio frequency interference*   RFI from a plethora of broadcast and communication systems is severe at these frequencies. The high speed digital data handling in the receiver system, in the correlator, and in a variety of analysis computers may also be sensitive to overloading unless specific mitigation steps are taken. The topic of RFI data handling is too multi-faceted to deal with in this introduction. Suffice it to say that, despite RFI, low frequency arrays are producing excellent science and new ones are being planned, including SKA-low.

*Computational load*   The computational loads are high in all the synthesis imaging regimes. This is due to the size of the data sets to be handled, the necessity to remove various forms of RFI, the complexity of the data corrections required, including the effects of non-isoplanicity, and finally the size of the final images to be made. The imaging of wide fields can be very time consuming but the science potential unlocked means that many astronomers are taking up the challenges of low frequency work.

### 11.6.2 Low Frequency Arrays

Some dish-based arrays (e.g. JVLA, GMRT; Section 10.1) are used for low frequency work. Here we describe three general-purpose phased arrays.

*LOFAR*   The international Low-Frequency Array is centred in The Netherlands (maximum baselines $\sim 100$ km) with (in 2017) external partner stations in Germany, France, UK, Sweden, Poland, and Ireland (maximum baselines 1500 km). LOFAR was the first and is the largest of the modern low frequency phased array telescopes. It operates in two bands: LOFAR-low (10–80 GHz) uses low-cost droop dipoles and the field of view covers the entire sky; LOFAR-high (120–240 GHz) is built up from 'tiles' made up from $4 \times 4$ bow-tie dipoles (Section 8.1.3), providing a field of view (FWHM) $\sim 30°$. A typical LOFAR station consists of 96 low-band antennas and 48 high-band antenna tiles. There are 18 'core' stations and 18 'remote' stations within the Netherlands and (currently) eight international stations. Each station can produce one or more beams, which can be cross-correlated with the equivalent beams from other sites to form an aperture-synthesis array. The latter system relies on broadband optical fibre data links, with the digital correlation carried out via software in a high performance computer rather than in purpose designed electronics (see also van Haarlem *et al.* 2013.)

*MWA.*   The Murchison Widefield Array has been developed by an international collaboration, including partners from Australia, Canada, India, New Zealand, and the USA. It is located in Western Australia near the planned site of the future SKA-low telescope. It consists of 2048 dual-polarization bow-tie dipoles (Section 8.1.3) optimized for the frequency range 80–300 MHz. As in LOFAR-high they are arranged in tiles of $4 \times 4$ dipoles, giving a field of view of 25° at 150 MHz. The majority of the 128 tiles are distributed within a $\sim 1.5$ km core region so as to provide high imaging quality at a resolution of several arcminutes (see also Tingay *et al.* 2013).

*LWA*   The Long Wavelength Array has been developed by the University of New Mexico and a consortium of US partners. There are two sites: in New Mexico close to the JVLA and in Owens Valley, CA (run by Caltech). At each site there is currently a single 'station' consisting of 256 linearly polarized crossed dipole elements, sensitive to the range 10–88 MHz and distributed over a 100 m diameter area (see also Ellingson *et al.* 2013).

*Special purpose arrays*   Other low frequency arrays being built at the time of writing (2018) include: the Hydrogen Epoch of Reionization Array (HERA); the Hydrogen Intensity and Real-Time Analysis eXperiment (HIRAX); the Canadian Hydrogen Intensity Mapping Experiment (CHIME); the TianLai Project, China.

## 11.7 Further Reading

*Supplementary Material* at www.cambridge.org/ira4.

Reid, M. J., and Honma, M. 2014. Micro-arcsec radio astrometry. *Ann. Rev. Astron. Astrophys.*, **52**, 339.

Schuh, H., and Behrend, D. 2012. VLBI: a fascinating technique for geodesy and astrometry. *J Geophys.* **61**, 68.

Schuh, H., and Böhm, J. 2013. Very long baseline interferometry for geodesy and astrometry. In: Xu, G. (ed.), *Sciences of Geodesy – II*, p. 339. Springer.

(TMS) Thompson, A. R., Moran, J. M., and Swenson, G. W. 2017. *Interferometry and Synthesis in Radio Astronomy*, 3rd edn. Springer (available on-line).

# Part III

## The Radio Cosmos

# 12

# The Sun and the Planets

We start our brief review of the cosmos as seen by radio with our solar system, in which there are sources both of thermal and of non-thermal radio waves (see Chapter 1). The visible surface of the Sun is a thermal emitter, but its surrounding atmosphere is the seat of violent and dynamic events which are characterized by their non-thermal radio emissions. Some of the planets also, including Earth, radiate non-thermally at long radio wavelengths from their surrounding atmospheres, while the surfaces of the outer planets may be used as standards of thermal radiation to calibrate millimetre-wave telescopes such as ALMA.

Before the advent of radio astronomy in the mid-twentieth century, the Sun was regarded as a non-variable star apart from the cycle of sunspots on its visible surface and the varying shape of the corona, visible only at total eclipses. It was recognized that the terrestrial aurorae were caused by solar activity, but the extension of the corona to interplanetary distances was invisible and could only be revealed by radio observations. The activity of the Sun is now seen to originate in its magnetic field, which is variable on a wide variety of spatial and temporal scales; it is a driver of dramatic events in the corona and extends far out into the solar system.

## 12.1 Surface Brightness of the Quiet Sun

The Sun is a star of spectral type G, which places it near the middle of the main sequence (although the great majority of stars are less massive than the Sun). The solar mass $M_\odot$ is $2 \times 10^{31}$ kg and its radius is $7 \times 10^5$ km. The temperature of its surface, the photosphere, is 5770 K, as determined bolometrically. Its visible spectrum is approximately that of a black body at that temperature, cut throughout by absorption lines originating in the chromosphere, which lies immediately above the photosphere.

Solar radio emission originates above the photosphere, at heights depending on wavelength. Emission at millimetre wavelengths is mainly from the chromosphere, and the millimetric Sun is similar to the visible Sun; it approximates to a disk at a temperature around 6000 K, somewhat hotter than the photosphere. At metre wavelengths the quiet Sun is larger, brighter, and more variable; here we are observing the solar corona rather than the chromosphere. The temperature of the corona is around $10^6$ K. Figure 12.1 gives a simplified view of the optical and radio solar spectrum, including the very variable long-wavelength radio emission from the active Sun.

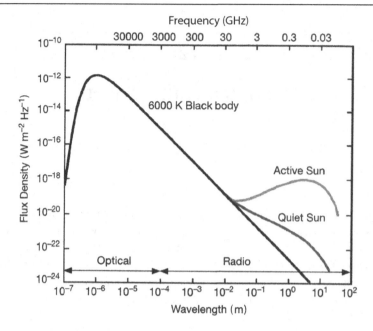

Figure 12.1. Schematic of the flux density of the Sun showing active and quiet phases in the radio band (Jet Propulsion Laboratory, courtesy of C. Ho).

The millimetre-wavelength emission from the chromosphere is also very variable, tracing the complexities of small-scale magnetic fields; these are energetic structures which are involved in the excitation of the upper chromosphere and corona. Following the details requires angular resolution comparable with optical, which is becoming available in observations by ALMA. Rapid single-dish scans by one element of the ALMA array, using a range of wavelengths, are shown in Phillips *et al.* (2015). The Nobeyama Radioheliograph, a synthesis array comprising 84 dishes each 80 cm in diameter, arranged on a 490 m east–west line and a 220 m north–south line, daily produces maps with a resolution of a few arcseconds at 34 GHz and 17 GHz (wavelengths 1 and 2 cm).

At long wavelengths, where the corona is optically thick, the Sun appears larger as well as brighter. At decimetre wavelengths the corona can be seen outside the photospheric disk as a *limb brightening*, where the long line of sight through the hot corona produces a higher brightness temperature, while at metre wavelengths the disk is lost behind the bright corona. The dramatic difference between the emissivity of the quiet Sun at decimetre and at millimetre wavelengths is due to the difference in opacity of the corona, which emits and absorbs by the free–free mechanism (Chapter 2). At intermediate (centimetre) wavelengths a ray path from the photosphere will pass through both the chromosphere and the corona, and the brightness temperature depends on the opacity and temperature along the whole ray path. The effect on the flux density from the Sun at a time of sunspot minimum is shown in Figure 12.1. Zirin *et al.* (1991) showed that, for frequencies $\nu$ between 1.4 and 18 GHz, $T_b$ can be modelled as the sum of two components: a constant contribution from the upper chromosphere $T_{chrom}$ at $11\,000$ K and an added component from the corona at $10^6$ K that is proportional to $\nu^{-2.1}$, the frequency dependence of the optical depth of the corona (see Section 7.5). The combination gives

$$T_b = A\nu_{GHz}^{-2.1} + T_{chrom}. \tag{12.1}$$

At sunspot minimum Zirin *et al.* found that $A = 140\,000$ K, giving $T_b = 70\,000$ K at 1.4 GHz and 11 000 K at 10.6 GHz.

The corona is variable in shape and extent over the sunspot cycle; there is no ideal symmetric quiet Sun,[1] but Figure 12.2 (McLean and Sheridan 1985) shows the increase in diameter at lower frequencies. The brightness temperature at the centre of the disk is approximately $(8-10)\times10^5$ K over this range of frequencies. At the lowest reliably observable frequencies, around 30 MHz, the radiation originates further out in the corona and the brightness temperature falls to around $5 \times 10^5$ K.

The high temperature of the solar corona is also observable in X-rays and in optical emission lines with a high excitation temperature. From these spectral lines, and from the radio and X-ray brightness temperatures, it is clear that the whole corona, extending out to several solar radii, is at a temperature of between 0.5 and $3 \times 10^6$ K.

As seen in Figure 12.2, the active Sun can be overwhelmingly bright at metre wavelengths, reaching a flux density of over $10^4$ Jy. Apart from its intrinsic interest, this intense radiation can occasionally dominate the sky and seriously affect all observations at long radio wavelengths.

## 12.2 Solar Radio Bursts

Powerful bursts of radio from the Sun were discovered in 1942 by J. S. Hey as interfering noise in radars operating at wavelengths of several metres.[2] Paul Wild in CSIRO (Australia) constructed the first radio spectroheliograph in 1949, and recorded dynamic spectra displaying a wide variety of bursts at metre wavelengths, indicated in Figure 12.3. The spectrum often shows two narrow peaks harmonically related and sweeping downwards over the whole recorded band, within a few seconds (Type III) or a few minutes (Type II). The source was later found to be located far from the Sun's surface, at a distance increasing with time. The interpretation is clear: there is a cloud of energetic particles ejected from near a sunspot and travelling outwards (for Type III) at a speed typically $0.25c$, i.e. up to $10^5$ km s$^{-1}$.

The two peaks are interpreted as the local plasma frequency $\nu_p$ and its harmonic $2\nu_p$. The excitation of oscillations in the plasma, known as Langmuir oscillations, is due to electrons with sub-relativistic energy streaming out along an open magnetic field line. A visible solar flare is often observed near the base of this field line. The stream of electrons, and the excitation of Langmuir oscillations, continues to large distances, as has been observed by radio receivers on interplanetary spacecraft, notably by the twin spacecraft STEREO (Reiner *et al.* 2004), stationed at distances of around 1 AU (the Earth–

---

[1] Colour images of the radio Sun can be found in http://www.astro.umd.edu/~white.

[2] See the detailed early history of radio astronomy *Cosmic Noise* by W. Sullivan.

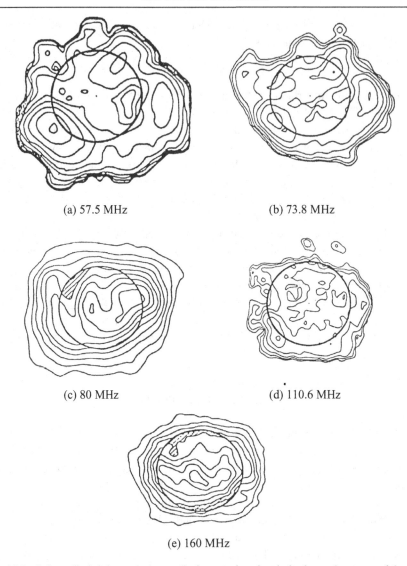

(a) 57.5 MHz                               (b) 73.8 MHz

(c) 80 MHz                                 (d) 110.6 MHz

(e) 160 MHz

Figure 12.2. Solar radio brightness at low radio frequencies; the circle shows the extent of the optical disk (McLean and Sheridan 1985).

Sun distance) from both Earth and Sun. The downward sweep of frequency extends below 1 MHz to 100 kHz or lower. A particle detector on STEREO also measures directly the ambient electron density, demonstrating the identity betwen the plasma frequency and the Type III peak (for the plasma physics involved in the coherent radio emission see the review Bastian *et al.* 1998). Figure 12.4 shows a useful model of the ambient electron density, plotted as the corresponding plasma frequency, at distances extending to 1000 solar radii ($\sim$ 4 AU) (Mann *et al.* 1999); this model is indicative only, as the electron density is variable and its distribution is controlled by magnetic field lines (Zucca *et al.* 2014).

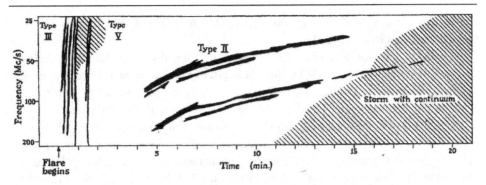

Figure 12.3. Dynamic spectra of solar radio bursts. This idealized diagram shows the timescales of various types; Type III and Type II show the characteristic sweep from high to low frequencies, with second harmonics (Boorman *et al.* 1961).

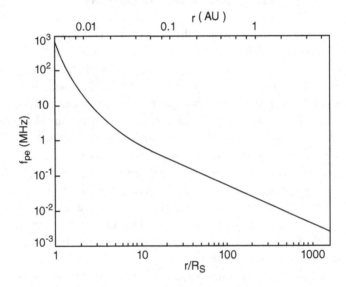

Figure 12.4. A model of the electron density in the heliosphere, plotted as the plasma frequency $f_{pe}$ (Mann *et al.* 1999); $R_S$ is the solar radius.

The huge release of energy producing a solar flare and the Types II and III bursts is due to a catastrophic rearrangement of the magnetic field over a sunspot group, known as a reconnection event. Electrons accelerated in this event travel downwards, towards the photosphere, as well as upwards to excite the radio emission shown in Figure 12.3. The downwards movement can occasionally be seen as a reverse frequency sweep in metre- and centimetre-wavelength radio emission, but it is also observed at X-rays and at millimetre wavelengths. Emission at these shorter wavelengths is incoherent; at millimetre wavelengths it appears to be synchrotron or gyrosynchrotron. (Gyrosynchrotron radiation occurs at low harmonics of the gyrofrequency.)

The original characterization of bursts at metre wavelengths was by Wild *et al.* (1963). An account of metre-wavelength radio bursts is given in McLean and Labrum (1985), and at decimetre wavelengths in Isliker and Benz (1994). A review of Type III radio bursts is found in Reid and Ratcliffe (2014).

Solar radio emission at metre and decimetre wavelengths is monitored regularly at Owens Valley Solar Array (OVRA). Here an interferometric array of 1.8 m dishes covers the frequency range 1–18 GHz. The antennas are fed by log-periodic arrays. LOFAR and the other low frequency arrays are actively engaged in studies of solar radio bursts.

The richness of the various solar radio emissions has excited the interest of a widespread network of observers, many using similar equipment and coordinating their observing times to form a complete patrol (Benz *et al.* 2005). This network, known as CALLISTO, uses a standard low-cost receiver, sweeping from 45 MHz to 870 MHz, that is capable of recording large solar radio bursts with only a broadband dipole or a single log-periodic array.

### 12.3  Coronal Mass Ejection (CME)

In common with many stars the Sun is the source of a continuous, although variable, outward flow of charged particles. This is the solar wind, which impinges on the Earth and all other planets. It is an ionized plasma, consisting mainly of electrons and protons with energies of order 1 keV and originating in the corona. The solar wind may be monitored by observing its effect on quasars seen through it at various distances from the Sun. This 'interplanetary scintillation' was the phenomenon exploited by Antony Hewish and Jocelyn Bell in observations which led to the discovery of pulsars (see Chapter 15).

The solar wind does not normally reach the Earth's atmosphere, as it is deflected by the terrestrial magnetic field, but occasionally, and particularly at times of solar magnetic activity, a discrete and more energetic cloud is ejected from the Sun. This coronal mass ejection (CME) may reach the Earth's atmosphere, exciting an aurora and a noticeable magnetic disturbance, a geomagnetic storm. The solar wind originates in the whole of the outer corona, but the CME occurs above a sunspot group and involves a massive release of energy from the local sunspot magnetic field.

The typical magnetic field over a sunspot group is a loop joining two spots with opposite polarities, with a strength of some thousands of gauss. This closed field expands outwards; near its base oppositely directed fields are forced together, forming the shape of a capital omega, $\Omega$. A merger of the opposite fields (magnetic reconnection) detaches a large body of plasma, with its enclosed magnetic field, and releases energy stored in the field, which drives the CME outwards. Depending on the direction of ejection, the CME cloud may reach the Earth some hours or days later.

The CME and Type III radio bursts are both associated with catastrophic releases of energy, but are distinct and different events occurring in different parts of the sunspot magnetic field. The electron clouds exciting radio emission travel along open field lines, which may be adjacent to the closed field lines of the CME. Type III bursts are more usually associated with visible solar flares, while a CME may be observed as a visible cloud outside the limb of the Sun.

It seems likely that many stars may, like the Sun, emit bursts of radio waves (it should be noted that transient radio sources are difficult to observe, requiring multi-beam synthesis arrays with rapid response). Whether the bursts which have been observed are to be regarded as similar to solar radio bursts, but with an even greater intensity, was discussed by Bastian (1994).

## 12.4 The Planets

At millimetre wavelengths the surfaces of the outer planets, Venus, Jupiter, Saturn, and Neptune are sufficiently stable and reliable sources of thermal radiation to be used as standard references for calibrating radio telescopes such as ALMA and the spacecraft Planck (Hafez *et al.* 2008); for example the surface temperature of Saturn is 130 K, depending slightly on its slowly changing distance from the Sun. The atmospheres of these planets may also affect the observed brightness temperatures; for example Saturn has an atmosphere containing a variable amount of ammonia, which affects the spectral distribution of brightness temperature (van der Tak *et al.* 1999). The $CO_2$ atmosphere of Venus keeps the whole surface at temperatures over 700 K (the 'greenhouse effect'), and at millimetre wavelengths the radio emission also has a component from cooler clouds in the atmosphere.

At longer (decametre) wavelengths, the radiation from several planets, and particularly Jupiter, is dominated by intense and variable non-thermal radio from the planetary atmosphere. The first detection of Jupiter was made at 40 MHz (Burke and Franklin 1955) using an experimental Mills Cross built by two of the present authors, FG-S and BFB, at the Carnegie Institute of Washington; the frequency was later found to be at the upper end of a band of intense radiation from the polar regions of the planet. There is also an equatorial radiation band, which has a broader spectrum. Figure 12.5 indicates the three sources of Jupiter radio emissions and gives a typical spectrum.

The intense non-thermal decametre radiation at frequencies below 40 MHz arises from electrons streaming into the polar regions along magnetic field lines. Their radiation frequency is determined by the strength of the field; their energies are sub-relativistic, in the keV to MeV region, giving gyrofrequency radiation rather than synchrotron. (For

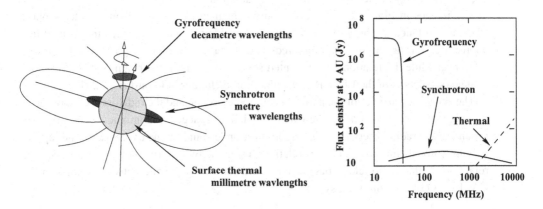

Figure 12.5. The three sources of radio from Jupiter (adapted from Girard *et al.* 2016). Thermal radiation from the surface dominates at millimetre wavelengths, gyrofrequency at long (decametre) wavelengths, and synchrotron from radiation belts at metre wavelengths.

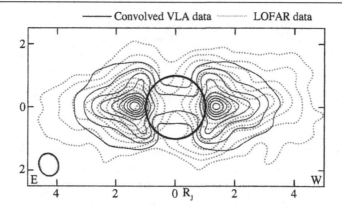

Figure 12.6. Synchrotron radiation from Jupiter's radiation belts, imaged by LOFAR at 150 MHz (dotted lines) and by VLA at 5 GHz (from Girard 2016). The telescope beamwidth is shown at the left.

non-relativistic electrons the gyrofrequency in a field of $B$ gauss is the cyclotron frequency $\nu_{\mathrm{cyc}} = 2.80\ B$ MHz). The electrons are accelerated by interaction with the solar wind, at a distance of 20 to 50 Jupiter radii ($R_J$). This very intense polar radiation occurs in short bursts on timescales of minutes down to milliseconds, those on the shortest timescale being related to the position of Jupiter's moon Io. The complexity of the polar bursts can be seen in recordings made over the band 10 to 33 MHz by three telescopes in France, Ukraine, and the USA (Imai *et al.* 2016).

The synchrotron radiation from the equatorial radiation belts is observed over a wide band from around 30 MHz to 30 GHz. The electron energies are in the MeV range, mildly relativistic, so that the spectrum is wide. Figure 12.6 shows the similar appearance of the radiation belts observed by LOFAR at 150 MHz and by the VLA at 5 GHz (Girard *et al.* 2016; Santos-Costa *et al.* 2009). Synchrotron radiation has a broad spectrum which is a smoothed version of the energy spectrum of the radiating electrons (see Chapter 2); the differences in the brightness distributions shown in Figure 12.6 indicate that the electron energies are lower in the outer regions of the radiation belt.

The images in Figure 12.6 were of course obtained by aperture-synthesis telescopes. There are technical problems to be overcome in making images of planets; they may move appreciably during the observations, requiring careful definition of positional frames of reference, while the relation of the wanted source to others in the beam or sidelobes may change (see details of the LOFAR data reduction for Jupiter in Girard *et al.* 2016).

The energetic particles of the auroral cyclotron emission and the radiation belt have been investigated directly *in situ* by the Cassini and Juno spacecaft. Continued monitoring by ground-based radio observations is essential for understanding the dynamics of the rapidly varying coronal emissions and their relation to solar activity; fortunately this only requires relatively simple radio telescopes, which could form a worldwide network similar to the CALLISTO solar monitoring system.

## 12.5 Further reading

*Supplementary Material* at www.cambridge.org/ira4.

# 13

# Stars and Nebulae

Radio emission has been observed from stars of all types along the main sequence, including the low-mass and relatively cold M-dwarfs (Güdel 2002). In some cases this is thermal emission, mainly from extended atmospheres or from the surfaces of the largest giant stars. Figure 13.1, which is the radio version of the Hertzsprung–Russell diagram, shows the range of stars detectable at lower frequencies. The brightest cm-wave stellar emission is non-thermal, arising from high energy electrons accelerated by a stellar magnetic field. The extreme example is the beamed radio emission from pulsars (Chapter 15), which is easily detectable with single-dish radio telescopes. Thermal emitters become more detectable at higher frequencies, especially for cooler stars with large angular diameters such as long-period variables on the asymptotic giant branch (AGB), extending to the upper right in Figure 13.1. They are in the transition from red giant to planetary nebula. Even so, almost all radio stars are observable directly only at cm wavelengths, using telescopes with μJy sensitivity coupled with high dynamic range and angular resolution. As we have seen in earlier chapters, this demands the complex and sophisticated aperture-synthesis arrays which are now available over the entire radio spectrum.

## 13.1 Thermal Radio Emission from Stars

Stars with observable thermal radio emission are either very close to us or present a large angular size, which usually means a relatively nearby post-main-sequence giant or supergiant, a young stellar object (YSO), or an active star surrounded by hot gas or dust.

For a star or extended disk with surface temperature $T$, subtending a solid angle $\Omega$, the flux density $S$ to be measured is $S = 2(kT/\lambda^2)\Omega$ (see Eqn 5.13). This practical limit is easily remembered as giving the minimum temperature $T_{min}$ detectable with a radio telescope whose minimum detectable flux in janskys is $S_{min}$; for a star with circular disk diameter $\theta$ arcsec, observed at wavelength $\lambda$ cm,

$$T_{min}\theta^2 = 1800\lambda^2 S_{min}. \tag{13.1}$$

Current radio telescopes can detect stars like the Sun out to a few parsec distance. The star $\alpha$ Cen A is the closest Sun-like star, at 1.34 pc, with a surface temperature $T_{eff} \sim 5800$ K and diameter 0.008 milliarcsec (mas) at optical wavelengths. This implies $S \sim 20$ mJy at 1 mm wavelength and 0.007 mJy at 5 cm. It has indeed been detected by ALMA in all

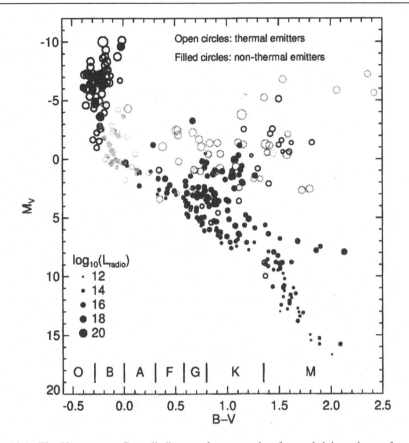

Figure 13.1. The Hertzsprung–Russell diagram for a sample of stars brighter than a few μJy at frequencies below ∼ 20 GHz, including those where the emission arises from close binary interaction (courtesy of S. White.)

observed bands. The SKA phase 1 will detect α Cen A and B at cm wavelenths and the next closest, τ Ceti (0.001 mJy at 5 cm, 3.6 pc), and the full SKA will detect more. Hotter stars are intrinsically brighter; the nearest (2.64 pc), the A-type Sirius A, with a temperature almost $10^2$ K, reaches ∼150 μJy at 33 GHz, with a blackbody spectrum, but the population of stars falls steeply as the mass increases.

Some of the coolest stars are brightest in the radio owing to their large diameters. The stellar surface of a cool star, which is not fully ionized, can be loosely defined as the surface where it becomes optically thick at a given frequency. The optical photospheric diameter of an AGB star is typically 200 times that of the Sun; such huge stars have very extended atmospheres and the outer layers are cooler. Thus, at lower frequencies the atmosphere becomes optically thick at progressively larger and larger radii, favouring detectability. The red supergiant Betelgeuse (α Orionis) has a photospheric surface temperature of 3600 K and optical angular diameter of 0.045 arcsec. Between 1.6

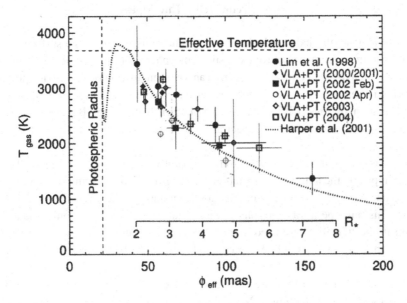

Figure 13.2. The broken lines show the optical photospheric radius and temperature of Betelgeuse (α Ori). The dotted line is the gas temperature profile predicted by the model of Harper *et al.* (2001). The brightness temperature measurements are based on observations from 43 to 1.6 GHz, left to right, compiled by O'Gorman *et al.* (2015).

and 43 GHz its flux density is proportional to $\nu^{1.33}$ with size $\propto \nu^{-0.36}$. Figure 13.2 shows observed brightness temperatures and diameters, with a semi-empirical model for the ionization balance and optical depth. The scatter is due to variability as well as uncertainties.

Resolving the disk of Betelgeuse requires long-baseline interferometers. Richards *et al.* (2013) used eMERLIN at 6 GHz to measure a $\sim$ 200 mas disk (about five times the optical diameter), total flux density $\sim$ 2 mJy with small positive and negative fluctuations of order 10%. At 338 GHz, two spots were also apparent, on a disk of diameter 58 mas, total flux density 545 mJy, measured using ALMA (O'Gorman *et al.* 2015). It is tempting to relate these to mass loss and suggestions include convection cells or magnetic effects leading to heating and local uplift; indeed, Betelgeuse has shown a molecular plume apparently related to a spot. Betegeuse is more than ten times the mass of the Sun, but solar-mass M-type and carbon-rich long-period variable stars, typically at a few tens of mJy at a similar distance (200 pc) can also be resolved by ALMA, and often also show spots. The somewhat hotter K stars have smaller radii but are also detectable if close enough, e.g. Antares (a supergiant at 170 pc) or Aldebaran (slightly more massive than the Sun, at 20 pc). A handful of the closest low-mass cool stars have aleady been detected at cm wavelengths at the mJy level, and an SKA survey could extend this to hundreds.

## 13.2  Circumstellar Envelopes

Although it is generally hard to detect thermal radio emission from the surfaces of stars, some comparatively cool stars have extensive atmospheres of dust and ionized gas whose thermal radiation is easily measurable, as can be seen in Figure 13.2. The process of condensation of gas and dust within nebulae can also be observed in some detail, especially in star-formation regions. (SFRs) The richest-known SFR is Sgr B2, near the galactic centre. Here there are thousands of spectral lines observable from hundreds of molecular species.

Radio emission is often observable from comparatively cool stars near the end of their evolution, typically the long-period variables such as Mira Ceti and some supergiants. Their common characteristic is a high rate of mass loss, amounting to between $10^{-8}$ and $10^{-4}$ $M_\odot$ per year. This must be a transient phase; their total mass is typically only about 1 $M_\odot$. They are in the transition from red giant to planetary nebula. Many are strong infrared emitters, appearing in the satellite survey catalogues from IRAS onwards (NASA/IPAC Infrared Science Archive, Caltech).

The intense stellar wind blowing out from these cool stars contains material from the interior dredged up by convection. Not far from the surface this material can condense, forming molecules and dust at a temperature of about 1000 K. Radiation pressure from the star then drives the dusty wind outwards to form a dense cloud more than ten times larger than our solar system. These stellar winds are believed to be feeding chemically enriched material to the interstellar medium.

Radio emission from such a circumstellar dust cloud often follows a blackbody spectrum, with spectral index $\alpha \approx -2$. Free–free radiation from ionized gas, such as the extended corona of $\alpha$ Ori (Figure 13.2), may also be present. The free–free component may be distinguished by a spectral index closer to zero; for an optically thin homogeneous plasma $\alpha = -0.1$. The pattern of flow in the stellar wind may be deduced from combined infrared and high-resolution radio mapping, which may show whether the flow is uniform in time and direction. For example, Wright and Barlow (1975) analysed the case of free–free radiation from a uniform and isotropic outward wind, in which the density falls inversely as the square of radial distance. For each line of sight through the sphere of plasma the flux density depends on the optical depth as

$$I(\nu, T) = \int_0^{\tau_{max}} B(\nu, T) \exp(-\tau)\, d\tau. \qquad (13.2)$$

Integration over the whole sphere then gives a total luminosity with spectral index $-0.66$ for infrared and $-0.6$ for radio (the difference is due to the Gaunt factor; see Section 2.6). If the radial structure can be resolved, there may be departures from the simple inverse square law of particle density. The radial distribution may then be related to the history of emission over a period of order $10^5$ yr; a discrete episode of emission, for example, would give a hollow shell. Structure of this kind is more easily seen in the brighter, non-thermal, maser emission which occurs in many circumstellar envelopes.

Much more extensive and energetic atmospheres are generated around the components of many binary star systems, which are the seats of novae and visible flares; many are also X-ray binaries. In many cases there are outflows with complex structure, and in some there are obvious interactions with a surrounding nebula or the interstellar medium.

The radio emission is often much more powerful than the thermal free–free emission of a hot corona; its intensity, spectrum, and polarization then indicate gyrosynchrotron or synchrotron radiation. Here there is a striking resemblance to the planet Jupiter, in which the radio emission over most of the radio spectrum is mediated by the planet's magnetic field, both in generating high energy particles and in providing a radiation mechanism. Beamed coherent emission is also observed; Hallinan *et al.* (2007) shows the light curve at 8.4 GHz for an M9 dwarf; periodic bursts occur at intervals of 1.96 hr, evidently the rotation period of the star, like that of a slow pulsar.

Spectral line emission from neutral hydrogen H I is observed from the extended atmospheres of a variety of evolved stars. These large-diameter sources of H I are observable with single-aperture radio telescopes, notably by the Nançay Radio Telescope (NRT; Gérard and Le Bertre 2006). Resolving their detailed structure requires aperture-synthesis telescopes; a combination of both types of telescope is needed to provide the full picture. Diep *et al.* (2016) presented combined observations with the NRT and the synthesis telescopes IRAM, JVLA, and ALMA, including spectral lines from both H I and CO.

## 13.3 Circumstellar Masers

In many molecular clouds, and particularly in the mass outflow from red giant stars, we observe the astonishing and beautiful phenomenon of maser radio emission. Masers can occur when the energy levels of a molecular species are populated in a non-Boltzmann distribution (see Chapter 3). Very small energy differences are involved; for radio emission the factor $h\nu/k$ is usually only of order 0.1 K. The molecular cloud contains material and radiation with several different temperatures, and a population inversion may be due either to collisions or to radiation. Several different molecules, each with several different masering transitions, may be observed in such stars; some of the strongest and most informative are the OH, $H_2O$, $CH_3OH$ (methanol), and SiO masers. Isotopic varieties, such as $^{28}SiO$, $^{29}SiO$, and $^{30}SiO$ can be observed; these are obviously useful in determining relative abundances but they may also be useful in determining temperatures and velocities when the more prolific lines are saturated.

Many other masers are observed in interstellar clouds; the distinction between circumstellar masers, originating in outflow, and interstellar masers, where material is condensing to form stars, may not always be clear, as some stars may simultaneously accrete material in their equatorial regions and expel a wind from their poles. Condensing material may form a disk, whose dynamics may be investigated by the Doppler shift of maser lines. In these cooler regions more complex molecules may be involved; Minier *et al.* (1999) found that methanol ($CH_3OH$) maser lines at 6.7 and 12.2 GHz traced Keplerian motion in disks in three such star-forming regions. The most powerful masers are observed in other galaxies; these are the megamasers, an example of which is described in Section 13.6. There is extensive literature on the subject of masers; see, for example, the review by Elitzur (1992) and the textbook by Gray (2012).

### *13.3.1 Silicon Oxide*

The energy levels involved in the SiO masers are comparatively simple since the molecule is diatomic. The principal radio line, at 43 GHz, is between rotational levels $J = 1-0$ in a vibrationally excited state. Silicon oxide masers have been detected in many rotational transitions up to at least $J = 15-14$, at 646.431 GHz (Baudry *et al.* 2018). The non-Boltzmann distribution, giving an excess population in the higher level, is due to collisions with neutral molecules (the maser 'pump'), in a region at a temperature of about 1000 K to 2000 K and a total particle density of about $10^9$ to $10^{10}$ cm$^{-3}$. Silicon oxide is itself a minor constituent in this dense cloud, which occurs within a few stellar radii of the surface. The maser cloud velocities in the maser emission extend only over less than 15 km s$^{-1}$, and the maser lines are correspondingly narrow.

This SiO maser region is more of an extended stellar atmosphere than a stellar wind. The maser emission is patchy, as seen by the several different components in a typical spectrum: in the star R Cas the separate components have been resolved spatially and are clearly seen over an area not much larger than the disk of the star's surface (Phillips *et al.* 2003). The intensities of the components vary greatly. The SiO molecule is non-paramagnetic, with only a small magnetic moment; nevertheless the circular polarization due to Zeeman splitting can be observed in several stars, indicating fields of some tens of gauss. Such a field is strong enough to control the dynamics of the stellar atmosphere, since the energy density $B^2/8\pi$ in such a field is greater than the particle energy density $nkT$. Strong linear polarization is often observed; this is less easily explained than the circular polarization, but must be related in some way to the structure within the SiO clouds.

The SiO masers do not extend beyond about five stellar radii above the surface. At this distance SiO becomes incorporated into dust particles; these are then accelerated outwards by radiation pressure, carrying the gas with them and forming the stellar wind.

The proper motion of maser components is difficult to measure, since the emission is usually variable on a timescale of months only, and also because very high angular resolution is necessary. Boboltz *et al.* (1997), using the VLBA, showed that the SiO masers round the star R Aquarii were moving inwards with a velocity of around 4 km s$^{-1}$; this is a star which is accreting, rather than creating an outward stellar wind.

### *13.3.2 Methanol*

Spectroscopically, methanol (methyl alcohol), $CH_3OH$, is a complex organic molecule with a complex set of energy levels. More than 20 lines have been detected at radio wavelengths. A line at 4.4 cm (6.7 GHz) is one of the brightest maser lines; but most are observed between 2 mm and 2 cm. Herbst and van Dishoeck (2009) pointed out that methanol, along with other complex organic molecules, might be found in the early stages of stellar evolution; this was proved in ALMA observations, at wavelengths around 1 mm, of a planet-forming disk (Walsh 2017).

### 13.3.3 Water

The $H_2O$ masers, like the SiO masers, are frequently found in Mira variables. They appear further from the star, at distances of 6–30 stellar radii. They are also found in active *star-formation regions* (SFRs) and particularly powerful examples, often called megamasers, are found in certain external galaxies such as NGC 5128. The most prominent line, at 22 GHz, is a transition between rotational states designated $6_{16}-5_{23}$. Excitation is by collisions with molecules; the temperature is of order 750 K, and the particle density is of order $10^8$ $cm^{-3}$. The total energy output is often phenomenal: the brightest $H_2O$ masers include, within a few kHz, more energy than the total bolometric output of the Sun.

The extension of receiver and interferometer techniques into the millimetre and sub-millimetre bands has allowed the observation of numerous transitions between rotational states both in SiO and in $H_2O$; for example, the 650 GHz $J = 15-14$ transition in SiO and the $J = 5-4$ at the higher vibration level $v = 4$ are observable (Menten and Young 1995).

## 13.4 The Hydroxyl Masers

Further out again from the star, the outflow from a red giant is established typically as a symmetric radial flow with a velocity of 10–20 km $s^{-1}$. The OH masers are observed at distances of $10^{16}-10^{17}$ cm, more than ten times greater than for $H_2O$ masers. They occur in a well-defined and comparatively thin shell, where the ambient ultraviolet light from the Galaxy is sufficiently strong to dissociate the dust. The density and temperature are of order 1000 $cm^{-3}$ and 450 K. The OH molecule is paramagnetic, with a dipole moment $10^3$ greater than those of SiO and $H_2O$, so that Zeeman splitting can be observed even though the magnetic field at the large radial distance is only of order some milli-gauss.

More than 1000 such OH stars have been observed. The strength of the OH masers is related to the rate of outflow in the stellar wind; the mass loss of $10^{-8}$ solar masses per year in a short-period Mira variable leads to a weaker maser emission than in a supergiant with an outflow of $10^{-4}$ solar masses per year. Since the maser emission occurs in the older population of stars, they are, like SiO masers, a useful dynamical probe for studying the structure of the Galaxy (see Chapter 10).

Four transitions are observed at wavelengths near 18 cm; these are all transitions within the ground state, which is split by lambda doubling and again by hyperfine splitting (Chapter 3). The excitation is due to infrared radiation at around 35 μm wavelength, which selectively excites the upper level of the doublet. In stars, the strongest maser action is found for the 1612 MHz line, especially in the stars with the greatest rates of mass loss. The masers are generally found to be saturated, with an intensity corresponding to a maser amplification by a factor of order $e^{20}$.

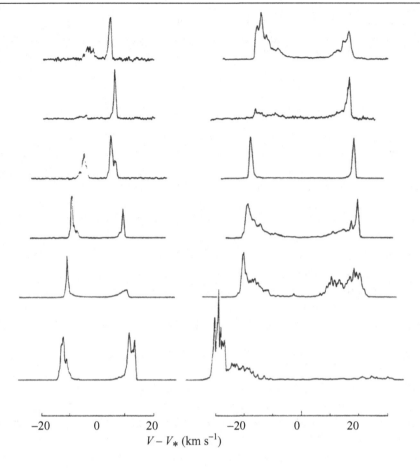

$$V - V_* \text{ (km s}^{-1})$$

Figure 13.3. Double OH maser lines from expanding shells. The lines are split by the different Doppler effects on the front and back of an expanding shell (Cohen 1989).

The thin expanding-shell structure of circumstellar OH masers is demonstrated by the structure of the dominant spectral line at 1612 MHz. Figure 13.3 shows examples of double and single lines; the double lines are due to Doppler shifts in the front and back of an expanding shell. The whole spectrum typically covers a 30 km s$^{-1}$ range of Doppler shifts, while the individual components are typically only 0.1 km s$^{-1}$ wide. A maser attains its full brightnesss only if its line of sight traverses a sufficiently long region with the same Doppler shift. If there is a radial gradient in the wind, this long line of sight occurs preferentially at the edges of the cloud; a ring of masers is then seen at the radial velocity of the star. If there is no gradient, the line may be strongest in the centre of the cloud, on the line of sight to the star itself; in this case, both the front and the back may be seen at velocities separated by twice the expansion velocity. This is seen in the maps of OH masers around the star OH 127.8 (Figure 13.4); here the maps at different frequencies across the line profile effectively show a series of cross-sections of the expanding cloud, the lowest and highest frequencies originating in the back and the front respectively.

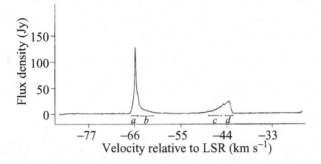

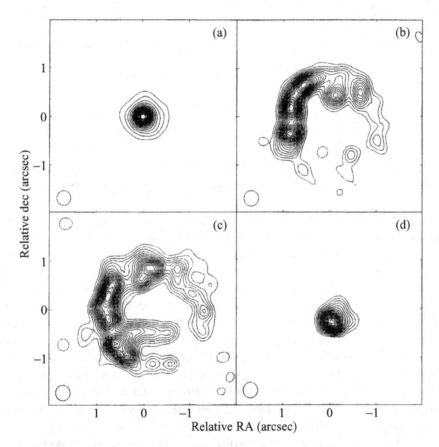

Figure 13.4. MERLIN maps of OH masers around the star OH 127.8. Four velocity ranges (a–d) have been selected from the spectrum (above) and separately mapped (below) (Booth *et al.* 1981).

For variable stars the 1612 GHz maser brightness follows the stellar phase. The radiating shell can be so large that the stellar radiation takes weeks to reach the OH shell. This means that the blue-shifted maser peak brightens around 10–100 days before the red-shifted peak, giving a measure of the shell radius (Engels *et al.* 2015).

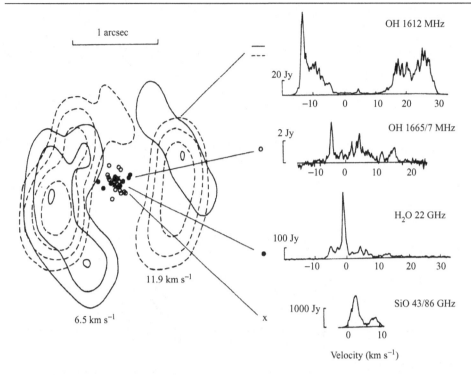

Figure 13.5. The OH masers in relation to the SiO and $H_2O$ masers in the circumstellar envelope of the star VX Sgr, showing their locations and their spectra (Chapman and Cohen 1986).

Circular polarization is frequently observed in OH masers, indicating magnetic field strengths of order 1 milligauss. Extrapolated back to the star surface, this would be a field of some tens of gauss, in agreement with the fields deduced from the SiO and $H_2O$ masers. Zeeman splitting is easily observed in OH masers, since the OH molecule is paramagnetic; a field of 1 milligauss may split a line by more than its intrinsic bandwidth. The $H_2O$ and $CH_3OH$ molecules are non-paramagnetic, and linear rather than circular polarization is usually observable. Field strengths of some tens of milligauss are observed in the in-falling molecular gas of star-formation regions; examples are the OH masers in Ceph A (Cohen *et al.* 1990) and methanol masers in W3 (OH) (Vlemmings *et al.* 2006).

The relation of the OH masers to the SiO and $H_2O$ masers in a single star, VX Sgr, is shown in Figure 13.5 (Chapman and Cohen 1986). Besides demonstrating the general picture set out in the preceding sections, this map shows that some complications remain to be understood. In particular, the outflow may not be spherically symmetrical, and the outward velocity gradient is not yet fully explained by radiation pressure on dust.

### 13.5 Classical Novae

The continuum radio emission from a quiet star like the Sun is dramatically outshone by the free–free radiation from the expanding stellar envelope resulting from a classical nova explosion. The 'nova' itself is not a new star; it is a white dwarf which is accreting material from a normal-star companion in a close binary system. The white dwarf itself is a star which has collapsed after completing its hydrogen burning, and is now mainly composed of helium. Accretion brings a new supply of hydrogen to its surface, forming a degenerate and unstable layer. A mass of about $10^{-4}$ $M_\odot$ may accumulate before the layer ignites. An increase in temperature in the degenerate hydrogen brings no increase in pressure; the hydrogen layer therefore explodes rather than expanding smoothly, and the whole layer then blows outwards as an expanding shell with a velocity of some hundreds or a few thousand km s$^{-1}$. The optical emission lasts a few days; the radio emission is from an outer shell which is ionized by intense ultraviolet light. This radio emission grows and decays over a typical period of some years. It is thermal radiation, which may be understood in terms of the theory of free–free radiation set out in Section 7.5, including both optically thick and optically thin regimes.

The geometry of the expanding shell is important: the optical depth may vary across different lines of sight. Consider first the radio emission from a simple model: a cube of ionized gas expanding with velocity $v$, initially optically thick and becoming thin as the density decreases. As in H II regions, the temperature is constant at around $10^4$ K, so that in the optically thick phase the flux density from the thermal radiation varies with time $t$ as the area of the source, i.e. as $v^2 t^2$, and with frequency $\nu$ as $\nu^2$ (spectral index $\alpha = 2$). In the later, optically thin, phase the total radio emission depends only on the density, and it falls as $t^{-3}$; the spectrum is flat with index $\alpha = -0.1$. The transition occurs later for lower radio frequencies; novae usually show a series of radio peaks which follow sequentially from high to low frequencies. More realistic models of shells may have tangential lines of sight which are optically thick, while those near the centre are thin. The transition from thick to thin therefore occurs over some time, and the spectral index may change more slowly than in the simple cube model. The details depend on the geometric thickness of the shell and on the fall of temperature as the shell expands (Seaquist 1989).

Figure 13.6 shows the radio emission at frequencies between 1.8 and 17.5 GHz from the nova V959. The eMERLIN observations at 5.0 and 5.7 GHz resolve the shell of the nova (Healy *et al.* 2017), allowing a model to incorporate changes in optical depth with radius and with expansion, giving models which fit the 5.0 GHz and 5.7 GHz light curves. It is remarkable that such a model accounts so well for the emission over such a wide frequency range. Interpretation is, however, more complex than might be seen in this figure, since the shell does not expand symmetrically; furthermore, the radio spectra indicate that there may be some synchrotron radiation in addition to thermal.

Radio images of the expanding shells of several novae are becoming available. For example the expansion of the thermally emitting shell has been observed for Nova V723 Cas at 5 GHz, as shown in Figure 13.7. The optically thick phase, with an expanding nearly uniformly bright disk and flux density increasing as $t^2$, lasted for almost 1000 days; the optically thin shell was less symmetrical, and after four years the fall in flux density was still not as steep as the expected $t^{-3}$.

In Nova Cygni 92 the angular diameter of the shell was resolved only 80 days after the outburst (Pavelin *et al.* 1993); the brightness temperature eventually reached 45 000 K, higher than expected from radiative excitation but still consistent with thermal radiation from shock-excited gas. The expanding shell is seen in a series of MERLIN 6 cm maps

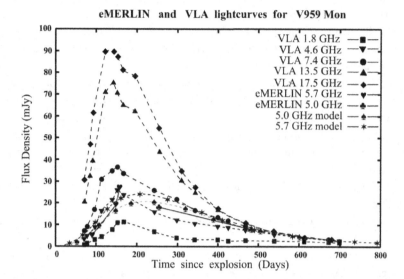

Figure 13.6. Radio light curves for V959 Mon; eMERLIN 5.7 and 5.0 GHz; multifrequency light curves VLA (Healy *et al.* 2017).

(Figure 13.8) by Eyres *et al.* (1996). The shell is far from symmetrical; this may reflect an anisotropic explosion or it may occur when the expanding cloud collides with structure in the surrounding interstellar medium

In other types of nova the radio spectrum and high brightness temperatures indicate non-thermal radiation; this may be due to high energy particles in the outflow itself. This may also occur for a classical nova; Figure 13.9 shows the remains of Nova Persei 1901, in which we now see synchotron radiation to one side of the site of explosion, where the expanding cloud is colliding with an invisible nebula. The high energy electrons responsible for this non-thermal emission are accelerated in shock fronts at this collision.

## 13.6 Recurrent Novae

The best observed example of a small class of novae which recur at intervals of several years is RS Ophiucus. Its outburst in 2006 was the sixth since it was first observed in 1898. The increase in brightness at outburst is dramatic; optically, it typically increases by six magnitudes while the radio emission, which was undetectable before the 2006 outburst, rose rapidly to intensities that could only be explained in terms of non-thermal radiation. The lowest radio frequencies were the last to be observed; the GMRT was used to detect emission at 610 MHz on day 20 after the outburst and at 325 MHz on day 38 (Kantharia *et al.* 2007). These novae are associated with symbiotic binaries; as in the classical novae, they consist of a white dwarf accreting from a red giant. The difference is primarily that the outburst occurs after a smaller accretion of hydrogen; this may be related to the large difference in the binary periods, which are 230 days for RS Oph and 0.6 days for a typical classical nova.

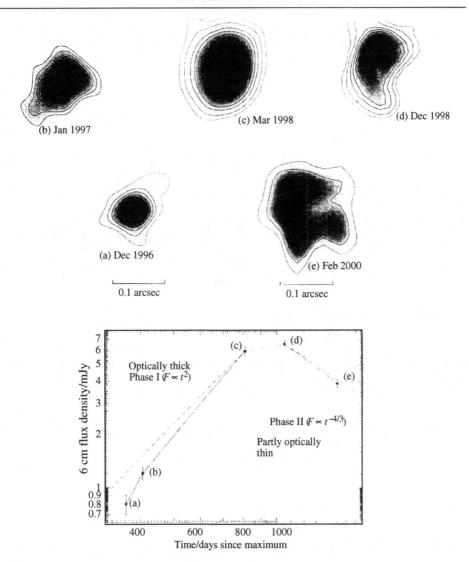

Figure 13.7. The thermal radio emission at 6 cm wavelength from the expanding shell ejected by Nova V723 Cassiopeia, showing an expanding optically thick disk becoming distorted and partially thin after 800 days (MERLIN observations, courtesy of T. O'Brien).

The cloud of material resulting from the thermonuclear explosion in RS Oph is ejected with a velocity of 4000 km s$^{-1}$. Without any interaction with other circumstellar material, this cloud might be expected to radiate free–free radio emission corresponding to a temperature of about $10^4$ K, as for a classical nova but with a much lower intensity because of the lower mass. Instead, in the 2006 outburst, intense radio emission was observed, growing for the first 100 days as the cloud expanded. Furthermore, aperture-synthesis observations using MERLIN, VLA, VLBA, and EVN were all started within two weeks (O'Brien *et al.* 2006), producing maps which gave an angular diameter showing that the

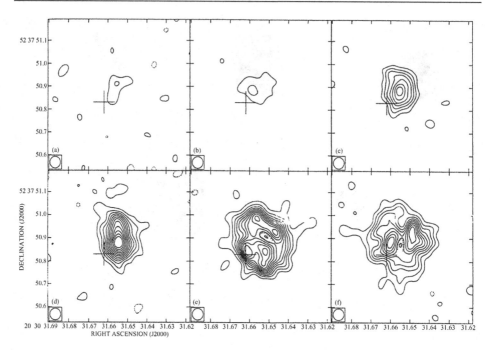

Figure 13.8. The expanding shell of thermal radio emission from Nova Cyg 92, mapped using MERLIN at 6 cm (Eyres *et al.* 1996).

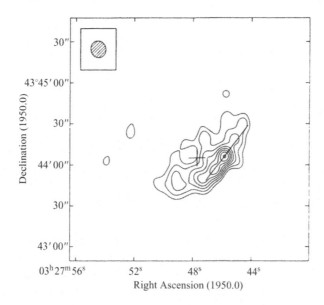

Figure 13.9. Non-thermal radio emission from the remains of Nova Persei 1901 (Reynolds and Chevalier 1984). The emission marks where the ejected shell interacts with the surrounding medium. The site of the nova itself is indicated with a cross; a more accurate position can be found in Seaquist *et al.* (1989).

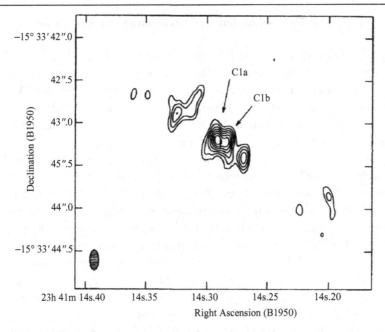

Figure 13.10. MERLIN image at 5 GHz of the symbiotic star R Aquarii. The star is located in the double source C1, and the ejected clouds are strung out over three arcseconds (Dougherty *et al.* 1995).

brightness temperature was at least $10^7$ K. This high temperature, and the radio spectrum, must be interpreted as due to synchrotron emission resulting from shock excitation of the cloud at a collision with circumstellar gas, probably in the form of a continuous wind from the red giant. Soft X-rays were also observed from this outburst; the spectrum again indicated an origin in high temperature gas derived from the collision.

The radio maps of several of these recurrent novae show ejected clouds which are very different from the isotropic model that we described for the classical novae. The MERLIN map of R Aquarii (Figure 13.10) shows clouds aligned along an axis, and a thermally emitting core (Dougherty *et al.* 1995). A later map using the VLA (Mäkinen *et al.* 2004) showed an outward movement of these individual clouds. This is our first example of many types of radio source in which the radiating material is in the form of twin jets; in this case, as in many others, they probably leave the star along the rotation axis. Well-collimated jets, rapid intensity fluctuations, and discrete structure within the jets suggest that the clouds in the jets are travelling outward at high velocity. We shall see in the next section that there are indeed outward velocities approaching the velocity of light in the X-ray binaries; in these very energetic objects we observe synchrotron radio emission. There are also, however, high velocity jet sources which are thermal radio emitters. An example is the radio jet in the Herbig–Haro objects HH 80-81 (Marti *et al.* 1995). The actual source may be hidden behind a dark molecular cloud which is the seat of recent star formation. Structure in the jet, which is 5 parsecs long, was observed at intervals of several years using the VLA at 3.5 cm wavelength. Proper motions in the range 70–160 mas $yr^{-1}$ were easily observed over this time, giving outward velocities of 600–1400 km $s^{-1}$.

## 13.7 Non-Thermal Radiation from Binaries and Flare Stars

Binary star systems with a condensed star as one component include some of the most spectacular radio sources in the Galaxy. There are however many examples among the main sequence stars of binaries which show strong radio emission. The well-known star Algol ($\alpha$ Persei) is quoted as a type for binaries consisting of late-type main sequence stars, although it is actually a triple system. The binary system CC Cas is in the same class of radio stars; it consists of a pair of interacting massive O stars. Algol itself has a continuous radio emission of about 0.01 Jy, detected at 2.7 and at 8 GHz, which rises rapidly by a factor 30 or more within hours at the time of a flare. The RS CVn variables, which include AR Lac and UX Ari, have more frequent flares, of the order of one per day; here the rise time may be only one hour, and there may be rapid fluctuations on timescales as short as five minutes. These flare stars are typically $10^{2-4}$ times more luminous than the brightest solar outbursts. The high intensity and high circular polarization indicate gyrosynchrotron radiation, with magnetic field intensities of order 100 gauss in the emitting regions.

The angular diameters of some of the Algol-type variables have been measured using global VLBI. (Algol itself, the triple system, was mapped and evaluated by Peterson *et al.* 2011.) Typically the source diameter is $10^{12}$ cm, comparable with the diameter of the binary orbit. The brightness temperature may reach $10^{10}$ K. Higher brightness temperatures are observed for some of the red dwarf variables; in a flare from YZ CMi the brightness temperature reached $7 \times 10^{12}$ K. Radiation at this level must be from a coherent source (see the discussion of synchrotron self-Compton effects in Sections 2.8 and 16.3.10); the nearest analogy may be the even brighter radio emission from pulsars.

## 13.8 X-Ray Binaries and Microquasars

More than 50 radio emitting X-ray binaries are known in our Galaxy, many of which have resolved jets; others are flat-spectrum compact sources which are assumed to be the self-absorbed cores of jets. Some produce spectacular outbursts. Cygnus X-3 is one of the most remarkable: it is a high mass X-ray binary (HMXB), which is also detectable as a gamma-ray source. The binary period is 4.8 hr. The compact object (a neutron star or a black hole) orbits a blue supergiant Wolf–Rayet star. The radio emission from Cyg X-3 is complex and rapidly variable; it has been resolved using real-time electronically connected global VLBI at 5 GHz (Tudose *et al.* 2010).

The peak of the radio emission occurs in time sequence from high to low radio frequencies, as might be expected from almost any emission mechanism for disturbances travelling outwards from an accretion disk surrounding a compact object. Such travelling disturbances are indeed seen in the form of twin jets; they move outwards with velocities variously estimated as 0.3 to 0.6 of the velocity of light. Observations with milliarcsecond resolution show a velocity of $0.63c$ and precession in the jets with a period of five days (Miller-Jones *et al.* 2004). The effects of such high velocities, which are encountered in quasars as well as in these sources within our Galaxy, require a relativistic analysis, which we give in Section 13.9 below. Synchrotron radiation from an expanding spherical shell, or from a sector of such a shell, may be analysed in a similar way to thermal radiation from the classical novae (Section 11.9). If a cloud initially containing a power-law distribution $KE^{-\gamma}dE$ of electron energies expands adiabatically to radius $r$, then $r^3 KE^{-\gamma}dE = $ constant and, at time $t$,

$$K(t) = K_0 \left(\frac{r}{r_0}\right)^{-(\gamma+2)}. \tag{13.3}$$

At the same time, however, the magnetic field will decrease during the expansion, and the geometry of the expansion has to be taken into account. A model by Marti *et al.* (1992) shows that the 1972 outburst of Cyg X-3, which was observed at eight different radio frequencies, can be accounted for by such a model (Figure 13.11). Synchrotron radiation alone is not, however, sufficient to account for the whole of these observed flux densities; in the initial stages of expansion, when the cloud is optically thick, the spectral index is lower than expected from synchrotron self-absorption, indicating free–free absorption from a larger mass of thermal electrons.

Finally, perhaps the most spectacular of these relativistic twin jet galactic sources is the binary SS433. This is observed as an optical as well as an X-ray and radio source; the optical emission includes spectral lines, showing that the jets contain hadronic material. There are twin oppositely directed jets along an axis which precesses over a period of 162.5 days. Discrete knots travel out along this precessing axis with velocity $0.26c$, as found from Doppler shifts of the optical spectral lines. The radio emission has been mapped over angular scales from 10 milliarcsec to several arcminutes, including VLBA measuremens at 1.6 GHz (Jeffrey *et al.* 2016). Figure 13.14(a) (Spencer *et al.* 1993) shows the travelling knots as they leave the centre; here the overall scale is 200 milliarcsec. The effect of precession is seen in Figure 13.14(b), covering 10 arcseconds, where the solid line traces out the spiral trajectory as seen projected on the sky. Between these two angular scales the brightness temperature falls from $10^8$ K to $10^5$ K; this must be synchrotron radiation, with the particle energies derived entirely by the initial ejection from the neutron star and the ballistic geometry determined solely by the precession of the star.

Many observations, including a long series by the X-ray spacecraft INTEGRAL have been made to determine the parameters of the binary star itself (see the review Fabrika 2004). It is often referred to as a microquasar, and appears to be unique in our Galaxy.

## 13.9 Superluminal Motion

The effects of superluminal velocities in astrophysical jets were first analysed in relation to the twin jets of extragalactic quasars (Chapter X). These jets remain narrowly collimated over distances of up to a megaparsec, and condensations within the jets leave the central source with velocities close to the velocity of light. The relativistic velocities may be observed as apparent velocities greater than $c$, depending on the aspect of the jet; these are the *superluminal* velocities. It is remarkable that this cosmic phenomenon is also seen, in miniature, in the X-ray binaries.

The first galactic superluminal source, GRS 1915+105, was discovered by Mirabel and Rodríguez (1994). Figure 13.12 shows maps made over a period of only 13 days. Four

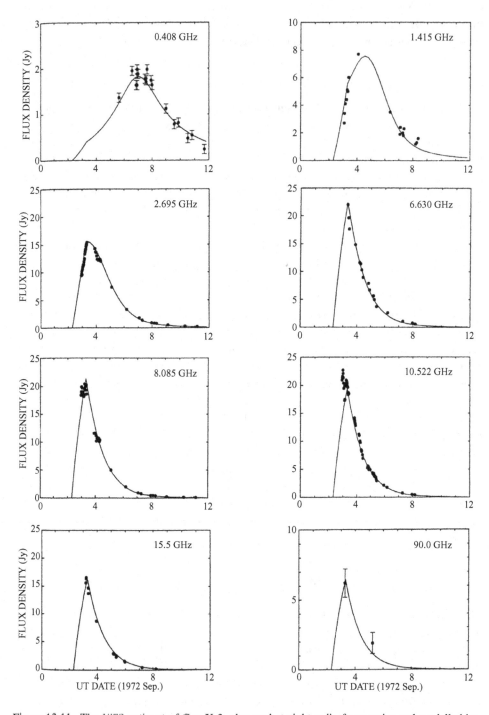

Figure 13.11. The 1972 outburst of Cyg X-3, observed at eight radio frequencies and modelled by Marti *et al.* (1992).

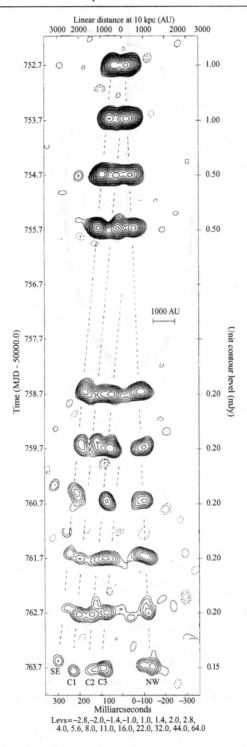

Figure 13.12.  A series of 5 GHz MERLIN maps of the superluminal X-ray binary GRS 1915+105 (Fender *et al.* 1999).

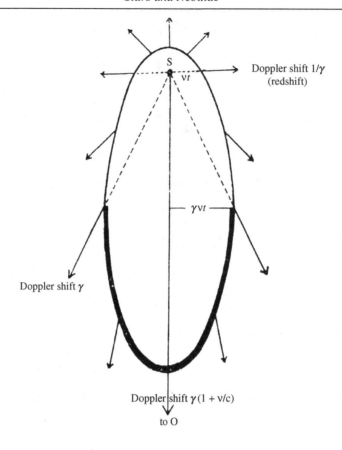

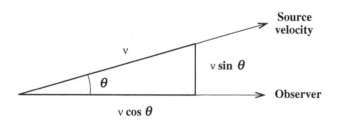

Figure 13.13. The geometry of superluminal motion. A relativistically expanding shell expanding isotropically with relativistic velocity $v$, seen by a distant observer (Rees 1966).

separate condensations are seen to leave the source and move outwards; the three to the left are moving with a proper motiom of $23.6 \pm 0.5$ mas per day, which corresponds to a velocity of $1.5c$, while the weaker one on the right is apparently moving with velocity $0.6c$.

An explanation for such supcrluminal velocities was proposed by Rees in 1966. He considered a spherical shell expanding isotropically with relativistic velocity $v$, as in Figure 13.13. The locus of radiating points as seen by the observer is a spheroid. Radiation from the front of the shell is seen by a distant observer to be Doppler shifted by a factor $\gamma(1 + v/c)$, where the relativistic factor $\gamma$ is equal to $(1 - v^2/c^2)^{-1/2}$, while the apparent velocity at the observed edge of the source is $\gamma v$, which may be greater than $c$. Consider now a discrete radiating source moving out from the central source with velocity $v = \beta c$ along a line at angle $\theta$ to the line of sight (Figure 13.13). In the time interval $\tau$ between two observations the sideways motion is $v\tau \sin\theta$. In this time the source moves a distance $v\tau \cos\theta$ towards the observer, so that the observed time interval is reduced to $\tau - \beta \cos\theta$. The apparent velocity is then $c$ times a 'superluminal' factor $F$, where

$$F = \beta \sin\theta (1 - \beta \cos\theta)^{-1}. \tag{13.4}$$

The maximum effect on the apparent velocity is at angle $\theta_m$, given by $\sin\theta = \gamma^{-1}$, $\cos\theta_m = \beta^{-1}$. The apparent velocity is greater than $v = \beta c$ for angles in the range $2\beta/(1 + \beta^2) > \cos\theta > 0$ and the maximum superluminal factor $F_{max}$ is given by

$$F_{max} = \gamma\beta. \tag{13.5}$$

Relativistic beaming increases the flux density; it also Doppler shifts the frequency of the radiation by a factor $\delta = \gamma^{-1}(1 - \beta \cos\theta)^{-1}$, so that the observed flux density depends on the emitted spectrum. If the emitted spectrum is $S_0(\nu) \propto \nu^\alpha$, the observed flux density is

$$S(\nu) = S_0 \left(\frac{\nu}{\delta}\right) \delta^{3-\alpha}. \tag{13.6}$$

This applies to an isolated discrete source; if there is a continuous stream, the apparent density of sources along the stream is increased by a factor $\delta^{-1}$ and the observed flux scales as $\delta^{2-\alpha}$.

If in the case of GRS 1915+105 we assume that the velocity of the receding jet (on the right in Figure 13.12) is the same as that of the three approaching jets (on the left) then a complete solution for $\theta$ and $v$ can be found from the two proper motions $\mu_{app}, \mu_{rec}$, since

$$\beta = \frac{\mu_{app} - \mu_{rec}}{\mu_{app} + \mu_{rec}}, \tag{13.7}$$

$$\tan\theta = 1.16 \times 10^{-2} \left(\frac{\mu_{app}\mu_{rec}}{\mu_{app} - \mu_{rec}}\right) d, \tag{13.8}$$

where $d$ is the distance of the source in parsecs (Fender *et al.* 1999). For this source we obtain $\beta = 0.9$ and $\theta = 66°$. A later revision (Punsly and Rodriguez 2016) reduces the velocity to around $\beta = 0.3c$.

## 13.10  H II Regions

The classic H II regions are discrete ionized clouds surrounding very hot O-type stars. A typical cloud might be approximately spherical (a *Strömgren sphere*), several parsecs across, and bounded by a well-defined ionization front, inside which there is an equilibrium balance between ionization by ultraviolet light from the star and recombination. Hydrogen atoms spend some hundreds of years in the ionized state, and some months in the neutral

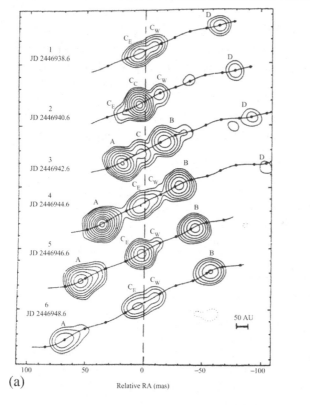

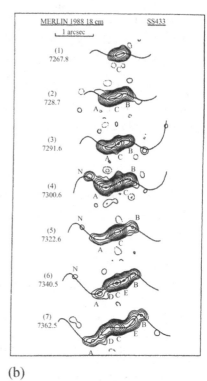

(a)                                                              (b)

Relative RA (mas)

Figure 13.14.  Radio emission from the X-ray binary SS433: (a) knots as they leave the centre; (b) the effect of precession (Spencer *et al.* 1993).

state. They are maintained at a temperature of 8000–10 000 K, above which the region loses energy very rapidly through the excitation of oxygen ions (see Section 2.6). Lower energy photons penetrate beyond the limit of the H II region, and can excite or ionize atoms and molecules in a *photo dissociation region* (PDR) surrounding the Strömgren sphere. The PDR contains a rich mixture of atomic and molecular species, as in the example of Figure 3.3.

The major part of the radio spectrum observed from H II regions is the continuum from free–free emission, in which the electron is unbound before and after collision with an ion.

Figure 13.15 shows the free–free emission from a group of H II regions known as W3. The radio spectrum from such regions follows the classic pattern of optically thin bremsstrahlung (spectral index $-0.1$) above a frequency of order 1 GHz; at lower frequencies the spectrum progressively approaches the optically thick case in which the spectral index becomes $+2.0$ (see Section 2.3). At millimetre and submillimetre wavelengths (Ladd *et al.* 1993) the H II regions are less conspicuous, while the 20 micron emission is concentrated on regions of star formation (Wynn-Williams *et al.* 1972). The ratio of radio recombination line emission to free–free emission provides a useful measurement of temperature, since it depends on the ionization ratio in the H II region. The

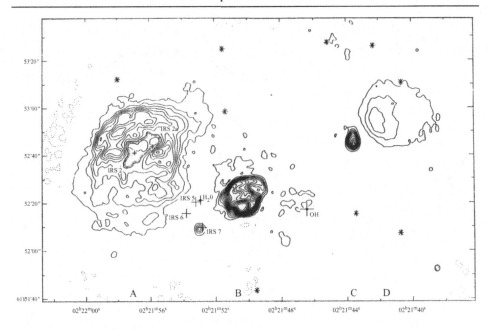

Figure 13.15. The free–free emission at 5 GHz from the W3 group of H II regions. The contour interval is 255 K (Harris and Wynn-Williams 1976).

emissivity of the line radiation, like the continuum free–free radiation, is proportional to the square of the electron density; the ratio depends on radio frequency $\nu$, quantum number $n$ and temperature $T_e$ as

$$\text{line/continuum} \propto \nu^{2.1} n^{-1} T_e^{-1.15}. \tag{13.9}$$

Typical temperatures determined from this ratio are between 8000 K and 10 000 K. The interpretation of this ratio is, of course, less certain when conditions within the H II region are not uniform; this is the case for the high-density 'compact' H II regions, such as W3. The ratio of line to continuum radiation is also more difficult to interpret for diffuse sources away from the galactic plane; here the background radiation is primarily synchrotron, and the ratio of line to continuum is less meaningful.

Large H II regions are detectable optically in the plane of the Galaxy for distances of several kiloparsecs. Georgelin and Georgelin (1976) showed that they follow the spiral-arm pattern much as can be seen in other galaxies. This delineation of the spiral structure can be extended through the whole Galaxy using radio recombination lines, since the Galaxy is transparent at radio wavelengths; Wilson *et al.* (1970) mapped all the bright H II regions of the Milky Way in 109 $\alpha$ recombination-line emission.

## 13.11 Supernova Remnants

All stars evolve and eventually die, and most leave small, condensed, fossil remains, typically a white dwarf. The death of stars with masses greater than $\sim 8 \, M_\odot$ is catastrophic and spectacular: the core of the star becomes a white dwarf whose mass reaches the limit

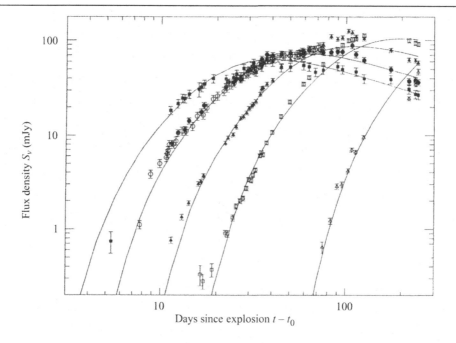

Figure 13.16. The growth of radio emission from SN 1993J in the galaxy M81, over the first 200 days. From left to right the curves show the flux density at 1.3, 2, 3.6, 6, and 20 cm (van Dyk *et al.* 1994).

of stability, and which then collapses within a second to form a neutron star or a black hole. A similar collapse can occur when a white dwarf in a binary system accretes mass from its companion star (Section 12.15). The energy released in the collapse drives infalling matter outwards in a violent explosion, which disrupts the surrounding remnants of the star, producing the visible supernova which can outshine its entire parent galaxy for several weeks (see Burrows 2000 for a brief review). The supernova phenomenon is an ideal example of multi-spectral astronomy. The explosion itself is essentially a phenomenon for the visible spectrum and for neutrino astronomy; radio astronomy takes over progressively as the expanding cloud of debris becomes more diffuse. Radio emission has, however, been observed from supernovae, notably SN 1987A in the Large Magellanic Cloud and SN 1993J in the galaxy M81. At the earliest stage the emission is probably from the expanding supernova debris, but it is the subsequent interaction with the surrounding ISM that produces the shell-like supernova remnants (SNR) which are the long-lasting memorials to the death of the star. The evolution over the first 200 days of the radio emission from SN 1993J in the galaxy M81 is shown in Figure 13.16. Short wavelengths are observed first, and the spectrum evolves rapidly towards longer wavelengths. This is synchrotron radiation from an expanding shell round the supernova. The expansion was measured by Bartel *et al.* (1994) and shows a linear growth to 250 microarcsec diameter over 250 days. The growth of the supernova remains over the subsequent 10 years (Figure 13.17) was mapped using global VLBI at 5 GHz (Marcaide *et al.* 2009).

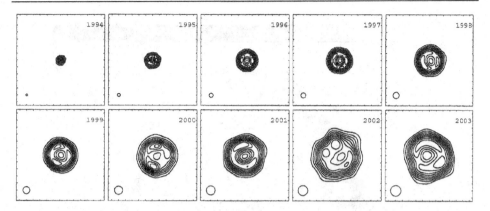

Figure 13.17. The expanding remnant of SN1993J in the galaxy M81, over the first 10 years, mapped by global **VLBI** at 5 GHz. The circular beam used to reconstruct each map is shown in the lower left corner. Contours are in steps of 10% of peak emission. The scale marks are at milliarcsecond intervals (Marcaide *et al.* 2009).

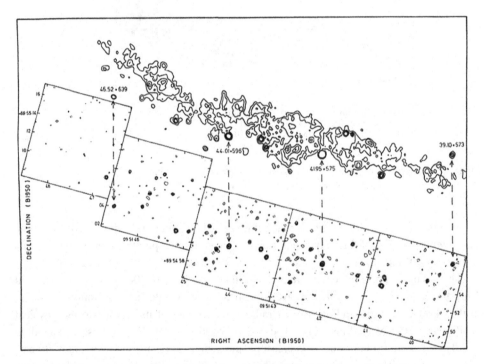

Figure 13.18. Supernova remnants in the starburst galaxy M82 (Muxlow *et al.* 1994).

The evolution of a young SNR at a later stage has been seen in the star-burst galaxy M82 (see Chapter 16), which contains at least 40 discrete radio sources which are identified as SNRs (Figure 13.18) (Pedlar *et al.* 1999; Muxlow *et al.* 1994). The expansion of one

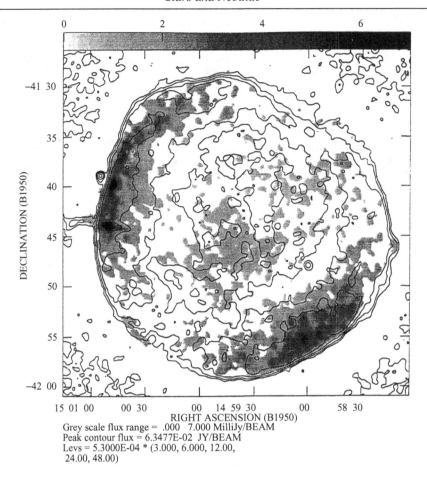

Grey scale flux range = .000  7.000 MilliJy/BEAM
Peak contour flux = 6.3477E-02  JY/BEAM
Levs = 5.3000E-04 * (3.000, 6.000, 12.00,
   24.00, 48.00)

Figure 13.19. Radio emission at 1370 MHz from the supernova remnant SNR 1006. The grey scale shows the polarized component (Reynolds and Gilmore 1986).

of these was followed for 11 years, when it had reached about 50 milliarcseconds across. The angular velocity of this expansion is about one arcminute per year, corresponding to a velocity of $10^4$ km s$^{-1}$ and a birth date in the early 1960s (see also Beswick 2006).

The ages of the observable radio remnants extend to at least 100 000 years (see Raymond 1984 for a review). If they all behaved identically, we would have a complete picture of the development and fading of their radio emission and of its relation to the physical conditions in the expanding cloud. The 300-year-old remnant Cas A is still expanding rapidly, but its radio emission is decreasing; the same has been found for the 400-year-old remnants SNR 1604 (Kepler) and SNR 1574 (Tycho). Two 1000-year-old remnants are also observed to be expanding: the Crab Nebula (1054) and SNR 1006 (Figure 13.19). The remains of much older SNR can also be observed: IC 443, the Cygnus Loop (Dickel and Willis 1980), and the Vela Nebula. These are probably between 10 000 and 100 000 years old. An invaluable working catalogue of SNRs is maintained by D. A. Green at http://www.mrao.cam.ac.uk/surveys/snrs/.

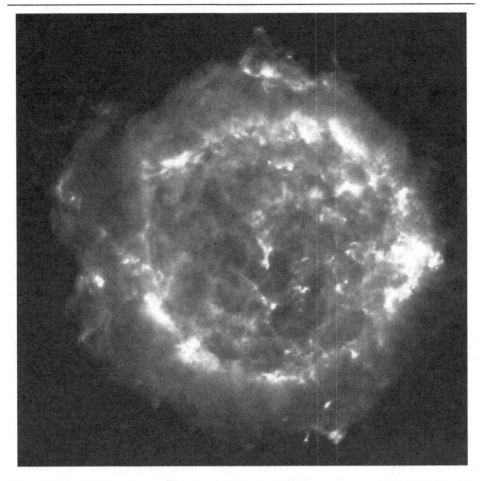

Figure 13.20. Radio emission from the supernova remnant Cas A mapped with the VLA. The overall diameter is 5 minutes of arc. (Image courtesy of NRAO/AUI. Credit: L. Rudnick, T. Delaney, J. Keohane and B. Koralesky, image composite by T. Rector.)

The supernova remnant Cas A, which is about 300 years old and is the youngest SNR in the Galaxy, is the brightest radio object in the sky (apart from the Sun), but was not noticed optically until radio observations drew attention to it. It contains filaments moving radially outward at 6000 km s$^{-1}$; others have been nearly stopped by collision with the surrounding ISM. The spherical shell shows where the ISM has been compressed by the supernova (Figure 13.20). The synchrotron radio emission from this and similar shells shows linear polarization neatly arranged round the circumference, indicating where the interstellar magnetic field has been swept up and compressed; a good example is 3C10, shown in Figure 13.21.

The common feature of all these various SNRs is the catastrophic release of energetic particles, with a total energy of order 10$^{51}$ erg, expanding into the ISM with a velocity of order 10 000 km s$^{-1}$. Although the mean free path is high, this blast wave is coupled

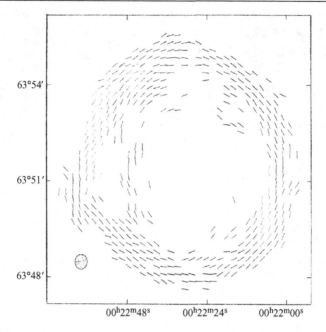

Figure 13.21. Polarized radio emission on the circumference of the supernova remnant 3C10. The map shows the intrinsic polarization angle distribution as derived from observations at 6 and 21 cm, correcting for Faraday rotation (Duin and Strom 1975).

to the interstellar gas by the ambient magnetic field, and the gas is swept up with it. Up to the point where the swept-up mass is comparable with the mass of the ejecta, the expansion is a simple unhindered explosion (Sedov–Taylor expansion); after this point, the shape and speed of the expansion depend on the density and distribution of the interstellar gas and its magnetic field. Parts of Cas A are still expanding at 6000 $\mathrm{km\,s^{-1}}$; the older SNRs appear to have been slowed down by interstellar gas. The shapes of IC 443, the Vela Nebula, and the Cygnus Loop seem to be determined mainly by collisions with interstellar clouds; these remaining wisps are moving at speeds of only some tens of $\mathrm{km\,s^{-1}}$. The effect on the clouds is best seen at millimetre and sub-millimetre wavelengths: for example, van Dishoeck *et al.* (1993) showed that the outer parts of the IC 443 shell contain shocked molecular gas emitting both lines, such as those from CO, and radio continuum.

There is, however, a distinctive feature of the Crab Nebula (see the Supplementary Material), which is shared by a minority of others. Over the wide spectrum through which it can be observed, the emission comes from a filled sphere rather than an expanding shell; such SNRs are known as *plerions*. The Crab Nebula appears as a tangled web of filaments, with some optical line emission but mainly synchrotron radiation. The continuum follows a power law, with unusually low spectral index, extending to X-rays. The lifetime of electrons sufficiently energetic to radiate X-rays is only a few years; therefore there must be an energy supply within the nebula. This is, of course, the Crab Pulsar, which we discuss in Chapter 15.

Although the Crab Nebula is expanding at a lower rate than the Cas A SNR, an extrapolation of its expansion back to its origin in AD 1054 shows that the expansion has not been slowed by sweeping up the surrounding interstellar gas and dust; on the contrary, it has accelerated. Again, the magnetic field required for its synchrotron radiation, around 100 microgauss, is hard to account for either as a remnant of the original stellar magnetic field or as a swept-up interstellar field. The pulsar provides the explanation for all these distinctive features. Only young SNRs can be expected to show these characteristics, but it is unclear what proportion of young SNRs will show the same evidence of an active pulsar. The 400-year-old Tycho and Kepler SNRs are already behaving very much like the old nebulae such as IC 443 and the Cygnus Loop.

## 13.12  Further Reading

*Supplementary Material* at www.cambridge.org/ira4.
Gray, M. 2012. *Maser Sources in Astrophysics*. Cambridge University Press.

# 14

# The Milky Way Galaxy

Our home Galaxy, the Milky Way, is only one of the enormous number of galaxies, perhaps $10^{11}$, that make up the Universe. Galaxies come in many forms and sizes and these can be classified, to a first approximation, by their optical appearance; the main division is between spirals and ellipticals. Radio observations of the H I line first proved that the Milky Way is a spiral; both line and continuum observations have contributed to our understanding of its structure, and especially of the interstellar medium.

## 14.1 The Structure of the Galaxy

The Milky Way has stars as its most visible component, but radio astronomy is usually more concerned with the interstellar medium (ISM). The ISM is composed principally of hydrogen and helium, with a trace of heavier elements ocurring in both atomic and molecular form and in larger aggregates referred to generically as dust. In addition there is a third, more tenuous, component, the high energy medium, composed mostly of energetic particles, principally protons and electrons, with energies extending well beyond $10^{18}$ electronvolts. These pervade the Galaxy and are contained within it by a magnetic field of a few microgauss;[1] they are detected on Earth as cosmic rays. Finally, it has become clear that our Galaxy, in common with galaxies in general, is embedded in a much larger halo of *dark matter*, detectable only through its gravitational influence on the stars and the ISM. Radio observations contribute to the study of all these components.

Spiral galaxies such as the Milky Way are distinguished by a flat disk, with a variously developed system of spiral arms and a central bulge of stars which may have a bar-like structure linked to the spiral arms. The dust and gas of the ISM, and the young stars presently forming from the ISM, are strongly concentrated on the plane of the disk, typically with an axial ratio of about 100 : 1. The stellar disk, comprising older stars, is a flattened ellipsoid, but thicker, with an axial ratio of around 10 : 1. The galaxies are designated S0 if they have no discernable spiral arms, and Sa, Sb, Sc for those with increasingly open spiral structure; the designation S becomes SB for barred spirals. The central bulge is most prominent in Sa

---

[1] Cosmic rays which have been detected in the energy range $10^{19}$ to $10^{20}$ electronvolts cannot be contained by the galactic magnetic field, and may come from extragalactic sources.

Figure 14.1. An image of the Milky Way made with the Gaia spacecraft using a Hammer equal-areas projection. The brighter regions correspond to higher concentrations of stars within a given pixel. The Magellanic Clouds can be seen in the lower right quadrant. (Credit: European Space Agency.)

spirals, and least in Sc spirals; the spiral arms are clearly delineated in Sa and Sb spirals but are rough, loosely wound, and irregular in the Sc's. In spiral galaxies, the oldest stars are found in a less conspicuous halo surrounding the galaxy, together with *globular clusters*, which are found in an approximately spherical distribution around the central region. Our Galaxy is classified as SBb, mainly on the basis of H I spectral line observations, which we present in this chapter.

The Milky Way has two satellite galaxies, the Large Magellanic Cloud (LMC) and Small Magellanic Cloud (SMC), both classified as irregular, having no obvious spiral or elliptical structure. Both are gravitationally bound to the Milky Way and are presently at a distance of roughly 45 kiloparsec (kpc),[2] six times the Sun's distance from the galactic centre. Our Galaxy, in turn, is a member of the Local Group, a gravitationally bound system of 30 or so galaxies, all closer than a megaparsec, the largest of which are the Milky Way and the well-known Andromeda Spiral, M31. The Local Group is a relatively small cluster but is a member of a much larger group called the Local Supercluster, which contains both field galaxies and clusters of galaxies, several richer than the Local Group by orders of magnitude. Its extent is not well defined but is more than 40 megaparsecs.

A striking image of the Milky Way in visible light is shown in Figure 14.1. It was made from measurements of the positions and fluxes from almost 1.7 million stars by the Gaia

---

[2] The astronomical unit of distance, the parsec, is based on the measurement of distance by parallax due to the Earth's orbital motion: 1 parsec (pc) = $3.08 \times 10^{16}$ m = 3.26 light years.

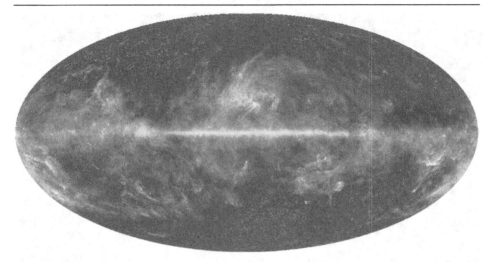

Figure 14.2. A multi-wavelength composite all-sky image from the Planck spacecraft (reproduced in colour in Supplementary Material) at a resolution of 5 arcminutes. In regions towards the top and bottom of the image, away from the diffuse emission from the Milky Way, the cosmic microwave background is visible. (Credit: European Space Agency.)

spacecraft. The galactic centre is in the middle and the galactic plane runs centrally across the image.[3] The dark regions arise from the absorption of the light of more distant stars by intervening interstellar dust and gas. The two bright objects below the plane are the Large and Small Magellanic clouds. There is little indication in this image of the three-dimensional structure of our Galaxy.

At microwave and mm wavelengths the appearance of the sky changes. The image in Figure 14.2 is a combination of maps made from all nine frequency channels (ranging from 30 GHz to 857 GHz) of the Planck spacecraft (see the Supplementary Material for a colour version). The lower frequencies are sensitive to synchrotron emission from relativistic electrons moving in the large-scale magnetic field of the Milky Way and to free–free emission from warm or hot interstellar ionized gas. The higher frequencies are increasingly dominated by thermal (quasi-blackbody) emission from the same cold interstellar dust that absorbs the visible light. The dust particles are distributed over a wide range of sizes from a few hundred microns down to several microns. Along the galactic plane, spectral line emission at 115 GHz from carbon monoxide is also significant.

At longer radio wavelengths the appearance of the sky changes again, as can be seen in the 408 MHz map in Figure 14.3 (Remazeilles *et al.* 2015). Dust plays no role and the image is completely dominated by synchrotron radiation. Large-scale loops and arcs can be seen extending away from the galactic plane. These mark the shells of ancient supernova

---

[3] The galactic coordinate system, longitude $l$ and latitude $b$, has been defined by the International Astronomical Union, and is based on radio H I measurements. The definition as given in Allen's *Astrophysical Quantities* approximates the plane of the Galaxy, with the origin ($l = 0°, b = 0°$) chosen close to the actual centre, the central massive black hole.

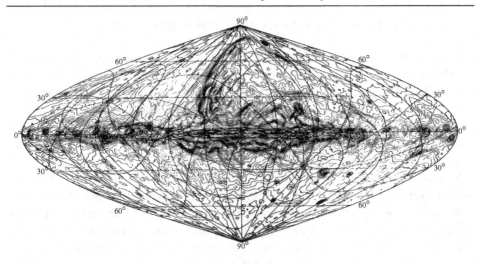

Figure 14.3. An all-sky contour map at 408 MHz at a resolution of 56 arcminutes (Haslam *et al.* 1982; revised). (Courtesy of C. Dickinson, Jodrell Bank Centre for Astrophysics.)

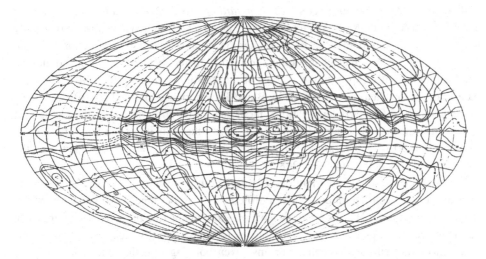

Figure 14.4. An all-sky contour map at 30 MHz. The contours of brightness temperature are labelled in units of 1000 K (Cane 1978).

remnants; in the case of the most prominent, the North Galactic Spur, the stellar progenitor exploded only a few hundred parsecs away from the Sun. The same broad features can be seen in the lower-resolution radio map made at 30 MHz (wavelength 10 m) in Figure 14.4, although synchrotron self-absorption confines our view to relatively local regions of the Milky Way. At the time of writing the most recent low frequency sky mapping has been made with the Long Wavelength Array in New Mexico covering the frequency range 40–80 MHz (Dowell *et al.* 2017).

## 14.2 Galactic Rotation: The Circular Approximation

Establishing the structure and dynamics of the Milky Way is not straightforward because our observing point in the solar system lies within the body of the rotating Galaxy; in addition dust obscuration limits the use of stars for the task. However, radio waves pass unimpeded through the dust, and the radial velocities of cool neutral hydrogen (H I) clouds, determined from the Doppler shifts of the 21 cm spectral line (see Sections 3.2 and 3.6), enable us to build up a picture of our spiral Galaxy as if we were looking from above the plane.

A simplified picture of the galactic disk, in which the stars and gas in the ISM rotate about the galactic centre (GC) in circular orbits, is shown in Figure 14.5(a). Towards the centre the rotation periods are shorter and the angular velocities $\Omega$ are higher. The result is *differential rotation* and, in a given direction, clouds at different distances $r$ from the Sun will have different radial velocities. In any given direction the observed H I spectrum is the sum of the emission from all distances. Distinct spectral peaks are often seen indicating denser gas at particular radial velocities; these peaks suggest the presence of spiral arms.

Before using the observed radial velocities to determine the galactic structure they must first be converted to the observer's *local standard of rest* (LSR) and this involves several corrections. The Earth–Moon system is rotating about a barycentre (with a velocity usually too small to be important); the Earth rotates on its axis (the daily correction is about 1 km s$^{-1}$); the Earth is in orbit about the Sun (with velocity about 30 km s$^{-1}$); and the Sun itself is moving with respect to the average of the stars in its vicinity (the LSR). The main parameters of the Galaxy, as reviewed by Bland-Hawthorn and Gerhard (2016), are known through several different determinations, including VLBI observations of the parallax of masers near the galactic centre (Honma *et al.* 2018): the velocity of the Sun with respect to the LSR is $30.24 \pm 0.12$ km s$^{-1}$, the distance to the GC $R_0 = 8.1 \pm 0.1$ kpc, and the rotational velocity at the Sun $V_0 = 248 \pm 3$ km s$^{-1}$.

To determine the spatial distribution of hydrogen in the disk one needs to establish the distances to the emitting clouds. This is difficult but can be done if we know the Milky Way's *rotation curve*; H I observations have played a major role in its determination. The key is to recognize that there are some places for which the distance of a cloud can be established from simple geometry. For any given pointing direction and hence galactic latitude $l$ the radial velocity $V_r$ reaches a maximum at the tangent point, i.e. for the cloud at T in Figure 14.5(a).[4] Other clouds with the same orbital radius have only a component of their velocity along the line of sight. At the tangent point we can write

$$R = R_0 \sin l \qquad\qquad (14.1)$$

and

$$V = V_r + V_0 \sin l. \qquad\qquad (14.2)$$

---

[4] The locus of the tangent points is a circle whose diameter is a line drawn from the Sun to the GC, but near the centre the motions become non-circular.

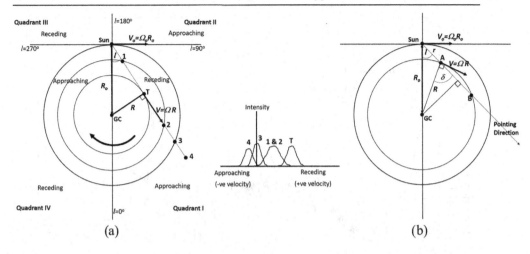

Figure 14.5. (a) Idealized plan view of the galactic disk with the galactic quadrants I–IV marked. The stars and gas are in circular orbits about the galactic centre (GC). The angular velocity increases towards the GC; hence, outside the solar circle objects fall behind the Sun and inside the circle they move ahead. Towards the GC at $l = 0°$ and the anticentre at $l = 180°$ the radial velocity is zero and the H I line emission is centred at the rest frequency. For other pointing directions positive (receding) and negative (approaching) radial velocities are seen, depending on whether the gas is inside or outside the solar circle. For the pointing direction shown, the radial velocity will be a maximum at the tangent point T. The schematic spectrum shows discrete contributions from other clouds; note that from points 1 and 2 the contributions overlap and the hence the distance is ambiguous. (b) Geometric construction for determining the distance $r$ to a cloud – see the main text for further details.

Armed with a knowledge of the Sun's orbital parameters, the galactic longitude of the pointing direction, and the maximum radial velocity in that direction, we can plot the rotation curve $V$ against $R$. The result, discussed in more detail below, is that $V$ changes relatively little with $R$ and, for an idealized model, one could assume that $V = \Omega R = \Omega_0 R_0$. We are now in a position to convert the radial velocity of a spectral peak into the distance of the emitting gas.

Consideration of Figure 14.5(b) shows that at galactic longitude $l$ the relative velocity of a cloud at point A with respect to the Sun will be

$$V_r = \Omega R \sin \delta - \Omega_0 R_0 \sin l \qquad (14.3)$$

and, since[5]

$$R \sin \delta = R_0 \sin l, \qquad (14.4)$$

we arrive at the fundamental equation

$$V_r = R_0 (\Omega - \Omega_0) \sin l \cos b. \qquad (14.5)$$

[5] From the sine theorem, $\sin l / R = \sin(180° - \delta)/R_0 = \sin \delta / R_0$.

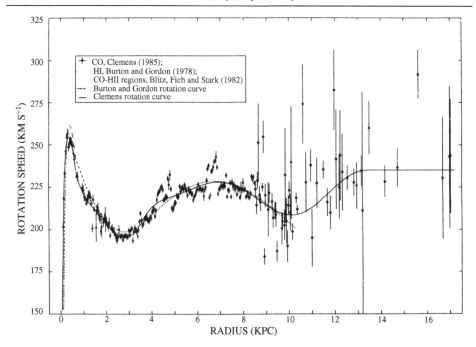

Figure 14.6. The rotation curve of the Galaxy, from observations of CO and H I emissions inside the solar radius, and from CO complexes associated with H II regions of known distance (Clemens 1985).

The extra factor $\cos b$ allows for sources away from the plane, i.e. at galactic latitude $b$. Using Eqn 14.5, $\Omega$ can be calculated for the cloud; then, from the rotation curve ($\Omega R$ against $R$, hence $\Omega$ against $R$) one can determine $R$ and so establish the Sun–cloud distance $r$. However, a problem remains: there is an ambiguity in the distance determination for clouds within the solar circle in quadrants I and IV: given $\Omega$ and $R$ there can be two possible locations. Thus in Figure 14.5(b) the cloud could be at A or B. Outside the solar circle there is no ambiguity. The ambiguity may be resolved from size alone with the more distant clouds appearing smaller on average. In this manner the spiral structure of the Milky Way (see Section 14.3 and Figure 14.8) was first determined.[6]

The current state of knowledge of the rotation curve of our Galaxy, the Milky Way system, is summarized in Figure 14.6. The inner part of the rotation curve is derived from H I and CO data, using the tangent points as in Figure 14.5(b). The curve beyond the solar circle uses optical data from observation of H II regions and open clusters; the scatter is probably caused by measurement uncertainties and will improve with time. Inside $R = 4$ kpc there are systematic differences between the velocities observed in regions I and IV, due to non-circular motions; Figure 14.6 shows averages. The simplest fit to the observations is a straight horizontal line from $R = 4$ kpc, that is, a constant velocity equal to the rotational velocity at $R_0$. This remarkable result is not peculiar to the Milky

---

[6] More recently distance ambiguities have been resolved using pulsar absorption lines (Chapter 15) or identification with known objects such as H II regions or molecular clouds.

Way; it is a common phenomenon, observed in other spiral galaxies both in the radio observations (Wevers *et al.* 1986) and in observations of optical emission lines (Rubin *et al.* 1985). In a Keplerian system, with a point mass at the centre, the tangential velocity would diminish as $R^{-1/2}$; a model of the Milky Way in which the mass is the sum of the observable luminous matter, the gas, and the dust gives a slower fall with radius, but there is no match possible to the nearly constant velocity from 4 kpc outward shown in Figure 14.6.

If the visible Milky Way resided in a singular isothermal sphere of matter, a constant rotational velocity, independent of radius, would be expected. Since the same phenomenon is observed in many other galaxies, this has led to the conjecture that galaxies are often embedded in spherical mass distributions larger than the visible galaxy, although the matter is not luminous. There is other evidence that *dark matter* is common: clusters of galaxies show dynamical behaviour that demands far more mass than can be accounted for by the observed luminous matter. Furthermore, gravitational lensing (Chapter 16) gives evidence that the observed multiple images suggest the presence of dark matter. This has led to the conclusion that the Universe contains dark matter that cannot be detected except by the influence of its gravitational field. It cannot be neutral hydrogen or ionized hydrogen, but it might be condensed objects, too small to become luminous stars, commonly referred to as massive compact halo objects (MACHOs). Two searches for such objects, the MACHO and OGLE projects, using the gravitational lensing of background stars, have so far failed to show their existence (as we write in 2018) if they are in the size range from the mass of Jupiter to the mass of a brown dwarf (0.001 to 0.05 solar masses). Another option, at the fundamental particle scale, conjectures that the dark matter consists of weakly interacting massive particles (WIMPs). Despite the difficulty in understanding the nature of dark matter, the evidence for it in our Milky Way, and many other massive galaxies, and other evidence for dark matter, is so compelling that its existence has to be granted.

The size of the halo of the Milky Way is not known, but it has to be limited by tidal interactions with other galaxies; Fich and Tremaine (1991) suggested a nominal halo radius of 35 kpc, and a total mass of the galaxy, including the halo, of

$$M_G = 3.9 \times 10^{11} \left( \frac{\Theta}{220 \text{ km s}^{-1}} \right)^2 \left( \frac{r_{max}}{35 \text{ kpc}} \right) M_\odot. \tag{14.6}$$

A detailed discussion of the circular approximation is given in Burton (1988).

## 14.3 Spiral Structure

The beautiful patterns of other spiral galaxies have radio counterparts; these can be seen both in the radio continuum and in the spectral lines from hydrogen and common molecules such as CO. This should be expected, because the optical spiral arms are outlined by star-forming regions, and the stars have formed from the ISM. This is illustrated in Figure 14.7, which shows the CO distribution in M51, constructed by aperture-synthesis measurements with the Owens Valley Millimeter Interferometer, superimposed on H$\alpha$ emission

CO on Hα

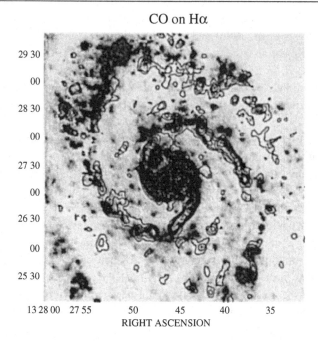

RIGHT ASCENSION

Figure 14.7. The spiral structure of M51, seen in CO line emission superposed on Hα emission (Rand and Kulkarni 1990). The contour intervals are at 4.5, 7.8, 13.5, 22.5, 31.5, and 40.5 Jy km s$^{-1}$ beam$^{-1}$.

that traces the H II region distribution. There is a practical problem to be overcome in making such maps of external galaxies, since their distance and consequent small angular size requires aperture-synthesis interferometers to obtain the necessary angular resolution. Since the molecular line emissions are weak and the interferometers have a small filling factor, long observing sessions are needed to obtain the necessary signal-to-noise ratio.

We can obtain much stronger signals from the ISM of our own Milky Way, and the fine details and small-scale motions are more easily observed, but we are immersed in its structure and the overall patterns are not traced easily. Nevertheless, the spiral structure of the Milky Way can be deduced, under the circular approximation, by taking the observed line profiles such as those used to construct Figure 14.9, and transforming them by use of the fundamental Eqn 14.5 into a face-on view of the Milky Way. Figure 14.8 is the celebrated result obtained by Oort *et al.* (1958), who combined the single-dish hydrogen observations taken in the Netherlands and in Australia to delineate the spiral structure of our Galaxy. This representation is not easily improved upon, since there are large-scale departures from circular velocity.

The observations from which the structure of the galactic plane can be found are shown in Figures 14.9 and 14.10. These are spectra of the line emission from H I (Burton 1988) and CO (Dame and Thaddeus 1985). The CO data were taken with two identical 1.3 m telescopes in the northern and southern hemispheres, so providing complete coverage of

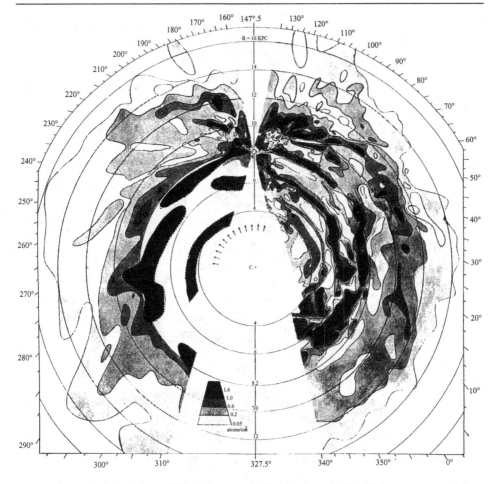

Figure 14.8. The Oort, Kerr, and Westerhout map of the spiral arms of the Galaxy, constructed from a circular model (see Figure 14.5), showing the maximum densities projected on the galactic plane.

the galactic plane.[7] (Because of the shorter wavelength of the the CO 1–0 line, the angular resolutions of CO surveys are similar to those for the H surveys.) The contours show the brightness temperature, as in the hydrogen plot, and the same cautions concerning optical depth apply. Along the entire galactic circle, low-velocity material can be seen, although its distribution in CO is patchy, since much of the CO occurs in giant molecular clouds. A similar velocity map can be drawn using hydrogen recombination lines (Alves *et al.* 2015).

Several patterns show up clearly in both diagrams. There is a clear envelope, reaching its most extreme velocities at galactic longitudes $l = \pm 20°$, that relates to the rotation of the galaxy. The most obvious departure from the general overall sweep of velocities due to galactic rotation is at longitudes within 5° of the centre, which indicates that the Galaxy should be classified as a barred spiral.

---

[7] A more detailed map of CO emission, by Dame *et al.* (2001), is shown in the Supplementary Material.

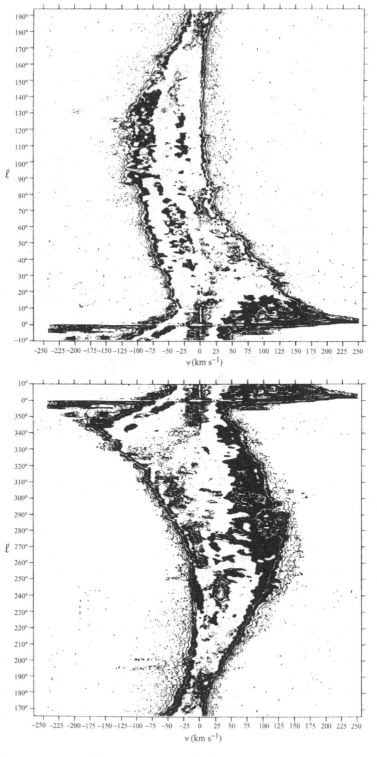

Figure 14.9. The spectrum of H I emission along the plane of the Galaxy. The velocity–longitude contours represent the neutral hydrogen intensities along the galactic equator, $b = 0°$ (Burton 1988).

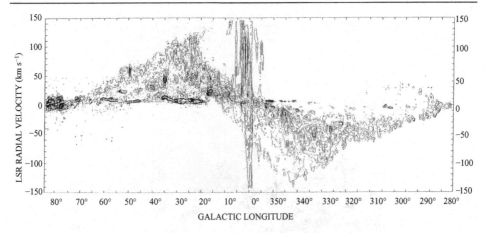

Figure 14.10. The spectrum of CO emission along the plane of the Galaxy, integrated over galactic latitudes −3.25° to +3.25° (Dame *et al.* 1987).

Barred spirals (types SBa, SBb, and SBc in the Hubble–Sandage scheme) form a particular morphological class, and nearly half of all spiral galaxies may possess a bar, according to a study by Kormendy and Norman (1979). Milky Way maps of molecular line emission, such as Figure 14.11, show a concentration about 3° long at the centre containing about 10% of the total molecular gas content of the whole Galaxy. Velocities are available from CO and other molecular measurements and from OH maser stars that belong to the older population (mostly Mira-type variable stars). Within a bar there must be large radial components of velocity; these are indeed observed, showing that the Milky Way has a bar, with its long axis extending to 2.4 kpc and making an angle of about 20° to the Sun–centre line, the nearer end being in the northern direction (Blitz *et al.* 1993).

A common feature of many spiral galaxies is a *warp* in the otherwise flat galactic plane, seen at large distances from the centre. This is observed in our Galaxy and others through H I line emission; a clear example is the warp in NGC 3741 (Gentile *et al.* 2007). These features are due to the gravitational attraction of other galaxies. An analysis by Binney (1992) showed that, like the Magellanic Stream, the warp in our Galaxy may have been due to the gravitational attraction of the LMC in a close encounter with the Milky Way.

The 21 cm hydrogen line from the Galaxy can be detected over the whole sky but mapping it with good angular resolution is a major task, which has only recently been achieved by the HI4PI collaboration (Ben Bekhti *et al.* 2016), which combines observations with the 64 m Parkes and 100 m Effelsberg telescopes, each using multiple beams and multichannel spectrometers, producing a huge data base for research into galactic structure.

## 14.4 The Galactic Centre Region

The dynamics of the central regions of the Milky Way are dominated by the gravitational field of the bar, which contains a mass of $(1-3) \times 10^{10}$ $M_\odot$. The effect of the black hole, which has a much smaller mass (Section 14.5 below), is only important close to the centre.

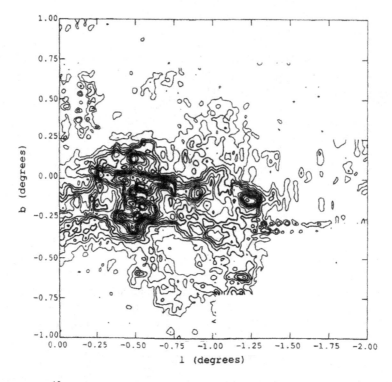

Figure 14.11. The $^{12}$CO, $J = 1-0$ emission from the central molecular zone of the Milky Way. The line emission is integrated over the velocity range $-60$ to $-190$ km s$^{-1}$ (Uchida *et al.* 1994).

Several features of the velocity maps presented in Figures 14.9 and 14.10 have been interpreted as discrete entities; for example an anomalous feature which crosses the galactic centre with a systematic velocity of $-55$ km s$^{-1}$, and extending from longitude 338° has been interpreted as a tangent point to an arm with radius 3 kpc: this was referred to as the '3 kpc expanding arm', but it should now be regarded as part of the bar. At larger distances from the centre, the 10–20 km s$^{-1}$ asymmetry of velocities between positive and negative galactic longitude is probably also an expression of the effect of the quadrupole gravitational component of the bar.

Within 2° of the centre there is a more clearly defined feature, the central molecular zone (CMZ), part of which is seen in the CO line intensity map of Figure 14.11. Within this feature there are markedly high velocities, far greater than the normal galactic rotational velocity.

The CMZ is a ring with radius 180 pc containing a high concentration of molecular species; it is comparable with the giant molecular clouds found elsewhere in the Galaxy, but with higher density and temperature (typically 70 K). A striking feature in the intensity map is the extenson to latitude $b = -0.8°$, which appears to be part of a ring with radius 75 pc.

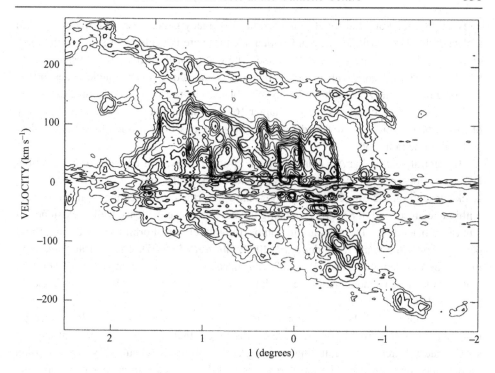

Figure 14.12. Velocities observed in $^{12}$CO in the galactic centre region, from a cut across the central molecular zone at galactic latitude $|b| \leq 0.1°$. Contours are at 1 K intervals (Binney *et al.* 1991).

High velocities unrelated to the general galactic rotation are observed in the molecular gas in the CMZ. Figure 14.12 shows the CO spectrum against galactic longitude in a cut across the CMZ at galactic latitude $b = +0.333°$, with velocities extending from $-135$ km s$^{-1}$ to $+165$ km s$^{-1}$. Structures which may be derived from these plots are described in Morris and Serabyn (1996). The CMZ has been extensively studied at millimetre wavelengths, for example by APEX (Atacama Pathfinder Experiment) and SMA (Sub-Millimetre Array), combining single-beam and interferometer techniques in mapping and spectrometry (Kauffmann 2016).

Within the CMZ and approaching the galactic centre itself, the dynamics are increasingly determined by the gravitation and energetics of the black hole.

## 14.5 The Black Hole at the Galactic Centre

The central region of the Milky Way is highly disturbed, as seen for example in surveys by Altenhoff *et al.* (1979), Handa *et al.* (1987), and Law *et al.* (2008). The strong central source, Sgr A*, has long been known as a variable non-thermal radio source, probably at the dynamic centre of the Milky Way, and has now been established as a black hole (see a review by Melia and Falcke 2001). The nuclei of many galaxies are believed to contain a black hole; as we will describe in Chapter 16, these are the centres of some of the most dramatic radio phenomena in the Universe. The centre of the Milky Way system, while not showing the violent events exhibited by quasars and active galactic nuclei, does indeed contain a black hole. Although our Milky Way black hole is a weak radio source compared with the quasars which we discuss in Chapter 16, it is surrounded by an astonishing variety of phenomena which are revealed by radio observations both in the continuum and in molecular line radiation.

The central source, Sgr A*, is hidden from optical observation by massive obscuration, but infrared observations with the ESO VLT telescopes (Gillessen *et al.* 2009) and the 10 m Keck telescope at 2.2 μm (Ghez *et al.* 2005) have nevertheless plotted the orbits of 28 stars within one arcsecond of Sgr A*, producing the most decisive evidence for the existence of a massive black hole. The stars are in compact highly elliptical orbits, one with an orbital period of only 6.5 years (Schödel *et al.* 2002; Ghez *et al.* 2005). The central gravitating mass is established at $(4.3 \pm 0.4) \times 10^6$ solar masses, and it is so compact that it must be a black hole. This is undoubtedly the black hole whose existence was surmised by Lynden-Bell and Rees (1971).

The angular size of the radio source Sgr A* has been measured by VLBI at a range of wavelengths (Figure 14.13). At wavelengths longer than 3.5 mm the apparent angular size is due to interstellar scattering, varying as $\lambda^2$, a clear indicator of scattering in an ionized medium; however, the measurements at 1.4 and 1.25 mm appear to resolve the source (Doeleman *et al.* 2008). At the shortest wavelength the linear size inferred from these observations corresponds to only four Schwarzschild radii for a black hole of four million solar masses. The combination of the infrared astrometry and the millimetre-wave VLBI observations (Section 11.4.1) establishes beyond a doubt that Sgr A* is a black hole.

## 14.6 The Spectrum of the Galactic Continuum

Jansky's first observations showed that the background radiation at 20 MHz had to be associated with the Galaxy, but could not be coming from stars such as the Sun. This is synchrotron radiation. Figure 14.14 (Cane 1979) shows the average brightness temperature at high galactic latitudes for frequencies below 1 GHz. Above 10 MHz the spectrum is well fitted by a power law of the form

$$T_b = 3 \times 10^5 \left(\frac{\nu}{10\ \mathrm{MHz}}\right)^{-2.55} \quad \mathrm{K}. \tag{14.7}$$

Figure 14.14 also shows the specific intensity $I_\nu$ (related to the brightness temperature by the equation $I_\nu = 2kT/\lambda^2$; see Chapter 2), fitted by a power law of the form

$$I_\nu = 9 \times 10^{-21} \left(\frac{\nu}{10\ \mathrm{MHz}}\right)^{-0.55} \quad \mathrm{W\ m^{-2}\ Hz^{-1}\ sterad^{-1}}. \tag{14.8}$$

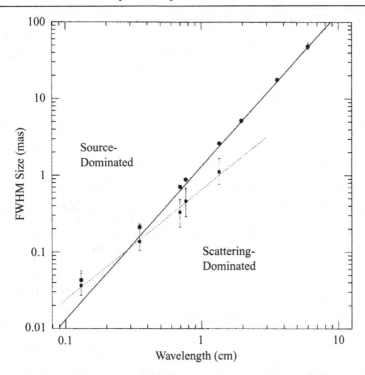

Figure 14.13. Wavelength dependence of the angular size of the radio source Sgr A* (Doeleman *et al.* 2008). The sizes measured at wavelengths 3.5 mm and 1.25 mm lie above the extrapolated scattering line, showing that the source itself is resolved at these short wavelengths. The square points and the dotted line represent the deduced source diameter in the absence of scattering.

The steep-spectrum synchrotron rises to a brightness temperature of about $1.6 \times 10^7$ K at 1 MHz. Above 1 GHz the spectrum continues to fall; above 10 GHz the synchrotron radiation falls below the 3° cosmic background (see Chapter 17 for details).

The mean index $\alpha$ (which conventionally refers to specific intensity) is $-0.55$ in this spectral region.[8] The spectrum is distinctly curved, changing smoothly from a spectral index of $-0.4$ at lower frequencies to $-1.0$ above 1 GHz (Reich and Reich 1988; Davies *et al.* 2006). Below 3 MHz the high latitude spectrum falls steeply; this is obviously an interesting, although badly observed, region of the spectrum. Unfortunately the terrestrial ionosphere makes it very difficult to observe at such low frequencies, while satellite observations from above the atmosphere necessarily have very low angular resolution and are not much help in sorting out the reason for the dramatic turnover in the spectrum.

Among the discrete galactic sources that are superimposed upon the galactic continuum there are many compact sources such as stars (Chapter 13) and pulsars (Chapter 15); these contribute little to the measured sky brightness. Supernova remnants and H II regions, on the other hand, show prominently in the continuum maps, mainly at low galactic

---

[8] Note that in many earlier texts the opposite sign is used, so that $I_\nu \propto \nu^{-\alpha}$.

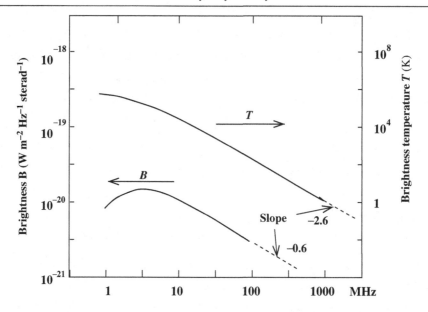

Figure 14.14. Brightness temperature and specific intensity of the Galaxy at high galactic latitudes (after Cane 1979).

latitudes. The supernova remnants exhibit non-thermal spectra and typically have high surface brightness and spectra that follow an inverse power law with frequency, with spectral indices not greatly different from the galactic background. The surface brightness can be high; the youngest supernova remnants have surface brightnesses in excess of $10^{-18}$ W m$^{-2}$ Hz$^{-1}$ ster$^{-1}$ at 100 MHz, corresponding to a brightness temperature of 300 000 K.

The H II regions are strong thermal sources: they are ionized gas clouds, heated and ionized by the ultraviolet radiation from young stars. Their temperature is limited to below 10 000 K by the efficient cooling effect of forbidden-line radiation from ionized oxygen. The radio spectrum of an H II region is that of a plasma which may be optically thick or thin. The Orion Nebula, M 42, is a typical example of a moderate-size H II region; its low frequency flux density is nearly proportional to $\nu^2$, following the Rayleigh–Jeans law closely. This means that the brightness temperature is constant (at about 9000 K), the plasma is optically thick, and the observed brightness temperature is a good measure of the kinetic temperature. At higher frequencies the spectrum is nearly flat, resembling the classical optically thin bremsstrahlung spectrum discussed in Chapter 2. If the spectrum could be expanded, it would show discrete spectral lines which are the hydrogen atom recombination lines.

At low frequencies, optically thick H II regions obscure the more distant bright non-thermal sources and reduce the brightness temperature on the plane. This can be seen in a survey of the northern sky at 22 MHz by Roger *et al.* (1999). The more widespread fall in the spectrum below 3 MHz, seen in Figure 14.14, may be due to absorption in more local ionized hydrogen, subtending a larger solid angle; this is examined further in Section 14.12.

Delineation of the sources of non-thermal radiation in the disk is a difficult problem owing to this absorption at low frequencies and the comparative brightness of the thermal sources at high frequences.

## 14.7 Synchrotron Radiation: Emissivity

Interpreting the synchrotron radiation of the Galaxy in terms of electron energies and magnetic field strength requires a three-dimensional model. In other spiral galaxies we see a thin disk containing the spiral arms with, in many cases, an extended halo which is apparent when the galaxy is seen edge on. We make the same distinction between disk and halo for the radio emission from our Galaxy, following Beuermann *et al.* (1985) who distinguished between a thin disk, containing a mixture of ionized and neutral hydrogen with supernova remnants, and a thick disk which merges smoothly with the high-latitude radiation. Radiation from the thick disk, or halo, comprises 90% of the total galactic emission at 408 MHz, which is $9 \times 10^{21}$ W Hz$^{-1}$. The thick disk has a full equivalent width of 2.3 kpc in the vicinity of the Sun (galactic radius $R = 8$ kpc), increasing to 6.3 kpc beyond $R = 12$ kpc. The observed brightness temperature $T_b$ at 408 MHz can now be assigned to an emissivity which contributes 7 K kpc$^{-1}$ in the thick disk at $R = 8.5$ kpc, increasing to 31.5 K kpc$^{-1}$ at $R = 4$ kpc (see Reich and Reich 1988). We now follow the analysis of synchrotron radiation in Chapter 2 to obtain the flux of cosmic ray electrons and the strength of the magnetic field.

There have been some reasonably direct measurements of the strength of the general galactic magnetic field (Section 14.10), but these have been mainly confined to the ionized regions of the thin disk. Again, there have been measurements of the flux of electrons at cosmic ray energies arriving at the Earth, but this may not be representative of the flux in the thick disk. The only measure for both the magnetic field and the electron flux in the thick disk is from the synchrotron emissivity itself. The radio emissivity depends on both quantities, but they can be separated if it can be assumed that the energy density of cosmic rays is in equilibrium with the energy density of the magnetic field. This assumption of *equipartition* is also made in the interpretation of other radio sources, and we briefly outline the analysis.

Note that the assumption of equipartition is a common hypothesis, but is not supported by rigorous theory. Nevertheless, it may well be a valid assumption if quasi-equilibrium conditions hold.

From Eqn 2.50 the total emitted power $P$ from $N$ electrons with relativistic energy $\gamma$ in a field $B$ varies as

$$P \propto N\gamma^2 B^2. \qquad (14.9)$$

We use the approximation that radiation at frequency $\nu$ originates in electrons with relativistic energy $\gamma$, where

$$\nu \propto \gamma^2 B. \qquad (14.10)$$

Equating the energy densities $E_{cr}$ of the cosmic rays and of the magnetic field $E_B$ then gives a value for the magnetic field which depends on the radiated power $P$ at frequency $\nu$ as (see Miley 1980):

$$\mathcal{B} \propto \left(\frac{P}{\nu}\right)^{2/7}. \tag{14.11}$$

Miley applied this to the various components of radio galaxies, emphasizing the assumptions in the analysis; he found for example fields of $10^{-5}$ gauss in hot spots. For the Milky Way, Heiles (1995) used the volume emissivity of the Galaxy in terms of brightness temperature at 400 MHz as 7.3 K kpc$^{-1}$, to derive a magnetic field strength of 7.4 microgauss. Note that in Eqn 14.11 this is a moment average which is weighted towards higher values.

It is gratifying to find that the value of the field found in this way for the Milky Way agrees well with other measurements, which we discuss in Section 14.10. The assumption of equipartition does not, however, apply in some other galaxies, where the magnetic field has been found from Faraday rotation measurements to be an order of magnitude higher than the equilibrium value. The calculation of the total energy density may nevertheless be useful, since this does not depend critically on exact equipartition. Spencer (1996) showed that in a typical discrete source the total energy varies by less than an order of magnitude when the ratio of magnetic and particle energy is between 0.01 and 100.

## 14.8 The Energy Spectrum of Cosmic Rays

The electrons responsible for the galactic synchrotron radiation mainly have energies in the range 1–10 Gev. They can be observed directly as cosmic ray electrons from balloon- and spacecraft-borne spectrometers, for example by the spacecraft PAMELA (Menn *et al.* 2013). At low energies the flux may be reduced by the deviation in the solar magnetic field (Moskalenko and Strong 1998), but the energy spectrum of cosmic rays can be extended to extremely high energies by observations of extensive cosmic ray showers produced by nuclear collisions in the high atmosphere. The most abundant high energy particles are protons. Helium nuclei have a flux that is about an order of magnitude less, while cosmic ray electrons have a flux that is about two orders of magnitude lower than the proton flux. Heavier nuclei are also present, but with still lower fluxes.

The mechanism by which high energy nuclei and electrons are accelerated is partly understood, starting with the work of Fermi (1949). Recognizing that the ISM is permeated by magnetic fields, and that the medium is not uniform but clumped into clouds, Fermi showed that charged particles could be accelerated by colliding with these magnetized clouds. In the collision the cloud acts as a body of enormous mass, and its random motion gives it an enormous effective temperature. By the same second-order collisional process by which gases come to equilibrium, the particles increase in energy until the magnetic field cannot prevent them from passing through the cloud. The shock waves of supernovae expanding into the magnetized ISM can also accelerate particles, this time as a first-order process. A survey of cosmic ray techniques and knowledge, together with references, is given in Longair (1994).

Above an energy of about 1 GeV the flux of particles reaching the Earth is reliably measured, with little influence from the magnetic field entrained in the solar wind. Between 5 GeV and 50 GeV the flux of electrons follows a simple power-law spectrum $S = S_0(E/E_0)^{-\beta}$, with index $\beta \approx 3.1$; the spectrum steepens at higher energies. The spectral index $\beta$ for electrons with energies around $10^{9-10}$ eV is estimated at 2.5. The theory of synchrotron radiation, outlined in Chapter 2, shows that radiation from cosmic ray electrons with such an energy power law is expected to have a power-law spectrum, with index $\alpha = (1 - \beta)/2$, that is $-0.75$. This agrees well with the observed spectral index near the galactic pole, $-0.4$ at low and $-1.0$ at high radio frequencies (Section 14.6). The curvature of the radio spectrum may indicate that the spectral index of the electron energy spectrum continues to fall at lower energies, where no direct measurements are possible, but note that the observed radio spectrum at low frequencies is probably affected by free–free absorption in ionized hydrogen.

The remarkably good agreement between the values of magnetic field found from the assumption of equipartition and using the measured electron cosmic ray flux shows that the energy densities of the field and the cosmic rays are similar (although this may not be so for other galaxies). We discuss below the configuration of the fields in the Milky Way and other galaxies, which comprise both organized and turbulent components.

## 14.9 Polarization of the Galactic Synchrotron Radiation

The observation of a high degree of polarization in the metre-wave galactic background provided the final proof that synchrotron radiation is the source of the continuum radiation. The polarization of the synchrotron radiation from most of the sky has been mapped with high angular resolution at 1.4 GHz (Uyaniker *et al.* 1999; Landecker *et al.* 2006). The angular resolution is sufficient to distinguish practically all the discrete sources; these can then be removed from the maps, leaving a map of polarization which shows a combination of large-scale order and small-scale complication. The level of polarization is generally no more than 10%, and careful calibrations are needed to keep the level of spurious polarization introduced by the antenna and receiver to below 1%. As described in Appendix 4, the characteristics of the telescope must be evaluated as a Mueller matrix; an example is the survey by Wolleben *et al.* (2006), using a 26 m reflector to survey the northern sky at 1.4 m wavelength.

From Chapter 2, the degree of linear polarization of synchrotron radiation in a well-ordered field should be up to 70%. Such a high degree of polarization would normally be reduced by the superposition along a line of sight through the Galaxy of sources with different polarization position angles; in several directions, however, the magnetic field is sufficiently well ordered for the polarization to exceed 10%. An excellent example, shown in Figure 14.15, is centred on $l = 140°, b = 10°$ (Brouw and Spoelstra 1976). The two maps show observations of the same region at 1411 MHz and at 465 MHz; the highly polarized region is more disordered at the lower frequency, and the position angles of the vectors have opened out like a fan. This is due to Faraday rotation in the ionized interstellar medium; at the centre of the polarized region the line of sight is perpendicular to the magnetic field, and

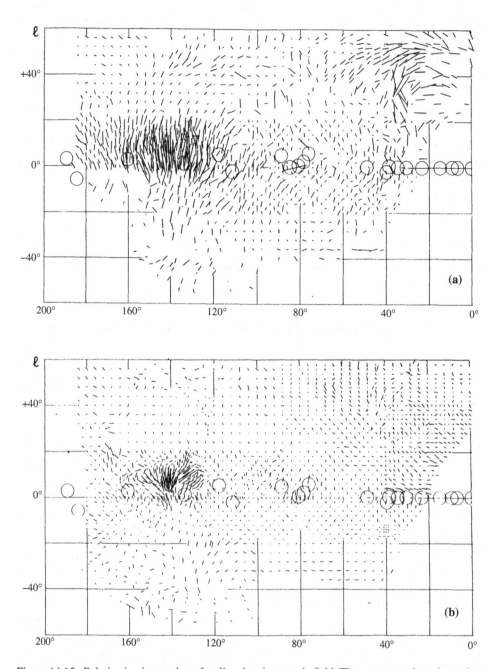

Figure 14.15. Polarization in a region of well-ordered magnetic field. The two maps show the region centred on $l = 140°$, $b = 10°$, observed at (a) 1411 MHz, and (b) at 465 MHz (Brouw and Spoelstra 1976).

on either side there is a component along the line of sight, rotating the vectors in opposite directions. At the edges of this region the degree of polarization is reduced because of the Faraday rotation occurring within the source; this is often a more important reason for the low degree of polarization in a synchrotron source than the topology of the field itself.

## 14.10 Faraday Rotation: the Galactic Magnetic Field

There is detailed structure in the polarized galactic radiation, even at high galactic latitudes and away from the galactic centre. The magnetic field apparently has ordered and randomly oriented components which are approximately equal, and the observed polarization is reduced and confused by the combination of the random superposition of Faraday rotation and differently orientated sources. Burn (1966) showed that if $N$ equal sources are seen in the same line of sight, with random orientations, the observed polarization is reduced by a factor of $N^{1/2}$; he also analysed the 'back-to-front' effect of Faraday rotation in a uniform emitting slab. If there is a total rotation measure $RM$ through the emitting slab then the polarization is reduced by a factor sinc$(RM\lambda^2)$. The polarization of the galactic background is so reduced by this effect that the large-scale structure is not apparent from polarization measurements alone. A detailed analysis of depolarization effects can be found in Sokoloff *et al.* (1998).

A better view of the large-scale structure is obtained from the Faraday rotation of polarized radiation from extragalactic sources. Faraday rotation is proportional to the product of electron density and the magnetic field component along the line of sight; it is also dispersive, so that it can be measured as a difference in polarization position angle at adjacent frequencies (Chapter 2). It obviously includes rotation within the extragalactic source itself, but an average over several sources can be used to find the galactic component in a region of sky. The distribution of rotation measures of extragalactic sources is shown in Figure 14.16(a), where positive and negative values are distinguished by solid and open circles (Han *et al.* 1999). A remarkable degree of organization emerges: for example, almost all values of $RM$ between galactic longitudes $0° < l < 135°$, latitudes $-45° < b < 0°$ are negative, while those in the opposite direction (near $l = 225°$) are positive. This is a strong indication of an organized field direction, extending well outside the spiral arm structure of the galactic plane. This component of the field is toroidal, in opposite directions above and below the plane.

Closer to the galactic plane the magnetic field has a more complex structure, which is associated with the spiral arms. Here the rotation measures of extragalactic sources are more difficult to interpret, as they are integrated along a line of sight which may cross several arms. There is less such confusion in the rotation measures of pulsars, which are located within the Galaxy and involve shorter paths.

The Faraday rotation of the highly polarized radio emission from pulsars provides our most detailed information on the magnetic field in the thin disk of the Galaxy. Along the line of sight to a pulsar, the radio pulses are delayed in the ionized interstellar medium; this delay can be measured by the difference in pulse arrival times at different radio frequencies. This dispersion in travel time is proportional to the total electron content along the line of sight. A model of the electron distribution can then give the distance of the pulsar. Since Faraday rotation is proportional to the product of electron density and the magnetic field component along the line of sight, the ratio of rotation measure (*RM*) to dispersion measure (*DM*) gives $\mathcal{B}_L$, the component of the interstellar magnetic field along the line of sight. In the usual units for *RM* (rad m$^{-2}$) and *DM* (pc cm$^{-3}$),

$$\mathcal{B}_L = 1.232(RM/DM) \quad \mu G. \tag{14.12}$$

The field obtained in this way is an average along the line of sight, and the average is weighted by the local electron density. It should also be remembered that the ionized regions of the galactic plane comprise only about 10% of the volume of the thin disk, containing a magnetic field that may not be representative of the whole of the disk. Nevertheless the surveys of pulsar rotation measures which are now available are invaluable in delineating the magnetic field structure and its relation to the spiral arms.

Figure 14.16(b) shows the distribution of measured *RM*s for pulsars more than 8° away from the galactic plane; this shows the same simply organized high-latitude field and agrees well with Figure 14.16(a)

Closer to the galactic plane, the lines of sight to extragalactic sources pass through a system of spiral arms, in which the magnetic field has a more complicated structure than the toroidal field above and below the galactic plane. Here the dispersion measures of pulsars are invaluable because of their shorter lines of sight. Figure 14.17 (Han 2007) shows a compilation of rotation measures of low-latitude pulsars, projected onto the galactic plane, with rotation measures from extragalactic sources around the periphery. The diagram is centred on the galactic centre, and the Sun is located towards the top; galactic longitudes are shown. The locations of three spiral arms are shown; the evidence for these depends mainly on the pulsar rotation measures. On the basis of a similar compilation, Brown *et al.* (2007) showed that the magnetic field follows the spiral arms in the inner Galaxy, clockwise in the Sagittarius–Carina arm and counterclockwise in the Scutum–Crux arm. Similar reversals have been proposed for the outer arms; Figure 14.18 is a sketch of the arms as delineated by electron density, with the magnetic field directions shown by arrows (Heiles 1995).

In addition to the rotation measures described above, evidence for the magnitude and structure of the galactic magnetic field is obtained from the intensity and polarization of the galactic synchrotron radiation, and, within some discrete sources, from Zeeman splitting of maser line radiation. Our location within the galactic plane requires a synthesis of all the evidence to proceed with a model involving a toroidal field out of the plane and a spiral arm field within it. The connection between these two systems is not understood.

There are both large- and small-scale deviations from this pattern. On small scales within the arms there is a random component of magnitude equal approximately to the organized field; this is the same random field as is seen in the thick disk as structure in the polarization of the synchrotron radiation. Between the arms, where the large-scale field is small, the random field predominates.

Most estimates of the strength of the galactic magnetic field are in the range 3–8 μG, comprising roughly equal turbulent and organized components. The organized component is larger towards the galactic centre, falling to 2 μG near the Sun. The origin of the field

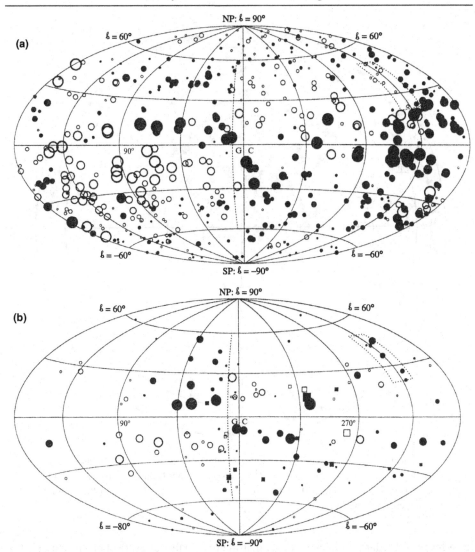

Figure 14.16. Rotation measures of (a) extragalactic sources (note the division between positive and negative values indicated by the dotted line at $l = 8°$); (b) pulsars at high galactic latitudes $|b| > 8°$. The area of the symbols is proportional to $|RM|$, ranging from 5 to 150 rad m$^{-2}$. The solid and open circles represent positive and negative $RM$ values respectively. (Courtesy of C. Han and the International Astronomical Union.)

is much debated. A 'primordial' field, built in from the early Universe, may play a role but that is still a conjecture, unsupported by observations or rigorous theory. Possibly the large-scale field is derived from stellar magnetic fields via supernova explosions. The development of the organized field would then be due to a dynamo action of the large-scale differential velocities in the Galaxy and the turbulent field would then be degrading to large-scale organization, the reverse of many other turbulent phenomena (see Beck 1996

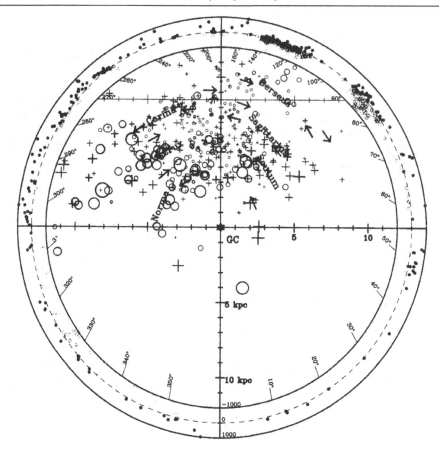

Figure 14.17. Rotation measures of pulsars and extragalactic sources with galactic latitude $|b| < 8°$. The size of the symbols is proportional to $RM^{1/2}$; the plus symbols have positive $RM$ and the circle symbols have negative $RM$ (Han 2007). A more recent version from Han *et al.* (2018) can be found in the Supplementary Material.

for a review). On larger scales it seems that major dynamical events both in the galactic centre and, at larger galactic radii, in supernova explosions, may drive the field into quite different configurations. This is most readily seen in other galaxies, where we have a distant perspective rather than the confusing view from inside. In many other galaxies the overall organization of the magnetic field is evident from the polarization of synchrotron radio emission. Figure 14.19 shows the orientation of the large-scale magnetic field in M51, where the field lines follow the spiral structure; in contrast the field lines in NGC 4631 (Figure 14.20) show a radial pattern which is believed to be due to an outward flow of ionized interstellar gas (Golla and Hummel 1994). Comprehensive reviews of magnetic fields in galaxies have been published by Kronberg (1994), Wielebinski and Krause (1993), and Beck (2016); see also Wielebinski and Beck (2005). See also Han (2017) for a review of interstellar and intergalactic magnetic fields .

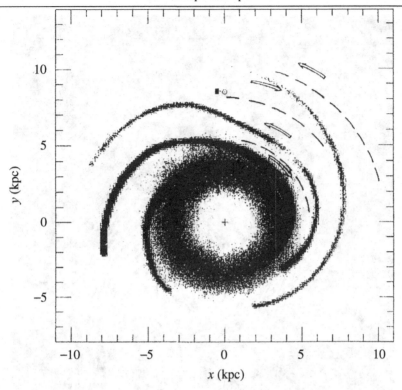

Figure 14.18. Magnetic field directions (arrows) and reversals (broken lines; Rand and Lyne 1994), and electron density $\langle n_e \rangle$ (grey scale; Taylor and Cordes 1993) (from Heiles 1995).

## 14.11 Loops and Spurs

The structure which can be seen away from the plane of the Galaxy, especially at low radio frequencies as in Figure 14.3, has no counterpart in stars or any other visible object. The most obvious feature in this high-latitude structure is the North Galactic Spur, which leaves the plane at $l = 30°$ and reaches to above $b = 60°$. It is best described as part of an approximately circular loop, with diameter $116°$, centred on $l = 329°$, $b = 17°$. This is also known as Loop I; two similar features are designated as Loops II and III. All are believed to be the outer shells of supernova remnants.

The radio emission from Loop I has the steep spectrum characteristic of synchrotron radiation. It is strongly linearly polarized; as expected, the degree of polarization is greater at higher frequencies where the effects of depolarization are least. The magnetic field direction is mainly along the line of the loop, indicating that the field has been swept up by an expanding shell. There is, however, hydrogen-line emission from the outer part of the shell, indicating an enhanced density of cool gas, and also enhanced X-ray emission from within the shell, indicating a very-high-temperature region. This suggests that hot gas within a bubble is compressing the cold interstellar medium as it expands (Heiles *et al.* 1980), and that this region already has an abnormally high density. The hydrogen

Figure 14.19. The orientation of the large-scale magnetic field in M51, from observations of the polarization of radio emission at high radio frequencies (Neininger 1992).

line radiation shows Zeeman splitting, giving a value of 5.5 μG for the magnetic field (Verschuur 1989). The effects of this field can also be seen in Faraday rotation of radio sources beyond the loop. Loop I is now regarded as a *superbubble*, the effect of a series of supernova explosions in an active region; the prominent shell is due to the most recent of these explosions (Wolleben 2007). The distance of the shell is probably about 100–200 pc. It is not seen below the plane; this is presumably due to the location of its supernova progenitor above the plane, and to a sufficient concentration of mass in the plane to halt expansion in this direction.

These loops are evidently local and comparatively short-lived features of the Galaxy. Supernova explosions may nevertheless be an important source of energy in the ionized gas which extends more than 1 kiloparsec from the plane, and the effects of an individual explosion may last as long as $10^6$ yr. The only direct evidence of the supernova itself would be the presence of a neutron star in the right place and with the right age. Such evidence

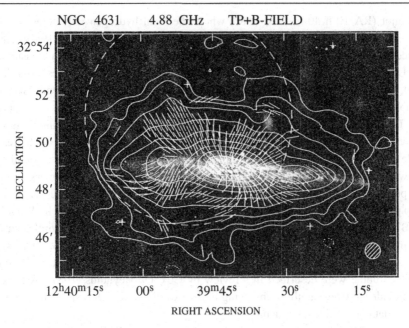

Figure 14.20. The radial magnetic field around NGC 4631 (Golla and Hummel 1994).

is given by the gamma-ray source Geminga. This is one of the three brightest gamma-ray sources in the sky, the other two being the Crab and Vela pulsars. Almost 20 years after the discovery of Geminga, it was found to be a strong X-ray source which is pulsating with a period of 237 ms; the same periodicity was then found in the gamma-ray observations. Although no radio pulsar can be seen, Geminga is evidently a neutron star. Its age is about 300 000 years, as determined from its slowdown rate. It is visible as a very faint object with a very large proper motion; even at a distance of only 100 pc its speed must be over 100 km s$^{-1}$. Its position and age are at least consistent with its identification as the collapsed centre of the explosion which produced Loop I.

## 14.12 The Local Bubble

The loops and spurs of the preceding section are obvious examples of the inhomogeneity and variety of the interstellar medium. Less obvious, although probably very similar, is a bubble-like structure which surrounds the Sun. This 'Local Bubble' is a region with dimensions of order 100 pc, with a comparatively low density of hydrogen, around $5 \times 10^{-3}$ cm$^{-3}$, and a high temperature, around $10^{6}$ K. The product of density and temperature is a pressure; this is reasonably in equilibrium with the surrounding regions.

Above galactic latitude $60°$, the column density of neutral hydrogen drops more rapidly than it should for a plane-parallel atmosphere, and this may be evidence for most of the high-latitude gas being ionized (see Dickey and Lockman 1990 for a summary of the evidence). In particular, there is a high-latitude 'hole' in the ISM, located in the constellation

Ursa Major (RA 10 h 40 m, Dec +58°) where the local hydrogen falls below a column density of $4 \times 10^{19}$ cm$^{-2}$. This region was studied in detail by Jahoda *et al.* (1990). The best evidence for the Local Bubble comes from combining soft X-ray data with measurements of the Ly$\alpha$ absorption in nearby O and B stars. The X-ray emission comes from the hot, ionized, gas, while the absorption depends on the line-of-sight density of neutral hydrogen. The bubble boundary is defined by a column density of $10^{19}$ hydrogen atoms cm$^{-2}$ (Paresce 1984). The temperature is given by the X-ray brightness. Its origin is uncertain; it may have been formed by outflowing hot gas from a group of young stars, or it may be the result of a single supernova explosion. If it is a supernova remnant, its age would be around $10^5$ yr and it would have occurred within 100 pc of the Sun. We have already remarked that the X-ray pulsar Geminga might have originated in a supernova which created Loop I; it is equally possible instead that this was the origin of the Local Bubble (see Gehrels and Chen 1993; Bignami *et al.* 1993). Only a general conclusion can be drawn: the complex structure of the interstellar medium, as seen in the radio maps of Figures 8.3 and 8.4, was at least partly formed by the transient effects of supernova explosions. The extent and outline of the Local Bubble were described in a review by Cox and Reynolds (1987). A large new body of evidence from satellites observing in the extreme ultraviolet will provide far more extensive data on such local structures.

## 14.13 Further Reading

*Supplementary Material* at www.cambridge.org/ira4.

Arp, H. 1966. Atlas of peculiar galaxies. *Astrophys. J. Suppl.*, **14**, 1.

Beck, R. 2016. Magnetic fields in spiral galaxies. *A & A Rev.*, **4**, 24.

Beck, R., and Wielebinski, R. (eds.) 2005. *Cosmic Magnetic Fields*. Lecture Notes in Physics, vol. 664. Springer.

Buta, R. J. 2013. Galaxy morphology. In: Oswalt, T. D., and Keel, W.C. (eds.), *Planets, Stars and Stellar Systems, Volume 6: Extragalactic Astronomy and Cosmology.*

Cox, D. P. 2005. The three-phase interstellar medium Revisited. *Ann. Rev. Astron. Astrophys.*, **43**, 337–385.

de Vaucouleurs, G., and de Vaucouleurs, A. 1964. *Reference Catalogue of Bright Galaxies.* Texas University Press.

Fall. S. M., and Lynden-Bell, D. (eds.) 1981. *The Structure and Evolution of Normal Galaxies*. Cambridge University Press.

Longair, M. S. 1994. *High Energy Astrophysics*, Volumes I and II, 2nd edn. Cambridge University Press.

Pacholczyk, A. G. 1970. *Radio Astrophysics: Nonthermal Processes in Galactic and Extragalactic Sources*. Series of Books in Astronomy and Astrophysics, Freeman.

Sandage, A. 1961. *The Hubble Atlas of Galaxies*. Carnegie Institute of Washington.

Sandage, A. 1994. *The Carnegie Atlas of Galaxies*. Carnegie Institute of Washington.

Verschuur, G. L., and Kellermann, K. I., 1988. *Galactic and Extragalactic Radio Astronomy*. Springer.

# 15

# Pulsars

## 15.1 Neutron Stars

In 1934 Baade and Zwicky suggested that the final stage of evolution of a massive star would be a catastrophic collapse of its core, leading to a supernova explosion and leaving a small and very condensed remnant, a neutron star. This theory was immediately successful in explaining observations of supernovae and their remnants, especially the Crab Nebula, which was identified as the remains of a supernova observed in the year AD 1054. Neutron stars seemed, however, to be hopelessly unobservable; they would be cold, only about the size of a small asteroid, and many light years from the Earth. More than 30 years later, neutron stars were discovered by both X-ray and radio astronomers. The original X-ray discoveries were intense thermal sources of radiation; these were neutron stars in binary systems, where the heating is due to the accretion of matter from a binary companion. Thermal radio emission is not detectable from neutron stars, but many radiate an intense beam of non-thermal radiation, rotating with the star and detected as a radio pulse as it sweeps across the observer. This pulsed radiation was later discovered in X-rays and gamma-rays in many pulsars. Other categories of high energy sources such as anomalous X-ray pulsars (AXPs) and soft gamma-ray repeaters (SGRs) are now known to be closely related to the radio pulsars. The intriguing and complex behaviour of pulsars not only relates to the condensed-matter physics of the star itself but also to many aspects of stellar evolution, galactic structure, and gravitational physics.

Early radio telescopes often used receivers with integration times of many seconds, too long to detect the periodic short pulses from pulsars. The discovery of pulsars (Hewish *et al.* 1968) was the result of a deliberate measurement of the rapid intensity fluctuations of radio sources due to scintillation in the solar corona. A large receiving antenna array was used, at the unusually long wavelength of 3.7 m; the sensitivity and time resolution of this system were sufficient to detect individual pulses at intervals of 1.337 s from the pulsar which we now designate PSR B1919+21 (PSR originally stood for pulsating source of radio and now generally for all pulsars; the numbers refer to its position in Right Ascension and Declination).[1] Pulsar signals are weak, and usually only detectable by adding together many

---

[1] The positions of most astronomical objects are designated in either 1950 coordinates (B) or 2000 coordinates (J). The convention is for all pulsars to have a J designation and retain the B designation only for those published prior to about 1993.

individual pulses. In searching for pulsars the technique has been to look for a long and very regular sequence of pulses, adding the pulses typically in a long effective integration time of some tens of minutes. Since the period is unknown, the signal must be detected originally by a process of Fourier transformation of the raw detected signal. When single pulses are strong enough to be detected individually, they often prove to be variable in intensity and structure; their radation mechanism is not understood.

Pulsars require large radio telescopes for their discovery and detailed study. The total known in 2017 exceeds 2500, including those discovered by their X-ray or gamma-ray emission. A full catalogue is maintained by The Australian Telescope National Facility (ATNF).[2]

This chapter provides an introduction to several aspects of pulsar research, concentrating on observations and their analysis.

### *15.1.1 Neutron Star Structure*

After the exhaustion of nuclear fuel has led to the gravitational collapse of a normal star from its equilibrium state, there are two stable condensed states short of total collapse to a black hole; these are represented by the white dwarf stars and the neutron stars. In both, a total collapse under gravitation is prevented by the pressure of degenerate matter: electrons in a white dwarf and neutrons in a neutron star. Each can exist only over a limited range of mass; adding mass to a white dwarf can lead to a collapse into a neutron star. The properties of each are determined solely by their individual masses.

The degenerate neutron gas of a neutron star behaves as though there are only static forces between the neutrons: the equation of state contains no reference to temperature. Theoretical models for neutron stars show that they can exist in only a small range of mass; the precise range depends on the uncertain equation of state but the models allow a range of 0.2 to 2.5 solar masses $M_{\odot}$ (Lattimer and Prakash 2001). A larger mass would lead to further collapse into a black hole. Several neutron star masses have been measured accurately from the characteristics of binary systems; they range from 1.2 to 2.0 $M_{\odot}$. Predicted radii lie within a small range of 10 to 12 km; observational evidence from the intensity of thermal X-rays from the surface of some pulsars agrees well with this prediction (see the review by Ozel and Freire 2016).

Figure 15.1 shows the structure of a typical model neutron star with mass 1.4 $M_{\odot}$. The outer part forms a solid crust about 1 km thick, whose density at the surface is of order $10^6$ g cm$^{-3}$, similar to that of a white dwarf. The density increases rapidly within the outer crust, which is a rigid and strong crystal lattice, primarily of iron nuclei. At higher densities electrons penetrate the nuclei and combine with protons to form nuclei with unusually large numbers of neutrons. This continues until, at a density of about $4 \times 10^{11}$ g cm$^{-3}$, the most massive of these nuclei become unstable and free neutrons appear. These neutrons form a superfluid which, as we will see from the rotational behaviour of the pulsar, can move independently of the solid crust. The core itself contains only neutrons with a small equilibrium proportion of electrons and protons; it is superconducting and superfluid. The possibility that a central solid core might exist depends on the equation of state; the existence of some

[2] www.atnf.csirc/research/pulsar/psrcat

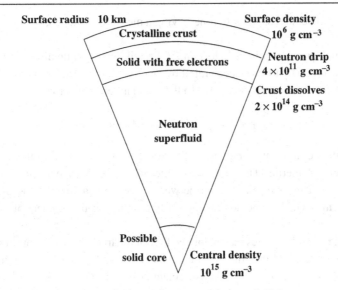

Figure 15.1. A section through a typical model neutron star, mass 1.4 $M_\odot$.

pulsars with masses close to 2 $M_\odot$ favours 'hard' equations of state in which this solid core is unlikely to exist.

Pulsars have very large surface magnetic fields, typically between $10^8$ and $10^{14}$ gauss ($10^{4-10}$ T). Both the rapid rotation and the high magnetic fields of young pulsars reflect their origin in the collapse of a stellar core in a supernova explosion; similar values would be seen if a normal star like the Sun were shrunk to the size of a neutron star, and in the process the angular momentum and the total magnetic flux were conserved. The external magnetic field, like the Earth's magnetic field, is largely dipolar.

The strong magnetic field, combined with the rapid rotation, generates very strong electric fields outside the neutron star. These fields pull charged particles from the surface, forming a dense and highly conducting *magnetosphere* (Goldreich and Julian 1969). The rotation of the magnetic field forces the magnetosphere into co-rotation with the solid star out to a large distance where co-rotation would be at the velocity of light. The magnetosphere is therefore contained in a *velocity-of-light cylinder*. Outside this cylinder, the rotating magnetic field becomes a classical dipole radiator, which carries away energy from the rotating star.

## 15.2 Rotational Slowdown

The rotational periods of the majority of pulsars lie in the range 100 ms to 3 s. *Millisecond pulsars* constitute a distinct category with rotational periods less than about 20 ms; the shortest known is 1.4 ms. Longer periods, up to 11 s, are found in the *magnetars*, a group observed mainly in X-rays containing (in 2017) only four which are known to emit radio pulses. All periods are measurable sufficiently accurately to give a rate of change; almost all are observed to be increasing with time, giving a measure of the rate of loss of rotational kinetic energy.

This rotational energy is primarily lost either directly by magnetic dipole radiation or by an outflow of energetic particles accelerated by electromagnetic induction. For a magnetic dipole the rate of loss of energy is related to the angular velocity $\omega$ by

$$\frac{d}{dt}\left(\frac{1}{2}I\omega^2\right) = I\omega\dot\omega = \frac{2}{3}M_\perp^2\omega^4 c^{-3}, \tag{15.1}$$

where $M_\perp$ is the component of magnetic dipole moment orthogonal to the spin axis and $I$ is the moment of inertia. The moment of inertia is almost independent of the mass, because there is an inverse relationship between mass and radius; $I$ is usually taken to be $3 \times 10^{44}$ g cm$^2$ (which coincidentally is nearly the same as the moment of inertia of the Earth).

This relation probably applies reasonably well even if part of the energy outflow is carried by energetic particles, and it is therefore customary to use Eqn 15.1 to assign a value of dipole moment to each pulsar from measurements of the rotation rate $\nu = \omega/2\pi$ and its derivative $\dot\nu$ (or period $P$ and $\dot P$). The dipole moment $M$ is usually quoted as a value $B_0$ of the polar field at the surface at radius 10 km for an orthogonal dipole, giving the widely used relation

$$B_0 = 3.3 \times 10^{-19}(-\dot\nu\nu^{-3})^{1/2} \quad \text{gauss.} \tag{15.2}$$

The slowdown may be expressed as the power law

$$\dot\nu = -k\nu^n, \tag{15.3}$$

where $k$ is a constant and $n$ is referred to as the *braking index*; if the slowdown follows Eqn 15.1 the index $n = 3$. If we assume that a pulsar has an angular velocity which is initially very high, and a constant magnetic field, its age $\tau$ may be found by integrating Eqn 15.3 to give

$$\tau = -(n-1)^{-1}\nu\dot\nu^{-1} = (n-1)^{-1}P\dot P^{-1}. \tag{15.4}$$

For a braking index $n = 3$, this calculated age becomes $(1/2)(\nu/\dot\nu)$; this is often quoted as $\tau_c$, the *characteristic age* of the pulsar. Given an initial rotation rate $\nu_i$, the age becomes

$$\tau = -(n-1)^{-1}\nu\dot\nu^{-1}[1 - (\nu/\nu_i)^{n-1}]. \tag{15.5}$$

A direct measurement of $n$ is attainable only if a stable value of the second differential $\ddot\nu$ can be measured. Differentiation of Eqn 15.3 gives

$$n = \frac{\nu\ddot\nu}{\dot\nu^2}. \tag{15.6}$$

A useful value of $\ddot\nu$ can be found only for young pulsars; even for these it requires a long run of timing measurements to allow for any irregular rotational behaviour, and few results have been obtained. Characteristic ages should be interpreted as actual ages only with some caution, since measured braking indices differ considerably from $n = 3$, ranging from 1.4 to 3.

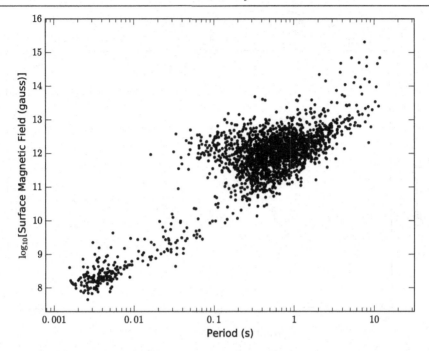

Figure 15.2. The polar magnetic field strengths of pulsars and other neutron stars. (Data from the ATNF Catalogue, courtesy of Xiaojin Liu, Jodrell Bank Centre for Astrophysics.)

## 15.3  Magnetic Dipole Moments

Figure 15.2 shows the distribution of polar field strengths $\mathcal{B}_0$ in all pulsars, calculated from measured values of $P$ and $\dot{P}$ according to Eqn 15.2. The majority, usually referred to as the *normal* pulsars, have values between $10^{11}$ and $10^{13}$ gauss. Those with long periods and the largest fields, up to $10^{15}$ gauss, are the *magnetars*; those with the shortest periods and the lowest fields, between $10^8$ and $10^9$ gauss, are the *millisecond* pulsars.

The normal evolution of pulsars in this diagram, according to the slowdown law of Eqn 15.1, should follow a horizontal track to the right, eventually reaching a remarkably well-defined boundary known as the *death line*. The progress across the diagram appears to be not so simple, and we discuss in Section 15.11 the possibility that the magnetic dipole field is not constant. It is remarkable that the distribution of millisecond pulsars in this diagram follows the same death line, even though their magnetic fields are a factor $10^4$ lower and they have followed a different evolutionary path, which we discuss later.

## 15.4  Rotational Behaviour of the Crab Pulsar

The slowdown in rotation rate of the Crab Pulsar is shown in Figure 15.3. Over 45 years, during which time the rotation rate fell from 30.2 Hz to 29.7 Hz, this pulsar completed more than $4 \times 10^{10}$ rotations; despite the rotational irregularities described below, the observations

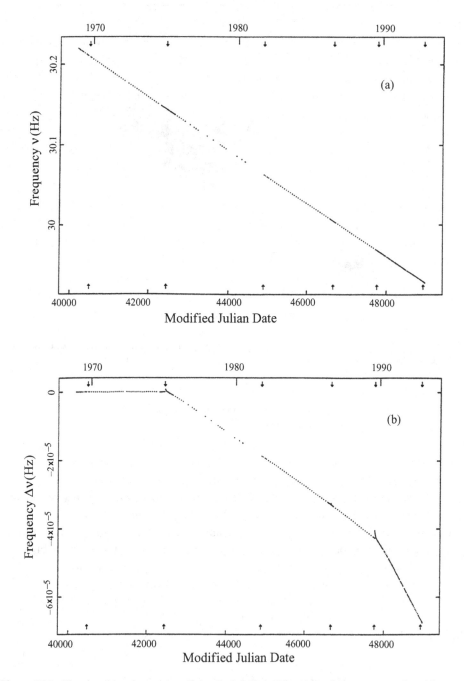

Figure 15.3. The slowdown in rotation of the Crab Pulsar: (a) as recorded over more than 45 years; (b) on an expanded scale after subtracting the initial slowdown rate; the glitches show as steps in the slope, corresponding to increases in the slowdown rate. The transient steps in frequency are insignificant on this scale (Lyne *et al.* 2015a).

give an almost complete account of all rotations over this time. The characteristic age, which is found, using Eqn 15.4, from the average slope of the plot in Figure 15.3(a), can be measured to a precision of 1% in a single day. Measured in this way the characteristic age is 1250 yr, which may be compared with the actual age of 960 yr since the supernova explosion was observed (1054 AD). The difference is a warning that Eqn 15.4 assumes a braking index $n = 3$, which differs from measured values when they are available; furthermore, present-day slowdown rates are only indications of actual age and not infallible measures of it.

The expanded plot of Figure 15.3(b), in which the mean slope is removed, allowing the vertical scale to be multiplied by 5000, shows steps in rotation rate known as *glitches*. At a glitch in the Crab pulsar, the rotation rate increases by a step of order 1 in $10^9$, typically accompanied by a step increase in slowdown rate of order 1 part in $10^4$. These steps have no significant effect on the determination of characteristic age.

A small curvature in the plot of Figure 15.3(b) provides a measurement of the value of the second differential $\ddot{\nu}$; this gives a value for the braking index $n = 2.50 \pm 0.01$. Why this differs from 3.0, as expected from the expression for magnetic dipole radiation, has been much discussed. A possible explanation is that there is a substantial loss of angular momentum in energetic material streaming out from the magnetic poles (Lyne *et al.* 2015a); see also Section 15.11.

## 15.5 Glitches in Other Pulsars

Much larger steps in rotation rate and in slowdown rate occur at glitches in some other pulsars. The Vela pulsar has a series at about three-year intervals, with steps in rotation rate $\nu$ of order 1 in $10^6$ and in $\dot{\nu}$ of order 1 in $10^2$, as shown in Figure 15.4. With non-linear recoveries between each glitch this pattern dominates the rotational behaviour of this pulsar; nevertheless there is a long-term trend in slowdown rate, which gives a measurement of $\ddot{\nu}$ and hence the characteristic lifetime (Espinoza *et al.* 2017). The glitch itself occurs within a few milliseconds; the recovery is slow, lasting many days. Other pulsars, such as B1758-23, similarly show large glitches repeating at intervals of several years, but with no recovery at each step in rotation rate (Shemar and Lyne 1996).

Espinoza *et al.* (2011) showed that glitch activity is least in both very young pulsars (like the Crab) and old pulsars. The Vela pulsar, which appears to be approximately 20 000 years old, is in the active 'middle-aged' category. Millisecond pulsars, in this context, are old and inactive; most behave like ideal clocks (apart from the small and well-characterized slowdown). Prior to 2017 only two glitches had been observed in millisecond pulsars (McKee *et al.* 2016).

### 15.5.1 Superfluid Rotation

The explanation of glitches involves some remarkable physics. The steps in rotation rate show that at least 2% of the moment of inertia is attributable to a separate component

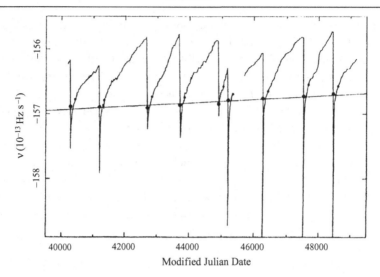

Figure 15.4. The slowdown rate $\dot{\nu}$ of the Vela pulsar over 25 years, showing a series of glitches at approximately three-year intervals, with incomplete recoveries. The long-term change in $\dot{\nu}$ marked by the straight line can be used to calculate the braking index (Lyne *et al.* 1996).

of the neutron star; the long time constant of the exponential recoveries observed after a glitch shows that this component is a superfluid. The rotation of a superfluid is abnormal; it is expressed as vortices, each of which contains a quantum of angular momentum. The area density of vortices is proportional to the angular velocity. As the rotation slows, the area density of the vortices must reduce by an outward movement. There is, however, an interaction between the vortices and the lattice nuclei, so that the outward flow is impeded by the pinning of vortices to the crystal lattice (Alpar *et al.* 1989). A glitch indicates a sudden unpinning and release of vortices, transferring angular momentum to the crust. Vortices which remain pinned to the crust take no part in the slowdown; this effectively removes part of the rotational moment of inertia and allows the pulsar to slow down faster.

## 15.6 Radio, Optical, X-Ray, and Gamma-Ray Emission from the Magnetosphere

Pulses from the Crab pulsar can be detected over an astonishingly wide spectrum, from radio through gamma-rays to optical. Over the whole of this spectrum there is a widely spaced double pulse structure (Figure 15.5). At long radio wavelengths there is also a *precursor*; other radio components appear at centimetric radio wavelengths; Hankins *et al.* (2015) identified a total of seven distinct components. Each is a cross-section of a narrow beam of radiation from a distinct location within the magnetosphere.

The strong magnetic field ($\sim 10^{12}$ gauss at the stellar surface) completely dominates all physical processes outside the neutron star. The force of the induced electrostatic field acting on an electron at the surface of a rapidly rotating pulsar like the Crab pulsar exceeds gravitation by a very large factor, which would be as much as $10^{12}$ if there were no

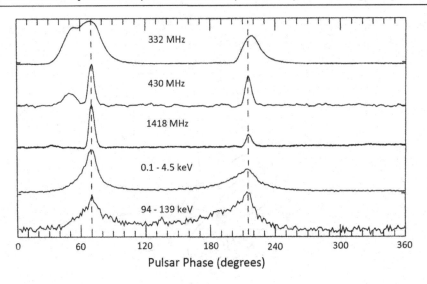

Figure 15.5. The profile of pulses from the Crab pulsar. The main pulse and interpulse are seen over the spectrum from radio to gamma-rays. The precursor component is observed only at low radio frequencies. (See Moffett and Hankins 1996 for references to the observations.)

conducting atmosphere. The magnetic field remains approximately dipolar within the light cylinder, i.e. out to a radial distance $r_c = c/\omega$. Within this cylinder the strong magnetic field allows charged particles to move along but not across the field lines. Figure 15.6 shows an equatorial region in which the dipole magnetic field lines close within the velocity of light cylinder and polar regions where field lines extend through the light cylinder. These open field lines originating near the poles allow energetic particles to escape; these particles are then able to energize a surrounding nebula such as the Crab Nebula (see Chapter 13).

The closed equatorial region contains a plasma which, unusually in astrophysics, has a high net charge. The net charge density generates a static field $\mathcal{E}$ which cancels the induced electric field due to the rotating magnetic field, so that

$$\mathcal{E} + c^{-1}(\omega \times \mathbf{r}) \times \mathcal{B} = 0. \tag{15.7}$$

This corresponds to a charge density in the plasma, where the difference in the numbers of positive and negative charges is

$$n_- - n_+ = \omega \cdot \mathcal{B}(2\pi ec)^{-1}. \tag{15.8}$$

Between the equatorial and polar regions there is a vacuum gap (the *outer gap*) in which there is a very high electric field. This field can accelerate charged particles along the magnetic field lines, reaching very high energies. The observer's line of sight can cut across beamed emission from the two such regions in each rotation, giving pairs of pulses widely separated by up to 180°. This is the broadband emission seen mainly in X-rays and gamma-rays, although it may extend into radio, as in the Crab pulsar. There

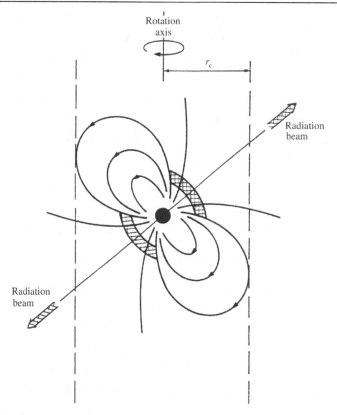

Figure 15.6. A pulsar magnetosphere.

is also a high electric field in an *inner gap* over the magnetic poles, which accelerates particles from the neutron star surface into the polar plasma. The radio emission in the majority of radio pulsars originates above the inner gap. The Crab pulsar shows beamed radiation from both regions (Figure 15.5): the double pulse is the outer gap emission, and the *precursor* appearing at low radio frequencies is emission from above the polar gap.

## 15.7 Polar Cap Radio Emission

Most pulses from radio pulsars originate above the magnetic poles, where electrons and positrons are accelerated by the strong induced electric field. Although the radiation mechanism is not understood (see Section 15.12 below), the radio emission almost certainly results from the curved motion of these particles, which are tightly constrained to follow the magnetic field lines, so that an individual emitting region produces a narrow component in the observed radio pulse. This polar cap radio emission is confined to small regions, whose distribution is reflected in the sometimes complex pulse shapes. The observed pulse shapes

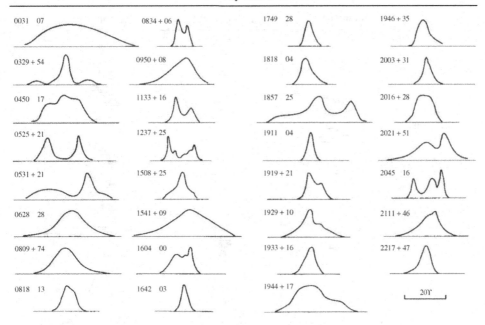

Figure 15.7. A selection of integrated pulse profiles.

are therefore a combination of a beamwidth inherent in the radiation process and a spatial distribution of emitters. The examples in Figure 15.7 show simple pulse shapes, which may be interpreted as beams from individual sources, and multiple sub-pulses which are from distributed sources.

In many pulsars a sequence of single pulses shows sub-pulse components which drift across the profile over a time of several pulse periods; this *pulse drifting* is regarded as a lateral movement of individual sources around and within the polar cap (Figure 15.8). In some pulsars the track of this movement appears to be closed, so that the same pattern of excitation can recur after an interval considerably longer than the time for a sub-pulse to cross the width of the pulse profile. This was interpreted by Deshpande and Rankin (2001) as a pattern of excitation rotating continuously round the polar cap, as in a carousel.

For those pulsars where the axes are nearly perpendicular, pulses may be observed from both magnetic poles (Figure 15.7), while for those in which the rotation and magnetic axes are nearly aligned the observer must be located close to the rotation axis; in this case the observed radio pulse may extend over more than half the pulse period.

## 15.7.1  Polarization

Strong linear polarization, which is observed in both the radio and the optical emission (polarimetry is becoming possible in X-ray but not yet in gamma-ray observations), provides valuable clues to the geometry of the emitting regions. In a simple radio pulse

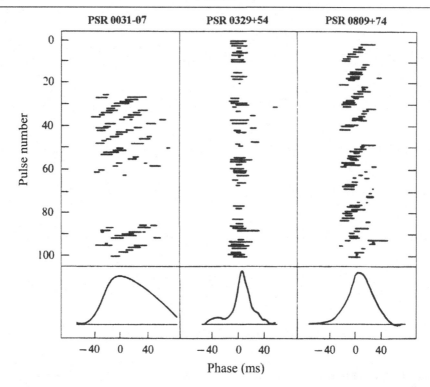

Figure 15.8. Pulse drifting in the PSRs B0031−07, B0329+54, and B0809+74. Successive pulses contain sub-pulses that occur at times that drift in relation to the average pulse profile. All three pulsars also show the effect of pulse nulling, when sequences of pulses are missing (Taylor and Huguenin 1971).

(Figure 15.9) the plane of polarization swings monotonically through an S-shape. This was interpreted by Radhakrishnan and Cooke (1969) as the successive observation of narrowly beamed radiation from sources along a cut across the polar cap; it may alternatively be inherent in the beam from a single source. In either case the plane of polarization is determined by the alignment of the magnetic field at the point of origin, so that the sweep of polarization can be related to the angle between the magnetic and rotation axes, and their relation to the observer. Lyne and Manchester (1988) showed in this way that the angles between the axes are widely distributed (it has been suggested, but with little evidence, that this distribution can be related to an evolution during the lifetime of a pulsar).

Many pulsars show a high degree of circular polarization, which often reverses during the pulse. A compilation of pulse shapes and polarizations from 600 pulsars by Johnston and Kerr (2017), using integration over many pulses, shows the degree of circular and plane polarization, with position angle, through the pulse. (Note that in pulsar astronomy the convention in RH and LH circular polarization is unfortunately opposite to the IAU convention, which is generally adopted in radio astronomy, including interferometry and aperture synthesis. See van Straten *et al.* (2010).)

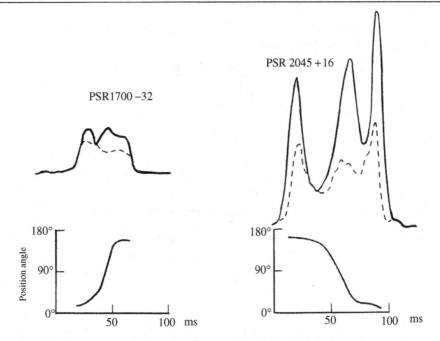

Figure 15.9. The swing of the plane of polarization across an integrated radio pulse. The rate of swing depends on the relation between the line of sight and the axes of rotation and of the magnetic field.

### 15.7.2 Individual and Integrated Pulses

The radio pulses vary erratically in shape and amplitude from pulse to pulse; however, the integrated profile obtained by adding some hundreds of pulses is usually reproducible, stable over at least several decades, and characteristic of an individual pulsar (Figure 15.10). In many pulsars the width of the integrated profiles varies with radio frequency. A tendency for larger width at lower frequency has been interpreted in terms of diverging field lines over the magnetic poles, as can be seen in Figure 15.6; the suggestion is that the lower frequencies are emitted higher in the magnetosphere, where the field lines diverge through a wider angle. A simple geometrical model then leads to an estimated height of emission; for a long-period pulsar this is about $300(\nu_{MHz}/300)^{-1/2}$ km. Following the field lines down to the surface, the region in which the excitation originates must be only about 250 metres across.

Since the radius of the velocity-of-light cylinder is determined by the angular velocity, the scale of the magnetosphere in pulsars with periods from milliseconds to several seconds must range over a ratio of 5000 to 1. Nevertheless the angular width of the integrated profiles, and the radio spectra, vary remarkably little over this range (Kramer *et al.* 1999). The millisecond pulsars have broader profiles, especially at lower radio frequencies; at high frequencies the only difference between the radio characteristics of the two classes is in luminosity, which is lower by a factor of ten in the millisecond pulsars. At the other end of the scale, many slowly rotating pulsars such as the 8.5 s pulsar (Young *et al.* 1999)

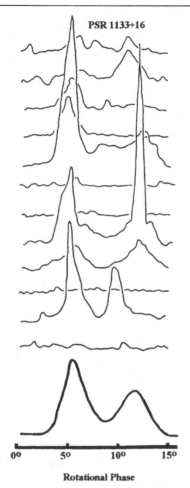

Figure 15.10. Individual and integrated pulses from PSR 1133+16.

have a narrow beamwidth of only around 1°: there is no explanation for this, and it suggests that there may be a substantial population of very slow pulsars, most of which cannot be detected at all because their narrow beams never point towards us.

### 15.7.3 Nulling, Moding, and Timing Noise

In many pulsars the variability from pulse to pulse extends to a complete gap in emission lasting for a number of pulses, as seen in pulse numbers 63 to 86 in PSR B0031-07 (Figure 15.8). This *pulse nulling* may extend for long periods: there are pulsars which disappear for many hours or even for many days. In pulsars with several components there may be sudden changes in the observed integrated profile, switching between different patterns of mean excitation; this behaviour is known as *moding*. These changes in pulse profile are instantaneous, occurring at intervals of days, months, or years (Lyne *et al.* 2010); they are accompanied by small switches in slowdown rate.

In the analysis of pulse arrival times, intended to derive a rotation rate and its derivative, there is usually a residual component that is generally referred to as *timing noise*. It seems likely that this is a manifestation of the same instability as moding, but at a low level and often on a short timescale. Examples of such timing residuals from observations over many years are shown in Figure 15.11; these display various characteristics, often a periodicity of some years, which are not apparent in shorter data sets (Lyne *et al.* 2010).

Pulse nulling, which may be regarded as an extreme form of moding, occurs over a wide range of timescales. In some examples only a few pulses, or even a single pulse, may be followed by a much longer null. An example is PSR J1929+1357, which only produces observable pulses for a few minutes at irregular intervals of several days (Lyne *et al.* 2017). Isolated single radio pulses with characteristics of pulsars have been observed and later identified as pulsars when further isolated pulses from the same location were observed; the intervals between these were found to be multiples of a precise period. These were designated RRATs (rotating radio transients) (McLaughlin *et al.* 2013); they appear to be part of the general population of pulsars, constituting around 3% of the known pulsars (Karako-Argaman *et al.* 2015).

It is often difficult to distinguish these extreme cases of nulling from the so-called FRBs (fast radio bursts). Isolated impulses often occur in radio observations, and are rightly attributed to manmade radio interference. However, an impulse detected in the Parkes survey for pulsars was unusual; it displayed a dispersion in arrival time characteristic of a distant pulsar. This discovery (Lorimer *et al.* 2007) was the first of a new category now known as fast radio bursts. They are distinguished from RRATs by their large dispersion measures, which are larger than expected from a pulsar located in the Milky Way (Keane and Petroff 2015), indicating an extragalactic origin. Identifying an associated optical object, which might give a distance based on a redshift, is difficult, since the precise location of a single transient cannot easily be found. One of the FRBs, however, was followed by an afterglow lasting six days; this was found to be in a galaxy with redshift $z = 0.5$ (Keane *et al.* 2016). Another, more certain, identification was made possible when a repeating FRB was found (Spitler *et al.* 2016), allowing a very accurate VLBI position and angular diameter measurements (Chatterjee *et al.* 2017). The source was identified with a faint galaxy at redshift $z = 0.19$. The physical nature of the source of such FRBs is unknown, and it may be as revolutionary as the discovery of pulsar radio emission itself: the bursts are hugely linearly polarized and have an unprecedentedly high and variable rotation measure (Michilli *et al.* 2018).

## 15.8 Magnetars

Magnetars are the group of neutron stars at top right in Figure 15.2, with the largest known magnetic fields. The history of their discovery starts in 1972, when a powerful burst of gamma-rays was detected by several spacecraft, notably the Venera 12 space probe (Mazets *et al.* 1979). The different times of arrival at the separate spacecraft gave a location in

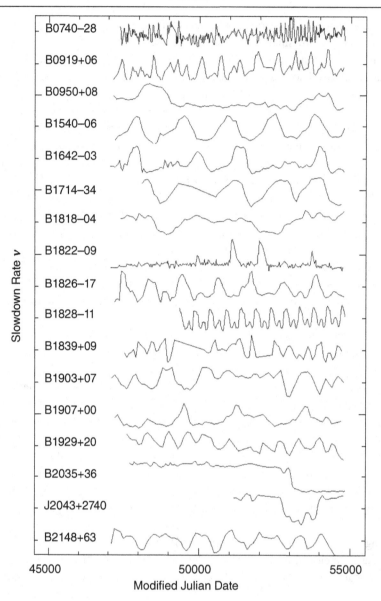

Figure 15.11. Timing noise from 17 pulsars. (Courtesy of Andrew Lyne, Jodrell Bank Observatory.)

a supernova remnant in the Large Magellanic Cloud, believed to be about 10 000 years old. Within the burst there was a modulation with period 8 seconds. In 1992 Duncan and Thomson proposed the name 'magnetar' for a rotating neutron star with a magnetic field up to $10^{15}$ gauss, accounting for a rapid spin-down from the short period of a typical pulsar.

Some gamma-ray sources were found to be repeating, and these SGRs were also identified with X-ray bursts (XRBs), one of which showed a periodicity of 7.5 seconds (see Kaspi and Beloborodov 2017 for references to the discovery). In 2017 there were 23 known magnetars (Olausen and Kaspi 2014). Of these, four have been detected as radio pulsars, the most interesting of which is PSR J1745−2900. This magnetar is only 2.5 arcseconds from Sag A\* at the galactic centre and is probably in orbit around the central black hole of the Galaxy. It is, however, erratic in both intensity and periodicity, making its orbit difficult to track in any detail.

The name magnetar now encompasses all neutron stars known to be rotating with periods between 2 and 11 seconds and with magnetic fields between $10^{13}$ and $10^{15}$ gauss. They are evidently evolving rapidly, but their irregular behaviour prevents any measurement of braking index. The intensity of their radiation cannot be explained by the usual magnetic dipole slowdown, and the source of their energy is usually attributed to magnetic field decay.

## 15.9  X-Ray Binaries and Millisecond Pulsars

The discovery of a pulsar with the astonishingly short period of 1.6 milliseconds (Backer *et al.* 1982) showed that the limitations of the early search techniques might be concealing a large population of short-period pulsars. Searches using techniques sensitive to shorter periods soon revealed a distinct population of *millisecond pulsars* (MSPs), with periods mainly below 20 milliseconds. There seems to be a lower limit of 1.4 ms on the period of MSPs (although this may still be an effect of observational selection). New discoveries of MSPs are frequent; for example Scholz *et al.* (2015) reported timing observations of five MSPs found in a survey at Arecibo; four of these are in binary systems and one is solitary. There were at that time 160 known MSPs distributed throughout the Milky Way galaxy, and 130 within globular clusters. (For a general introduction to millisecond pulsars see Lorimer 2008.)

The explanation of this new class of millisecond pulsars, and their linkage to the main population, came from X-ray astronomy. In 1971 thermal X-rays from a powerful source, Cen-X3, were observed to be modulated with a period of 4.8 s; furthermore, a periodic Doppler shift in the pulse interval showed that the source was in a binary system (Giacconi *et al.* 1971). More than 100 such X-ray sources are now known, distributed throughout the Galaxy. All are neutron stars in binary systems. Most of the companions are young stars with masses between 5 and 20 $M_\odot$, which are evolving through the red giant phase towards a supernova collapse, and whose remnant core would probably become another neutron star. Others, the low-mass X-ray binaries (LMXBs), have companions with masses about 1 $M_\odot$, which are evolving towards collapse to a white dwarf.

The atmosphere of the companion star in these binary systems has expanded beyond the Roche limit, at which matter can flow out under the gravitational attraction of the neutron star. Accretion on the surface of the neutron star has two effects: heating and transfer of angular momentum. The heating is concentrated in hot spots above the magnetic poles; the observed modulation occurs because rotation brings the hot spots in and out of view. The transfer of angular momentum, derived from the binary system, accelerates the rotation of the neutron star to milliseconds. The star only becomes a pulsar when the accretion ends: there is an evolutionary sequence, from X-ray binary to millisecond pulsar. This transition has actually been observed: PSR J1023+0038, a 1.7 ms pulsar in a binary with orbital period 4.8 hr, switches between two modes in which it is alternately a radio pulsar and an XRB (Archibald *et al.* 2009; Patruno *et al.* 2014; Bassa *et al.* 2014.)

## 15.10  Binary Millisecond Pulsars

Figure 15.12 shows the observed orbital Doppler shift for two examples of binary pulsars. PSR 0655+64 is an MSP in an orbit with very low eccentricity; in contrast the young pulsar B1820-11 is in a highly eccentric orbit. Most binaries are millisecond pulsars; about 80% of MSPs are in binary systems. All those with small eccentricities have a low-mass white dwarf companion; slow accretion by the neutron star from the extended atmosphere of a main sequence star as it evolves to become a white dwarf tends to circularize the orbit. Those with high eccentricity usually have a neutron star companion; these appear to be the outcome of a second supernova explosion in a binary system. Solitary millisecond pulsars may have evolved from either LMXBs or HMXBs; in the case of the HMXBs the binary may have disrupted at the time of the supernova explosion, while for the LMXBs the white dwarf may have been evaporated by intense radiation from the pulsar itself. The latter scenario is supported by observations of several pulsars with very-low-mass companions, in which an occultation occurs over a large part of the binary orbit; this is attributed to a cloud of ionized gas streaming away from the white dwarf.

### 15.10.1  Gamma-Ray Pulsars

Gamma-rays from the Crab and Vela pulsars were first detected by the SAS-2 and COS-B spacecraft, and EGRET found several more discrete gamma-ray sources. When the Fermi Large Area Telescope (LAT) was planned there was little expectation that there would be many more pulsars whose spectra extended into gamma-rays. The LAT, however, has not only detected gamma-ray pulses from known radio pulsars but has found that the radiation from many previously known gamma-ray sources is pulsed, adding substantially to the catalogues of both normal and millisecond pulsars. The detection of gamma-ray pulses meets a difficulty unknown to radio astronomy, due to quantization of the high energy radiation. Gamma-ray photons arrive at intervals very much greater than the periods of even the slowest pulsars. Given the pulse period, it becomes clear after the arrival times of some hundreds of photons have been measured that they are arriving only within strictly periodic pulse intervals. Even when the pulse period is not known from radio observations, it may be discovered from the intervals between photon arrival times, which must be at integral multiples of the pulse period.

The LAT telescope on the spacecraft Fermi has detected more than 200 gamma-ray pulsars, most of them millisecond pulsars and young pulsars (see a review by Smith

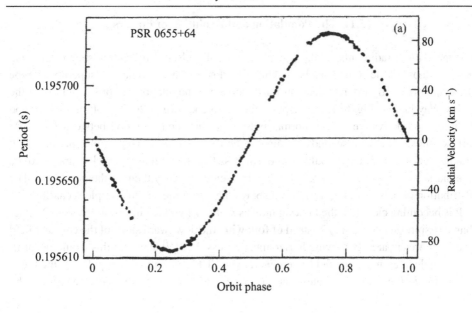

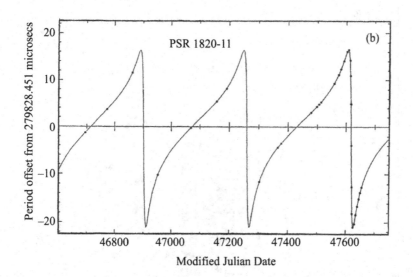

Figure 15.12. Doppler shifts of pulse periods in binary pulsars: (a) orbit with small eccentricity (PSR 0655+64; Damashek *et al.* 1982); (b) orbit with large eccentricity (PSR 1820-11; Lyne and McKenna 1989).

*et al.* 2017). There is also a growing number discovered as gamma-ray pulsars with no detectable radio emission; the distinction between these categories is related to their geometry and their magnetic field strengths (Rookyard *et al.* 2017).

## 15.11  The Population and Evolution of Pulsars

The majority of radio pulsars, the so-called 'normal' pulsars, are neutron stars originating in the collapse of the cores of isolated white dwarf stars. These events, the supernovae Type II, occur sufficiently often to account for the observed population of normal pulsars in the Milky Way galaxy. Figure 15.13 displays the pulsars known in 2017, plotting their period $P$ and period derivative $\dot{P}$. The normal pulsars are in a concentration between $P = 100$ milliseconds and $P = 8$ seconds. Assuming their rotation is slowing as in Eqn 15.3, with braking index $n = 3$, they would all move across this diagram along tracks parallel to the line marked $10^{12}$ gauss; this evolution would eventually carry them to the *death line*. The distribution of pulsars in Figure 15.13 displays the difficulties in this simple scenario.

It is becoming clear that the braking indices $n$ of most young pulsars are close to 2 rather than 3 (Espinoza *et al.* 2017). Instead of following the downward slope of the constant field line, they are apparently moving horizontally across the $P/\dot{P}$ diagram; this would occur if, for example, their magnetic fields were *increasing*. The concentration of pulars in the lower part of the $P/\dot{P}$ diagram indicates that this horizontal flow must eventually be replaced by

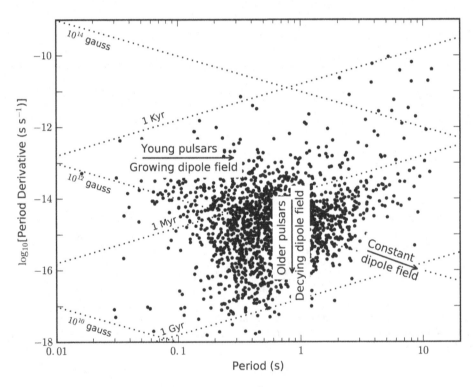

Figure 15.13. Pulsar evolution: the flow across the $P/\dot{P}$ diagram. The known normal (non-recycled) radio pulsars are located according to their periods and first period derivative. Arrows show the directions of evolution of young pulsars with low braking index $n$ and older pulsars with high braking index $n$ moving towards lower slowdown rates. (Data from the Australia Telescope National Facility (ATNF) Catalogue, courtesy of Xiaojin Liu, Jodrell Bank Centre for Astrophysics.)

a nearer-vertical downward flow; this could be due to a *decreasing* field strength. Both are physically possible; the increasing field could be due to the resurfacing of a field buried beneath the highly conducting surface of the neutron star, followed eventually by an ohmic decay.

The magnetars, at the top of the $P-\dot{P}$ diagram, are short-lived pulsars, characterized by their long periods and large slow-down rates. It is not clear whether they have a different origin from that of the normal pulsars, or whether some normal young pulsars evolve into magnetars (Espinoza *et al.* 2011).

## 15.12 The Radiation Mechanism

The intensity of the radio emission from pulsars shows at once that it cannot be thermal in any sense; short pulses can only come from compact sources, so the brightness temperature must be very high, exceeding $10^{30}$ K in some cases. Such high brightness can only be produced by the coherent motion of charged particles; if $N$ particles move together, they radiate an intensity proportional to $N^2$ rather than $N$. (For a general discussion of coherent radiation mechanisms, see Melrose 2017). Furthermore, the radio pulses show a very high degree of polarization, both linear and circular, which on occasion may approach 100%; this cannot be explained in terms of synchrotron or curvature radiation. The radio emission is therefore coherent, and the association of a particular frequency with a definite radial distance shows that this distance is determined by a resonance in the plasma of the magnetosphere. Melrose argues that a two-stage process is involved, in which the coherence derives from bunching in an unstable stream of particles and the radiation is a resonant coupling at a critical density to a propagating mode directed along a field line. The linear polarization of the radiation is then similar to that of curvature radiation.

The original acceleration of the particles takes place near the surface of the polar cap, in a cascade process. In this cascade, as suggested by Sturrock (1971), electrons or positrons are accelerated to a high energy and radiate gamma-rays via curvature radiation; these gamma-rays then create electron and positron pairs as they encounter the strong magnetic field, and the new particles are accelerated to continue the cascade. The optical and other high energy radiation from the Crab and Vela pulsars, in contrast, can be accounted for as incoherent curvature or synchrotron radiation from individual high energy particles streaming out along field lines in the outer gap at the edge of the polar cap.

Pulsars are most often and most easily observed at radio frequencies in the range 100 MHz to 2 GHz. Some spectra have been measured over a wide range, from 20 MHz to 80 GHz. Their integrated radio emission has a continuous steep spectrum, with spectral index $\alpha \sim -1.4 \pm 1$ (Bates *et al.* 2013); a survey by LOFAR shows that many spectra turn down at low frequencies, below around 100 MHz (Bilous *et al.* 2016).

The spectral characteristics of pulsars cannot be related directly to the energy spectrum of the emitting particles, as can be done for incoherent sources of synchrotron radiation (Chapter 5).

The distribution of intensity of pulses from a small number (around 1%) of the younger pulsars includes so-called 'giant' pulses, which may be ten or a hundred times above the average. These may contain a structure that is very narrow both in duration and in radio frequency (Hankins *et al.* 2016), indicating an origin from regions only a few metres across.

## 15.13 Pulsar Timing

A rotating neutron star acts as a massive cosmic flywheel; its pulsed emission, locked to the embedded magnetic field, allows us to regard its rotation as a precision clock. Timing the arrival of pulses is informative both for the behaviour of individual pulsars and for the characteristics of propagation through the interstellar medium. It also provides remarkably accurate positions and transverse velocities, measured as parallax and proper motion. For pulsars in binary systems it can provide information on the orbits, with such accuracy that general relativity is required for their characterization, with the effect that the masses of both components are measurable with remarkable precision.

Pulsar timing requires a high signal-to-noise ratio, which is achieved using wide receiver bandwidths and by the addition of many individual pulses to form an integrated profile. The pulse arrival time varies over the large frequency bandwidths, owing to dispersion in the interstellar medium (Chapter 4). This may be allowed for by dividing the wide band into many sub-bands, each acting independently, and compensating for the dispersion by delaying the individual outputs; the necessary *de-dispersion* delay is of course different for each pulsar. Further compensation for dispersion is usually achieved within each sub-band in a process of *coherent de-dispersion*; this requires a Fourier transform of the incoming signal to produce a spectrum, which can then be delayed and transformed back to restore the signal (see the supplementary material). Coherent de-dispersion requires a fully digitized signal (Appendix 3), requiring computer power that increases rapidly with increasing signal bandwidth. Digital sample rates must be at least twice the bandwidth of the signal in each channel; it is therefore usual to transform the signal frequencies downwards by a heterodyne, so minimizing the necesary rate of sampling.

A timing observation for a pulsar involves an integration for several minutes, building up a pulse profile that can be fitted to a standard profile, giving an arrival time at the radio telescope. Timing may achieve an accuracy of 10 nanoseconds, which is astonishing considering that the pulse may have been travelling for thousands of years before it reached the radio telescope. The distance travelled in 10 nanoseconds is only 3 metres, so it is essential as a first step in timing analysis to take account of the position of the telescope on the rotating Earth at the time of the observation, reducing pulse arrival times to a time of arrival at the centre of the Earth (the geocentre). The next stages of analysis must allow for the motion of the Earth in the solar system, and any motion of the pulsar in a binary system or across the sky.

Timing the pulse arrival times as the Earth orbits the Sun yields positions with milli-arcsecond accuracy. Figure 15.14 shows the approximately sinusoidal annual variation of arrival times due to the travel time across the Earth's orbit. This variable delay, known as the Römer delay, is at a maximum of 500 seconds for a source in the plane of Earth's orbit (the

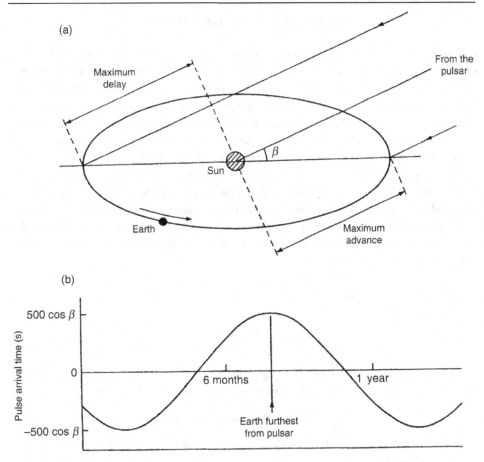

Figure 15.14. The Römer delay.

ecliptic); the pulsar position is found from the amplitude and phase of the annual variation. The orbit must be defined precisely in an *ephemeris*, which takes account of the Earth's orbital ellipticity and the influence of other planets.

A further delay, the Shklovskii delay, is due to the transverse velocity, if any, of a pulsar. Following a straight path (and not the curved path which would keep the distance constant) the distance increases, giving a delay which increases with time as $v^2t^2/2dc$, where $v$ is the transverse velocity of the pulsar at distance $d$. The effect is observed as a change in period derivative $\dot{P}$.

Another variable delay, which appears in general relativity, is the Shapiro delay, related to the distortion of space–time in the gravitational field of the Sun. This delay is combined with the Römer delay in the timing analysis; it is a small effect in the solar system, but as we will see it is very important in binary pulsar systems. For a detailed look at pulsar timing and other pulsar observing techniques, see the *Handbook of Pulsar Astronomy* by Lorimer and Kramer (2005).

## 15.14 Distance and Proper Motion

For many nearby pulsars the parallax (giving distance) and the proper motion can be measured by interferometry, independently of pulse timing. The most accurate measurements are made by comparing the positions of the pulsar and an adjacent quasar, when only differential measurements need be made. Assuming that the position of the quasar is related to a fundamental frame, the pulsar's position may be found in this way to sub-milliarcsecond accuracy. Continued timing observations over a year or more provide measurements of distances through parallax, while their proper motion is also measurable for many pulsars.

Distances are also obviously available for those pulsars which are identified or are associated with visible objects, such as those in supernova remnants. Some distance information is available for pulsars close to the galactic plane, which are observed through neutral hydrogen clouds (see Chapter 14). These H I clouds absorb at wavelengths near 21 cm; the actual absorption wavelength depends on their radial velocity and hence gives an indication of the distance of an individual pulsar.

For most pulsars the only measurement of distance is provided by the frequency dispersion in the pulse arrival times. The *dispersion measure* (DM) is observed as a delay $t$, usually quoted as

$$t = \frac{\mathcal{D} \times \mathrm{DM}}{\nu^2}, \tag{15.9}$$

where the frequency $\nu$ is measured in megahertz. The *dispersion constant* $\mathcal{D}$ is the product of electron density and distance. With the usual units of $\mathrm{cm}^{-3}$ and parsecs,

$$\mathcal{D} = 4.149 \times 10^3 \quad \mathrm{MHz}^2 \ \mathrm{pc}^{-1} \ \mathrm{cm}^3 \ \mathrm{s}. \tag{15.10}$$

Given a model of the distribution of electron density in the interstellar medium, the DM found from Eqn 15.9 immediately gives the distance. For nearby pulsars an average electron density $n_e = 0.025 \ \mathrm{cm}^{-3}$ pc may be used, with an accuracy of around 20%; for example the Crab pulsar with DM = 57 is known to be at a distance of about 2 kpc. At distances greater than 1 kpc it is, however, necessary to take account of the structure of the Galaxy; a model distribution by Taylor and Cordes (1993), revised by Yao *et al.* (2017), is generally used for such distance determinations.

## 15.15 Binary Radio Pulsars

In 1993, the Nobel Prize in Physics was awarded to Russell Hulse and Joseph Taylor of Princeton University for their 1974 discovery of a double neutron star binary system, PSR B1913+16. By observing this system for several years they found a steady decrease in the orbital period agreeing exactly with Einstein's prediction of energy loss by gravitational radiation (Figure 15.15). This dramatic demonstration of the power of binary pulsars in demonstrating and measuring the effects of general relativity opened a new field of relativistic astrophysics.

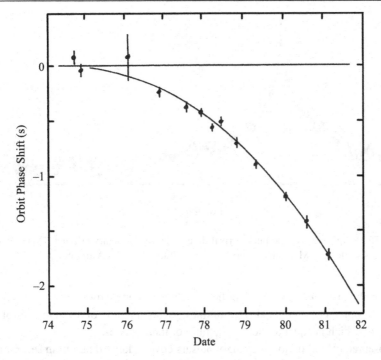

Figure 15.15. The loss of energy by gravitational radiation shrinks the orbit of the binary of PSR B1913+16, decreasing the orbital period, as shown by the earlier pulse arrival time (Taylor and Weisberg 1989).

### 15.15.1 The Analysis of Binary Orbits

Measurement of pulse arrival times around a binary orbit is in effect a measurement of *range* which provides much greater accuracy than the velocities available from Doppler shifts measured in optical astronomy. An orbit is first characterized by five Keplerian parameters: orbital period $P_b$, eccentricity $e$, major axis $a_{psr}$ projected on the line of sight $x_{psr}$, time of periastron $T_0$, and longitude of periastron $\omega$. The inclination $i$ to the line of sight, and the masses of the two components become available only in the next, post-Keplerian, stage of analysis; however, the period and orbit diameter already yield a mass function $f$:

$$f = \frac{(M_C \sin i)^3}{M_T^2} = \frac{4\pi^2}{T_\odot} \frac{x_{psr}}{P_b^2}, \tag{15.11}$$

where $M_T$ is the total mass and

$$T_\odot = \frac{GM_\odot}{c^3} = 4.925\,490\,947 \tag{15.12}$$

is the solar mass in units of time.

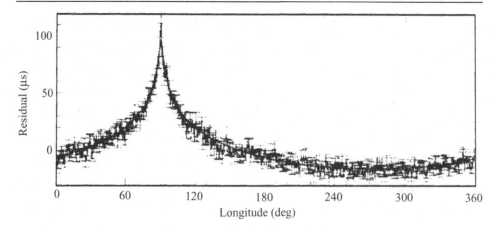

Figure 15.16. Shapiro delay. The pulse arrival time varies as the binary pulsar J0737-3039 orbits its companion. (Courtesy of M. Kramer, Max Planck Institute for Radio Astronomy.)

When the binary companion is identified (usually as a white dwarf) it may be possible to observe a Doppler shift in optical spectral lines, giving the ratio of the velocities of the two components, which is equal to the inverse ratio of their masses.

The observed orbital periods of binary pulsars cover a large range from several hours to 50 years (Lyne *et al.* 2015b).

### 15.15.2  Post-Keplerian Analysis

The precision of timing in many millisecond binary pulsars requires general relativistic theory for analysis. Five new post-Keplerian (PK) parameters are involved:

1. the precession of the elliptical orbit, measured as $\dot\omega$;
2. the Einstein delay, due to gravitational redshift and time dilation in the orbit;
3. the decay of the orbit, measured by Hulse and Taylor as a change in orbital period $\dot P_b$;
4. the Shapiro delay, due to distortion of propagation in space–time by the gravitation of the companion. An example of this delay is shown in Figure 15.16 (see Kramer *et al.* 2006). Two parameters are involved: the range $r$ and the shape $s$ of the delay curve.

Combining the Römer delay and the Shapiro delay yields the full geometry of the orbit, giving the orbital period, the size and ellipticity of the orbit and its attitude in relation to the line of sight. The Shapiro delay yields the mass of the companion star.

In one of the double neutron star systems, J0737-3039, which has the short orbital period of 2.4 hours, both components were observed to be pulsars (Lyne *et al.* 2004). The origin of this system is clearly understood: the older star, with a rotation period of 23 milliseconds, has been spun up by accretion from a massive main sequence companion, which subsequently evolved in a supernova collapse, leaving a younger normal pulsar which now has a period of 2.8 seconds. This double pulsar system is

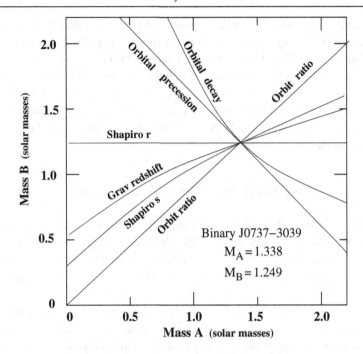

Figure 15.17. The masses of the two pulsars comprising the binary J0737-3039 (after Kramer *et al.* 2006).

a rich field of astrophysics: as well as displaying the same loss of orbital energy via gravitational waves (as in PSR B1913+16), it happens that the plane of the binary orbit is close to our line of sight, so that an interesting and complex occultation occurs in every orbit.

With the orbits of both pulsars independently determined from timing observations, the characteristics of the binary system are overdetermined. The agreement of the independent measurements of the mass of both components demonstrates the correctness of Einstein's theory of general relativity. This is illustrated in Figure 15.17. Here the various factors determining the masses $M_A$ and $M_B$ of the two pulsar components J0737-3039A and J0737-3039B are shown as tracks which intersect at the actual masses. The observations giving the parameters of each track are: (i) the ratio of the orbit amplitudes, which gives the ratio of the masses; (ii) the rate of precession of the orbits, which for this binary is 17° per year; (iii) the rate of change of the orbital period; (iv) the two Shapiro parameters $r$ and $s$; (v) the combined effect of gravitational redshift and time dilation.

Kramer *et al.* (2006) showed this mass–mass diagram with track widths indicating the accuracy with which they were measured. The remarkable precision with which all tracks coincide is a demonstration both of the correctness of general relativity theory and of the analysis of the orbital parameters.

Several further tests of general relativity have been proposed; for example, the coupling between pulsar spin and orbit in a binary has already demonstrated, and if a pulsar–black-hole binary can be discovered, there is a possibility that this coupling may reveal the spin of the black hole. The many ways in which pulsar timing can probe general relativity were set out by Kramer in his 2016 George Darwin Lecture (2017).

An internationally coordinated system of pulse timing and analysis is essential, especially for the gravitational wave programme to be described in Section 15.17. All the basically geometric elements are combined in the computer program TEMPO, which originated in Princeton (Taylor and Weisberg 1989), and the newer TEMPO 2 (Hobbs *et al.* 2006), which is used for all radio telescopes.

## 15.16 Searches and Surveys: The Constraints

Figure 15.18 was compiled at a celebratory meeting *50 Years of Pulsars* at Jodrell Bank Observatory in 2017. The concentration at the Milky Way suggests that the searches have explored the distribution throughout our Galaxy, but the limited sensitivity of surveys has largely confined detections to the nearer half of the Galaxy. Observational selection, which hid the millisecond pulsars from the early observers, must always be borne in mind when the luminosity and spatial distributions of pulsars in the Galaxy are to be described. Besides the limitations in instrumental sensitivity related to antenna size, receiver noise, bandwidth, integration time, and of course the inverse square law, which limits the distance at which a pulsar can be detected, there are effects of propagation in the ionized interstellar medium to be considered. The most obvious is the frequency dispersion in pulse arrival time, described in Chapter 4. This spread limits the time resolution of the receiver by lengthening the pulse. The effect is overcome by restricting the bandwidth of the receiver or by splitting the receiver band into a number of sub-bands, each with its own detector. Detection of a weak signal from a pulsar with a given value of DM then requires the outputs of the separate

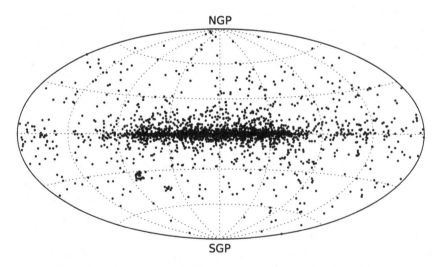

Figure 15.18. The distribution of the first 2500 pulsars over the sky. (Data from the ATNF Catalogue, courtesy of Xiaojin Liu, JBCA.)

channels to be added with appropriate time shifts; if a search is to be made for pulsars with unknown values of DM, there must be a succession of trials with different values of DM.

The sharp frequency dependence of propagation effects forces the searches for all but nearby pulsars to be conducted at frequencies around 1 GHz or above. Unfortunately this means that for a sufficiently large single-aperture radio telescope the beam area is small; for the major search at 1.4 GHz using the Parkes telescope the beam diameter was only 13 arcminutes. Each beam area is usually observed for an integration time of 30 minutes, so that to search the whole sky with a single beam would take over 40 years of continuous observations. As described in Chapter 8, the Parkes telescope was equipped with a multiple feed system, so that 13 independent beams could be used simultaneously, each with separate receivers for the two hands of circular polarization (Manchester *et al.* 1996); even so, a reasonably complete coverage of the sky requires many years with such a system. The alternative is to use a multi-element system such as LOFAR, with large primary beams combined electronically to give many synthesized beams simultaneously.

A severe constraint on the rate at which a survey can be conducted is the massive task of data handling. The detected output of each channel must be sampled through a long integration time at a rate several times faster than the shortest period pulsar to be discovered. The set of samples at different radio frequencies must then be added with delays appropriate to some value of dispersion measure; the sum must then be Fourier transformed and the resulting periodogram searched for significant signals. The process must then be repeated for a series of values of the dispersion measure. The length of each process increases rapidly with the integration time, so a balance has to be struck between survey time, sampling interval, range of dispersion measure, and computation time. The observational constraints on any large survey inevitably impinge mainly on the sensitivity to pulsars with the shortest periods and the largest dispersion measures.

Binary systems, in which the observed pulse period may be changing rapidly, may be missed in a survey with a long integration time. To preserve the sensitivity of observations with long integration times, it is necessary to search for a changing periodicity. An approximately linear rate of change is the best that can be accommodated in this additional search, leaving the possibility of missing some very-short-period binaries.

## 15.17 Detecting Gravitational Waves

Given all the corrections in Sections 15.13 and 15.15.2, observers can address questions beyond the rotational behaviour of the pulsar itself, and search for one more geometrical effect, that of gravitational waves. The short burst of gravitational waves from the collapse of a binary neutron star system has already been observed by LIGO and Virgo (Abbott *et al.* 2017); much weaker, but much slower, waves are being sought from an effect of cosmic inflation. If this occurred, and there was some asymmetry in the expansion, there should be a background of gravitational waves, with low (nanohertz) frequencies, which might affect the relative arrival times of pulses from pulsars at different locations across the sky.

Several observatories have undertaken a series of such observations lasting many years, either in networks or singly. A set of millisecond pulsars, selected for their simple pulse profiles and their rotational stability, are being monitored to accuracies of order 10 nanoseconds. Three networks of observatories, known as pulsar timing arrays (PTAs), are currently engaged in such a cooperative programme, based in North America, Parkes (Australia), and Europe. These are coordinating their procedures and observing programmes to form an International PTA. See for example Wang and Mohanty (2017) on the possibility of observing gravitational waves with the SKA and other large radio telescopes.

## 15.18 Further Reading

*Supplementary Material* at www.cambridge.org/ira4.

Lorimer, D. R., and Kramer, M. 2005. *Handbook of Pulsar Astronomy.* Cambridge University Press.

Lyne, A. G., and Graham-Smith, F. 2012. *Pulsar Astronomy,* 4th edn. Cambridge University Press.

# 16

# Active Galaxies

Although the radio emission from our Milky Way galaxy dominates the radio sky, it would not be a conspicuous object at the distance of any nearby galaxies. As we described in Chapter 1, the discovery and identification of the powerful radio sources designated Centaurus A, Virgo A, and Cygnus A, and especially the high redshifts of the quasars, opened up a dramatic new arena of astrophysics focussed on the accretion of material onto a central supermassive black hole (SMBH) and the release of gravitational energy. Objects in which this takes place have become known as active galactic nuclei (AGN) and are the main subject of this chapter.

Most galaxies have essentially the same mixture of components, including, we now know, SMBHs in their nuclei. However, their appearance can differ greatly depending on whether the SMBH is accreting material and has become 'active' and/or the current rate of star formation is enhanced for some reason. The fact that AGN constitute only a minority of all galaxies implies that the accretion phase occupies a relatively brief period in their history; this is reflected in the statistics of the large numbers of radio sources now observed at intensities many orders of magnitude below those of the discovery observations. We will devote the second part of this chapter to these statistics. In the first part we are mainly concerned with active galaxy phenomena as revealed by high-resolution radio telescopes,[1] starting with the star-forming galaxies (SFG), in which activity is widespread rather than SMBH-driven.

## 16.1 Star-Forming Galaxies

A wide-ranging review of the radio emission from normal galaxies such as the Milky Way was given by Condon (1992). The emission is mainly synchrotron radiation from electrons with cosmic ray energies, diffusing throughout a galaxy and radiating as they are accelerated in a magnetic field. These high energy electrons have been produced in the remnants of supernova explosions of massive ($>8$ $M_\odot$) stars. In some galaxies the supernova rate is higher than the one per century in the Milky Way; the closest such galaxy is M82 (distance

---

[1] The radio images in this chapter are a small sample of the available material, reproduced here in monotone. Further images are to be found in colour in the Supplementary Material. We also refer the reader to the websites for Leahy *et al. Atlas of DRAGNs* and the VLBA MOJAVE survey images of Lister *et al.* (2018).

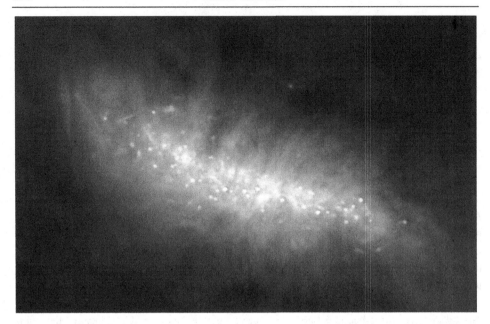

Figure 16.1. The starburst galaxy M82; the discrete sources embedded in the diffuse emission are supernova remnants; see also Figure 13.18. (VLA Image Gallery. Credit: Josh Marvil (New Mexico Institute of Mining and Technology/National Radio Astronomy Observatory (NRAO)), Bill Saxton (NRAO/Associated Universities, Inc./National Science Foundation).)

3.5 Mpc), in whose central region the supernova rate is about ten times higher. Figure 16.1 shows a radio image of this central region, where individual supernova remnants can be identified as the discrete sources amidst the diffuse interstellar background emission (see also Figure 13.18). Observations of M82 with eMERLIN confirm the identification with remnants of Type II supernovae, since many can be seen to be shells expanding at $\sim 10\,000$ km s$^{-1}$. Illustrative images are shown in the Supplementary Material.

The spectrum of M82 is typical of star-forming galaxies. Figure 16.2 shows the region extending from low radio frequencies to the far infrared (FIR). Three separate components are identified by the broken-and-dotted, broken, and dotted lines: synchrotron, free–free, and thermal. The free–free radiation makes a significant contribution at millimetre wavelengths (near 100 GHz); at lower frequencies the spectrum is dominated by synchrotron radiation, while in the FIR it is dominated by thermal radiation from dust.

After many such galaxies had been studied, a remarkably close relationship between their overall radio and FIR luminosities emerged (Helou *et al.* 1985). The correlation shown in Figure 16.3 is one of the tightest in all astronomy and holds over a wide range in radio luminosity ($L_{1.4\,\mathrm{GHz}}$ from $10^{18}$ to $10^{23}$ W Hz$^{-1}$) and for a wide diversity of galaxy types. The fundamental cause is the common origin of the radio and FIR emission in star-formation activity. Ultraviolet emission from hot young (O-type) stars heats the interstellar dust. These massive stars are short-lived and undergo supernova explosions, which provide the energetic electrons for the synchrotron radiation. In both cases it is the total energy input that is significant, while factors such as the strength of the interstellar

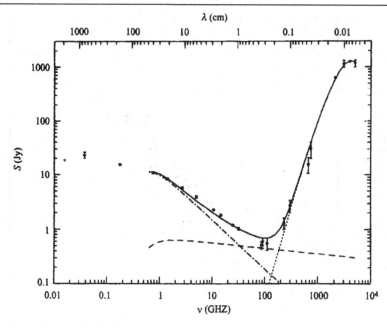

Figure 16.2. The radio and infrared spectrum of the starburst galaxy M82. The broken-and-dotted, broken, and dotted lines show the components of the emission from synchrotron, free–free, and thermal radiation mechanisms (Peel *et al.* 2011).

magnetic field have little effect. All that is required is that a simple proportion of the energy supplied by the young massive stars should be converted without significant loss into the overall radio and FIR emission. This being said, the tightness of the correlation is not completely understood. Nonetheless, given that the FIR emission is directly related to the formation rate of young stars, the tight radio–FIR correlation enables the radio luminosity to become an excellent proxy for the start-formation rate in a galaxy. Ultraviolet light, being another indicator of young stars in galaxies, might seem to provide a more direct way of distinguishing galaxies with enhanced star-formation rates. However, absorption of the ultraviolet light by dust within the galaxy is a severe limitation; in contrast the radio emission suffers no absorption. Details of the relations between the ultraviolet, radio, and FIR emission from normal galaxies are given in the review by Condon (1992).

Galaxies with high star-formation rates are termed 'starburst' galaxies (SBGs) because the enhanced star-forming activity can only be sustained for a limited period ($10^7 - 10^9$ years), after which the huge amount of gas required to support the burst becomes depleted. These objects often have a disturbed appearance in optical and/or infrared images and are generally associated with mergers or interactions between galaxies. Such interactions are usually obvious in nearby galaxies, and in fact our example star-forming galaxy M82 is a small-scale starburst in which the activity has been triggered by a tidal interaction with its larger companion galaxy M81.

Galaxies lying at the top end of the radio–FIR correlation are termed ultra luminous infrared galaxies (ULIRGs). Their profuse infrared emission compared with that of the

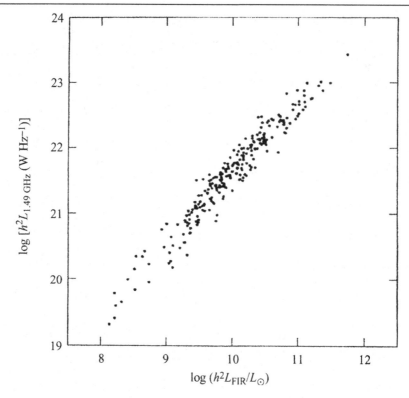

Figure 16.3. The relation between the radio and the FIR emissivities of normal galaxies (Condon *et al.* 1991).

Sun ($L_{IR} \simeq 10^{11}-10^{12}$ L$_\odot$) is a result of the exceptionally large numbers of young O-type stars heating the interstellar dust. An archetypical example, and the closest to the Earth (distance $\sim 75$ Mpc), is Arp 220. This ULIRG is over ten times more luminous in the FIR than M82, and its activity is due to the merger of two gas-rich galaxies. The current supernova rate of a few per year has been directly determined via the observation of new discrete radio sources appearing across the face of the galaxy (Lonsdale *et al.* 2006). The star-formation rate, $>100$ M$_\odot$ yr$^{-1}$, inferred from the infrared luminosity is already much higher than in M82, but there are more extreme examples dubbed 'extreme' or even 'hyper luminous' infrared galaxies. For a brief overview of starburst galaxies from a radio perspective see Muxlow *et al.* (2006).

The study of star formation over the history of the Universe is a subject of much current research. The first generation of mm-wave and submm-wave telescopes such as the JCMT (see Section 8.10.3) discovered a population of dusty galaxies, often at high redshift, and the dust and molecular-gas content can now be studied with powerful interferometers such as the JVLA and ALMA (see the review by Carilli and Walter 2013) coupled with infrared data from spacecraft. It has recently become clear that some SMBH-powered AGN are also undergoing a burst of star formation and that there can be a close connection between the star-formation history of such galaxies and the energy released by the AGN. We will return to this in Section 16.6.

## 16.2 Active Galactic Nuclei

Radio galaxies and quasars belong to an overall family of active galactic nuclei (AGN) powered by 'a central engine', at whose heart is an SMBH of mass $10^6-10^{10}$ M$_\odot$, sometimes dubbed a 'monster'. Padovani *et al.* (2017) estimated that $\sim 1\%$ of all galaxies exhibit high-level AGN activity at any one time and perhaps 10% show evidence of low-level activity. The defining characteristic of an AGN is its broadband continuum emission, which extends over a much wider spectral range than that of a normal galaxy. Some rare objects, the blazars, emit over the complete range from radio ($10^8$ Hz) to ultra-high-energy (TeV) gamma-rays ($\sim 10^{28}$ Hz), although most AGN spectra cut off in the X-ray band before $\sim 10^{19}$ Hz. The total energy output of an AGN can reach $10^{41}$ W ($>10^{14}$ L$_\odot$), which immediately indicates that the energy source is something other than the nuclear fusion which powers the $\sim 10^{11}$ stars in a massive galaxy. We shall see that this enormous energy output is generated from within a relatively tiny region of $\sim 1$ pc$^3$.

Active galactic nuclei have been intensively studied since the 1960s and have been classified in many different ways.[2] We concentrate here on the properties of *radio-loud* AGN,[3] defined as objects with $L_{1.4\text{GHz}} > 10^{23}$ W Hz$^{-1}$, whilst recognizing that their physical interpretation *a fortiori* requires the consideration of data from all wavelengths. While luminous AGN were first discovered as a result of radio observations, only a minority ($\sim 10\%$) of AGN are 'radio-loud'. We tie in with the 'radio-quiet' ones in Section 16.7. Obtaining redshifts, and thereby distances and luminosities, is pivotal for AGN studies. The main route has been via optical emission line spectra (see Section 16.4.1), but optical wavelengths suffer absorption by intervening dust and so obtaining redshifts for complete samples of radio sources is difficult. Millimetre/sub-mm-wave spectroscopy, in particular with ALMA, has opened a new route to the redshifts of distant dusty galaxies.

Three main classes of radio-loud AGN are recognized:

- *Quasars* These exhibit a wide range of radio structures and spectra and are bright continuum emitters across the EM spectrum; they have characteristic strong emission lines in the ultraviolet and/or optical. Most quasars are not radio-loud.
- *Radio galaxies* Their radio structures are generally more extended and have steeper spectra than those of quasars; they also exhibit a broader range of non-radio properties, some having strong emission lines and others not.
- *Blazars* There are two sub-classes, the BL Lacs and the flat-spectrum radio quasars (FSRQs).[4] Their radio structures are always dominated by compact components and they are very bright continuum emitters right up into the gamma-ray band. BL Lacs show flat, often featureless, optical spectra but the FSRQs show strong emission lines. Both classes are highly variable and often highly polarized.

We can provide only a simplified account of the phenomenology and physical understanding of radio-loud AGN. In doing so we have leaned heavily on some excellent reviews,

---

[2] In the review by Padovani *et al.* (2017) 51 different class acronyms are listed.
[3] A radio-loud AGN was originally defined as one having a ratio of radio flux density to optical flux density $R \geq 10$. We adopt a definition based on absolute radio luminosity.
[4] Sometimes called optically violently variable (OVV) quasars.

including Heckman and Best (2014) and Padovani *et al.* (2017), who discuss the AGN population as a whole, while Tadhunter (2016) focusses on radio-loud AGN. Other reviews and useful books are identified in the appropriate sections below.

Before describing their properties it is useful to sketch briefly the physical picture underlying our current understanding of radio-loud AGN. We provide more details in Section 16.5. The energy is generated by mass accretion onto the SMBH ($< 10^{-4}$ parsec scales) within the 'central engine' (parsec scales). Two antiparallel highly collimated and highly relativistic jets are launched from close (sub-parsec scales) to the SMBH; *strong radio sources are fundamentally linked with such jets*. Outside the nucleus the jets interact with the medium through which they drive; in this process ambient gas may be entrained. Initially the jets pass through the interstellar medium of the host galaxy (tens of kiloparsec scales), beyond which they encounter the ionized gas of the intergalactic medium and their energy is dissipated in giant 'lobes' (hundreds of kiloparsec scales). If the angle to the line of sight is small, 'relativistic beaming' (Sections 13.9 and 16.3.9) causes the forward pointing jet to appear much brighter than the backward pointing one. Orientation also plays a major role in the visibility of other emitting regions in the central engine owing to the screening effect of a surrounding dusty torus (tens of parsec scales). At some angles all regions can be seen but when the torus is viewed side-on most are hidden.

## 16.3 Radio-Loud AGN

There is a wide variety of radio source types, ranging from massive double sources that are over a megaparsec across to bright objects completely dominated by structure on the scale of tens of parsecs or less; the reasons will become clear as we proceed. They share common underlying features, outlined above, and radiate predominantly via the synchrotron mechanism (but note Section 16.5.1). We now describe their properties and their physical interpretation, starting with the largest. An excellent set of images, together with much explanatory material and appropriate references, can be found in *An Atlas of DRAGNs* (double radiosources associated with galactic nuclei).[5]

### 16.3.1 Classification of Extended Radio Sources

The enormous double-lobed structure straddling the parent galaxy seen in Cygnus A (Figure 16.4) turned out to be a general characteristic of radio-loud AGN. Cygnus A is $\sim 140$ kpc across but some radio galaxies can be over ten times larger. As the quality of radio interferometric images improved in the 1970s other common features of their structures were discerned. There was invariably a compact 'core' coincident with the centre of the host galaxy and often one or more thin 'jets' apparently emanating from it.

---

[5]  The *Atlas* is compiled and maintained by P. Leahy, A. Bridle, and R. Strom and can be found at http://www.jb.man.ac.uk/atlas/.

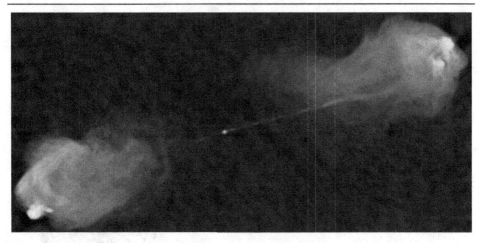

Figure 16.4. The FR II radio galaxy Cygnus A (Perley *et al.* 1984).

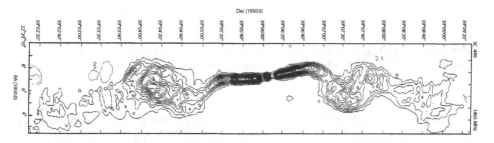

Figure 16.5. 3C449: an FR I radio galaxy, imaged at 21 cm by the VLA (Perley *et al.* 1979).

Fanaroff and Riley (1974) were able to discern two distinct patterns. In the most powerful sources ($L_{1.4GHz} > 10^{25}$ W Hz$^{-1}$) such as Cygnus A the 'outer lobes' are edge-brightened and show 'hot spots'. This type of structure was dubbed FR II. In lower-luminosity objects ($L_{1.4GHz} = 10^{23-25}$ W Hz$^{-1}$) the twin jets and lobes merge together and are not brightened at the outer extremities. Such structures were dubbed FR I and an archetypical example, 3C449, is shown in Figure 16.5.

### 16.3.2 *FR II Sources*

The theory of powerful double radio sources as being continuously supplied by energy from a central object was developed by the Cambridge group in the 1970s (e.g. Scheuer 1974 and references therein), essentially to account for FR II sources. The theory was well developed before many of the beautiful radio images shown in this chapter had been made. The book by De Young (2002, see the Further Reading section) provides a 'one-stop' update to this classical picture.

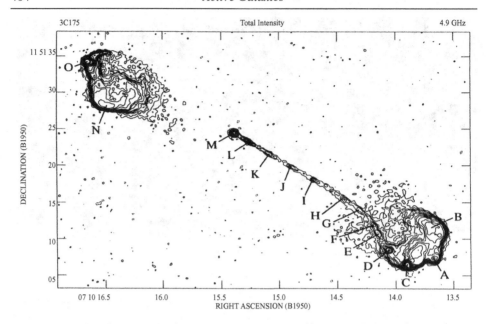

Figure 16.6. The FR II quasar 3C175 mapped at 5 GHz with 0.4 arcsec resolution: M is the central feature; the other labels indicate hot spots in the jet and lobes (Bridle *et al.* 1994).

In both Cygnus A and and 3C175 (Figure 16.6), a single thin jet can be seen stretching from the core to one of the lobes. In fact deeper images show that the other lobe is also linked to the centre by a faint jet. The jets can be amazingly well collimated over vast distances (hundreds of kpc and more). The detailed map of 3C175 shows that the radio brightness of the jet varies very little over its length. The symmetry of the outer lobe structures strongly suggests an intrinsic symmetry to the energy flow, and so relativistic beaming (Section 16.3.8) is the likely cause of the jet brightness asymmetry; the backward-directed jet can be attenuated to the extent that it falls below the map noise level. The implication is that the bulk flow remains relativistic and supersonic right out to the outer lobes. Looking again at 3C175, the small bright patches along the jet represent turbulent interactions with the surrounding interstellar and intergalactic medium. These interactions clearly do not cause the jet to dissipate much of its energy before it is deposited at the outer extremities of the lobes. A high degree of polarization of the radio emission, reaching 50% in some cases, suggests that a well-organized magnetic field may be involved in the stabilization of the jet but the physical processes involved in jet collimation remain poorly understood (e.g. Romero *et al.* 2017). Jets in FR IIs are generally straight; where large wiggles or bends are seen, this can be attributed to a projection effect in jets directed fairly close to the line of sight.

The radio lobes develop when the jets are stopped by the ram pressure of the diffuse intergalactic gas; at this point hot spots of radiation develop. The energy deposited at the hot spot flows back, evacuating a cavity, which we observe as radio lobes. The lobes contain fewer high energy electrons and consequently have steeper radio spectra than the jets. The

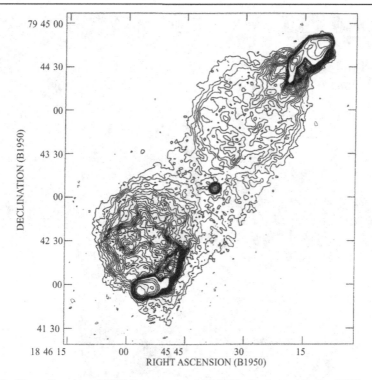

Figure 16.7. The radio galaxy 3C390.3, mapped at 1450 MHz (Leahy and Perley 1995).

entire cavity is therefore seen best at lower radio frequencies: Figure 16.7 shows a VLA map of 3C390.3 at 1450 MHz (Leahy and Perley 1995), in which a jet is almost invisible but the hot spots, cavities, and older parts of the lobes are prominent. While the average flows in the lobes are sub-relativistic, and so their brightness is unaffected by beaming, the hot spot fluxes may be beamed to some extent (e.g. Leahy and Perley 1995).

### 16.3.3 FR I Sources

In the lower-luminosity FR I sources such as 3C449 (Figure 16.5) the less powerful jets entrain material and become subsonic closer to the nucleus. This leads to a more diffuse structure with no obvious concentration at the outer edges as is found in FR II sources. Another consequence of entrainment is that jets in FR Is tend to look more symmetrical as they are decelerated quickly by the external medium and beaming effects are much reduced.

The nearby massive elliptical galaxy M87 provides an excellent example of an FR I source which, along with Centaurus A, has been more extensively observed than any other radio galaxy. It also exhibits some generic features of radio-loud AGN.

M87 was one of the first powerful radio sources to be identified and was originally known as Virgo A. The top left panel in Figure 16.8 shows an optical image of the galaxy taken with the Hubble Space Telescope. A jet, whose projected length is $\sim 1.5$ kpc or $\sim 5000$ light years, can be seen extending radially from the centre of the galaxy, and this is shown

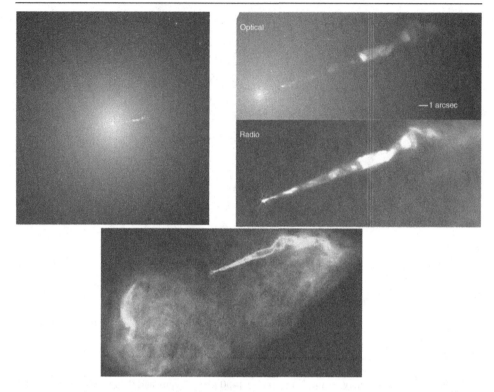

Figure 16.8. (top left) An HST optical image of the elliptical galaxy M87. The jet can be seen extend-
ing radially from the centre. (Credit: NASA, ESA, and the Hubble Heritage Team (STScI/AURA).
Acknowledgment: P. Cote (Herzberg Institute of Astrophysics) and E. Baltz (Stanford University)).
(top right) Optical (HST) and radio (VLA) images of the jet whose extent (projected) is $\sim 1.5$ kpc.
(Credit: J. Biretta; HST.) (bottom) A VLA image showing part of the extended lobe structure
(NRAO/AUI, courtesy of F. Owen).

in greater detail in the pair of images at top right; one of these is an HST optical image
and the other is a radio image taken with the VLA.[6] These two images show a remarkable
similarity, immediately suggesting that, like the radio, the optical radiation is due to the
synchrotron mechanism. The ordered structure in the jet is indicative of shocks. Only one
large-scale jet is seen, probably owing to the effects of Doppler boosting. The bottom panel
of Figure 16.8 is a VLA image of some of the extended radio structure; with no sign of
outer hot spots it is characteristic of an FR I source.

M87 has been extensively studied in the radio. In view of the relative closeness of the host
galaxy ($\sim 16$ Mpc) there is particular interest in imaging with VLBI arrays at millimetre-
wavelengths, where the physical scales probed in the inner jet can be as small as $\sim$ ten
times the radius of the central SMBH (the Schwarzschild radius $r_s = 1.8 \times 10^{13}$ m or
1000 light seconds for a mass of $6 \times 10^9$ M$_\odot$). Walker *et al.* (2018) and Kim *et al.* (2018)

---

[6] See Sparke and Gallagher (2007), *Galaxies in the Universe: An Introduction*, 2nd edn, Chapter 9.

describe the most recent results and give extensive references to previous work on M87. Walker *et al.* (2018) present results from 17 years of observations with the VLBA at 43 GHz; their latest radio image is shown in the top panel of Figure 16.9, in which the projected jet is traced over $\sim 25$ mas (i.e. $\sim 2$ pc or $\sim 6$ light years). It is clearly edge-brightened and near its base has a wide opening angle, which narrows down beyond a few milliarcseconds; a faint counter-jet can also be seen. Proper motions reveal that bright regions can move 'downstream' at apparent relativistic velocities of $\geq 2c$, but as a whole the jet propagates non-ballistically at significantly lower speeds; this is best seen in a 17 year movie.[7] Currently the highest resolution studies of the inner jet are by Kim *et al.* (2018) using the GMVA (Section 11.4.11) at 86 GHz; their image of the central $\sim 3$ mas is shown in the bottom panel of Figure 16.9. In these innermost regions the outer edges of the jet are expanding parabolically and the base is only $\sim 4-5.5\, r_s$ in extent. For a detailed astrophysical interpretation of these extraordinary images the reader should consult the original papers.

The less powerful jets in FR I radio galaxies can be distorted if the galaxy is moving with respect to its surroundings; this phenomenon is seen in NGC 1265 (Figure 16.10(a)). Such structures form a morphologically defined class of 'tailed' radio sources and invariably occur in clusters of galaxies. Depending on their exact shape they have been dubbed at various times 'head–tail', 'narrow-angle tail' and 'wide-angle tail'. A remarkable twin-source structure is seen in 3C75 (Figure 16.10(b)). Here jets emerge from two accreting black holes in the same galaxy (NGC 1128). The galaxy is moving at $\sim 1000$ km s$^{-1}$ through the cluster gas and jets on both sides are swept in the same direction (towards the northwest).

### 16.3.4 Core–Jet Sources

Not all extended radio sources fit simply into the FR I or FR II class. A well-known source with an anomalous radio structure, shown in Figure 16.11(a), is the first quasar to be identified, 3C273. It has a bright core and a highly asymmetric one-sided jet extending over 20 arcsec to the south-west. The radio jet is shown in more detail in Figure 16.11(b), where it is superposed on an HST image which reveals a corresponding optical jet. Deep radio images by Perley and Meisenheimer (2017) reveal additional steep-spectrum low-brightness emission to the north and west of the jet and compact core. Since the major axis of the radio sources does not, in general, lie in the plane of the sky a plausible (but not necessarily correct) interpretation is that we are seeing a radio source viewed in a direction where a 'backward' lobe lies behind the forward-pointing jet.

### 16.3.5 Repeated Outbursts

The circumstances in which the centre of a galaxy is triggered into becoming an AGN are, as yet, not fully understood. The duration of AGN activity is also uncertain, but

---

[7] On R. C. Walker's website http://www.aoc.nrao.edu/~cwalker/M87/index.html.

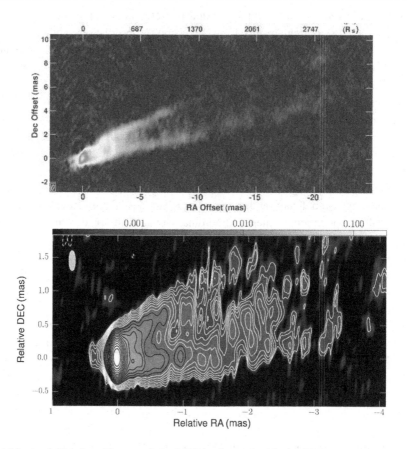

Figure 16.9. (top) The first 25 mas of the M87 jet imaged with the VLBA at 43 GHz (Walker *et al.* 2018); the faint emission is better seen in a colour rendition (see the Supplementary Material). (bottom) The inner 3 mas of the jet imaged with the GMVA at 86 GHz (courtesy of J.-Y. Kim and T. Krichbaum).

there are clues in several radio galaxies which exhibit *episodic* activity. Figure 16.12 shows the radio image at 1.4 GHz of the giant radio galaxy J0116-473, which has a central unresolved core and two superposed double-lobed sources, each of which has the appearance of a standard radio galaxy. The larger double is more diffuse, with bright edges, and is not fed with any obvious jets. The smaller double appears to be connected to the central object and is aligned on an axis which would be the axis of a larger jet, now non-existent, which might have powered the larger double. Other sources exhibit similar behaviour. The obvious interpretation of such double–double radio galaxies is that the activity of the central source is episodic within a timescale similar to the lifetime of the lobe structure, estimated at around $10^8$ yr (Komissarov and Gubanov 1994).

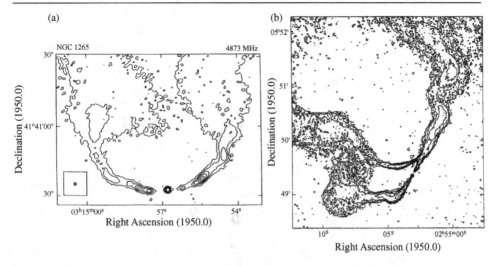

Figure 16.10. Two 'tail' sources: (a) distorted twin jets from NGC 1265 (O'Dea and Owen 1986); (b) the radio galaxy 3C75, showing two sources of radio jets within one galaxy (Owen *et al.* 1985).

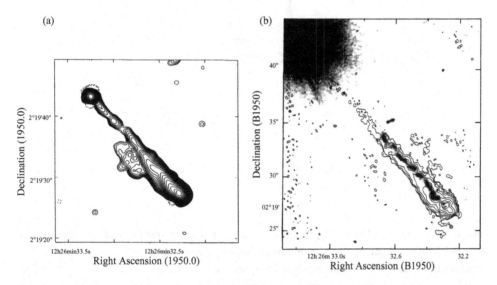

Figure 16.11. The quasar 3C273, an asymmetric core–jet source: (a) at 408 MHz (Davis *et al.* 1985); (b) at 1660 MHz, combining images from MERLIN and the VLA, superposed on a grey scale image from the Hubble Space Telescope and shown at double the scale (Bahcall *et al.* 1995).

## *16.3.6 Synchrotron Emission from Extended Sources*

The radio emission from outer jets and extended lobes is optically thin synchrotron radiation. A selection of radio-loud AGN spectra is shown in Figure 16.13; when the lobes dominate (as here in the case of 3C123 and the low frequency section in 3C84) the spectrum is *steep*, with spectral indices $\alpha$ distributed in a fairly narrow range around $-0.8$. The jets

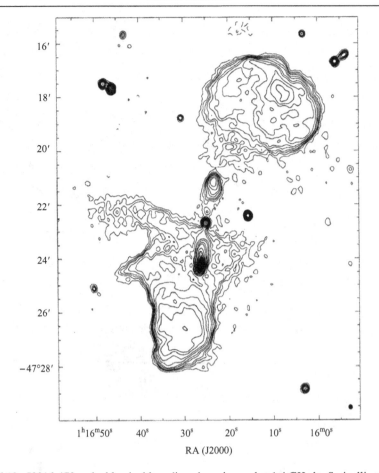

Figure 16.12. J0016-473, a double–double radio galaxy, imaged at 1.4 GHz by Saripalli *et al.* (2002, 2003), providing evidence for episodic activity.

also show a similar behaviour although there are gradients in spectral index in these outer jets and lobes indicative of electron energy losses. There is a range of inferred physical conditions in the lobes but the electron energies are of order $< 1-10$ GeV ($\gamma < 10^3$ to $\sim 10^4$) and the magnetic fields are in the range 1–10 microgauss. The total energy in the lobes can be found from the reasonable assumption of equipartition between particle energy and magnetic field energy; it reaches over $10^{53}$ J in the most luminous radio galaxies. Recalling Section 2.7.1, the relation $\alpha = (p-1)/2$ between the index $\alpha$, where $S(\nu) \propto \nu^\alpha$, and the electron energy spectrum $N(E) \propto E^{-p}$, we find an index $p = 2.6$, similar to that of cosmic ray electrons in our Galaxy.[8] The more unusual behaviour in these spectra is discussed in Section 16.3.8.

---

[8] Note the opposite sign convention in use for cosmic ray particles.

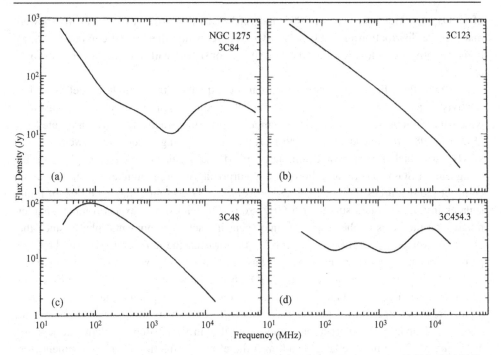

Figure 16.13. Typical spectra of radio galaxies (3C84, 3C123) and of quasars (3C48, 3C454.3) (Kellermann and Owen 1988).

### 16.3.7 Sub-Galactic-Scale Radio Sources

Radio surveys in the 1990s began to find some compact sources, with smaller than galactic dimensions, and in some cases < 1 kpc in size, and with spectral turnovers ranging from hundreds of MHz to >10 GHz. The spectra are peaked because the radio components are small enough to be self-absorbed (see also Section 16.3.8). Many such sources are now known, with various sizes, shapes, and spectra. They are usually designated according to their radio spectra as compact steep spectrum (CSS) or gigahertz peaked spectrum (GPS) objects. Reviews were given by O'Dea (1998) and by Snellen and Schilizzi (2002).

A closely related class, the compact symmetrical objects (CSOs) (Wilkinson *et al.* 1994), was discovered in the Caltech–Jodrell-Bank (C–J) VLBI surveys at 1.6 and 5 GHz (Taylor, G. B., *et al.* 1994; Henstock *et al.* 1995 and references therein). The class-defining object, 2352+495, has a structure resembling a low resolution image of an FR II radio source but its size, $\sim 150$ pc, is about $10^3$ times smaller. Further light was shed when 2352+495 was observed with higher VLBI resolution and at several epochs (Owsianik *et al.* 1999); one of the images is shown in Figure 16.14. The resemblance to extended FR II radio galaxies was confirmed and the separation of the two well-defined hot spots was seen to increase from 49.1 to 49.4 milliarcseconds over 10 years (Figure 16.14(b)). Dividing the hot spot separation by the expansion rate of 21 microarcseconds per year gives a separation velocity $\sim 0.3c$ and an age of only $1900 \pm 250$ years. Similar ages and velocities, or in some cases upper limits, have been found for a significant number of CSOs (e.g. Gugliucci *et al.*, 2005 and references therein). Gugliucci *et al.* showed that the age distribution in CSOs peaks at <500 years and that the jet components have higher relativistic velocities than the hot spots, consistent with passage down a cleared channel.

It is now thought that all these types can be explained by a single model in which relativistic jets encounter a dense interstellar medium. An evolutionary scenario in which the various sub-galactic sized sources evolve into each other was presented by O'Dea (1998). Readhead *et al.* (1996) proposed a unifying model for powerful radio sources in which CSOs evolve into large FR II radio galaxies (which might now be designated LSOs). There are, however, a surprisingly large number of these young compact radio sources. For example Polatidis *et al.* (1999) found that about 8% of radio sources in a flux-limited survey at 5 GHz are CSOs. One would expect a much smaller number on the basis of the length of time spent in each developmental phase, since the expansion velocities seem to remain similar. If, however, the radio emissivity in the CSO phase is much higher than in the FR II phase then the CSOs would be overrepresented despite their youth. Alternatively, the CSO phenomenon may be, in part, episodic in nature (see Section 16.3.5) and some of them might die young before having the chance to mature into large-scale FR II sources. Regardless of their future development, their sizes and their lack of polarization demonstrate that in these sources relativistic jets are encountering dense interstellar gas and must therefore be affecting their environment (see e.g. Bicknell *et al.* 2018).

### 16.3.8  Compact Radio Sources

Compared with the lobe-dominated FR I and FR II sources the jet-dominated radio sources such as 3C273 (Figure 16.11) also have brighter compact cores. In other sources, predominantly quasars and blazars, the core dominance can be greater and in some cases there is little or no sign of extended (arcsec-scale) emission even in high dynamic range maps. These are highly beamed sources (see the next section), where a jet is pointing close to the line of sight. Their radio spectra are strongly affected by synchrotron self-absorption (Section 2.7.2 and Figure 2.12). In practice the simple core spectrum with a peak at a radio frequency $\nu_c$ (the 'turn-over frequency') is never observed, since the emitting regions are not uniformly bright. In some cases, as in the quasar 3C48, there is a smooth distribution in the sizes of the emitting regions and the transition from optically thin to optically thick takes place gradually – the result is a curved spectrum. More often, however, compact source spectra are *flat* (practically defined as $\alpha < 0.5$) over a wide frequency range. This behaviour is exemplified by the FSRQ blazar 3C454.3 (a bright gamma-ray source), whose spectrum exhibits several peaks indicative of distinct components of different sizes and turn-over frequencies; together these produce a composite spectrum which is roughly flat. In Figure 16.13 the 'bump' in the high frequency section of the 3C84 spectrum shows that, in addition to extended lobes, this source also has a bright self-absorbed core. The flux densities of compact flat-spectrum sources are often found to be variable (see Section 16.5.1).

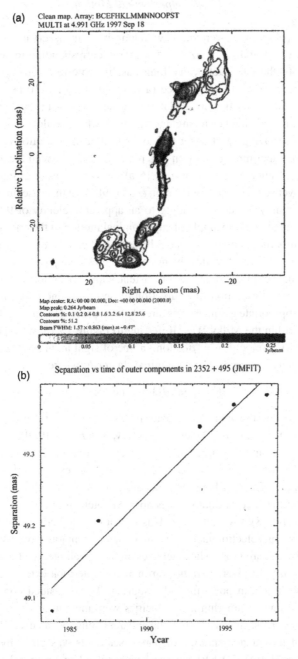

Figure 16.14. (a) A global VLBI image of 2352+495, a compact symmetrical object with $z = 0.238$. Three components are resolved into a core and double-jet structure closely resembling an FR II radio galaxy. In a series of similar observations over 10 years, the angular separation of the hot spots was found to increase linearly, as shown in (b) (Owsianik *et al.* 1999).

### 16.3.9 Superluminal Motion

The brightest compact sources were found with long-baseline observations in the late 1950s and early 1960s, which indicated sizes < 0.1 arcsec corresponding to sub-galactic physical dimensions. In the late 1960s very long baseline interferometry (VLBI) came into operation. At a resolution $\sim 1$ milliarcsec (a physical size of $\sim 10$ pc at $z = 1$) it was soon noticed that large flux density variations in the cores of bright radio quasars were accompanied by marked changes in angular structure, which could be interpreted as being due to components moving apart faster than the speed of light (Cotton *et al.* 1979; Cohen *et al.* 1977). Such 'superluminal' motion was put beyond doubt with the advent of reliable VLBI images (see Section 10.10). Figure 16.15(a) shows the classic example of the core of 3C273, mapped over several years by Pearson *et al.* (1981). In this case, the angular velocity was $0.76 \pm 0.04$ mas yr$^{-1}$, corresponding to an apparent velocity of $9.6 \pm 0.5c$. Figure 16.15(b) shows another early example of superluminal motion in the quasar 3C345 (Biretta *et al.* 1986); by this time techniques had advanced sufficiently to show that components could accelerate and change direction as they moved down the jet.

The superluminal jet phenomenon is consistent with conventional physics and can be understood as relativistic motion in a direction close to the line of sight (e.g. Blandford *et al.* 1977). The appropriate geometry has already been presented, in Chapter 13, in relation to radio sources within the Milky Way. Referring to Figure 13.13, we repeat the result for a source moving with velocity $v = \beta c$ along a line of sight at angle $\theta$ to the line of sight. The apparent velocity is $c$ times a 'superluminal' factor $F$, where

$$F = \beta \sin\theta (1 - \beta \cos\theta)^{-1}. \tag{16.1}$$

The maximum effect on the apparent velocity is at $\sin\theta = \gamma^{-1}$, where $\gamma$ is the relativistic factor $\gamma = (1 - \beta^2)^{-1/2}$. For the typical observed values $F = 5$–$10$, the minimum required value of $\gamma$ is about 7, and the largest effect is seen for line-of-sight angles $\theta = 5°$–$20°$. The effect is still important at $\theta = 45°$, where $\gamma = 7$ gives a superluminal velocity of $3c$. The corresponding Doppler boost of flux density compared with that from a stationary source depends on the spectral index but is approximately $\gamma^3$ when the velocity effect is a maximum, and up to $8\gamma^3$ when $\theta = 0$. This increase by a factor of up to 1000 or more gives rise to a powerful selection factor, in which the large majority of observed bright flat-spectrum radio sources are those whose jet axes are nearly aligned with the line of sight.

Following the early discoveries, radio astronomers undertook extensive VLBI observations of parsec-scale jets in powerful radio sources. The one-sided core+jet structure is ubiquitous and a range of superluminal velocities were measured; a typical value is $\sim 5c$ although in some sources, such as 3C279, the apparent velocity can exceed $15c$. A comprehensive review of the phenomenology of parsec-scale jets was given by Zensus (1997). More recently the MOJAVE VLBI surveys[9] have aimed to provide long-term systematic monitoring of relativistic motion in jets, focussing on bright, highly variable, blazars. Lister *et al.* (2018 and references therein) provide an overview of the programme.

---

[9] See the images in https://www.physics.purdue.edu/MOJAVE/.

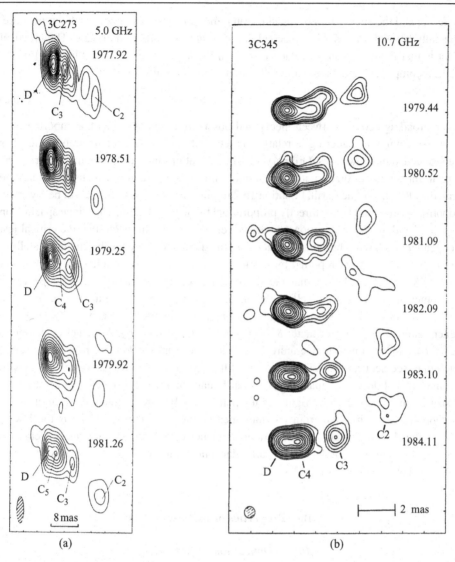

Figure 16.15. Superluminal motion in the cores of: (a) 3C273 (VLA at 5 GHz; Pearson *et al.* 1981); (b) 3C345 (VLA at 10.7 GHz; Biretta *et al.* 1986).

### 16.3.10 Brightness Temperatures in Compact Sources

Core radio components typically have an angular diameter of the order of one milliarcsecond, corresponding to a linear size of a few parsecs at moderate redshifts. This small angular size implies very high brightness temperatures in the range $10^{11}$–$10^{12}$ K. Following the discussion in Section 2.8, this observation is consistent with a calculation by Kellermann and Pauliny-Toth (1969), who showed that there is a brightness temperature limit for a static relativistic electron gas set by inverse Compton scattering. This is known as the *self-Compton temperature limit*. The ratio of the particle energy loss by inverse Compton scattering and the energy loss by synchrotron radiation increases rapidly with temperature, especially beyond $10^{12}$ K where second-order scattering starts to dominate. The practical effect is that there is a simple relation between the angular size $\theta$ (in milliarcsec) of the smallest component and its maximum flux density $S_m$ (Jy) at $\nu_c$ (GHz):

$$\theta \simeq S_m^{1/2} \nu_c^{-1}. \tag{16.2}$$

The broad agreement between theory and observation demonstrated the important role of inverse Compton scattering in relativistic jets, but the brightness temperature limit has been exceeded in a few cases. Ordinary VLBI observations have difficulty in measuring the angular diameters of objects with such high brightness temperatures, because baselines are limited by the size of the Earth. Equation 16.2 implies that the limiting size will be inversely proportional to $\nu_c$ and hence directly proportional to the wavelength. The fringe spacing, on the other hand, is also proportional to the wavelength, and it turns out that the critical test is dependent on having baselines greater than the diameter of the Earth. The first satellite-based observations were reported by Linfield *et al.* (1989); see Section 11.4. They found that, at a baseline of 2.3 Earth diameters, several sources exhibited brightness temperatures that somewhat exceeded the self-Compton limit. In principle the relativistic effects can explain this observation since the intrinsic brightness temperature can be boosted to a higher value by the Doppler factor $\delta = (1 - \beta^2)^{1/2}/(1 - \beta \cos \theta)$. A greater challenge to the $< 10^{12}$ K brightness temperature limit has come from the *RadioAstron* Space Radio Telescope (see Section 11.4). Observations of the quasar 3C273 (Kovalev *et al.* 2016) with baselines of 171 000 km (equal to over 13 Earth diameters) have revealed angular structure of size 26 microarcsec with brightness temperature $\sim 10^{13}$ K. Kovalev *et al.* argued that the Doppler factor inferred from previous measurements of the 3C273 jet is too low to produce the observed brightness temperature, and they called into question the conventional incoherent synchrotron model for this particular source. Further *RadioAstron* measurements will show how general this result is.

## 16.4 Other Properties of Radio-Loud AGN

### 16.4.1 Optical Emission Lines

Optical spectral lines are observed in emission from most radio-loud AGN. They are predominantly from a small number of atomic species, in some cases partly ionized.[10] In the visible band some of the brightest are also seen in H II regions, e.g. H (Balmer lines), CIII] and CIV, [NII], [OII], [OIII], and MgII. There are *broad lines*, with velocity widths of $\sim 10\,000$ km s$^{-1}$, and *narrow lines* with typical velocity widths of a few hundred km s$^{-1}$. Within these generalities the diverse emission line characteristics of AGN have been variously categorized by many authors over the years (see Tadhunter 2016 and

---

[10] The astronomical nomenclature is as follows: a neutral species is indicated by I, a singly ionized species by II, and a doubly ionized species by III, etc; lines with no brackets are quantum mechanically allowed, those with a single bracket are 'semi-forbidden' while two brackets indicate 'forbidden'.

Padovani *et al.* 2017). Here we give only a simplified version, focussing on radio-loud objects and following Tadhunter's (2016) nomenclature. The links between extended radio structure and optical line emission are summarized in Figure 16.16.

- *Broad line radio galaxies and quasars* (BLRG/Q)  Powerful radio galaxies and quasars exhibit both broad and narrow lines and strong continuum emission; their radio classification is predominantly FR II but some FR II sources show NLRG or WLRG spectra (see below).
- *Narrow line radio galaxies* (NLRG)  Some radio galaxies show only narrow lines accompanied by weak continuum emission; their radio classification is invariably FR II.
- *Weak line radio galaxies* (WLRG)  In many FRI radio galaxies the optical spectrum is dominated by the starlight of an elliptical galaxy, which shows marked stellar absorption features; only one or two weak lines may be seen.
- *No line objects*  Some blazars, in particular the BL Lac objects, show only a featureless continuum spectrum. These are not shown in Figure 16.16.

### 16.4.2 Spectral Energy Distributions (SEDs)

A characteristic of AGN is their broadband continuum spectra extending from radio to X-rays and sometimes gamma-rays; in all cases, however, the relative amount of power emitted in the radio band is tiny ($< 1\%$). Figure 16.17 shows a schematic SED associated with an FR II radio source. As we explain in Section 16.5.1, the radiation comes from a mixture of spatially separated regions in and around the central engine. In practice there is a wide scatter in the power radiated in the different bands due to: (i) differences in physical conditions; (ii) intrinsic variability including flaring episodes; and (iii) the orientation of the particular AGN with respect to the line of sight.

All AGN are intense emitters in the X-ray band (0.2–200 keV; $4 \times 10^{16} - 4 \times 10^{19}$ Hz) and so X-ray surveys are the most direct route to finding them. The reasons are that: (i) they are produced close to the heart of the central engine; (ii) they suffer relatively little from absorption by gas and dust, especially at higher energies; (iii) there is little contamination from other X-ray sources in the host galaxy. Radio-loud AGN are the *only* cosmic sources of energetic gamma rays from $\sim 100$ MeV and in some cases up to $>10$ TeV ($>10^{27}$ Hz). The Fermi spacecraft has made the greatest contribution to their study, see the review by Dermer and Giebels (2016). The most intense sources are both types of blazars, which, while rare in the AGN population as a whole, dominate the extragalactic gamma-ray sky.

### 16.4.3 Variability

Multiwavelength variability is a characteristic of AGN, particularly in blazars. Radio–mm outbursts, on timescales from a few days to a few weeks, can be ascribed to synchrotron emission from shock waves propagating down the inner relativistic jet (see Figure 16.20). In turn these can be ascribed to variations in the accretion flow rate in the central engine. In the radio band the Metsähovi and Michigan groups provided long-term monitoring of bright compact radio sources for several decades. Their joint results, primarily covering the range 4.8 GHz to 87 GHz (and for some sources to 230 GHz with data from other telescopes)

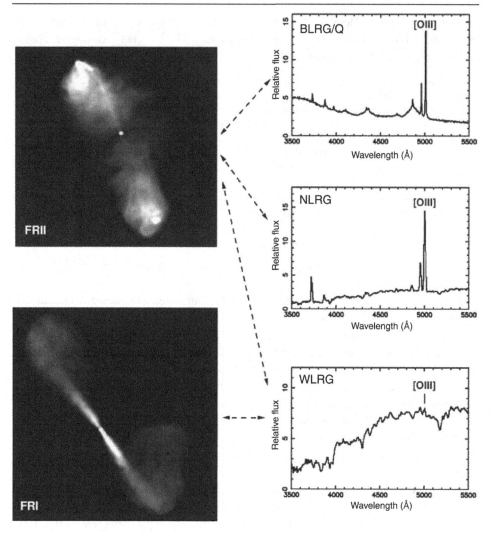

Figure 16.16. Classifications of radio-loud AGN. (left) Radio morphology: the FR II source is 3C98; the FR I source is 3C296. (right) Optical spectroscopy. The broken lines indicate the links between the classes. (Credit: Clive Tadhunter.)

on 90 sources were summarized by Hovatta *et al.* (2008). They found a complex range of variability timescales, with flaring events typically happening every few years. The twice-weekly Caltech 15 GHz monitoring programme[11] of >1400 blazars was begun in 2008. Variability takes place on timescales of less than a week up to many months, and, while there is no difference in the behaviour of BL Lacs and FSRQs, Richards *et al.* (2014) found a connection between gamma-ray emission and radio variability in FSRQs but not in BL Lacs. They attributed this to the higher Doppler factors in FSRQ jets. A few blazars

---

[11] www.astro.caltech.edu/ovroblazars.

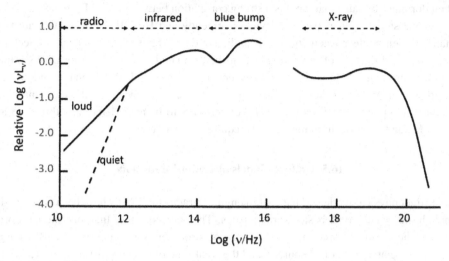

Figure 16.17. Schematic representation of a typical spectral energy distribution (SED) in an AGN associated with an FR II radio source. Galaxies linked to FR I sources do not show the the 'bump' at rest-frame ultraviolet wavelengths; blazars show an additional broad spectral hump above $10^{20}$ Hz (see Dermer and Giebels 2016, and Section 16.5.1 below).

show a specific link between radio and gamma-ray variations, with those in gamma-rays leading the radio by $\sim 100$ days (Max-Moerbeck *et al.* 2014). Their interpretation is that the gamma-rays come from the relativistic inner jet, and only when the opacity of the expanding jet has decreased sufficiently can radio emission be seen; this happens 'downstream' and hence at a later time (see Figure 16.20 and the discussion on the relativistic jet in Section 16.5.1).

At other wavelengths the fastest variations (hours to minutes) are seen in the X-ray band and these are thought to be associated with instabilities in the accretion disk close to the SMBH. Significant variations (say, a factor 2) with timescales $\Delta t$ limit the size of the X-ray emitting region to $\sim c\Delta t$, since the light travel time across the region has the effect of smearing out variations on shorter timescales. The timescales of $10^3 - 10^4$ seconds correspond to physical scales in the accretion disk comparable to the solar system.

### 16.4.4 Host Galaxies and SMBHs

Heckman and Best (2014) provided a panoramic overview of the relationship between host galaxies and SMBHs. A common thread connecting radio-loud AGN of all classes is that they are situated in massive elliptical galaxies. The total stellar mass can be estimated from the integrated starlight from the galaxy and the assumption of a typical mass-to-light ratio. Black hole masses can be obtained in a variety of ways. For nearby galaxies (mostly not AGN) the nuclear region can be resolved and its kinematics studied with optical spectroscopy; dynamical modelling then provides an estimate of the central mass dominated by the SMBH. By this means it has been shown not only that most galaxies harbour an SMBH (albeit dormant) but also that there is a strong correlation between the SMBH mass and the mass of the stellar 'bulge' (see the review Kormendy and Ho 2013). Most AGN are too distant for their nuclear kinematics to be studied in detail (but see a special case in Section 16.5.1) and so the mass of the SMBH has to be estimated by secondary means (Section 16.5.1). There is a strong correlation between radio power and SMBH mass, and so the most powerful AGN, and hence radio sources, are located only in the the most massive galaxies.[12] To power a radio-loud AGN there appears to be an SMBH mass threshold at $\sim 10^8$ $M_\odot$, and the most luminous objects require black hole masses up to $10^{10}$ $M_\odot$.

## 16.5 Unified Models of Radio-Loud AGN

Two distinct physical models of the central engine in radio-loud AGN have been developed, although they share some basic characteristics. The need for more than one model arose from the dichotomy between the two classes of strong-emission-line objects (BLRG/Q, NRLG), associated with FR II sources, and the weak-line objects (WLRG) associated with FR I sources. There was also the need to explain two classes of blazars with different optical-emission-line properties. The review by Urry and Padovani (1995 and references therein) described these models, which have become known as 'unified schemes'.

Since they produce radio-loud AGN, both types of central engine produce powerful relativistic jets. However, to anticipate the discussion to come, in the *radiative-mode* objects the emission lines are strong and the bulk of the broadband continuum radiation comes from a radiatively efficient accretion disk and its surroundings; the inner jet adds beamed radiation when it is pointing approximately towards us. Radiative-mode AGN only launch powerful jets in a minority ($\sim 10\%$) of cases and these are the most luminous objects (FR II class); the majority of radiative-mode AGN are *radio-quiet*. Lower-power (FR I class) radio-loud AGN are predominantly *kinetic-mode* objects; all kinetic-mode objects launch powerful jets, hence their name. The jets account for most of the broadband continuum radiation while the accretion disk is radiatively inefficient. In most cases the starlight from the galaxy can therefore be seen and the emission lines are weak.

### 16.5.1 Radiative-Mode AGN

The model is schematized in Figure 16.18, and right at the start we re-emphasize the point that the direction from which one views the central region of an AGN determines its observed properties.

**The SMBH**   The energy derives from the release of the gravitational potential energy of the surrounding material as it falls into the black hole. This idea had been conjectured since the discovery of quasars but it was Lynden-Bell (1969), in a seminal paper, who put the idea on the now-accepted footing. Using simple Newtonian arguments he showed that the energy

---

[12] With masses 3–20 times that of the Milky Way's mass, which is $\sim 6 \times 10^{11}$ $M_\odot$

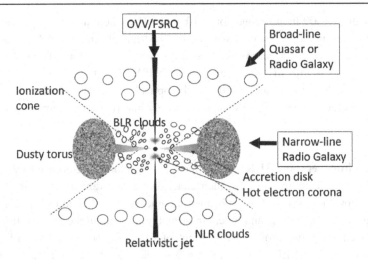

Figure 16.18. Highly schematic representation of the central engine in a radiative-mode AGN (not to scale). The appearance of the AGN, and hence its classification, depends upon the view angle. See the main text and Figure 16.20 for the wide range of physical scales involved.

available from the infall of a mass $m$ onto the last stable orbit of an SMBH was a significant fraction ($\sim 10\%$) of the rest mass energy;[13] this is an order of magnitude greater than is available (0.7%) from the nuclear fusion processes in stars. Despite the high conversion efficiency, the enormous energy output of quasars, $>10^{40}$ W, requires the conversion of $>10$ M$_\odot$ per year, with the most luminous objects demanding ten or even a hundred times this rate. Recognizing that this consumption rate was not sustainable over the lifetime of a galaxy, Lynden-Bell (1969) realized that many galaxies should harbour an inactive SMBH at their centre and this has proved to be the case. A pedagogic review of black holes in astrophysics was given by Bambi (2017).

**The accretion disk**  The infalling material may be gas from the interstellar medium or the intracluster medium, but it may also include stars that are disrupted as they pass within the Roche limit of the black hole. Angular momentum of the infalling material concentrates the infalling material to an accretion disk with steep gradients of angular velocity. Further collapse occurs through magnetically linked dissipation in the differentially rotating disk, and by turbulent dissipation. Gradually material spirals in towards the SMBH until it reaches the last stable orbit at a distance of order $\sim GM/c^2$. For a hole of mass $10^8$ M$_\odot$ this

---

[13] A simplified version of the argument goes as follows: material of mass $m$ moves from a great distance, where its potential energy PE = 0, towards mass $M$. Along the way it exchanges potential energy for kinetic energy. At a distance $R$ it has lost PE = $GMm/R$ but not all of this is available for radiation since the mass has angular momentum and becomes part of an orbiting accretion disk (see above). To maintain a stable orbit requires that the orbital KE = PE/2 (the virial theorem) and hence only $GMm/2R$ of energy is available to be radiated away during infall. If material is accreting at a rate $\dot{m} = dm/dt$ then the total luminosity $L = GM\dot{m}/2R$. The last stable orbit around a stationary black hole is three times the Schwarzschild radius (which defines the event horizon) i.e. $R = 6GM/c^2$ and hence $L = \dot{m}c^2/12$.

corresponds to $\sim 1000$ light seconds or $\sim 10^{-5}$ pc. The whole disk is only a few parsecs across so that it cannot be resolved with optical instruments (note that 1 pc corresponds to 1 milliarcsec even for a nearby AGN at a distance of 200 Mpc). The radiation from the disk is thermal but the disk is hotter close to the SMBH, emitting ultraviolet radiation, and cooler towards the outside, emitting in the optical and infrared bands. The observed spectrum is therefore a composite of approximately blackbody radiation over a range of temperatures.

**The broad-line region (BLR)**   The line-emitting clouds are ionized by ultraviolet radiation from the accretion disk but their exact disposition remains unclear. The broad-line widths show that the clouds must be moving quickly ($\sim 10\,000$ km s$^{-1}$) but some may be infalling, some outflowing, and others in orbital motion about the SMBH. It is possible to model the emission-line profiles with a range of different models giving similar profiles. Whatever its form, the BLR must be hidden from view by the dusty torus when viewed from the side (see Figure 16.18). The BLR's typical size, as deduced from the emission-line variability, is light months in low-luminosity objects and up to a few light years in luminous quasars.

**The narrow-line region (NLR)**   The NLR extends high above the disk, where it can be observed unobscured by the dusty torus. It can be spatially resolved in nearby low-luminosity AGN, where its morphology often shows a biconical structure with a broad opening angle. This suggests that the ionizing photon field is collimated by the dusty torus, as indicated in Figure 16.18. In these nearby AGN the dimensions are typically a few hundred pc; in bright quasars the NLR dimensions are inferred to be a few kpc. As expected with these dimensions, the line-emitting clouds are moving relatively slowly (at hundreds to $\sim 1000$ km s$^{-1}$) and, in contrast to the BLR, no clear variations of the emission lines are observed in response to large continuum variations.

**The mass of the SMBH**   Except in nearby galaxies the gravitationally influenced region around an SMBH is too small to resolve spatially (but see the Event Horizon Telescope, Section 11.4.11) and so the mass must be determined by secondary methods. The standard technique relies on 'reverberation mapping'. The time delay between variations in the ionizing continuum emission and variations in the strong broad emission lines is a measure of the light crossing time of the BLR and hence of its radius $R$. The mass $M$ of the SMBH is then determined, using simple Newtonian arguments, as $M \approx R\Delta V^2/G$, where $\Delta V$ is determined from the Doppler width of the broad lines emitted by clouds assumed to be gravitationally bound to the SMBH. For more details and the description of a data base of SMBH masses determined in this way, see Bentz and Katz (2015 and original references therein). Another way to set a lower limit on the mass is via the rate at which a radiating mass can accrete material. There is a battle between the outwards pressure from escaping photons and the inward attraction of gravity. For gravity to triumph the maximum luminosity for a given SMBH mass is $L \leq 3.8 \times 10^4 (M_{SMBH}/M_\odot)$ $L_\odot$.[14] This is known as the Eddington limit, $L_{Edd}$. By this means one can estimate SMBH masses in luminous AGN to be $> 10^8$ $M_\odot$.

**The dusty torus** Surrounding the thin accretion disk there is a thicker torus of accreting material, which is cooler and sufficiently opaque, owing to its dust content, to make the thin accretion disk and the BLR invisible when viewed from the side. There is, however, an excess of infrared radiation, indicating the presence of large dust clouds, and several AGNs contain huge $H_2O$ masers, analogous to those observed in molecular clouds around some stars in the Milky Way (Chapter 12). The torus is therefore depicted as a cool dusty molecular cloud whose inner radius is in the range 1–10 pc and whose outer radius is 50–100 pc. The amount of sky which the torus blocks can be judged from the shapes of some resolved NLRs, which are consistent with a cone of ionization of angle 60–90 degrees. Over half the sky must be blocked from view, as seen from the SMBH.

Direct observation of the torus in a typical AGN requires an angular resolution of the order of one milliarcsecond. Fortunately, this is available from VLBI; furthermore, the narrowness of the water maser lines may allow the dynamics of the torus to be revealed from the distribution of spectral components across the central regions of the AGN. This was achieved by Miyoshi *et al.* (1995) in the nucleus of the spiral galaxy NGC 4258 (M 106), where the water maser sources are seen to be strung along a line, and their velocities vary progressively along the line, towards us at one end and away at the other (Figure 16.19). They are convincingly represented as being located in a flattened inner part of the torus, which is rotating at velocities up to 900 km s$^{-1}$ round a massive central object. This has a mass of $\sim 40$ million solar masses within a radius of 0.13 pc (five light months) and hence is a most convincing candidate for a black hole.

The water masers of NGC 4258 illustrate well the origin of maser emission, which may be either the selectively amplified emission from a continuum source lying behind the maser or self-generated within the maser itself. Figure 16.19 shows the torus as a ring round a compact central source, which cannot itself be seen directly. The diagrammatic spectrum, shown below the torus, has components at velocities around $\pm$ 900 km s$^{-1}$, which derive from the edges of the rotating torus, where the line of sight through the torus has its maximum length. There is also a cluster of central components due to the amplified radiation from the central source, which shows a gradient of velocities across the central source. This allows the acceleration to be determined, and hence the central mass, as noted above.[15] The spin axis of the AGN is aligned with the radio jets which are seen at much

[14] The outward radiation flux over a sphere of radius $r$ is $L/4\pi r^2$; thus the momentum flux, i.e. the outward radiation pressure, is $L/4\pi r^2 c$. The Thomson cross-section for scattering by an electron is $\sigma_T$ and hence the outward force on an electron is $\sigma_T L/4\pi r^2 c$. The inward attractive force due to a mass $M$ on a proton–electron pair in an ionized plasma is $GMm_p/r^2$ (only the mass $m_p$ of the proton is important). In order to accrete, the luminosity $L \leq 4\pi Gcm_p M/\sigma_T$.

[15] Further VLBI observations (Herrnstein *et al.* 1999; Humphreys *et al.* 2013) have provided measurements of the lateral movement of the central components, giving their orbital velocities. Combining the angular size and the velocities gives a distance $7.6 \pm 0.23$ Mpc to NGC 4258, which is both precise and independent of all

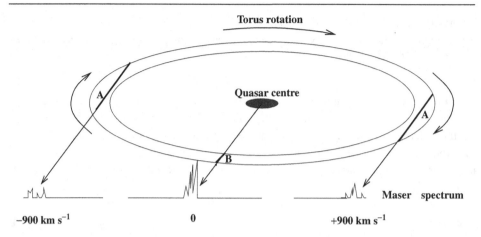

Figure 16.19. Maser emission from the inner regions of a torus in the AGN of the galaxy NGC 4258. The central maser, at the systematic velocity of the galaxy, is energized by the quasar core; the outer lines have sufficient path length to self-energize.

larger distances, about 500 pc, from the nucleus. The rotation axis of the whole galaxy is, however, at an angle of 119° to that of the nucleus and its jets. Such misalignment seems to be common; the behaviour of the AGN is often independent of the alignment of the main bulk of the galaxy.

**The relativistic jets**   Marscher (2006) provides an overview of relativistic jets in AGN, and the book *Relativistic Jets from Active Galactic Nuclei*, Boettcher *et al.* (2012) (see the Further Reading section) contains extensive reviews of the phenomena. Figure 16.20 (which complements Figure 16.18) shows a schematic view of the inner region of the central engine traversed by the jets. Only one jet is shown for simplicity and, indeed, only one jet is usually seen in the VLBI images, owing to differential Doppler boosting. The emerging blobs of emission, seen in VLBI maps of superluminal motion, may be discrete concentrations in moving material or shock fronts moving along a smoother stream. It is likely that the unresolved core in VLBI images is simply due to the base of the jet, where it becomes optically thick at a particular frequency. Boosted synchrotron radiation can extend from radio up into the X-ray band; emission at high X-ray and gamma-ray energies is associated with the inverse Compton scattering of photons by the ultra-relativistic particles in the jet (see also Sections 16.3.10 and 16.4.3).

Discussion of the complex astrophysics involved in jet production and collimation is beyond the scope of this book but, as first suggested by Blandford and Znajek (1977), the process is likely to be associated with magnetic fields anchored in the accretion disk and wound up by differential rotation. The plasma jet is encircled with a toroidal magnetic

other distance determinations. The Megamaser Cosmology Project Survey (e.g. Kuo *et al.* 2015 and references therein) has extended this programme to other objects to provide an independent measurement of the Hubble constant (see also Appendix 2).

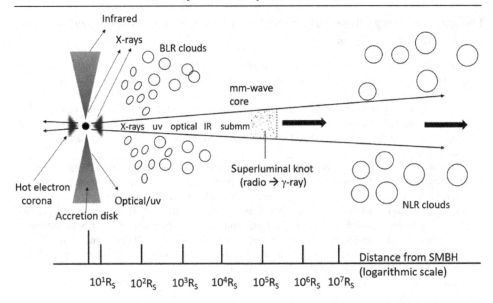

Figure 16.20. Highly schematic representation of part of the central engine in a radiative mode AGN, showing details of the relativistic jet. The scale is in terms of the Schwarzschild radius $R_S$ of the SMBH, $2GM/c^2 = 3 \times 10^{-5}$ pc for $M = 10^8 \, M_\odot$ (after Marscher 2006).

field, which confines it by hoop stress. This magnetic 'corset' may continue to play a role in confining the jet out to distances of hundreds of kpc, although the processes involved in the collimation of jets remain uncertain.

**Orientation effects**  As shown in Figure 16.18, seen edge-on, with the jet in the plane of the sky, the torus can obscure the accretion disk and the BLR, but not the NLR; hence we see NLRGs. The BLRG/Qs are viewed from an intermediate angle over the edge of the torus. From within the relativistic beaming half-angle of the jet axis, $\theta < 1/\gamma$, the beamed continuum from the jet reaches its maximum and from this direction we see FSRQs. A key piece of evidence, which demonstrates that BLRG/Qs and NLRGs are not two different species, was obtained by Antonucci and Miller (1985). They showed that in narrow-line objects, where the torus axis is surmised to be roughly edge-on, both continuum and broad-line radiation can be observed in polarized light. The interpretation is that this is radiation from inside the torus scattered by hot electrons in a 'corona' which lies above the torus. The high degree of polarization (15%–20%) is a characteristic feature of scattering.

   The supporting evidence for orientation differences within the populations is that radio galaxies are larger than quasars. Furthermore, when one jet is much brighter than the other, the lobe being fed by the brighter jet is always more highly polarized than the other lobe. This is the Laing–Garrington effect (Garrington et al. 1988), which can be understood as differential depolarization due to Faraday rotation: the radiation from the more distant lobe has to traverse a greater distance through the surrounding medium.

**The spectral energy distribution (SED)**   We are now in a position to outline a simplified interpretation of Figure 16.17.

– *Radio*   The entire radio source only accounts for $\sim 1\%$ of the total energy up to $\sim 10^{12}$ Hz.
– *Infrared*   The dusty torus reprocesses ultraviolet radiation from the accretion disk; this energy
  is reradiated in the infrared and dominates the spectrum from $\sim 1$ to $\sim 100$ microns ($3 \times 10^{14}$
  to $3 \times 10^{12}$ Hz). The inner parts of the torus are hotter and radiate at the shorter wavelengths.
  Orientation-dependent effects are evident in the fact that quasars are more luminous in the mid-
  infrared than radio galaxies, owing to Doppler-boosted synchrotron emission and dust extinction
  in radio galaxies.
– *Optical/ultraviolet*   The thermal spectra from the accretion disk, the ultraviolet from the hotter
  parts closer to the SMBH, and the optical emission from the cooler regions further out combine to
  produce a characteristic 'blue bump' peaking at $\sim 10^{15}$ Hz (the rest frequency).
– *X-rays*   Thermal radiation from the innermost parts of the accretion disk is boosted to X-ray (keV)
  energies by inverse Compton scattering off electrons in a hot compact (tens of Schwarzschild
  radii) 'corona' over the SMBH and the central part of the disk (e.g. Wilkins and Gallo 2015).
  Higher energy X-rays come from relativistically boosted inverse Compton radiation from the inner
  relativistic jet (Figure 16.20); see also Section 16.3.10.
– *Gamma-rays*   As described above, the emission is Doppler-boosted radiation from regions close
  to the base of the relativistic jet.

### 16.5.2  Kinetic-Mode AGN

A related but distinctly different physical model has been developed to account for the
weak-emission-line objects (WLRGs) typically associated with FR I sources. A schematic
picture is shown in Figure 16.21.

In the *kinetic-mode* objects the energy is still associated with infall onto an SMBH but
the accretion disk is different from that in the radiative-mode objects. It is thought to be

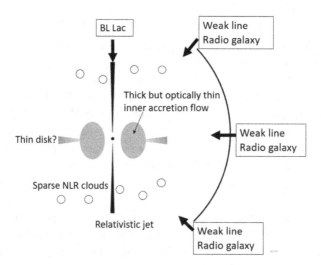

Figure 16.21. Highly schematic representation of the central engine in a kinetic-mode AGN
(not to scale) (after Heckman and Best 2014).

physically thicker but optically thin and therefore radiates only weakly; this explains the absence in WLRGs of the 'blue bump' in the spectral energy distribution in Figure 16.20. Most of the radiated energy from the nuclear region comes from the inner relativistic jets. It is not clear whether obscuring tori are present but they certainly are much less evident than in radiative-mode objects. There is little gas around the nucleus to give rise to ultraviolet or optical emission lines. A unifying feature is that the BL Lac objects can be interpreted as kinetic-mode AGN seen down the axis and therefore subject to maximum relativistic boosting of the jet emission. This emission therefore swamps the weak lines which can be seen at other orientations. This scheme is less clear cut than the radiative-mode scheme and there are certainly cases where the phenomenology is intermediate. Heckman and Best (2014) and Tadhunter (2016) provide an overview of the current picture.

## 16.6 Accretion Rates and Feedback

It is becoming clear that the underlying differences between FR I (kinetic-mode) and FR II (radiative-mode) radio sources is mainly due to their accretion rates into the central regions. In the radiative-mode the accretion rate is high whilst in the kinetic-mode it is low. The threshold between the two modes is diagnosed by $L/L_{\rm Edd} \sim 0.01$ (Section 16.5.1), where the luminosity includes the mechanical power in the radio jet. Radiative-mode objects are above the threshold, and kinetic-mode objects are below. This difference in accretion rate leads to the different structures inferred for the accretion disks. Many authors have drawn attention to the similarities with galactic X-ray sources driven by accretion onto stellar-mass black holes (e.g. Romero *et al.* 2017). Much of the physical picture developed for these sources, including the ability to switch between different accretion modes, can be transferred across to AGN.

It has also become clear that there are large-scale feedback processes associated with AGN. The feedback cycle involves the fuelling of the central engine coupled with the effect that the huge amount of energy it produces has on the evolution of the host galaxy and its surroundings. This topic was reviewed by Fabian (2012) and is also well covered in Heckman and Best (2014).

In radiative-mode objects the energy released is in the form of radiation, plus, in the minority which are radio-loud, the mechanical energy of the jets. In the more numerous kinetic-mode objects the energy transfer is dominated by the jets. The importance of the mechanical energy carried by radio jets was pointed out by Scheuer (1974). He showed that it is greater, by a factor of at least 100, than the radio energy that the jets emit. This is evident from the total particle and magnetic energy in the radio lobes, to which must be added the work done against the pressure of the external gas. Heckman and Best (2014 and references therein) showed that the inferred jet power $P_{\rm jet}$ is a function of radio luminosity, $P_{\rm jet} \sim 10^{37}$ W being an appropriate value for an intermediate-power radio-loud AGN ($L_{\rm 1.4GHz} = 10^{25}$ W Hz$^{-1}$).

The feedback cycle is clearest in the kinetic-mode (FR I) sources, in which the elliptical host galaxies are gas-poor. In such objects the jet power is absorbed in the intergalactic gas, which is detectable by its X-ray emission (see also Section 17.13). As an example,

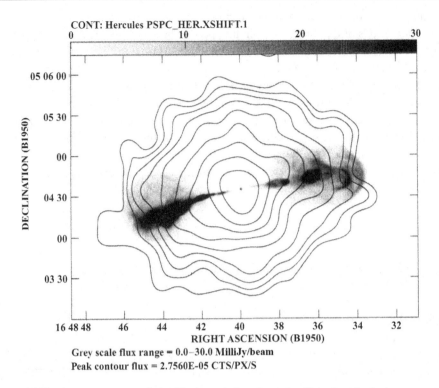

CONT: Hercules PSPC_HER.XSHIFT.1

Grey scale flux range = 0.0–30.0 MilliJy/beam
Peak contour flux = 2.7560E-05 CTS/PX/S

Figure 16.22. A contour map of the X-ray emission from the Hercules A cluster superposed on a grey-scale image of the radio galaxy at 18 cm wavelength. The hot gas emitting the X-rays extends through the cluster of galaxies, whose diameter is greater than 1 megaparsec (Gizani and Leahy 1999).

Figure 16.22 shows the radio galaxy 3C348, or Hercules A, superposed on the X-ray emission from a surrounding cluster of galaxies. The hot cluster gas cools, forming filaments which can move towards the galactic nucleus to act as fuel for the AGN. However, unless there is a compensating heating of the cluster gas many new stars would form from the cooling flows, and these are not seen. Heating by the jet flows approximately balances the cooling rate and a dynamical, if episodic, equilibrium is established. Powerful evidence for this process is provided by high-resolution X-ray images showing a rich variety of cavities and bubbles inflated by radio jets and lobes. The best studied case is the radio galaxy 3C84 and its X-ray halo (see the Supplementary Material).

Radiative-mode AGN present a different scenario. These high-accretion-rate AGN require a ready supply of cold gas from within their host galaxies to fuel them. Because star-formation and radiative-mode AGN both demand cold gas, radiative-mode AGN coexist within SFGs and SBGs. In this case the radiation from the central engine and the mechanical energy from the jets, to which must be added the radiation and outflows from massive stars and supernova remnants, can drive outflows. In the feedback cycle these outflows reduce the rate of star formation and the rate of accretion onto the SMBH. In powerful AGN the outflows can be directly imaged in the radio and/or mm-wave band via the spectral lines of H I or molecules. In the mm-wave regime the principal instruments are currently ALMA and NOEMA, while the JVLA and eMERLIN provide complementary information at longer wavelengths (e.g. Morganti 2017 and Fan *et al.* 2017).

## 16.7 Radio-Quiet AGN

As stressed by Padovani *et al.* (2017), a radio detection does not mean that an AGN is radio-loud. From a radio perspective such sources are defined by a luminosity $L_{1.4GHz} > 10^{23}$ W Hz$^{-1}$ and lie above the tight radio–FIR correlation discussed in Section 16.1. The majority of AGN are typically a thousand times less radio-luminous than radio-loud AGN and so are loosely dubbed *radio-quiet*, although a better term would be "radio-faint". Some authors use the terms 'jetted' and 'non-jetted' to emphasize the physical difference between radio-loud and radio-quiet AGN.

**Radio-quiet quasars (RQQs)**   About 90% of quasars are radio-quiet but their broad-band spectral energy distributions are similar to those of radio-loud quasars and they also show the same broad and narrow emission lines. The implication is that the physical situation in the central engine is essentially the same in the two classes. Radio-quiet quasars must be radiative-mode objects which, for a reason which is not understood, do not produce powerful relativistic jets. The latter is consistent with the fact that none are gamma-ray sources – in fact RQQs can be described as 'gamma-ray silent' and radio-faint (Dermer and Giebels 2016). What generates the relatively weak radio emission in RQQs remains a matter of continuing debate. For example, Leipski *et al.* (2006) showed that in many low-redshift RQQs a jet-like radio structure can be discerned, coupled with evidence of interactions with the surrounding medium. On the other hand, Kellermann *et al.* (2016) argued that the bulk of radio emission from RQQs comes from star formation in the host galaxy.

**Seyfert galaxies**   Spiral galaxies with unresolved bright optical nuclei were recognized in the early 1940s by Carl Seyfert but their significance was not appreciated for several decades since in other respects they look like normal spiral galaxies. Modern surveys reveal that 1%–2% of spirals can be classified as Seyfert galaxies. They come in two main types: Seyfert 1 galaxies show broad and narrow emission lines; Seyfert 2 galaxies show only narrow emission lines. Other than their radio and optical luminosity, little distinguishes Seyfert 1 galaxies from quasars (radio-loud or radio-quiet) or broad-line radio galaxies. Similarly, the spectra of Seyfert 2 galaxies resemble weaker versions of narrow-line radio galaxies. The clear implication is that these objects are low-luminosity radiative-mode AGN with the same orientation-dependent properties as were outlined in Section 16.5.1. It is likely that Seyferts are accreting more slowly onto smaller SMBHs ($10^6$ to $10^7$ M$_\odot$) and, while they do not develop powerful relativistic jets, weak jet-like structures can be seen in a minority of these objects (Nagar *et al.* 1999).

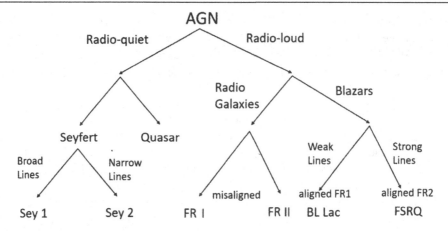

Figure 16.23. The main types of AGN based on observational criteria (after Dermer and Giebels 2016).

**LINERS and other spirals** About 30% of spiral galaxies exhibit low-excitation nuclear-emission-line regions (LINERs) which are weaker than in Seyferts. As with radio-quiet quasars there is a debate about the source of their activity. The most luminous probably host a low-luminosity AGN while the others are dominated by star formation (see Heckman and Best 2014). The nuclei of even apparently 'normal' spiral galaxies can emit excess radio power along with narrow emission lines: M 51 is a prominent example (e.g. Bradley *et al.* 2004), and in Chapter 14 we saw that even our own Milky Way galaxy has an 'anomalous' source, Sgr A*, at its centre.

### 16.8 Summary of AGN Phenomenology

All galaxies with masses comparable to, or larger than, the Milky Way are now firmly believed to host dormant or active SMBHs, and low-level nuclear activity is quite common. Powerful AGN, and even more so radio-loud AGN, are rare. The radio-loud objects are invariably found in elliptical galaxies while radio-quiet host galaxies are usually spirals. Figure 16.23 summarizes diagramatically the main types of AGN on the basis of obser-vational criteria while Figure 16.24 emphasizes the basic physical differences between them. Within the radiative-mode and kinetic-mode pictures a broad understanding of the different types of AGN and their properties has been reached. We have painted only a greatly simplified picture but one which should enable the interested reader to start tackling the research literature.

Many questions remain to be answered; for example, why do some massive galaxies host powerful radio sources (and hence launch relativistic jets) but other, apparently similar, massive galaxies do not? Why do radiative-mode AGN develop jets in only a minority of cases – and when they do why are there only the FR II type at the top end of the radio

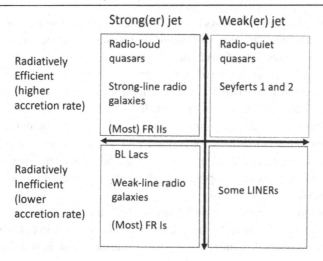

Figure 16.24. Summary of the physical differences between the two types of unified model (after Padovani *et al.* 2017).

luminosity range? How is it that two different physical circumstances, the radiative-mode and kinetic-mode models, produce very similar 'radio engines'? The answers will lie in a complex astrophysical interplay of SMBH mass, spin rate, and accretion rate and the availability of fuel. Finally, we note that the effect on the evolution of the host galaxy of the intense radiation and mechanical energy supplied by an AGN has become an intensive area of study (see also Section 16.9.5).

We now turn from the morphology and physics of individual radio sources to the statistics of their occurrence, extending to the very large numbers of faint sources which are now found in deep-survey observations.

## 16.9 Surveys, Source Counts, and Evolution

Sky surveys in the radio band have advantages over those carried out in other bands across the electromagnetic spectrum. The intensity of radio emission from a distant source is largely unaffected by absorption or scattering by the intervening medium and it is feasible to make a survey which is complete and has uniform sensitivity over a large region of the sky. A plethora of surveys for discrete continuum sources have been carried out over the 70-year history of radio astronomy. In the early years it was thought that they could have a major impact on cosmology.

### 16.9.1 Early Source Counts

Consider a static Euclidean universe uniformly populated with sources of the same intrinsic luminosity. The flux densities of the sources will fall with their distances $D$ as $D^{-2}$, while their numbers will increase as $D^3$. As a result the total number of sources $N(>S)$ detected above a flux level $S$, the *cumulative or integral source count*, will vary as $S^{-1.5}$ and hence a plot of $\log N(>S)$ versus $\log S$ will show a power law of slope $-1.5$. In this simple geometry the slope does not change even if there are sub-populations with different intrinsic luminosities (e.g. Wall and Jenkins 2012). The high luminosities of some radio sources identified in the 1950s and early 1960s (see Wall 1994) implied that, as the limiting flux density of a radio survey fell, the source counts would probe cosmologically significant distances. It therefore seemed that a $\log N$–$\log S$ plot could be a cosmological diagnostic when the observed slope deviates from the simple theoretical model. The concept seemed straightforward but astronomers' hopes for a definitive cosmological test were dashed, for a variety of reasons.

The first problem was confusion (see Sections 8.10.3 (the single-dish case) and 10.12.1 (the interferometer case)), which dogged the earliest radio source surveys. Strong sources, falling in poorly understood sidelobes, contributed spurious detections while closely separated weaker sources could be blended by the large reception beams. As a result the first-generation surveys produced widely differing power-law slopes depending upon the telescope used and the analysis method adopted (e.g. Hazard and Walsh 1959a). The disagreements led to a heated debate between different observing groups on the relevance of their results to cosmology and, in particular, to the steady-state cosmology for which a slope of $-1.5$ was predicted.[16] The Revised 3C Catalogue (3CR; Bennett 1962) finally provided a good measure of $\log N$–$\log S$ for strong sources but it contains only a few hundred sources of heterogeneous types and luminosities. While the slope was steeper than $-1.5$ it began to be realized (e.g. Longair 1966) that the main impact of the counts would not be as a cosmological discriminant but as a study of evolving radio source populations. Now, over 50 years later, surveys detect sources at least one million times fainter than 3CR and the counts cannot be described by a single slope; they are mainly determined by the evolution of different source populations (see Section 16.9.4) rather than by the geometry of the space in which they are located.

Before moving on to describe the modern counts two further issues must be addressed. The early debate over the interpretation and implications of source counts was based on simple Euclidean geometry. But, to interpret the counts of increasingly fainter and hence probably more distant sources, the equations of relativistic cosmology in an expanding universe must be used to calculate distances and volumes; account must also be taken of the effect of redshift on the observed flux density, which depends on the shape of the radio spectrum. The relevant relations are given by Condon (1988). The effects are powerful and the expected slopes rapidly become much shallower than $-1.5$, even for a Universe uniformly filled with non-evolving sources. To add emphasis to an earlier point, one needs to assume a geometry for the Universe in order to interpret the counts, as opposed to deriving cosmological constraints from the counts.

The second issue is that integral source counts obscure the interpretation (e.g. Jauncey 1967). Gathering sources into separate flux bins, with each fainter bin also containing the

---

[16] The interested reader can catch the flavour of this debate from: Hazard and Walsh (1959b); www.astro.phy
.cam.ac.uk;rahist.nrao.edu; Frater *et al.* (2013).

contents of all the stronger ones, makes the error bars non-independent. In addition any features in the integral counts will be smeared and thus less easy to recognize. These problems are obviated by the use of the *differential source count*, i.e. a plot defined by counting sources $n(S)$ in flux density bins from $S$ to $S + \Delta S$; in a static Euclidean universe, $n(S) \propto S^{-2.5}$. A steep slope like this is inconvenient for graphical presentation and so standard practice is to plot the so-called *Euclidean normalized source counts* $n_{norm}(S) = kS^{2.5}n(S)$, weighting down the large numbers of sources at low flux densities. If $n(S)$ happens to follow the slope for a static Euclidean universe over a range of fluxes, the points in $n_{norm}(S)$ will be distributed around a horizontal line (zero slope); thus 'Euclidean' has become shorthand for a zero slope in the normalized differential counts. The arbitrary normalization factor $k$ is often chosen so that $n_{norm}(S) = 1$ at $S = 1$ Jy; alternative choices for $k$ allow counts taken at different frequencies to be combined on the same plot.

### 16.9.2 Modern Source Surveys

The issues involved in making reliable discrete source surveys with single dishes and interferometers have been discussed by Gregory and Condon (1991) and Condon *et al.* (1998) respectively while Condon (2017) has summarized the lessons learned. de Zotti *et al.* (2010), Padovani (2016), Norris (2017b), and Simpson (2017) have compiled complementary lists of surveys and discussed their astrophysical implications. Here we mention a few of the more influential surveys from recent decades.

Almost the entire sky was covered, at a specific frequency (5 GHz) and with comparable quality, by the early 1990s. The north was covered by the 87GB survey (Gregory and Condon 1991) and its enhancement, the GB6 survey (Condon *et al.* 1994), while the southern sky was covered by the PMN survey (Griffith and Wright 1993). These surveys were made with single dishes (resolution $\leq 5$ arcmin), and at the catalogued flux levels (tens of mJy and above) they were operating well away from their confusion limits. At $S_{5GHz} = 25$ mJy the surface density of sources is $\sim 10^4$ sterad$^{-1}$ ($\sim 3$ deg$^{-2}$). Figure 16.25 shows how these discrete radio sources are distributed across the sky. The galactic sources, mainly H II regions, are seen as a thin line outlining the plane of the Milky Way; these are confined to within $5°$ of the plane. At higher latitudes nearly all the sources are extragalactic (with the exception of a few associated with stars). Their distribution appears to be isotropic and this has been strongly confirmed by subsequent deeper surveys (e.g. NVSS; see below).

In the 1990s interferometers began to dominate the task of cataloguing much fainter sources. By this time synthesized beam formation and its effect on the confusion limit of a survey was better understood. The most influential of this generation of surveys, NVSS (Condon *et al.* 1998) and FIRST (White *et al.* 2000) were both made with the VLA at $\sim 1.4$ GHz. The NVSS covered $> 80\%$ of the sky and catalogued 1.8 million sources. It used the most compact array of the VLA, providing a resolution of 45 arcsec, to capture radio sources with large angular diameters. The NVSS completeness limit is 2.5 mJy, at which point the surface density is $\sim 1.6 \times 10^5$ sources sterad$^{-1}$ ($\sim 50$

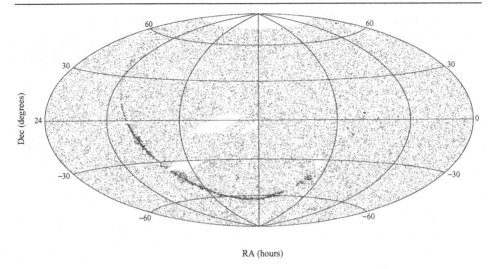

RA (hours)

Figure 16.25. All-sky map of radio sources from the combined surveys of Gregory and Condon (1991) and Griffith and Wright (1993). Apart from the thin line of sources associated with the Milky Way, the distribution is essentially uniform over the sky.

sources deg$^{-2}$). The FIRST survey used a more extended VLA configuration resulting in 5 arcsecond resolution and a sensitivity limit of 1 mJy. The survey area was chosen to overlap with that of the SDSS optical sky survey. In comparison with NVSS, FIRST obtains higher angular resolution on individual sources at the cost of resolving out and hence missing some sources with large angular structure. At the 1 mJy detection level there are $\sim$ 90 sources deg$^{-2}$. Updated catalogues for NVSS and FIRST are available on the Web.

With these and other large area surveys 'in the bank', attention turned to finding sources in much smaller areas of sky ($<$ 1 deg$^2$) but with flux densities as much as $\sim$ 100 times fainter than either NVSS or FIRST. The first survey of this generation centred on the Hubble Telescope Deep Field (HDF; Richards *et al.* 1998; Richards 2000); which allowed identifications for exceptionally faint optical objects. For this survey the VLA was used in its most extended configuration, at 1.4 GHz and 8.5 GHz, with integration for 50 hours to reach flux densities of 40 μJy at 1.4 GHz and 8 μJy at 8.5 GHz. The discussion of observing techniques by Richards (2000) is particularly illuminating as an example of the care needed when carrying out long aperture-synthesis integrations at the limit of what is practical at the time. Thus, just as in the early days, one must deal with confusing sidelobe effects both from strong sources outside the field of view and from mJy sources within the field of view, to which can be added differential effects across a wide frequency bandwidth and baseline-related offsets. Since then a variety of other deep fields have been surveyed, mainly with the updated JVLA. At the time of writing the most recent collection of references can be found in Vernstrom (2016b) and Simpson (2017). However, as new instruments come into operation new source surveys are being undertaken; see the Supplementary Material for details.

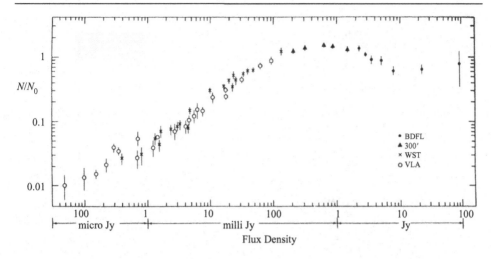

Figure 16.26. Counts ($N$) of radio sources, over a range of $10^6$ in flux density, compared with those expected ($N_0$) from a static Euclidean Universe with no evolution. The counts are presented as the ratio of the observed number density and the number density expected in a simple Euclidean universe with constant number density (Kellermann and Wall, 1987).

### 16.9.3 Modern Source Counts

The source counts are made up from a heterogeneous mix of radio sources, but at long radio wavelengths (below a few GHz) the steep spectrum sources dominate and this simplifies the interpretation. However, at long wavelengths beamwidths become large and confusion quickly sets in as fainter sources are observed. A compromise frequency of 1.4 GHz has therefore become the *de facto* standard for generic discussions.

Figure 16.26 shows the Euclidean normalised count at 1.4 GHz as it stood at the end of the 1980s. Except at the faint end (below $\sim 1$ mJy) little has changed since then. Following Kellermann and Wall (1987) the broad features in these counts can be summarized thus:

- *Sources stronger than* $\sim 3$ Jy: this is the flux range probed by the early counts (Section 16.9.1). At the brightest end the statistics are poor but the count is consistent with Euclidean after which the counts steepen. The cumulative surface density of sources is $\sim 50$ sterad$^{-1}$.
- *From* $\sim 3$ Jy *to* $\sim 0.1$ Jy: the counts are approximately Euclidean. Down to 0.1 Jy the cumulative surface density is $\sim 6500$ sources sterad$^{-1}$ ($\sim 2$ sources deg$^{-2}$).
- *At* $\sim 0.1$ Jy: there is a downturn (becoming flatter than Euclidean) and the downward slope ($n(S) \propto S^{-1.5}$) extends to $\sim 1$ mJy at which point the cumulative surface density has reached $\sim 300\,000$ sources sterad$^{-1}$ ($\sim 100$ sources deg$^{-2}$).
- *Below* $\sim 1$ mJy: the counts begin to flatten again.

Recent progress has, perforce, been at the faint end, essentially all the strong sources having now been counted. At the start of this century the results presented by Richards (2000), extending down to $\sim 40$ µJy, showed that the flattening off at the faint end had become clearer. Surveys over the next 15 years extended the counts down to $\sim 20$ µJy but below $\sim 100$ µJy there is significant scatter between the results of different surveys. The

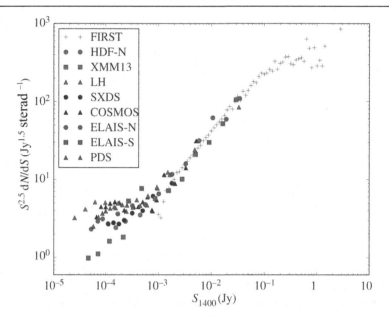

Figure 16.27. Euclidean-normalized source counts from deep radio surveys at 1.4 GHz compiled by Simpson (2017); see that paper for further details.

compilation made by Simpson (2017) shown in Figure 16.27 summarizes the situation close to the time of writing. Here we find echoes of the challenges encountered in the early days of counting sources, translated down in flux density by a factor of a million! The issues are well known but Vernstrom *et al.* (2016a) addressed them quantitatively. They found that 30%–70% of faint sources are missed or fit inaccurately if the sources are larger than the synthesized beam and that sources separated by less than the beam size are blended. It is likely that a combination of these and other effects, coupled with attempts to compensate, accounts for much of the scatter at the faint end of the counts.

The most reliable, unbiased, method for establishing the faint source statistics is the so-called *P(D)* analysis first introduced by Scheuer (1957). This approach does not deal with individual detections but treats the deviations $D$ in the map statistically. At its simplest the fluctuations in the map $\sigma_m$ are a quadratic combination of the thermal noise $\sigma_n$ and the fluctuations $\sigma_s$ from faint sources around or below this level, i.e. $\sigma_m^2 = \sigma_n^2 + \sigma_s^2$. This enables $\sigma_s$ to be determined and compared with models of the faint-source statistics. Unsurprisingly there are additional subtleties involved, which were discussed by Condon *et al.* (2012) and by Vernstrom *et al.* (2016b), whose derivation of the faint 1.4 GHz counts is shown in the Supplementary Material. The *P(D)*-derived count is roughly flat (Euclidean) from 1 mJy to $\sim 20$ µJy and fits through the middle of the scatter of the points derived from catalogued sources. However, at $\sim 20$ µJy the *P(D)*-derived count shows a marked downturn, with the slope ($n(S) \propto S^{-1.5}$) becoming similar to that between 1 and 100 mJy. At 1 µJy the predicted cumulative surface density is $\sim 2 \times 10^8$ sources sterad$^{-1}$ ($\sim 65\,000$ sources deg$^{-2}$).

### 16.9.4 Evolution of Source Populations

Lists of radio positions and flux densities at a given frequency are of limited use on their own. As we described earlier in this chapter, recognizing the different classes of radio sources requires a combination of radio morphology and spectra complemented with measurements in other wavebands. To recapitulate: radio sources are a mixture of SMBH-powered AGN and supernova-powered star-forming galaxies (SFGs), with some sources exhibiting both phenomena. At a radio luminosity $L_{1.4GHz} > 10^{23}$ W Hz$^{-1}$, essentially all sources are AGN but the radio-quiets ('unjetted') fall below this luminosity. The radio-loud AGN ('jetted') can be split into the extended FR Is ($L_{1.4GHz} < 10^{25}$ W Hz$^{-1}$) and FR IIs ($L_{1.4GHz} > 10^{25}$ W Hz$^{-1}$) plus the compact strongly beamed FSRQs and BL Lacs. The close links between the beamed and non-beamed classes were discussed in Sections 16.5.1 and 16.5.2. The SFGs exhibit a range of star-formation rates (SFRs) and, as discussed in Section 16.1, their radio luminosity is a proxy for the SFR and is strongly correlated with FIR luminosity.

The need is now to move from taxonomy to demography, i.e. to understand how these classes fit within evolving populations of radio sources. This is part of the broader task of understanding how an entire galaxy population evolves over the age of the Universe. To trace radio source evolution with cosmic epoch the demographer must: (i) establish a local radio luminosity function (LRLF), i.e. the space density of sources in a particular population as a function of luminosity at zero redshift; (ii) determine how this LRLF evolves with redshift. It is important to stress that the evolution relates to populations of sources, not to individual objects. Condon (1984, 1988) described the process in detail. There is a vast literature on the cosmological evolution of radio sources since Longair's (1966) seminal paper. For discussions of this complex topic the reader can consult the reviews by: Condon (1984); Kellermann and Wall (1987); Wall (1994); de Zotti et al. (2010); and Padovani et al. (2016). The analytical details are beyond the scope of this book, and here we simply sketch some basic features of the evolutionary picture which have emerged over the past 50 years.

The differential counts extend over eight orders of magnitude in flux density. Within the top two orders of magnitude the slope is either steeper than or equivalent to Euclidean. This must be seen in the context of the effects of the expanding Universe. As the redshift increases, the combination of source dimming and relativistic geometry forces the differential counts for a non-evolving population to be much flatter than Euclidean. As Longair (1966) pointed out, the fact that the counts for the strongest sources do not follow this trend must mean that these sources are evolving strongly and were more common in the past. In fact the comoving densities of powerful radio sources at $z = 1$ to $z = 2$ are more than 1000 times greater than the local densities. This rapid evolution must, however, switch off at somewhat higher redshifts, otherwise there would be too many faint sources. With this in mind we now look at specific regions in the counts:

– *The brightest few tens of sources* These are largely a mixture of low- and high-redshift AGN. The roughly Euclidean slope arises from a combination of the start of the evolution of the RLFs of the most powerful FR II sources and the local RLFs of the low-luminosity FR I sources.

- *The sources down to* 0.1 Jy    These are largely drawn from the most powerful FR II radio sources, whose population is undergoing very strong evolution with redshift. The evolution of powerful sources was confirmed by a closely related technique known as the *luminosity–volume* test, and also known as the $V/V_{max}$ test, introduced by Schmidt (1968) for quasars and by Rowan-Robinson (1968) for infrared sources. This test requires a complete sample with redshifts for each object but is free of the biasses which can affect the source counts. For each object the volume of space up to its distance is designated $V$, and the volume up to the distance at which the object would be just detectable (in either the radio or the optical or infrared) is designated $V_{max}$. The average of the ratio $V/V_{max}$ for many objects should, in a simple Euclidean universe (as in Section 16.9.1) be 0.5. Both for the quasars and for the infrared sources the average ratio was found to be $\sim 0.7$; this confirmed that they were strongly evolving on a cosmological timescale. Further discussion of the link between the source counts and the $V/V_{max}$ test is given in de Zotti *et al.* (2010).
- *The sources between* $\sim 0.1$ Jy *and* $\sim 1$ mJy    The relativistic effects which suppress the slope of the counts now prevail. The sources are still mainly radio-loud AGN but are dominated by low-luminosity FR I sources. This part of the population is at best only weakly evolving and objects are more uniformly distributed in redshift than are the most luminous sources.
- *The sub-*mJy *sources*    The fact that the differential slope again becomes Euclidean implies that a new strongly evolving population has emerged; these are the SFGs. It is already known that the SFR from optical and/or IR measurements grows strongly to $z = 1$–$2$ (e.g Madau and Dickinson 2014) and falls off at higher redshifts. The sub-mJy source population mix was discussed by Vernstrom *et al.* (2016a). Based on cross-comparisons with X-ray, optical, and infrared data they derive a mix of $< 10\%$ radio-loud AGN, $< 28\%$ radio-quiet AGN, and 58% SFGs, with 4% unclassified. Radio morphology is another route to classification since, in contrast with AGN, SFGs are more diffuse and lie within the outline of the optical galaxy. From 1.4 GHz MERLIN+VLA radio images of the Hubble Deep Field, Muxlow *et al.* (2005) were able to show that the proportion of starburst systems increases with decreasing flux density. Below 100 μJy over 70% of the sources are of the starburst type associated with major disk galaxies with $z = 0.3$–$1.3$. More recent eMERLIN+JVLA observations of the central 15 arcminute area of the well-studied GOODS-N field (eMERGE DR-1, Muxlow *et al.*, in preparation) have confirmed this trend from observations of a sample of 849 sources down to $\sim 9$ μJy. In addition, many luminous star-forming systems are found to exhibit bright nuclear starbursts embedded within regions of extended star formation, and a significant number of radio-quiet AGN are seen to be dominated by radio emission associated with star formation whilst AGN activity is found in other wavebands (for images, see in the Supplementary Material).

In summary, differential evolution results in the radio source counts being dominated by particular populations over different flux density ranges (for more details see Wilman *et al.* 2008 and Mancuso *et al.* 2017). These are signposts in the quest to understand the development of a galaxy population as a whole. One immediate clue is that SFGs, predominantly found in spiral galaxies, seem to share a similar evolutionary history with the radio-luminous FR II AGN hosted by massive ellipticals.

### 16.9.5 Future Continuum-Source Surveys

Disentangling the astrophysical phenomena which govern the coevolution of galaxies, massive stars, and SMBHs is a very active research area demanding large samples to be studied at multiple wavelengths. An increasingly important role for radio surveys is to provide unbiased samples of SFGs, not affected by dust obscuration, from which the SFR can be obtained from the radio luminosity. To provide the required surveys new radio facilities are coming on line: SKA1 is about to start construction (2020) and SKA2 is in prospect. Padovani *et al.* (2017) and Norris (2017a) have discussed the potential of these facilities and provide many references.

New surveys are possible, thanks to improvements in the survey speed *SS*:

$$SS \propto (A_{\text{eff}}/T_{\text{sys}})^2 \Delta\nu \, \Omega_{\text{FoV}}. \tag{16.3}$$

Here $\Omega_{\text{FoV}}$ is the solid angle of the field of view; otherwise, the parameters are as defined in Chapter 8. The solid angle $\Omega_{\text{FoV}}$ can be the beam of a dish equipped with a single feed, a combination of beams from many feeds, or one of the many beam combinations open to phased arrays. For example the JVLA is now undertaking a large-area survey, VLASS, in the band 2–4 GHz, taking advantage of the improvements in $T_{\text{sys}}$ and particularly in $\Delta\nu$ after the enhancement to the array completed in 2012. The VLASS survey will reach an rms brightness level of $\sim 70$ µJy per beam and, when completed in 2023, will produce a catalogue containing around 10 million sources. Using the new technology of phased array feeds (PAFs, see Section 8.2) the Australia SKA Pathfinder (ASKAP) can undertake previously infeasible surveys. The EMU Survey at 1100 MHz will cover a large fraction of the sky down to an rms brightness of $\sim 10$ µJy/beam and should generate a catalogue of 70 million sources. At metre wavelengths the now mature technology of aperture arrays (see Section 8.2) is enabling the MWA and LOFAR to make deep wide-area surveys. For more details see Norris (2017b).

As Norris (2017a) has pointed out, efficient handing of these large data sets and optimal extraction of the source information has passed beyond the capabilities of mere humans and will require the development of new machine-learning techniques. Then, after the radio sources have been classified and catalogued, techniques for efficiently cross-comparing them with enormous catalogues from other wavebands will be required. This involves new software technologies complementing the new hardware of PAF-enabled dishes and aperture arrays.

Finally we look forward to source surveys with the SKA, initially SKA1, in the late 2020s. Currently the vast majority of objects in optical surveys have no radio data, but SKA1-MID will begin to change this. A large fraction of the sky will be surveyed down to reliable detection levels of 20–30 µJy while the deepest selected-area survey will detect sources at 0.25 µJy (5σ). As the limiting flux level falls, radio measurements will pass beyond their present focus on exceptional objects in the Universe to the more mundane. For example the SKADS-simulated sky (Wilman *et al.* 2008) predicts that, at 1 µJy, more than 80% of sources will be SFGs, only $\sim 7\%$ will be low-luminosity AGN, and the remainder will be radio-quiet AGN; hardly any sources will be distant high-luminosity AGN. The SKA2 will have the potential to detect any galaxy similar to the Milky Way. With these high surface densities, subtle patterns may be discerned and, 70 years after hopes for cosmological tests were dashed, new cosmological tests will become possible (e.g. Jarvis *et al.* 2015). Furthermore, as deep multiwavelength surveys probe unexplored regions of observational 'parameter space', astronomers can anticipate rewards in the form of completely new phenomena (e.g. Wilkinson *et al.* 2004; Norris 2017a and references therein).

## 16.10 Further Reading

*Supplementary Material* at www.cambridge.org/ira4.

Boettcher, M., Harris, D. E., and Krawczynski, H. (eds.). 2012. *Relativistic Jets from Active Galactic Nuclei*. Wiley.

Condon, J. J. 2017. In *Proc*. Conf. on *The Many Facets of Extragalactic Radio Surveys: Towards New Scientific Challenges*, 20–23 October 2015, Bologna, Italy. Online at https://pos.sissa.it/267/004/pdf.

Condon, J. J., Broderick, J. J., Seielstad, G. A., Douglas, K., and Gregory, P. C. 1994. A 4.85 GHz sky survey. 3: Epoch 1986 and combined (1986 + 1987) maps covering $0°$ to $\leq 75°$. *Astron. J.*, 107, 1829.

De Young, D. S. 2002. *The Physics of Extragalactic Radio Sources*. University of Chicago Press.

Gregory, P. C., and Condon, J. J. 1991. The 87GB catalog of radio sources covering delta between O and + 75 deg at 4.85 GHz. *Astrophys. J. Suppl.*, 75, 1011.

Griffith, M. R., and Wright, A. E. 1993. The Parkes–MIT–NRAO (PMN) surveys. I – The 4850 MHz surveys and data reduction, *Astron. J.*, 105, 1666.

Padovani, P. 2016. The faint radio sky: radio astronomy becomes mainstream. *Astron. Astrophys. Rev.*, 24, 13.

Richards, E. A., 2000, The nature of radio emission from distant galaxies: the 1.4 GHZ observations. *Astrophys. J.*, 533, 611.

Richards, E. A., Kellermann, K. I., Fomalont, E. B., Windhorst, R.A., and Partridge, R. B, 1998. Radio emission from galaxies in the Hubble Deep Field. *Astron. J.*, 116, 1039.

White, R. L., Becker, R. H., and Gregg, M. D. *et al.* 2000, The FIRST bright quasar survey. II. 60 nights and 1200 spectra later. *Astrophys. J. Suppl.*, 126, 133.

# 17

# The Radio Contributions to Cosmology

## 17.1 The Expanding Cosmos

The most astounding achievement of observational astronomy is our ability to observe the Universe on the largest possible scale. We can examine its present state, its past history, and extrapolate into the future. Observational cosmology started with optical observations, but radio has proved to be the major contributor to our understanding of the Universe. The discovery of discrete radio sources, and their identification with extragalactic nebulae with high redshift (most notably Cygnus A) signalled that the radio sources might well serve as interesting cosmological probes. Attempts to use the statistics of the early source counts failed (see Chapter 16), but a new chapter in cosmology was opened when optical identifications from the 3C survey began. Two sources, 3C48 and 3C273, were located with sufficient accuracy to show that they were in an entirely new class of object, which became known as quasars. The discovery of very large redshifts in their spectra called attention to the importance of relativistic phenomena in the Universe and led to the discovery of black holes. Then, in 1965, the discovery of the cosmic microwave background (CMB) gave direct evidence of the 'Big Bang' model of the early Universe and laid the foundation of the quantitative interpretation of cosmic evolution.

In this chapter we give a brief review of the historical development of cosmology, as a basis for understanding the theoretical and experimental advances of the past century.[1] We then present the basic elements of cosmological theory,[2] introducing the main parameters of modern descriptions of the cosmos, which are now measured with remarkable precision by observations of the CMB. Our main concern is to describe the observations of the CMB in some detail, showing how the radiometry and polarimetry techniques of earlier chapters have been extended to astonishing levels of precision.

## 17.2 A Brief History

Two events during World War I marked the beginning of a new era in cosmology. Albert Einstein extended his theory of general relativity (GR) to include the action of gravity

---

[1] For a comprehensive historical account we recommend Malcolm Longair (2006), *The Cosmic Century*.
[2] Among the many accounts of the fundamentals we recommend Barbara Ryden (2017), *Introduction to Cosmology*, 2nd edn.

throughout the Universe, and in 1917 Vesto Melvin (V. M.) Slipher published the high radial velocities of 25 spiral nebulae on the basis of their optical spectra, the fruit of several years of work. Edwin Hubble determined distances by identifying Cepheid variables, publishing his famous paper 'A relation between distance and velocity among the extragalactic nebulae' (Hubble 1929), which caused a sensation among cosmologists. The Hubble velocity–distance relation became known as the recession of the nebulae, or, in more modern terminology, the expansion of the Universe. The standard cosmological model of a Universe of galaxies, expanding from an initial state of high density, became the focus of theoretical cosmology, notable contributions coming from the Russian mathematician Alexander Friedmann and the Belgian priest Georges Lemaître.

Observational cosmology contributed little more for three decades. Then in 1965 cosmology was completely transformed by the discovery of the cosmic microwave background (CMB) by Arno Penzias and Robert Wilson, which was immediately understood as radiation from the early Universe, as anticipated by Ralph Alpher and Robert Herman in 1948. The observed spectrum of this radiation peaks at millimetre wavelengths, corresponding to a temperature of 2.725 K (Alpher and Herman predicted a CMB with temperature of 5 K); it actually originated as thermal radiation at a temperature of $\sim 3000\,\mathrm{K}$ but is redshifted by a factor $z \approx 1100$. Theoretical models of the expanding Universe could now be tested against precise observations.

Mapping the structure in the CMB, which evolved into the galaxies and clusters of galaxies of the present-day local Universe, required radiometer observations of unprecedented accuracy; the structure has a maximum amplitude of only $10^{-5}$ of the average value of the CMB. This was a challenge which could be met only by the most sensitive observations; success came more than 25 years after the discovery of the CMB, initially from radiometers in high-altitude balloons and ultimately from spacecraft. The astonishingly accurate measurements of CMB structure by the three spacecraft COBE, WMAP, and Planck can now be interpreted in terms of a model Universe whose components are specified with accuracies of 1% or better; the model now requires the major components to be dark matter and dark energy, neither of whose physical natures are understood. The visible components, the baryonic matter in stars and interstellar and intergalactic gas, play only a minor part in the large-scale dynamics of cosmic expansion.

### 17.3 Geometry and Dynamics

Hubble's law relating redshift $z = \Delta\lambda/\lambda$ to distance $d$ could be understood within the Einstein–Friedmann GR cosmology: the recession of the nebulae was evidence for the expansion of space itself. This approach requires a broad smoothed-out view of the cosmos in which local structures such as individual galaxies and clusters of galaxies are ignored. The expanding Universe was a natural consequence of GR, in which the fundamental precept is the principle of equivalence: inertial mass and gravitational mass are the same thing. The gravitational force is then a property of geometric space.

The geometry of space is determined through Einstein's field equations by the distribution of matter and energy. Finding an appropriate solution or solutions to his formidable field equations was made easier by the observational fact that the Universe appears to be isotropic; it looks the same in all directions. Hubble worked hard to show that this was so for galaxies; more modern studies of radio galaxies and quasars (see Section 16.9), and *a fortiori* the CMB, have verified the isotropy out to redshift $z = 1100$. Furthermore, observations show that the Universe is homogeneous, with the same physical laws everywhere. In sum, we can adopt the *cosmological principle*: on a sufficiently large scale, the Universe presents the same aspect everywhere at a given epoch. This simplifying circumstance does not place the observer in a privileged location, but it does imply that the timescale of expansion is the same at all locations.

Smoothing out the galaxies and clusters of galaxies into a featureless fluid, the expansion with time is characterized by a scale factor $a(t)$, which is set equal to unity at the observer. The expansion applies to the wavelength of radiation, providing the explanation of Hubble's observation. In this expanding coordinate system, the redshift is interpreted as the expansion in wavelength as the coordinates expand. Consider a light wave emitted at time $t_{em}$ that we now observe at time $t_{obs}$ with redshift $z$. The observer at time $t_0$ observes the wave redshifted by $z = (\lambda_{obs} - \lambda_{em})/\lambda_{em}$. It follows that the scale factor of expansion is

$$a(t_{em}) = 1/(1 + z). \tag{17.1}$$

Although we do not know the time of emission $t_{em}$, if we have a model for $a(t)$ we can tie the redshift to that earlier time; other parameters follow. Provided that matter and radiation are uncoupled, the density of matter scales as $a(t)^{-3}$. The wavelength of a photon scales as $a(t)$; its energy scales as $1/a(t)$, so the photon energy density scales as $a(t)^{-4}$. This in turn implies that the cosmic background temperature scales as $1/(1 + z)$.

The expansion rate varies with time. It is usually expressed as a 'Hubble's law' applying at a particular time $t$:

$$H = \dot{a}(t)/a(t) \tag{17.2}$$

in which $H$ is the *Hubble parameter*, which may vary with time. Locally, at time $t_0$, it becomes the *Hubble constant $H_0$*.

The formal description of the geometry of space has an interesting history. The scale factor $a(t)$ was derived by Einstein in 1917, but it was for a static Universe. The Dutch astronomer and cosmologist de Sitter convinced Einstein that his model was unstable; gravity would, with any fluctuation, cause the Universe to collapse. Einstein's response was to add a new term to his field equations, the cosmological term, featuring a constant $\Lambda$, the cosmological constant, which was intended to oppose the attraction of gravity. Its action was, in effect, to introduce a force that grew with distance. The observation that the Universe is not static but expanding led Einstein to retract his proposal of the cosmological term as a mistake; it reappeared, however, several decades later, on the basis of observations of redshifts of distant galaxies.

The geometry of the Universe is specified by GR; the simplifying assumption of an isotropic, homogeneous, universe leads to a differential equation for the scale factor. This formalism is known as the Robertson–Walker metric (sometimes the names Friedmann and Lemaître are added.) The acceleration of the scale factor $a(t)$ is

$$\ddot{a}(t) = -[4\pi G a(t)/3](\rho + 3P/c^2) + \Lambda a(t)/3. \tag{17.3}$$

This reminds one of the acceleration that results from applying forces to a test particle at the surface of a sphere of radius $a$ and density $\rho$. The first term on the right shows the inward gravitational acceleration due to the matter in the sphere, with the associated energy represented by the pressure $P$. The matter density $\rho$ has two components, the baryonic density $\rho_b$ (the familiar constituents of atoms), and the dark matter density $\rho_d$, whose existence was postulated by Fritz Zwicky and proven by Ruben and Ford. The last term is Einstein's oppositely directed cosmological term, which grows with distance with a strength given by the cosmological constant $\Lambda$.

By integrating Eqn 17.3, remembering that the mass inside $a$ is constant, an expression is reached that looks like an energy equation (the terms have been rearranged to make the physics clear, and $a(t)$ is simply written as $a$):

$$\dot{a}^2/2 - (4\pi G a^2/3)\rho_{m,e} = \Lambda a^2/3 - c^2/R. \qquad (17.4)$$

On the left-hand side are the kinetic energy associated with the expansion and the gravitational potential energy, with $\rho_{m,e}$ representing the mass and energy associated with matter. The first term on the right is the energy associated with the cosmological term; the second term (a constant of integration) represents the energy associated with a curved universe with radius of curvature $R$. The energy associated with $\Lambda$ expresses the action of the repulsive cosmological term. It grows stronger with increasing distance, so it does work as the Universe expands. This stores up energy throughout the Universe, which is known as *dark energy*.

By setting $\Lambda$ to zero Einstein reduced the complexity of his model and was joined by de Sitter to frame the Einstein–de Sitter model. Following the usual convention of using the inverse of $R$, the *curvature constant* $\kappa$, in this model the simplified energy equation is

$$\dot{a}^2/2 - (4\pi G a^2/3)\rho_{m,e} = -\kappa c^2. \qquad (17.5)$$

This displays a key relation between the total energy and the curvature. If on the one hand the gravitational potential energy exceeds the energy of expansion, the system is bound and has positive curvature; it will expand to a maximum value, and then collapse to 'the big crunch'. On the other hand, if the kinetic energy of expansion is dominant, the curvature will be negative and the Universe will expand forever.

Now consider the special case when the kinetic and gravitational terms are equal. The Universe will have zero curvature: a flat universe. This allows one to define a critical density $\rho_c$. Since $H = \dot{a}/a$, Eqn 17.5 leads to the following form for the critical density:

$$\rho_c = 3H_0^2/8\pi G. \qquad (17.6)$$

In the Einstein–de Sitter cosmology, described by Eqn 17.5, this is the baryonic matter density that just closes the Universe. Its present value would be $10^{-26}$ kg m$^{-3}$, or six hydrogen atoms per cubic metre. Observations show that the baryonic matter density is more than an order of magnitude too small to close the Universe. Still, the Einstein–de Sitter model is valid for the early Universe. It shows that, before the $\Lambda$-term had any effect, the curvature of the Universe must have been small; otherwise it would have collapsed or expanded catastrophically before galaxies could form.

Since there are at least three contributors to the total mass–energy density, it has become common practice to adopt a quantity $\Omega$ that is the ratio of the total mass–energy content of the Universe and its total closure density. A flat universe would have $\Omega = 1$. We can also define components $\Omega_b$, $\Omega_{DM}$, and $\Omega_\Lambda$ representing the density of the baryonic, dark matter, and dark energy constituents of the total closure density. Analyses from WMAP and Planck agree that the approximate fractions are: baryonic matter 0.04; dark matter 0.25; and dark energy 0.71. The value of $\Omega$ is 1, within 1%, so the geometry of the Universe is close to being flat (this has an important bearing on what has become known as the inflationary scenario).

The Universe which we have described so far is known as the $\Lambda$CDM model;[3] the CDM stands for cold dark matter, which is described as cold since it does not manifest its presence by thermal radiation or kinetic energy. The model represents the Universe on a large scale as a featureless fluid, while in contrast the real Universe which we see around us is composed of structures such as galaxies and clusters of galaxies. This structure is supposed to have developed from tiny density fluctuations in the very early Universe, which grew through self-gravitation. At an intermediate stage these density fluctuations are detectable as small perturbations on the CMB; it is the amplitude and scale of these perturbations that are remarkably informative about the parameters controlling the expansion.

It is now clear that the cosmological term is non-zero; it is also clear that the curvature term is small, and may be zero. This leads to two outstanding problems to be addressed. First, there is the curvature problem: why is the curvature so small? Second, there is the simultaneity problem: why is the Universe so homogeneous, even for regions that have never been in contact with one another? These questions lie at the core of radio observations in cosmology today.

In the following sections we will describe the observations of the CMB which support the model of the Universe that we have briefly described above and allow such accurate measurements of its parameters.

## 17.4 The Early Universe: The CMB

The expansion of the Universe, prompted by Hubble's discovery, led immediately to GR models that shared a common property: the Universe was expanding from an initial state of extremely high density. In 1946 the nuclear theorist George Gamov proposed that the initial state was an immense ball of neutrons. This resulted in a detailed study of how the elements heavier than hydrogen, helium in particular, could be synthesized (Alpher, Bethe, and Gamov 1948; actually Bethe played no role but the playful Gamov added his friend's name so it could be read $\alpha, \beta, \gamma$). The Japanese astrophysicist Chushiro Hayashi then noted that it made more sense physically to have a mixture of protons and neutrons in

---

[3] An accessible derivation of the most useful equations in a flat $\Lambda$CDM using Newtonian mechanics is given in Condon and Matthews (2018).

thermal equilibrium with a surrounding sea of photons. Alpher and Herman picked up on his suggestion and carried out a correct calculation of helium synthesis. Steven Weinberg, in his little book *The First Three Minutes* (1977), gives an excellent account of both the physics and the historical background of these events.

The hot dense early Universe is not seen by radio or any other electromagnetic radiation. At first, as the Universe expands and cools, the mean free path of a photon in the dense plasma is so short that it cannot escape. At a temperature of about 3000 K, however, when a proton and an electron can combine to form a stable atom, the radiation field has become too weak to reionize the atom. This epoch is known as the EoD – the era of decoupling (sometimes incorrectly called the era of recombination). The age of the Universe at this point is about 350 000 years; the redshift $z$ is approximately 1100. The redshifted temperature of this cosmic fireball is now diminished to about 3 K. It exhibits a blackbody spectrum that peaks at about 1.6 mm wavelength, an optically thick screen that hides earlier events from view. Alpher and Herman predicted that it should be observable as a cosmic microwave background at a brightness temperature of about 5 K.

The discovery of the CMB did not derive from theorists' predictions. A measurement of such a low absolute temperature is an extraordinarily difficult task, requiring the mastery of cryogenic techniques, the best low-noise electronics, and extraordinary care to ferret out the many sources of extraneous noise. It is no surprise that the process started with the development of a new communication system, at the Holmdel Field station of Bell Laboratories (the place where Jansky, working on a communication system in 1933, discovered radio astronomy!). The group were working as part of a satellite communication project to bounce radio signals off the orbiting Echo balloon. Their task was to determine the sensitivity limits of the communication system.

The first measurements at Bell Laboratories were made in a band centred at 2.4 GHz ($\lambda = 12.5$ cm). The antenna was a truncated horn, a 'sugar scoop', rotatable about its longitudinal axis and designed to minimize unwanted noise from sidelobes (Crawford *et al.* 1961), and the receiver was a newly developed travelling-wave maser. The noise contributions from the antenna and receiver system were carefully measured, and the atmospheric contribution was found by measurements at elevations between horizon and zenith (Ohm 1961). The expected total noise temperature of the system, after measuring the contributions of all components, including the Earth's atmosphere, was 18.9 ± 3.0 K. The measured sky temperature was 22.2 K, only slightly above the instrumental and atmospheric contributions. Ohm assumed that the 3 K difference could be attributed to some small inaccuracy in the measurements, although we now know that this was in fact the first direct detection of the CMB.

In 1962 two radio astronomers, Arno Penzias and Robert Wilson, joined Bell Laboratories. They set out to measure the sky temperature with unprecedented accuracy, using the same horn antenna, with a new very-low-noise ruby maser receiver for 4 GHz ($\lambda = 7.5$ cm) and with detailed attention to the calibration and losses in the receiver system. The same 3 K background was established, after an exhaustive series of tests and measurements, as a background brightness of 3.5 ± 1.0 K. The reality of the CMB had been established, although this interpretation was not yet clear. A hundred kilometres to the west, at Princeton, Professor Robert Dicke had conjectured that there might well be a radiation background left over from the early Universe. (He was unaware of the 5 K prediction of Alpher and Herman.) Unbeknown to Penzias and Wilson, he had assembled a group and was well advanced in constructing experimental apparatus at the shorter wavelength of 3 cm. A mutual friend (BFB) put the two groups in touch. Following their conversation, a pair of papers appeared in the *Astrophysical Journal Letters* (1965), one by Penzias and Wilson giving their results, and the other by Dicke and Peebles outlining the theory.[4] Once the result was known, the Big Bang cosmology immediately became the standard model.

### 17.5 The Cosmic Dipole: The Coordinate Frame of the Universe

The first feature of the CMB to be discovered is known as the cosmic dipole. The CMB can be regarded as a sea of radiation through which the observer is moving due to the combined velocities of the rotation of the Earth and its orbital motion about the Sun, together with the velocity of the whole solar system. An observer travelling at velocity $\mathbf{v_0}$ through a radiation field at temperature $T_{\mathrm{CMB}}$ will observe, in a direction $\hat{\mathbf{r}}$, a Doppler-shifted blackbody spectrum with an apparent temperature shift $\Delta T_{\mathrm{CMB}}$ (the non-relativistic formula is sufficiently accurate):

$$\Delta T_{\mathrm{CMB}} = T_{\mathrm{CMB}}(\mathbf{v_0} \cdot \hat{\mathbf{r}})/c. \tag{17.7}$$

This Doppler-induced temperature shift caused by the solar motion through the CMB (duly corrected for Earth's rotation and orbital motion about the Sun) was first measured by a differential microwave spectrometer carried at high altitude in a U-2 research aircraft. The experiment was successful; the solar motion was detected at a level of one part in $10^3$ of the CMB background (Smoot *et al.* 1977).

Later spacecraft measurements show that the Doppler shift due to the Sun's velocity has an amplitude of $3.343 \pm 0.006$ mK, corresponding to a velocity of 363 km s$^{-1}$ toward galactic coordinates $l, b = 264°, 48°$ (Fixsen *et al.* 1996). This is a composite of the solar motion with respect to the local standard of rest, the motion about the centre of the Milky Way, the motion with respect to the local group, and (probably) the motion of the local group with respect to the local supercluster. This represents our motion within the coordinate frame of the Universe. The temperature difference across the dipole is now used as a standard in calibrating other CMB observations, such as those by the Planck spacecraft.

### 17.6 The Blackbody Spectrum of the CMB

The discovery that the CMB defined the rest frame of the Universe addressed a fundamental problem, but there were two basic questions that had to be settled. Was the spectrum of the CMB a true blackbody spectrum? At what level did intrinsic fluctuations of the CMB appear? The sky background is dominated at metre wavelengths by the integrated radiation from the Milky Way and from radio galaxies and quasars, while at shorter (infrared) wave-

---

[4] See Partridge (1995) for an account of the discovery and a more general review; a full account is given in the Nobel Prize addresses by Penzias and Wilson.

lengths it is dominated by extragalactic sources and by faint structures in the Milky Way. Happily, the microwave and millimetre-wave region between these two wavelength ranges contains the peak wavelength of the cosmic background radiation (CMB). The spectrum of the CMB can be traced to wavelengths as long as 50 cm but with decreasing accuracy, as it becomes lost behind the galactic synchrotron radiation (Sironi *et al.* 1990). Observing an accurate spectrum requires observations from above the terrestrial atmosphere, which was first achieved by the spacecraft COBE, launched in 1989.

The COBE spacecraft was furnished with three instruments: the far-infrared absolute spectrophotometer (FIRAS), the differential microwave radiometer (DMR), and the diffuse infrared background experiment (DIRBE). COBE was launched in 1989 into a circular orbit with an altitude of 900 km and an inclination of 99°; this orbit is 'sun-synchronous', precessing one complete cycle per year, so keeping the same orientation with respect to the Sun (Figure 17.2). The instruments always pointed away from the Earth, and approximately 90° away from the Sun, thus assuring a reasonably constant thermal environment; they were protected by their shielding and orientation from the powerful solar and terrestrial radiation. A general description of the satellite and its instruments was given by Boggess *et al.* (1992).

The task of FIRAS was to measure the spectrum of the background radiation by comparison with radiation from an onboard cold blackbody source, whose temperature could be adjusted to give near equality to the CMB. The instrument was a Fourier transform spectrometer cooled by liquid helium and covering the spectral range 1 cm – 0.5 mm. The detector was a bolometer. The very long integration times possible in such a space mission allowed unprecedented accuracies, quoted in millikelvins (mK).

The liquid helium for the FIRAS instrument lasted nine months, which was sufficient to measure the background brightness temperature over the wavelength range and to establish the isotropy of the background radiation (apart from the cosmic dipole described in Section 17.5). The power spectral density integrated over a large angular scale is shown in Figure 17.1, where the experimental points have such small uncertainties that they seem to lie exactly on the theoretical blackbody curve. A plot of the data together with the ideal Planck spectrum shows the deviation of the data from a 2.725 K blackbody curve as a function of wavenumber (wavenumber, the inverse of wavelength, is the natural unit for a Fourier transform spectrometer). This beautiful observation led to the award of a Nobel Prize (2006) to John Mather and George Smoot for their contributions to cosmology.

Although the CMB spectrum was expected to be much like that of a blackbody, there was no such reliable guidance for the level of CMB fluctuations. The task of the differential microwave radiometer (DMR) on COBE was straightforward: to detect the initial fluctuations or set as stringent a limit on them as possible. At some level there had to be fluctuations that would grow under the influence of gravity into the marked concentrations, such as stars and galaxies, that we observe today.

## 17.7 The Search for Structure

The visible Universe is clearly not a uniform, featureless, mass distribution. Stars, galaxies, and clusters of galaxies are major density fluctuations, and those fluctuations have to have

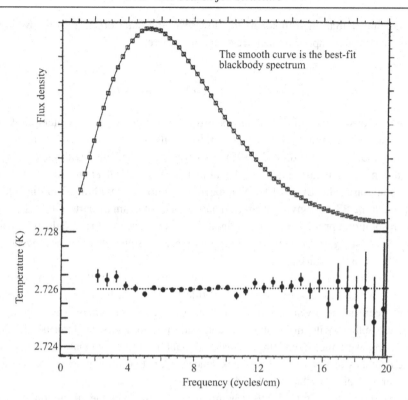

The smooth curve is the best-fit blackbody spectrum

Figure 17.1. Spectrum of the CMB from spectrometer observations in the COBE satellite (Mather *et al.* 1994). Deviations from an ideal blackbody spectrum are shown on a 400× magnified scale.

grown from small initial fluctuations in the early Universe. The differential microwave radiometer on COBE gave conclusive evidence that these density fluctuations existed, but their angular-size distribution had to be measured. Measurements of angular size began to be made at good sites, and balloon-borne radiometers offered promise, but satellite-borne radio telescopes and radiometers were clearly indicated for making observations of the necessary precision. In the meantime, the theory became more sophisticated, notably with the work of Sakharov, Zel'dovich, Peebles, and Tegmark. The key was to recognize that dark matter was more abundant than baryonic matter, and that the density fluctuations of dark matter would start first, dragging the baryonic matter with it. There is a difference between these two populations: there is no known dissipative force for dark matter, since it is influenced only by gravity, whereas baryonic matter does have dissipative forces, most notably electromagnetic. Its collapse is limited by the subsequent heating of the plasma, and it rebounds. Moreover, it is a resonant system and it rings with overtones that should be detectable. The mass motions are actually sound waves, travelling close to the speed of light. Therefore these are called acoustic fluctuations, the fundamental having an angular size of about one degree.

The task of detection was formidable: the dipole component (Section 17.5), which was discovered first, was only $10^{-3}$ of the CMB, and the ripples turned out to be more than

100 times fainter than the dipole. At this low level the main obstacle to be overcome is radiation from water vapour in the terrestrial atmosphere.

### 17.7.1  First Observations of Structure

The largest acoustic peak in the spatial spectrum was observed in the winters of 1993 and 1994 at Saskatoon; central Saskatchewan is noted for its cold, dry, winters. The analysis was reported in 1995 (Netterfield *et al.* 1995). This was fully confirmed with greater accuracy by a balloon-borne measurement in Antarctica in 1998 (Crill *et al.* 2003). This long-duration balloon flight, BOOMERANG, carried a radiometer with a 1.2 m diameter dish to an altitude of 38 km, giving a more detailed map of a small part of the sky with an angular resolution approaching 10 arcminutes. The existence of the one-degree peak was established; a survey covering the whole sky with greater sensitivity was now needed. This was clearly a task for spacecraft.

Sensitivity of order 10 microkelvins was needed in the face of the system noise of the radiometer, which is measured in kelvins. Wide bandwidths, long integration times, and excellent receiver stability are essential; in practice this has been achieved using comparison radiometers, either switching systems or correlation radiometers, as described in Chapter 6. Differences were measured either between signals from two antennas directed to widely separated patches of sky, or, in the Planck spacecraft, between an antenna and a cold load at a temperature close to that of the CMB.

Full-sky observations of the CMB structure were first made by the differential microwave radiometer (DMR, Figure 17.2) on the spacecraft COBE (Bennett *et al.* 1996). The angular resolution of 7° was limited by the size of the horn antennas. Later spacecraft missions, described below, achieved improvements in both sensitivity and angular resolution, requiring larger antennas.

These radiometers inevitably also observed radiation from the Milky Way, which is at a minimum at long millimetre wavelengths but which nevertheless contains structure on the same angular scales as the CMB. There are three main components of this unwanted foreground radiation: synchrotron, free–free, and thermal radiation, to which must be added a previously unknown 'anomalous microwave emission' (AME) at around 30–40 GHz, prominent in some parts of the Galaxy (see Figure 17.3 and the Supplementary Material). Fortunately these foregrounds may be distinguished by their different spectra, by comparing maps made at a range of wide-spaced frequencies. This was achieved by the COBE DMR, which observed at three frequencies and isolated CMB structure at an angular scale of 7°. The Wilkinson Microwave Anisotropy Probe (WMAP) used five frequency bands, from 22 GHz to 90 GHz, while Planck used nine frequencies, from 30 GHz to 857 GHz, providing a definitive separation of the foreground components and mapping the CMB structure over the whole sky.

The final COBE map was smoothed from the antenna resolution of 7.5° to 10°. Even so, there was not a sufficient signal-to-noise ratio in the map of $T(l, b)$ to show the CMB fluctuations directly;[5] the map was a speckle pattern which was a mixture of system noise

---

[5]  The quantities $l, b$ are spherical harmonic coordinates: see Appendix 5.

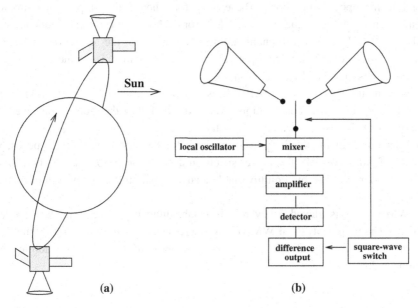

Figure 17.2. (a) COBE in orbit. The telescope axes were maintained perpendicular to the direction of the Sun; a full scan of the sky was achieved in one year. (b) The DMR radiometer detected the difference between signals from two horn antennas, using a Dicke switch system (Chapter 6).

and the cosmic signal, the two being of comparable amplitude. There were approximately 1600 independent pixels in the map, however, and the structure of the CMB cosmic signal was obtained by a statistical analysis (akin to the $P(D)$ analysis in Section 16.9), averaging over the whole sky.

The DMR results demonstrated that the angular power spectrum was clearly detectable at angular scales down to $7°$, with an rms of $30\,\mu K$, and showed that the angular power spectrum at large angular scales was approximately a power law, with a spectral index of $1.2 \pm 0.3$ consistent with the predicted Harrison–Zel'dovich spectrum (Harrison 1970; Zel'dovich *et al.* 1972). The angular resolution of COBE was, however, insufficient to show the expected peak in the spectrum at around $1°$, which required observations with larger-diameter telescopes.

### 17.7.2  Wilkinson Microwave Anisotropy Probe (WMAP)

The WMAP mission, launched in 2001, was built on the legacy of the DMR experiment. Thermal noise from the Earth, and to a lesser extent from the Moon, leaking in the sidelobes of the DMR antennas had contributed to the measurement uncertainty; to minimize sidelobe contamination, the WMAP satellite was placed in an orbit around the L2 solar Lagrange point, 1.5 million kilometres from the Earth, where it was maintained by small thusters. A pair of $1.6 \times 1.4$ m telescopes were mounted back to back to compare the sky brightness temperature in opposite directions. These were furnished with comparison radiometers using the correlation technique (Section 6.8.2) for each of five different frequency bands (23 GHz to 94 GHz), chosen to enable an accurate model of the Milky Way foreground contributions to be constructed. Dual-polarization feeds were included to measure the polarization properties of the CMB fluctuations.

An early reduction of the WMAP results, in 2013, produced a fluctuation map of the CMB in which noise contributions were so small that the fluctuations were real (unlike the COBE results, which were statistical). The improved angular resolution allowed derivation of the spatial power spectrum for the first four acoustic peaks, with accuracies of order 1%–2%. The power spectrum depends on six parameters (see Section 17.8), which were derived by a multivariable analysis, ushering in a new age of precision cosmology.

The WMAP results played a key role in establishing the current standard model of cosmology. The final analysis of WMAP's measurements was presented by Bennett *et al.* (2013) and the cosmological significance by Hinshaw *et al.* (2013). The satellite ceased operations in 2009.

### 17.7.3 Planck

The final stage (at least for some decades) in satellite observations of the angular structure was the spacecraft Planck, launched in 2009. Like WMAP, Planck was placed into an orbit round the distant Lagrangian point L2. With a larger telescope aperture ($1.9 \times 1.5$ m) and an array of detectors covering a $30 : 1$ wavelength range from 1 cm to 0.3 mm, and rotating at one revolution per minute, Planck covered the spatial spectrum out to $l = 2500$ (angular scale $\approx 0.07°$). Using only a single off-axis paraboloid, an array of horns for 74 detectors in nine frequency bands sensitive to frequencies between 25 GHz and 900 GHz was located in the focal plane. Transistor amplifiers (HEMTs) and pseudo-correlation radiometers (Chapter 6) similar to those used in WMAP were used in the three lowest frequency bands, centred on 30, 44, and 70 GHz; their performance is described by Mennella *et al.* (2011). Bolometers covered the six high frequency bands from 100 GHz to 857 GHz (wavelengths 3 mm to 350 microns) with both linear polarizations in the four lower bands (Lamarre *et al.* 2010). All measurements were made against a reference cold load at 4 K. Calibrations were also made against the thermal radiation from the cosmic dipole and several planets.

Planck operated for over four years, imaging the whole sky twice each year. For an overview of its products and scientific results, see the Planck Collaboration XIII (Adam *et al.* 2016) and updates in the Supplementary Material.

The wide spectral coverage of Planck enabled a detailed measurement of galactic and other foreground sources of radiation which can contaminate the primordial signal. Fortunately, these have mainly different spectra, so that observations at a range of frequencies, as made by WMAP and Planck, can distinguish between them (Dickinson 2016). Figure 17.3, from the Planck Collaboration X (2016), shows schematically the spectra of this foreground radiation from 10 GHz to 100 GHz in an area away from the plane of the Milky Way. Synchrotron radiation, which dominates the sky at low radio frequencies,

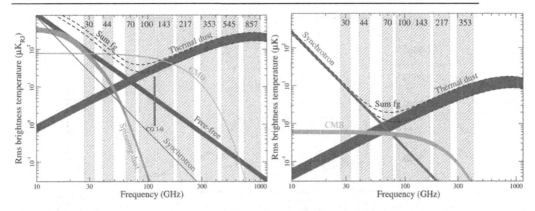

Figure 17.3. Spectra of foregrounds and CMB: on the left, the temperature and on the right, the polarization, showing the contributions from synchrotron, free–free, and dust. A 'spinning dust' component may be responsible for the increased total radiation between 20 GHz and 90 GHz, as shown. (Reproduced with permission from *Astronomy & Astrophysics*, ©ESO; original source ESA and the Planck Collaboration.)

makes a smaller contribution to the background temperature at 1 mm wavelength than free–free radiation from ionized gas and blackbody radiation from dust (familiar to observers in infrared). The temperature spectral indices ($T \propto \nu^\beta$) at GHz frequencies of these main contributors to the foreground, as used in the Planck analysis, are: synchrotron $\beta = -2.7$; free–free $\beta = -2.1$; thermal dust $\beta = 1.6$ (Davies *et al.* 2006). A further component, known as the anomalous microwave emission (AME) and attributed to spinning dust, is required to fit the observed spectrum, as shown in Figure 17.3. The foreground minimum is at around 70 GHz. These foreground components have been separately mapped, and are presented in the series of papers by the Planck collaboration. They are included in the Supplementary Material.

### 17.7.4 South Pole Telescope

The diameters of antennas on spacecraft are limited to around 1.5 m (apart from the possibilities of assembly in space, as in the James Webb Space Telescope), so that angular detail of the CMB structure that is finer than around 0.1° has required ground-based observations using telescopes with diameters around 10 m. Although Planck had measured the spatial spectrum of the CMB intensity over the whole sky, there remained a need for further measurements of its polarization and finer angular structure (the Sunyaev–Zel'dovich effect, discussed below). The outstanding example is the South Pole Telescope (SPT), situated at the Amundsen–Scott South Pole Station at an elevation of 1.8 km, where the atmosphere is very cold and dry, with a very low water vapour content. The SPT is a 10 m diameter Gregorian telescope (Carlstrom *et al.* 2011). It was originally equipped with a remarkable focal plane array of 966 horns feeding receivers at wavelengths 3, 2, 1.3 mm (95, 150, 220 GHz),[6] giving arcminute resolution over a degree-width field of view. The detectors for

---

[6] A new receiver array SPT-3G covering the same frequencies has 16 000 detectors; it started observations in 2018.

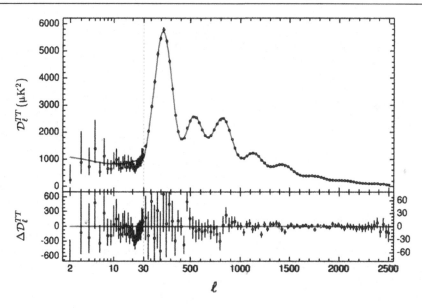

Figure 17.4. The angular spectrum of the CMB, as measured by the Planck spacecraft, showing the power spectrum of the temperature fluctuations at spherical harmonic *l*. Residuals with respect to a model spectrum are shown in the lower panel (note that the scale changes at *l* = 30, where the horizontal axis switches from logarithmic to linear). (Reproduced with permission from *Astronomy & Astrophysics*, ©ESO; original source ESA and the Planck Collaboration.)

the whole array are transition edge bolometers (Chapter 6). It should be noted that multi-beam operation is essential for such survey observations, since long integration times must be used; the factor $[(\Delta\nu)\tau_{\text{int}}]^{1/2}$ (Chapter 6) has to be typically $10^6$ to reduce a receiver rms noise of order 1 K to a sensitivity of 1 μK. The necessary integration times of more than 1000 seconds mean that a single beam of 1 arcminute would require over 3000 years to cover the whole sky.

## 17.8  The Derivation of Cosmological Quantities

The removal of foreground contributions to the intensity map of the CMB appears to be very successful, even at low galactic latitudes, as may be seen in the Planck map of the CMB in the Supplementary Material.

The spatial spectrum must be presented in terms applicable to the spherical sky. Instead of the spectral components or wavenumbers applicable to a rectangular map, the angular power spectrum of the brightness fluctuations is represented in terms of spherical harmonics (Appendix 5), with coefficients *l* relating to the angular scale. The spectrum is convention-ally plotted with ordinates $C_l$ proportional to temperature squared.

The power spectrum of the angular structure in the CMB, as measured by Planck, is shown in Figure 17.4, where $C_l$ multiplied by a customary scaling factor $l(l+1)/2\pi$ is plotted. The solid line in this plot is a fitted model based on the $\Lambda$CDM ($\Lambda$ Cold Dark Matter) model.

Further measurements of the CMB structure by spacecraft are unlikely to improve on the angular spectrum of Figure 17.4. The large error bars at the largest angular scales (i.e. at the lowest $l$ numbers) are inherent in the CMB structure itself, since there is a limited number of fluctuations across the whole sky (the so-called 'cosmic variance' limitation). Ground-based observations, using larger apertures, may provide more information on the finer angular structure. Most observational initiatives, however, are concentrating on measurements of the polarization, for which even greater sensitivity is required.

Six parameters were determined in the fitting process, characterizing the whole cosmological scene to 1% accuracy. These are: dark energy density, $\Omega_\Lambda$; the baryon density, $\Omega_b$; the dark matter density, $\Omega_d$; the age of the era of decoupling, $T_d$; the present age of the Universe, $T_u$; and the rms scale of the fluctuations, $\sigma_8$. The review by Hu and Dodelson (2002) gives a detailed discussion with references, and the elementary exposition by Tegmark (1995, see the Further Reading section) provides an introduction to the physical ideas. Several excellent illustrated tutorials are available on the web, e.g. those of Wayne Hu, NASA WMAP, and ESA Planck.

The $l$ values below the first peak are close to a Harrison–Zel'dovich scale-free spectrum, which is consistent with the inflationary scenario of Guth and others.[7] The first peak is an accurate indicator of the curvature of the Universe. Its location at $l = 220$ shows that the Universe has flat geometry. The second and third peaks depend strongly on the baryonic fraction $\Omega_b$. The present consensus is that the spatial geometry of the Universe is 'flat', i.e. $\Omega_\lambda + \Omega_d + \Omega_b = 1$ to an accuracy of about 0.4%. The time of the era of decoupling was at 370 000 years, and the age of the Universe is 13.8 billion years. In round numbers, the Universe is about 5% baryons and 25% dark matter, while dark energy is inferred to constitute 70% of the mass–energy of the Universe. A complete list of cosmological parameters may be found in Planck Collaboration Paper XIII (2016).

The inflationary scenario requires the Universe to have zero curvature, and the observed near-zero curvature gives strong support to this model. Since the inflationary scenario explains the simultaneity and homogeneity puzzles it is no longer a hypothesis, perhaps not quite a theory but a strong scientific construct. Dark energy remains a fact but, even though it is consistent with a universe with a non-zero cosmological constant, there are other ideas that need examination, and we need large-scale studies of galaxies and radio sources to see if their large-scale distribution is consistent with inflation.

## 17.9 Polarization Structure of the CMB

Measurements of the polarization of the CMB provide unique information on the early Universe. The subject is complicated and here we simply sketch in some top-level concepts.

---

[7] Cosmic inflation is a hypothetical rapid expansion of the very early Universe which provides an explanation for the observed flatness and isotropy. For an exposition, see Ryden (2017), Chapter 10.

For more details we refer the reader to the pedagogic exposition by Hu and White (1997)[8] and other references listed in the footnotes and the Supplementary Material.

The CMB is a blackbody radiator to one part in $10^5$ but below this level it exhibits a spectrum of fluctuations, as noted earlier. An ideal black body would be unpolarized, but the fluctuations generate locally anisotropic (quadrupolar, i.e. with peaks at multiples of 90°) radiation fields which generate patterns of linearly polarized radiation. These patterns depend on the form of the spatial fluctuation structure. In the simplest case, a brighter region illuminates the adjacent fainter regions and the radiation is Thomson-scattered, in the perpendicular direction, with linear polarization. This is analogous to the partial polarization of the sky due to scattered sunlight.[9] Fluctuations generated by the growing baryonic matter concentrations (density fluctuations) give patterns of polarization orientation that are gradient-like (curl-free) and hence analogous to the behaviour of electric fields; these are therefore called the $\mathcal{E}$-mode patterns.

There is another significant source of quadrupolar variations which produces different polarization patterns. Gravitational waves generated in the earliest phase of the Universe's history are shear waves, simultaneously squeezing and stretching space and everything within it in orthogonal directions. The patterns of scattered radiation have an $\mathcal{E}$-mode component but also a component with a different, odd, symmetry. The latter are curl-like (divergence-free) and hence analogous to the behaviour of magnetic fields; they are therefore called the $\mathcal{B}$-mode patterns.

To recapitulate, density perturbations generate only $\mathcal{E}$-mode polarization patterns whilst perturbations due to gravitational waves generate both $\mathcal{E}$-mode and $\mathcal{B}$-mode patterns. The polarization pattern across the sky is then a random superposition of these modes but the "handedness" of the $\mathcal{B}$-mode enables it to be statistically distinguished from the $\mathcal{E}$-mode. Measuring the relative strength of the $\mathcal{E}$- and $\mathcal{B}$- modes gives a fundamental insight into the very early Universe. Theory suggests that gravitational waves dominated when the Universe was only $\sim 10^{-16}$ seconds old, far earlier than the time of the first-three-minutes-of-matter dominance.

The observational challenge is formidable since discerning linear polarization patterns requires mapping at the μK level or better. The first hurdle to overcome is making reliable measurements untainted by subtle instrumental effects. The $\mathcal{E}$-modes have, however, been detected with good signal-to-noise ratio at $\sim 10\%$ of the intensity fluctuations; this is a strong validation of the overall picture of the CMB anisotropies. The next challenge is to measure the intrinsic $\mathcal{B}$-modes, which are expected to be at least ten times weaker than the $\mathcal{E}$-modes. In order to disentangle the $\mathcal{B}$-mode patterns one must also take account of the confusing effects of galactic foregrounds. Radiation scattered by dust grains which are preferentially aligned by galactic magnetic fields can mimic $\mathcal{B}$-mode patterns. The spectra in Figure 17.3 illustrate the problem of distinguishing the CMB emission from that of galactic dust in the range 100–150 GHz, where the best polarization measurements have been made. The escape route is that the cross-section for dust scattering is frequency dependent

---

[8] See also http://background.uchicago.edu/~whu/polar/webversion/polar.html
[9] http://background.uchicago.edu/~whu/intermediate/polar.html.

and so multifrequency mapping will enable the dust contribution to be subtracted.[10] A second 'foreground' is gravitational lensing by matter concentrations along the line of sight; lensing can turn an $\mathcal{E}$-mode into a $\mathcal{B}$-mode[11,12]. Theory suggests, however, that the peak of the intrinsic $\mathcal{B}$-mode angular spectrum should be at $l \sim 100$ and lower whilst the peak of the $\mathcal{B}$-mode spectrum due to lensing should be at $l \sim 1000$.

No intrinsic $\mathcal{B}$-mode polarization has yet (in early 2019) been detected although several observational campaigns are underway. Intrinsic $\mathcal{B}$-modes have been dubbed the 'smoking gun' diagnostic of inflation and so the prize for the first reliable identification is great.

## 17.10 The Transition to the Era of Reionization

Between the era of decoupling and the era of reionization lies an almost unknown Universe, often called 'the dark ages'. The faint fluctuations that we observe in the CMB grow as the Universe expands and, as they do so, fragmentation begins, ultimately becoming the stars, galaxies, and larger structures that we observe today. The era of reionization starts when the first Population III stars appear – the first stars composed mostly of hydrogen and helium. These hot, luminous, stars start to ionize the surrounding neutral gas and upset its thermal conditions. As time proceeds, the fractional ionization increases until the star is surrounded by an ionized 'bubble'. Eventually these bubbles merge, and the transition to a fully ionized plasma is complete. It may be possible to follow this process by observing neutral hydrogen at the large redshifts of the era over which this occurs.

The resulting brightness temperature of neutral hydrogen depends upon the state temperature, the gas kinetic temperature, and the local radiation temperature. The net result of the first infusion of ultraviolet radiation is to couple the state temperature to the gas temperature, which is lower than the background radiation temperature; thus an absorption line is produced. This will occur at a range of redshifts, and the absorption line will be correspondingly broad.

Two approaches are under way at the time of writing (2018), either searching for absorption in individual features of the background or looking with a wide beamwidth at an average over the whole sky. Individual features might be observable by existing arrays such as MWA and PAPER, which have published upper limits at frequencies in the range 120–140 MHz ($z \sim 10$). As in the first searches for structure in the CMB, individual features may not be detectable with sparse arrays (such as LOFAR), and the general structure must be sought through a three-dimensional power spectrum analysis.

The other approach is to look for the absorption feature in the spectrum of the whole sky. This is difficult because of the strong foreground radiation from galactic synchrotron emission, which however does have a smooth spectrum in contrast with the redshifted

---

[10] https://www.cfa.harvard.edu/~cbischoff/cmb/
[11] http://sci.esa.int/planck/51607-gravitational-lensing-of-the-cosmic-microwave-background-animation/
[12] http://background.uchicago.edu/~whu/Presentations/cmblens.pdf

hydrogen absorption. Such an effect was reported by Bowman *et al.* (2018) as a decrease in brightness, as measured by a single broadband dipole, of 0.5 K in a band 19 MHz wide at 78 MHz. The effect is only of order $10^{-4}$ of the synchrotron background; it was established only by meticulous calibration of the whole multifrequency receiver system, and it will require confirmation by similar observations elsewhere. The result is an average of many regions, covering a large area of the sky; the detection of individual regions will require a telescope with a large collecting area and many-channel frequency resolution. A planned instrument, HERA, should be able to do this, but its completion is some years away.

At a smaller redshift, around $z \sim 1$, there is expected to be a pattern in the distribution of matter that has grown from the baryon acoustic oscillations (BAO), which appear as structure in the CMB. The H I emission is mostly concentrated in galaxies and acts as a tracer of the total matter distribution. An observation of such structure, at frequencies of around 1 GHz and below (avoiding the local galactic hydrogen at 1.4 GHz), is in a frequency range where the sky is dominated by synchrotron emission from the Milky Way. The intention is to observe the frequency structure related to the hydrogen, in contrast with the smooth spectrum of the synchrotron radiation. The BAO structure that corresponds to the integrated emission from hundreds of galaxies should have an angular size of the order of $1°$ and an intensity of order $10^{-4}$ of the background. This is being attempted by several new telescope systems, which we illustrate with two different approaches to this demanding problem: the projects charmingly known as CHIME and BINGO. Both are designed to observe at redshifts of order $z = 1$, CHIME (the Canadian Hydrogen Intensity Measurement Experiment) at $0.8 < z < 2.5$ and BINGO (Baryon Acoustic Oscillation from Integrated Neutral Gas Observations) at $0.1 < z < 0.45$. The H I emission is distinguishable from the foreground by its comparatively narrow frequency structure, which must be recognized by using a receiver with multiple frequency channels with width around 1 MHz. The expected hydrogen signal is of order 100 μK, requiring very long integration times. High angular resolution is not necessary; the angular size of the dominant structure is expected to be around $1°$. Both systems provide many beams simultaneously, without which scanning the sky would be impractically slow; CHIME uses aperture synthesis in a close-packed array 100 m square (Chapter 9) while BINGO uses an aperture array of 50 horns in the focal plane of a fixed 40 m diameter parabolic reflector (Chapter 8). Both telescopes will deliver independent observations in several hundred adjacent frequency channels. As in many modern telescopes, the data rate will be huge. For details of CHIME, which started observations in 2017, see Newburgh *et al.* (2014); for BINGO see Battye *et al.* (2013). For a review of the progress up to 2017 in these and other H I mapping telescopes, see Kovetz *et al.* (2017).

## 17.11 The Sunyaev–Zel'dovich Effect

Starting from the primordial density fluctuations, dark matter has, under the action of gravity, aggregated on many scales. The dynamically less significant baryonic matter then congregates in the dark matter potential wells, there to form stars and galaxies. The highest-mass bound aggregates are clusters of galaxies, with total masses in the range $10^{14}$ to over

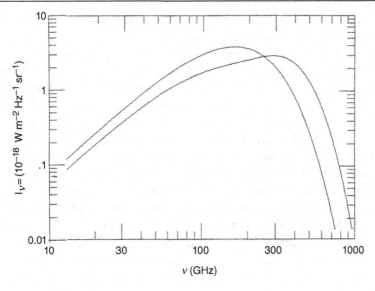

Figure 17.5. The Sunyaev–Zel'dovich effect, showing the shift of the CMB spectrum to higher frequencies (Birkinshaw 1999).

$10^{15}$ $M_\odot$ and containing hundreds or thousands of galaxies. Surrounding gas falling into such a cluster's gravitational potential well heats up to temperatures $10^7 - 10^8$ K (about 1–10 keV) and the electrons in the resulting hot intracluster plasma can be directly observed in the X-ray band via their free–free (bremsstrahlung) emission. They can also be indirectly observed in the radio band. Photons of the CMB travelling on a line of sight through a cluster interact with the high energy electrons and gain energy via the process of inverse Compton scattering. The overall effect is that the whole CMB spectrum is moved up in frequency, as shown in Figure 17.5; at the same time, the hot electrons give up a little energy in the scattering process.

At a frequency below 218 GHz ($\lambda = 1.4$ mm), the background temperature is reduced since some of the original photons have been shifted up in energy and hence to higher frequencies. The observable is, therefore, a local temperature decrement, in the range 100–1000 µK, centred on the cluster; at higher frequencies the cluster appears as a hot spot. This is the Sunyaev–Zel'dovich effect (SZE; Sunyaev and Zel'dovich 1980). On the scale of 1–10 arcmins the SZE provides a significant foreground to the CMB and must be taken into account when interpreting the intrinsic anisotropies. The SZE can be distinguished from CMB anisotropies because of its frequency-dependent brightness.

Measurements of the SZE, in combination with X-ray and optical observations, hold out much promise for astrophysics and cosmology. Birkinshaw (1999) and Carlstrom *et al.* (2002) provided comprehensive reviews of the fundamental physics and assessed the impact of the 'first generation' SZE measurements. The implications of the 'second generation' measurements were reviewed by Kitayama (2014). The SZE touches on a wide variety of subjects but most fall beyond the scope of this book. Among these, we note particularly the relation to X-ray observations.

The SZE and X-ray measurements of clusters are complementary probes of cluster physics. The SZE is proportional to $\int n_e T dl$, i.e. to the energy density and hence to the gas pressure integrated along the line of sight. On the other hand the X-ray emissivity is proportional to $\int n_e^2 T dl$, and hence the X-rays preferentially respond to the densest gas, typically near the centre of the cluster. Comparisons of SZE and X-ray images therefore contain information on the thermal structure of the intracluster gas. High-sensitivity high-resolution SZE images are now probing the internal substructure of selected clusters, for example using MUSTANG on the GBT with a 9 arcsec beam (e.g. Korngut *et al.* 2011) and NIKA on the IRAM 30 m with an 18.5 arcsec beam (e.g. Adam *et al.* 2017). Some of the shock-indicating substructure thus found may be associated with cluster mergers.

As a first-order approximation, however, the distribution of the intracluster gas can be assumed to have spherical symmetry and to be in hydrostatic equilibrium. Under the latter assumption the total mass of the cluster can be derived from a combination of the gas temperature (from X-ray spectral measurements of the free–free emission) and its density as a function of radius (from the X-ray surface brightness) – see e.g. Ettori *et al.* (2013). It is not surprising that there is a strong correlation between the gas temperature after infall and the cluster mass. The total mass of a cluster can also be derived from the velocities of the constituent (gravitationally bound) galaxies (Zwicky's original method) and from gravitational lensing (strong or weak). Given the temperature distribution, the SZE then provides one with a measure of the integrated *baryonic* mass of intracluster gas on the line of sight. The cluster baryon fraction can thereby be determined, and a broad picture has emerged: most of a cluster's mass ($\sim 85\%$) is in the form of dark matter and most of the baryons ($\sim 75\%$) are in the intracluster gas rather than in stars.

An important feature of the SZE for cosmological studies is that the surface brightness at a given frequency is essentially independent of redshift. This contrasts with the situation for X-rays (and indeed for most astrophysical sources), whose brightess dims as $(1+z)^{-4}$ owing to a combination of relativistic effects in an expanding universe. The SZE represents a fractional change in the brightness of the CMB whose energy density increases as $(1+z)^4$, negating the cosmological dimming.[13] Surveys using the SZE can therefore, with sufficient angular resolution, detect massive clusters at any redshift. However, since the typical size of a massive cluster, 5–10 Mpc, corresponds to $\sim 2$ arcmin at $z = 1$, angular resolution better than this is necessary to identify distant clusters.

The most extensive SZE surveys to date have been those by the Planck spacecraft and the South Pole Telescope. Planck has produced an all-sky catalogue of 1227 cluster and cluster candidates (Ade and *et al.* 2015). The SPT has produced an independent catalogue of cluster candidates from surveys of 2500 degree$^2$ of sky (Bleem *et al.* 2015). At the $5\sigma$ level there are 409 candidates, the great majority of which have been confirmed from ground-based observations. The two surveys complement each other and their detection

---

[13] A factor $(1+z)^3$ arises from the reduced volume, and photon energies are higher by $1+z$ at earlier epochs.

rate as a function of redshift illustrates the effect of matching the resolution to the cluster size. The Planck beam size is between 5 and 10 arcmin, depending on the frequency, so it is more sensitive to relatively nearby clusters. The beam of the SPT (1 to 1.5 arcmin depending on the frequency) is better matched to clusters at high redshifts. Thus, whereas Planck does not detect clusters with $z > 1$, the SPT catalogue contains several clusters with $z > 1.5$.

Since clusters are the largest bound structures, establishing their numbers and masses as a function of epoch is a powerful diagnostic of the emergence of structure in the expanding Universe. The amounts and distributions of dark matter (which enhances the rate of aggregation) and dark energy (which delays aggregration) play leading roles, but a variety of processes modify the primordial spectrum. To predict the evolution quantitatively, a set of complex inter-related effects must therefore be simulated; a seminal discussion, based on the $\Lambda$CDM model, is given in White *et al.* (1993).

Cluster demographics are particularly sensitive to a factor designated $\sigma_8$, the mass variance within a sphere of radius $\sim 10$ Mpc in the primordial density field. The largest clusters must have grown out of overdensities on this scale. When the current generation of SZE surveys was in the planning stage it was thought that $\sigma_8 \sim 1$ and at least ten thousand cluster detections were anticipated. In the event the numbers are closer to one thousand and favour $\sigma_8 \sim 0.8$; this is close, but not identical, to the value inferred from the Planck CMB data (see Ade *et al.* 2016).

Sunyaev–Zel'dovich effect surveys out to high redshifts are opening an unique window on cosmology. But extracting the requisite information demands multiwavelength studies: to understand better the cluster astrophysics; to measure reliable cluster masses; to measure cluster redshifts. These tasks all become more challenging at higher redshifts. Nevertheless it can be confidently expected that SZE-based cosmological studies have now reached only 'the end of the beginning'.

## 17.12 Gravitational Lensing

The bending of a light ray in the gravitational field of a massive body was first observed at the solar eclipse of 1919, when it provided a crucial test of general relativity (GR). In 1936 Einstein pointed out that, if a distant star were to be seen almost precisely behind a closer star, two ray paths would be possible through the gravitational deflection of the rays by the closer star, and it would be seen as a double image. This was observed in 1979 when Walsh *et al.* (1979) discovered that the radio source B0957+561 was identified with a close pair of star-like images on a sky survey plate; these turned out to be identical images of a single quasar. The two images, 6 arcsec apart, were found to have almost identical optical spectra, both with redshift $z = 1.41$; the separation of the two images was due to gravitational 'refraction' in a 'lensing' cluster of galaxies on the line of sight, at a redshift $z = 0.36$. Although conceptually it is easier to think of gravitational lensing as a refraction effect of gravity on a light ray, the correct analysis instead regards gravity as distorting space–time, in which light follows a redefined straight line (a geodesic) with velocity $c$. Nevertheless, it is permissible to state simply that the effect of a point mass is a deviation by an observable angle. There are two regimes in which such a deflection is observed.

– *Strong lensing*, with multiple images, distorted by the lensing process. Three categories of strong lensing can be distinguished:

　(i) macrolensing, where a foreground cluster of galaxies creates many images of distant galaxies, on a scale of tens of arcseconds;
　(ii) mesolensing, where a foreground galaxy images a distant quasar or galaxy, on a scale ranging from several arcseconds to a fraction of an arcsecond;
　(iii) microlensing, where the individual stars in a foreground galaxy, or in our own Galaxy, image a more distant object, on a scale of microarcseconds. Here the images can be confused, but the effect is usually manifested by scintillations of the lensed object.

　　Radio astronomy has had a significant impact on mesolensing by galactic masses, but not as yet on macro- or microlensing.

– *Weak lensing*, when a background object shows only a single, but distorted, image (Section 17.15).

### 17.13 Ray Paths in a Gravitational Lens

According to general relativity, the angular deflection $d\alpha$ in a ray path $dl$ due to a gravitational field $g$ is

$$d\alpha = 2c^{-2}g\,dl. \tag{17.8}$$

Integrating along a ray path close to a mass $M$, with impact parameter $b$, as in Figure 17.6, gives a total angular deflection

$$\alpha = 4MG/c^2b. \tag{17.9}$$

The geometry of this effect is qualitatively different from the geometry of a simple optical lens, in which the angular deflection is proportional to axial distance; here the deflection is *inversely* proportional to the axial distance, giving distorted images more like those of a mirage rather than a lens (the base and stem of a wineglass distort light rays in an analogous way). Note that the angle of deflection can be written $2R_S/b$, where $R_S = 2MG/c^2$ is the Schwarzschild radius for the mass $M$.

The lensing geometry is shown in Figure 17.6. The object, whether a quasar or a radio galaxy, is treated here as a point source. It is located in the object plane at distance $D_s$, and by convention an image plane is constructed centred on the deflecting mass, the lens, at distance $D_d$; the source would have been seen at an angle $\beta$ from the centre of the deflecting mass if there were no lensing effect. Two ray paths, on either side of the lens, give rise to the two images seen by an observer; these are seen projected on the image plane at an angle $\theta$ with respect to the axis joining the observer and the centre of mass. In this small-angle approximation, a simple geometrical construction, with the impact parameter $b = D_d\theta$, gives, using the point-mass deflection law of Eqn 17.9,

$$\theta = \beta + \left(\frac{D_{ds}}{D_dD_s}\right)\left(\frac{4GM}{c^2}\right)\frac{1}{\theta}, \tag{17.10}$$

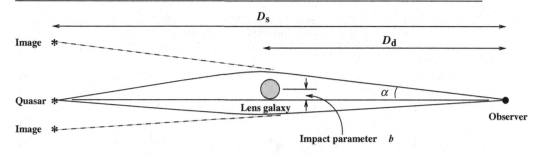

Figure 17.6. Ray paths in a gravitational lens in which the lensing galaxy is treated as a point mass. Two paths are shown; in the more general case there are three or five, depending on the geometry of the extended lensing galaxy.

where $D_{ds}$ is the lens-to-source distance. There is a particularly simple solution of Eqn 17.10 when the point mass and the source are lined up ($\beta = 0$): the source becomes a ring, generally referred to as an 'Einstein ring', of angular radius $\theta_E$:

$$\theta_E = \sqrt{\left(\frac{4GM}{c^2}\right)\left(\frac{D_{ds}}{D_d D_s}\right)}. \tag{17.11}$$

The Einstein ring radius in physical terms can be written $R_E = (2R_s D_*)^{1/2}$, where $D_* = D_s D_d / D_{ds}$ is a 'corrected distance'. A useful approximation applicable to a lensing cluster at a distance of the order of 1 Gpc with mass $M$ measured in units of $10^{11}$ solar masses is

$$\theta_E = \left(\frac{M}{M_\odot^{11}}\right)^{1/2} \quad \text{arcsec.} \tag{17.12}$$

Equation 17.10 is a simple quadratic equation, and the solutions for non-zero $\beta$ give one image outside the Einstein ring locus (the principal image), and one image inside the ring. The images are magnified; the closer the source is to the observer–lens axis, the higher the magnification. The magnification of a small source is the ratio of the observed solid angle $\delta\theta_1\delta\theta_2$ in the image plane to the solid angle $\delta\beta_1\delta\beta_2$ that the source would have subtended in the object plane if there had been no lens, where the angles are now expressed in two dimensions. When the lens mass distribution is circularly symmetric but extended, there will still be an Einstein ring for an aligned source, but the mass in Eqn 17.11 will be the lens mass contained within the ring.[14]

The text by Schneider *et al.* (1992) gives a comprehensive treatment of gravitational lensing, with a summary of the extensive literature and history of the subject. See also Mellier (1999).

---

[14] The essential features exhibited by simple lens–source configurations can be explored using the on-line simulator at www.jb.man.ac.uk/distance/frontiers/glens/gsim.html.

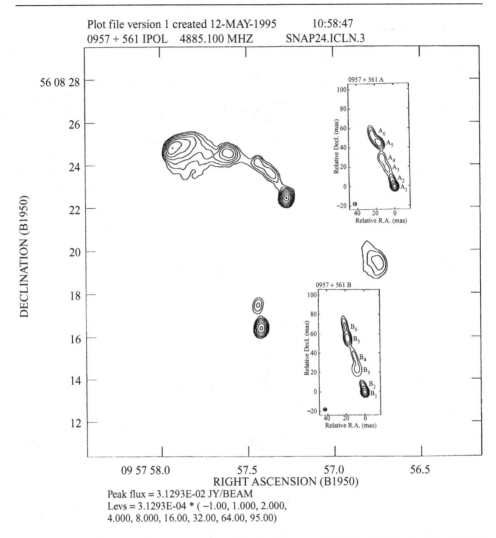

Plot file version 1 created 12-MAY-1995          10:58:47
0957 + 561 IPOL    4885.100 MHZ          SNAP24.ICLN.3

Peak flux = 3.1293E-02 JY/BEAM
Levs = 3.1293E-04 * ( −1.00, 1.000, 2.000,
4.000, 8.000, 16.00, 32.00, 64.00, 95.00)

Figure 17.7.  Radio map at 6 cm wavelength of the 'double quasar' B0957+561. The detailed (VLBI) maps of the two main components (Garrett *et al.* 1994) in the insets show similar, but not identical, structures.

### 17.13.1  Imaging by Extended Lenses

The 6 cm radio maps of the double quasar B0957+561 (Garrett *et al.* 1994), shown in Figure 17.7 (see also the Supplementary Material), demonstrate the complexities that arise in real cases. The northern source is the principal image, and the VLBI image clearly has a core–jet structure, but no such structure is immediately evident for the southern image. The source near the southern image is associated with the most prominent galaxy in the field, and it is that galaxy, the brightest in an extended rich cluster that covers the field, which exerts the largest lensing influence. There is, however, a considerable gravitational effect from the mass of the cluster as a whole, and from the dark matter that may be contained in it; finding the mass distribution from the details of such images is a complex task. Even an isolated galaxy acting as a lens is not a point mass and can seldom be treated as spherically symmetrical.

The gravitational-lensed systems discovered by radio telescopes are shown on the Supplementary Material website. The observed rate of strong lensing as observed in the CLASS radio survey (Browne *et al.* 2003) is consistent with the prediction of Turner *et al.* (1984) that a sample of 1000 radio sources will typically contain two to four lensed objects.

## 17.14 Lensing Time Delay

Except for the symmetrical system of an Einstein ring, the lengths of the ray paths corresponding to the various images in the strong-focussing regime will be different. Any variation in intensity of the source will therefore be observed at different times in the different images. There are two components in the observed delay, one due to this excess geometric path and the other due to the different gravitational time delays as the rays cross the gravitational potential well $\psi(\theta)$ of the galaxy (the Shapiro delay); in practical cases these two components are of comparable magnitude. Both must be scaled by the redshift $z$ of the lensing galaxy. Interpretation of an observed time delay therefore depends both on measured values of the distances and on an evaluation of the potential $\psi$ obtained by modelling the mass distribution in the lensing galaxy.

Time delays have been measured using radio observations of several lens systems. For example, a delay of $11.3 \pm 0.2$ days has been measured using VLA monitoring observations of the lens system B0218+357 (Biggs and Browne 2017). The delay is measurable using independent light curves of total intensity, polarized intensity, and position angle. The lensed image of this object also includes an Einstein ring, which helps in the construction of the geometry of the lensing galaxy. Using the angular separation of the two components (335 milliarcseconds) and the redshifts of the quasar, $z = 0.96$, and of the lensing galaxy, $z = 0.68$, an independent distance scale can be deduced; it gives a value of the Hubble constant $H_0 = 72.9 \pm 2.6$ km s$^{-1}$ Mpc$^{-1}$. Other examples are given in the Supplementary Material. The problem of deducing a value for $H_0$ is discussed in Jackson (2015).

## 17.15 Weak Gravitational Imaging

The gravitational lensing we have so far considered is in a *strong* regime, in which objects are close to the lensing mass. Outside this area there is a regime of *weak* gravitational lensing, in which small effects may be observed over a larger area of the sky. The effect of an individual lensing galaxy or cluster is to distort the spatial distribution of a number of distant sources and to elongate their images: the problem of interpretation is that their original distribution and individual shapes cannot be known, and only a statistical effect can be observed. The importance of weak lensing is that the lensing mass distribution, including dark matter, may be observed on a larger scale than for the strong regime; as we saw in Chapter 14, the dynamical behaviour of stars and interstellar gas suggests the existence of unobserved dark matter at large radii in many galaxies and clusters of galaxies.

For such observations a large observational survey is required. Most such work has so far been done optically, where techniques for imaging some millions of galaxies are available. Radio astronomy is developing a similar capability and has a special importance because radio surveys select sources with high redshifts, where cosmological effects become especially interesting. A survey mapping at least ten radio galaxies per square arcminute is needed: this will become possible when the Square Kilometer Array comes into operation.

## 17.16 Further Reading

*Supplementary Material* at www.cambridge.org/ira4.
Alpher, R. A., and Herman, R. 2001. *The Genesis of the Big Bang.* Oxford University Press.
Advancing astrophysics with the Square Kilometer Array. *www.skatelescope.org/books/.*

Challinor, A. 2004. In: *The Physics of the Early Universe.* Lecture Notes in Physics, vol. 653, p. 71. Springer.
Jones, B. J. T. 2017. *Precision Cosmology.* Cambridge University Press.
Longair, M. S. 2006. *The Cosmic Century.* Cambridge University Press.
Longair, M. S. 2008. *Galaxy Formation*, 2nd edn. Springer.
Peacock, J. 2000. *Cosmological Physics.* Cambridge University Press.
Rees, M. 2000. *Just Six Numbers.* Basic Books.
Ryden, B. 2017. *Introduction to Cosmology*, 2nd edn. Cambridge University Press.
Tegmark, M. 1995. In: *Proc. Enrico Fermi Course CXXXII*, Varenna.
Weinberg, S. 1977. *The First Three Minutes.* Basic Books.

# Appendix 1  Fourier Transforms

## A1.1  Definitions

A harmonically oscillating amplitude $a(t)$ can be represented by the complex quantity

$$a(t) = a_0 e^{i2\pi \nu t}. \tag{A1.1}$$

The complex modulus $a_0$ gives the absolute value of the amplitude $|a_0|$ and the phase offset $\phi$:

$$a_0 = |a_0| \, e^{i\phi}. \tag{A1.2}$$

The *phasor* $a_0$ rotates with angular frequency $\omega = 2\pi \nu$, and its projection on the real axis gives the physical amplitude $a(t)$. The linear addition of various components of a quantity such as the electric field in a radio wave is conveniently done by adding the phasors as vectors. Adding two components with identical phasors but with opposite signs of $\omega$ gives the real quantity directly.

The amplitudes $a_\nu$ associated with components at various frequencies $\nu$ can be super-posed to give general amplitudes that can represent any physically realizable form $f(t)$. Fourier theory relates $f(t)$ to the spectral distribution function $F(\nu)$ of these components by the following dual transformation theorem:

$$\text{if} \quad f(t) = \int_{-\infty}^{\infty} F(\nu) e^{i2\pi \nu t} d\nu$$

$$\text{then} \quad F(\nu) = \int_{-\infty}^{\infty} f(t) e^{-i2\pi \nu t} dt. \tag{A1.3}$$

The *spectral distribution function* $F(\nu)$ is the *Fourier transform* of $f(t)$. The frequency $\nu$ and the time $t$ are *Fourier dual* coordinates. Another example of such dual coordinates is the relation between the radiation pattern of an antenna and the current distribution in its aperture (Chapter 8). The reciprocal relation between such pairs of functions may be written $f(t) \rightleftharpoons F(\nu)$. Figure A1.1 shows some particularly useful examples.

The Dirac impulse function $\delta(t)$ has an especially simple Fourier transform. Since by definition it is zero everywhere except at the origin, but has integral unity, its transfer form is

$$F(\nu) = \int_{-\infty}^{\infty} \delta(t) e^{i2\pi \nu t} dt = 1. \tag{A1.4}$$

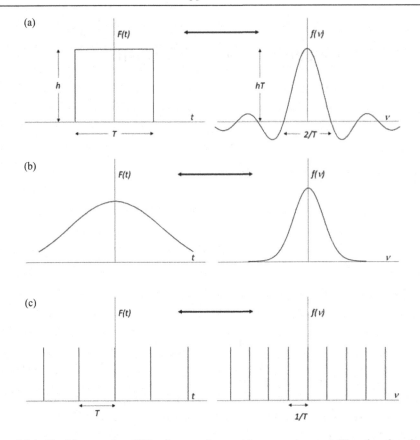

Figure A1.1. Useful examples of Fourier transforms: (a) square-step or $1/T$ gating function; (b) Gaussian; (c) periodic sampling function, spacing $T$, with transform spacing $1/T$. (See also Section 9.4 and Appendix 3).

Thus, a unit impulse has a flat spectrum of amplitude unity. If such an impulse occurs at a time $T$ other than zero, the spectral distribution still has all frequencies present with unit amplitude but with a frequency-dependent phase $2\pi\nu T$ (see also Section 10.8). The square-step or gating function $\Pi(t/T)$, with amplitude unity from $-T/2$ to $T/2$ and zero elsewhere, transforms into the 'sinc' function (Figure A1.1(a); see also Sections 7.2 and 9.4).

$$F(\nu) = \sin(\pi\nu T)/(\pi\nu T) \equiv \text{sinc}(\pi\nu T).\tag{A1.5}$$

The Gaussian function (Figure A1.1(b)) has the interesting property that its Fourier transform is also a Gaussian (see also Section 9.4)):

$$f(t) = e^{-(t/T)^2} \rightleftharpoons \frac{1}{\pi^{1/2}T}e^{-(\pi\nu T)^2} = F(\nu).\tag{A1.6}$$

The Gaussian function illustrates a general principle: when the characteristic time $T$ of a function is short, its spectrum is wide, and vice versa. Similarly, if a sinusoidal function is switched on and off at times $-T/2$ and $T/2$, its spectrum has a significant value over a bandwidth $\Delta\nu$ that is of order $1/T$.

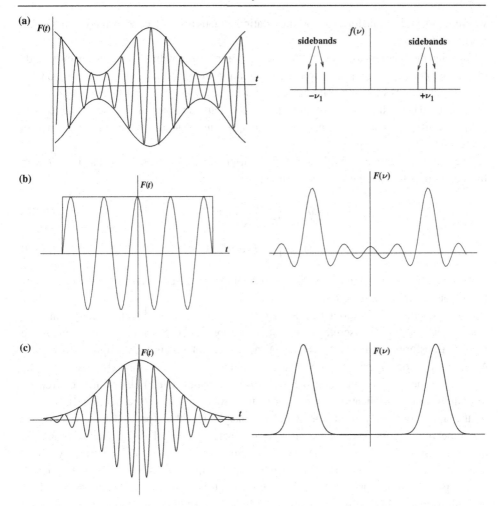

Figure A1.2. Fourier transforms of modulated waves.

Another self-similar pair is the periodic sampling function III(t), named by Bracewell and Roberts (1954) after the cyrillic character 'shah' (Figure A1.1(c)). If the samples are separated in time by $T$ then they are separated by $1/T$ in the transform domain.

Modulated waves, familiar to radio engineers, are encountered throughout physics. Typically, the amplitude of a cosinusoidal wave varies with time (Figure A1.2), either periodically as in the beat pattern (a) or aperiodically as in an isolated group of waves. If the modulating function is $g(t)$, so that the wave is $g(t)\cos(2\pi\nu_1 t)$, then its spectrum is

$$F(\nu) = \int g(t)\cos(2\pi\nu_1 t)\, e^{i2\pi\nu t}dt$$

$$= \int g(t)\frac{1}{2}[e^{i2\pi(\nu-\nu_1)t} + e^{i2\pi(\nu+\nu_1)t}]dt$$

$$= \frac{1}{2}G(\nu - \nu_1) + \frac{1}{2}G(\nu + \nu_1), \tag{A1.7}$$

where $G(\nu)$ is the transform of the modulating function $g(t)$ (see also the discussion in Section 6.6).

A cosine modulation, as in Figure A1.2(a), has a spectrum with 'sidebands' on either side of the components at $\pm\nu_1$. The top-hat modulation of (b) broadens the line components into sinc functions covering a frequency band inversely proportional to the length of the wave train. The Gaussian modulation in (c) has Gaussian spectral line components; this is the spectrum of a wave group, such as the spectrum of matter waves in a wave packet associated with a single particle.

There are alternative definitions of the Fourier transform. One form that is met with frequently uses the angular frequency, $\omega = 2\pi\nu$:

$$F(\omega) = \frac{1}{2\pi} \int_{-\infty}^{\infty} f(t) e^{-i\omega t} dt, \tag{A1.8}$$

$$f(t) = \int_{-\infty}^{\infty} F(\omega) e^{i\omega t} d\omega. \tag{A1.9}$$

Note the factor $1/2\pi$ in Eqn A1.8. The $\nu$-convention has the advantage of symmetry, while the $\omega$-convention is more compact.

The evaluation of a Fourier transform must frequently be done numerically. One can see that if a set of $N$ discrete values $f_i(t_i)$ is chosen, the numerical evaluation of the Fourier transform (Eqn A1.3) over the $N$ measurement points would appear to require $N^2$ multiplication operations to compute $N$ values of $F_i(\nu_i)$. Fortunately this is not the case, since there exist a number of *fast Fourier transform* (FFT) algorithms that reduce the number of multiplication operations to the order of $N \ln N$ (see Section 10.8). One of the most commonly used, the 'Cooley–Tukey' algorithm (previously developed by a number of authors, with Gauss as a forerunner) requires that the sample points be uniformly spaced; since in many practical situations, such as the aperture-synthesis technique (Chapters 9 and 10), the original data points are usually unequally spaced, an interpolation procedure is needed to obtain a regular grid of values. There are other FFT algorithms that do not require gridding, usually at the cost of greater demands on memory space.

## A1.2 Convolution and Cross-Correlation

The *convolution theorem* plays an important role in practical applications of Fourier methods, for example in formulating the effect of smearing by the antenna beam from a single telescope (Chapter 8) or on an aperture-synthesis map (Chapters 9 and 10). For a pair of functions $g(t)$ and $h(t)$ with Fourier transforms $G(\nu)$ and $H(\nu)$, their convolution is defined as

$$f(t) = \int_{-\infty}^{\infty} g(t') h(t - t') dt'. \tag{A1.10}$$

A convenient notation is $f(t) = g * h$, where the asterisk implies the convolution process. Note the reversal of sign in $h(t - t')$; note also that the process is commutative, i.e. $g * h = h * g$. The convolution process is shown in Figure A1.3, which shows the functions $g(t')$

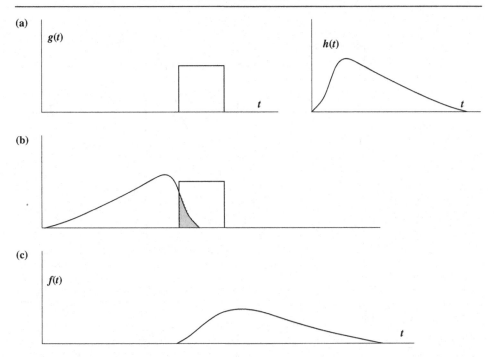

Figure A1.3. The convolution of two functions $g(t)$ and $h(t)$, shown in (a), evaluated (b) as their product (the shaded area) with $h(t)$ reversed. The convolution $f(t)$ is shown in (c).

and $h(t)$ with the reversed and shifted function $h(t - t')$. Their product is the shaded area, which is $f(t)$ at the time shift $t$. The convolution theorem states that the Fourier transform of $f(t)$ is simply the product of the Fourier transforms of the two convolved functions:

$$F(\nu) = G(\nu)H(\nu). \tag{A1.11}$$

The theorem is easily proven and has many practical applications. For example, a signal $g(t)$ may be an input signal to an amplifier or other linear device that has a response $h(t)$ to an impulsive input; the spectral distribution of the output is then the product of the spectra of the input signal and of the device. When a device is being considered in the frequency domain, $H(\nu)$ is known as the *transfer function*.

The convolution theorem may be used to facilitate the rapid calculation of Fourier transforms in many instances. For example, a triangle with unit height and base $B$ is the autocorrelation function of the gating function $\sqcap(t/T)$, as seen in Figure A1.4. The Fourier transform of $\sqcap(t/T)$ is sinc $\nu T$, and it follows immediately that the autocorrelation function of the triangle must have the Fourier transform sinc$^2\nu T$.

The *cross-correlation* of a pair of functions bears a close relation to convolution. For a pair of functions $f(t)$ and $g(t)$ the operation is often designated $f \otimes g$, with

$$f \otimes g \equiv \int f(t')g(t' - t)dt' \tag{A1.12}$$

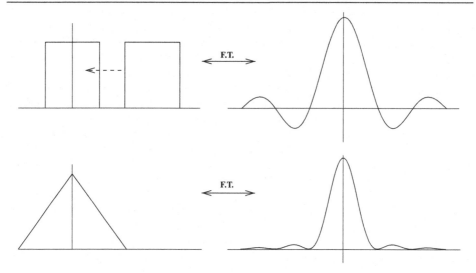

Figure A1.4. A autocorrelation of the gating function $\sqcap(t/T)$, indicated by the broken-line arrow at upper left, produces the unit triangle shown at the lower left. The respective Fourier transforms are shown on the right-hand side.

and thus

$$f \otimes g = f(t) * g(-t),\qquad\qquad(A1.13)$$

from which it follows that

$$f \otimes g = F(\nu)G^*(\nu).\qquad\qquad(A1.14)$$

Note the time reversal in Eqn A1.13; cross-correlation is not a simple commutative operation.

Cross-correlation has a particular significance in radio interferometry, where two signals $f(t)$ and $g(t)$ may be obtained from spaced antennas (see Chapters 9 and 10). The product $fg^*$, which has the dimensions of power, is then referred to as the *cross-power*, and the product $FG^*$ is known as the *cross-spectral power density* (see also Section 9.1).

The *autocorrelation function* $C(t)$ is defined by

$$C(t) \equiv \int_{-\infty}^{\infty} f(t')f(t' + t)dt'.\qquad\qquad(A1.15)$$

Note that there is no time reversal; the function is multiplied by itself at a variable time shift. The autocorrelation function is related to the spectrum of $f(t)$; for zero time shift it is simply its square (see Section 7.2). The autocorrelation function has a particular relevance in signals and systems, whose behaviour is often described in terms of *amplitudes* $e(t)$ (e.g. voltages, field strengths and currents), while the measurement process involves the *power* $p(t)$. Using a convenient choice of units we can write

$$p(t) = e^2(t).\qquad\qquad(A1.16)$$

The spectrum of a signal can be represented in Fourier terminology; the *spectral power density* $S(\nu)$ gives the power density in a given infinitesimal bandwidth $d\nu$. The total energy $E_{em}$ emitted over time is then

$$E_{em} = \int e^2(t)dt = \int_0^\infty S(\nu)d(\nu). \tag{A1.17}$$

Note that the distinction between positive and negative frequencies is no longer meaningful.

The spectral power density $S(\nu)$ is evidently proportional to $E(\nu)E^*(\nu)$, the product of the Fourier transforms of $e(t)$ and $e(-t)$. We will explore this relation by writing the self-convolution $e(t') * e(-t')$, so that the emitted energy $E_{em}$ may be expressed as

$$E_{em} = \langle e(t) * e(-t) \rangle_{t=0} = \int_{-\infty}^\infty e(t')e(t'+t)dt'. \tag{A1.18}$$

It follows from the definition of convolution (Eqn A1.10) that

$$\int_{-\infty}^\infty E(\nu)E^*(\nu)e^{-i2\pi\nu t}d\nu = e(t) * e(-t), \tag{A1.19}$$

so that, for $t = 0$,

$$E_{em} = \int_{-\infty}^\infty e^2(t)dt = \int_{-\infty}^\infty |E(\nu)|^2 d\nu. \tag{A1.20}$$

This is known as *Rayleigh's theorem* and is a generalization of the Parseval theorem for Fourier series. The integral over frequencies needs be taken over positive frequencies only, since $|E(\nu)|$ is symmetric in $\nu$, and so, for this physical case,

$$S(\nu) = 2E(\nu)E^*(\nu) = 2|E(\nu)|^2. \tag{A1.21}$$

The self-convolution $e(t') * e(-t')$ is symmetric in the time offset $t$ and when written with the reverse sign becomes the autocorrelation function $C(t)$, as in Eqn A1.15 above; that is,

$$e \otimes e \equiv C(t). \tag{A1.22}$$

The autocorrelation function has a most important property, known as the Wiener–Khinchin theorem (see Sections 5.1 and 7.2), which states that the Fourier transform of the amplitude autocorrelation is the power spectral density (with correction by a factor of 2 if only positive frequencies are treated, as in Eqn A1.20):

$$E(\nu) \otimes E^*(\nu) \rightleftharpoons C(t). \tag{A1.23}$$

This relation, summarized in Figure A1.5, provides the basis for autocorrelation spectrometry (Sections 5.1 and 7.2).

In radio astronomy the autocorrelation function is obtained from the product of a digitized signal and the same signal subjected to a variable delay. In practice many different delays may be used simultaneously, where the signal may be amplified coherently and split into many channels (see Chapter 10).

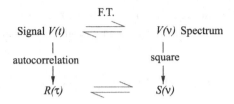

Figure A1.5. The relation between autocorrelation and Fourier transformation (see also Section 7.3).

## A1.3 Two or More Dimensions

The definitions and examples in this appendix have all been presented in the one-dimensional case, but since the Fourier transform is a linear operation it can be generalized to Cartesian coordinates in many dimensions. Given a function $f(\mathbf{x})$ of the $n$-dimensional variable $\mathbf{x}$, it will have a Fourier transform $F(\mathbf{k})$. The Fourier dual coordinate $\mathbf{k}$ is called the *spatial frequency* by analogy with $t$ and $\nu$ in the time–frequency case. A straightforward calculation shows that the fundamental Fourier inversion theorem, Eqn A1.3, becomes

$$f(\mathbf{x}) = \int_{-\infty}^{\infty} f(\mathbf{k}) e^{i2\pi \mathbf{k}\cdot\mathbf{x}} d^n k \tag{A1.24}$$

$$F(\mathbf{k}) = \int_{-\infty}^{\infty} f(\mathbf{x}) e^{-i2\pi \mathbf{k}\cdot\mathbf{x}} d^n x, \tag{A1.25}$$

thus defining

$$f(\mathbf{x}) \rightleftharpoons F(\mathbf{k}). \tag{A1.26}$$

## A1.4 Further Reading

*Supplementary Material* at www.cambridge.org/ira4.
Bracewell, R. N. 1978. *The Fourier Transform and Its Applications*, 2nd edn. McGraw Hill.
Oran, B. E. 1974. *The Fast Fourier Transform*. Prentice Hall.

# Appendix 2  Celestial Coordinates and Time

## A2.1  The Celestial Coordinate System

The Earth is a moving platform, and reference points in the Universe are not defined easily. There are no fixed stars in the Universe, and the very distant quasars must be used for astrometry and geodesy. The closest approximation we have to a universal coordinate system is defined by the cosmic microwave background. This nearly featureless radiation background, discussed more fully in Chapter 17, appears to be the remnant of the cosmic fireball. The sum of all motions of the solar system and the Milky Way galaxy in space causes a measurable Doppler shift in the background radiation, implying that our Milky Way galaxy has a resultant velocity of approximately 340 km s$^{-1}$ towards a point not far from the Virgo cluster. In theory, this could be used as the axis of a polar coordinate system, although the origin of the azimuthal angle would still have to be defined. In practice, the uncertainties in the measurement are so large that the idea is not feasible. The angular momentum axis of the Milky Way galaxy, and the line from the Sun to the galactic cen-tre, could be imagined as defining a nearly fundamental system, but neither direction is sufficiently well known as yet.

The positions and motions of bodies in the solar system thus provide the most practical coordinate reference system. The motions are complex, but are well defined and have been carefully analyzed. One starts with the rotation of the Earth, whose axis defines the orientation of a polar coordinate system as shown in Figure A2.1. Instead of using the polar angle, it has been conventional to use the co-polar angle, measured from the *celestial equator*. This angle is the *declination*, usually symbolized by $\delta$. The origin of the azimuthal angle is determined by the Earth's orbit about the Sun, which is given observationally by the apparent motion of the Sun along its path in the sky, known from ancient times as the *ecliptic*. The ecliptic has two intersections with the celestial equator. The *ascending node* or *vernal equinox* ($\Omega$) (the point of the spring equinox, when the mean Sun crosses the equator in an upward direction) defines the origin of the azimuthal angle, known as the *Right Ascension*, symbolized by $\alpha$, which increases in the counterclockwise direction about the celestial pole. Two further conventions can be noted: in celestial coordinates, north is taken in the direction of increasing declination, as in a normal terrestrial map, and east is taken to the left, in the direction of increasing Right Ascension. (This convention follows from mapping the terrestrial headings of north and east directly onto the celestial sphere.)

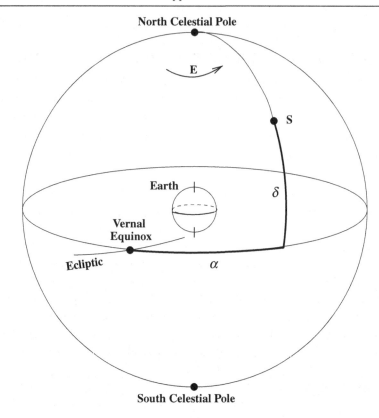

Figure A2.1. Polar coordinate system, based on the rotation of the Earth. Part of the ecliptic, the great circle defined by the apparent motion of the Sun against the background of the sky over a year, is shown intersecting the equator at the vernal equinox.

The *position angle* on the sky is defined from north through east, i.e. counterclockwise as seen from the Earth (see also the Supplementary Material).

Care must be exercised in defining the two great circles, the celestial equator and the ecliptic, that define the celestial coordinate system. The Earth's pole precesses in space because of the torque on the oblate Earth exerted by the Sun, Moon and planets, with a period of nearly 26 000 yr, or an angular rate of about 55 arcsec per year. There is also a nutational motion driven by periodic external torques with a more complex behaviour. The dominant term is a periodic motion of amplitude 9 arcsec and period 19 yr imposed by the Moon. The Earth's orbit, in turn, is perturbed by the other planets, so the ecliptic also varies in time. Even the Earth's rotation axis moves with respect to the Earth's crust, showing a 14 month variation (the *Chandler wobble*) with a fluctuating amplitude of approximately 0.1 arcsec, and a possible secular variation of about one millarcsecond per year. Because of the large number of time-varying corrections, therefore, the Right Ascension and declination of a cosmic body only have meaning when the *epoch* is specified for the equinox. The equinox of 1950.0 was commonly used, but has been superseded by 2000.0. Note that the equinox and pole determine the angular coordinate system, while 'epoch' refers to time; the question of timescale is discussed more fully in Section A2.2. A more accurately defined timescale, more accurate values of the astronomical constants, and an improved theory of the Earth's nutation are included in the new system, and to distinguish the old and new conventions the designations B1950.0 and J2000.0 have been adopted (B standing for Besselian year, the timescale defined for the former, and J for Julian year, the timescale for the latter epoch). Formal definitions and references are given in compact form in the *Astronomical Almanac*, jointly published by the UK and the USA (there are a number of other national counterparts). Further material can be found in the *Explanatory Supplement* and in Smart's *Spherical Astronomy* (1977). One should note that the determination of the celestial pole had traditionally been made by observing stars, but now observations of extragalactic radio sources have become the method of choice. Similarly, the ecliptic plane has traditionally been determined by observing planets and asteroids, but radio and radar measurements of planetary spacecraft have supplanted the optical methods.

The definition and realization of the International Celestial Reference Frame (ICRF) depends on accurate astrometric measurements extending over wide angles. Flamsteed at Greenwich achieved an accuracy of $10''$; nearly 200 years later most star catalogues based on optical observations are still accurate only to about $1''$. Very long baseline interferometry methods, which are more precise and more easily relate the positions of widely separated objects, now cover the celestial sphere with an accuracy of order 100 μarcsec, an improvement of four orders of magnitude (see Section 11.4.8). The astrometric satellite Gaia is producing a catalogue of millions of optical positions covering the whole sky and based on the IRCF, so that there is now a grid of well-determined positions in this frame to an accuracy of $< 50$ μarcsec (Liao *et al.* 2018).

Gaia is also measuring the distances of stars through their annual parallactic motion as the Earth (and the spacecraft) moves in its orbit round the Sun. The catalogues of the parallax and proper motion of over a billion stars that are being produced will transform our knowledge of the geometry and dynamics of the Milky Way galaxy.

## A2.2 Time

Space and time are the coordinates by which we label events. Ever since Minkowski, in 1907, wrote about the geometric interpretation of Einstein's special theory of relativity, the four-dimensional coordinate system of space–time has provided the framework for physics and astrophysics. Alternative schemes have been proposed from time to time, but no compelling experiments have yet dictated a change in this viewpoint. Time, in particular, can be measured with great precision using either gravitational clocks (via the dimensions of the solar system) or atomic clocks (via the frequency of photons from the ground-state hyperfine splitting of caesium-137). The rate of change of gravitational time with respect to atomic time is less than $10^{-12}$ per year; it is generally assumed, therefore, that time intervals are the same, regardless of the clock being used for the measurement.

The most accurate clocks in use today are probably the caesium atomic-beam devices, which achieve an absolute accuracy of at least one part in $10^{14}$ over a period of a day or longer. Hydrogen maser clocks have a *stability* of one part in $10^{15}$ over periods of hours to days but are more sensitive to magnetic, pressure, and thermal perturbation and thus may be less reliable as absolute standards. They have an important use in radio astronomy as stable frequency standards for very long baseline interferometry (VLBI), a subject treated in Chapter 11.

It now seems likely that a number of millisecond pulsars have a long-term rotational stability that is at least the equal of caesium clocks. If this feature of millisecond pulsars is to be used as a check on the smooth running of terrestrial time standards, it will be necessary to monitor the rotation of a set of millisecond pulsars over a period of several years. This is also a requirement of a search for gravitational waves, in which the relative times of arrival of pulses from different parts of the sky might reveal low frequency waves generated in the early Universe. Three networks, based in Europe, North America, and Australia, are monitoring the arrival times of millisecond pulsars for this purpose; their efforts are combined in an International Pulsar Timing Array (IPTA). The long sequences of recorded data are combined to search for ultra-low-frequency waves, with frequencies $10^{-9}$ to $10^{-8}$ Hz (Hobbs and Dai 2017).

Civil time is governed by the strong human desire to have noontime (1200 hours) occur at midday, and, at least on the average, for the year to start on the first day of January. There is the dual problem of specifying the time interval (the second) and the time (the elapsed interval since a specified origin for the time coordinate). Several conventions, agreed upon by international bodies, have been defined by the motion of the Earth in its orbit. By definition, the *tropical year* is the interval required for the Earth to travel from vernal equinox to vernal equinox (i.e. for the mean Sun to travel 360° along the ecliptic). The tropical year, in turn, defines *Ephemeris Time* (ET). This requires accurate measurement of the celestial equator (the instantaneous axis of the Earth's rotation) and of the plane of the ecliptic (the apparent motion of the mean Sun), since the intersections of these great circles define the equinoxes. Since the mutual perturbations of the bodies of the solar system are appreciable, an accurate accounting must be made of the variations of the system with time. By international agreement, the length of the tropical year for the epoch 1900.0, Jan 0 $12^h$ ET, was set at 31 556 925.9747 seconds. Currently, the repeatability of comparisons of intervals in this unit of time with atomic time is approximately one part in $10^{12}$ over several decades. It is generally accepted that atomic time and ephemeris (i.e. gravitational) time are identical; if they are not, the difference is evidently small.

Ordinary human activities are more strongly governed by the solar day, and the most common civil unit of time is *Universal Time* (UT), starting and ending at midnight, with respect to the transit time of the mean Sun at the location of the ancient transit-circle of the Royal Observatory at Greenwich. The most primitive definition, UT0, uses the instantaneous rotation of the Earth to define the time and is the time that would be derived at any instant from observing the Sun and stars. This is an imperfect clock, since the Earth does not rotate uniformly about its axis: it is a multipiece gyroscope, exchanging angular momentum between the core and the mantle, between the mantle and crust, and between the crust and atmosphere, with dissipative tidal torques from the Sun, Moon, and other bodies in the solar system.

The seasonal and secular variations are known with some accuracy, and when these corrections are applied, and successive improvements have been agreed on, better definitions of UT can be adopted. Polar motion corrections are applied to define UT1, which is then independent of the observer's location; UT2 is defined by applying corrections for seasonal variations in the Earth's rotation rate. The term UT is used when these distinctions are unimportant, and tables are given to relate the various timescales. Since caesium atomic-beam clocks are the most accurate devices currently in use, they provide the best time interval reference. By international agreement, photons from the ground-state caesium-137 hyperfine transition are defined to have an average of 9 192 631 770 oscillations per second, thus defining *International Atomic Time* (TAI). Such an atomic clock runs about 300 ns s$^{-1}$ slow with respect to UT (the offset was deliberately chosen to avoid confusion). In order to keep civil time in average agreement with solar time, and yet to have time intervals determined by the more precise TAI, the national time services generally use *Coordinated Universal Time* (UTC). Caesium clocks determine the short-term timescale, and 'leap seconds' are inserted at agreed times every few months to keep $0^h$ UTC in average agreement with UT. Corrections are listed in the *Astronomical Almanac*.

For most purposes, the chosen origin of time is $12^h$ on day 0 of the tropical year 1900, defining an origin at noon, 31 December 1899. Time is sometimes specified in *Julian days* (JD) with JD 0.0 starting at noon on 1 January 4713 BC (Julian calendar). To avoid very large numbers a modified Julian date (MJD) is usual, with origin arbitrarily chosen as 0 hr on 17 November 1858. Conversion tables are given in the *Astronomical Almanac*. The civil day 1 January 1990, for example, starts on JD 2 447 892.5. The choice of noon was made because the Sun was originally the best fundamental clock but, as techniques developed, the Moon, planets and stars replaced the Sun as the best reference bodies, and the Time Commission of the International Astronomical Union agreed to define midnight as the start of the astronomical day as of 1 January 1923 and, ever since, this convention has been followed when the time of observation is referred to in terms of year, month, day, hour, minute, and second. Julian days, however, still start at noon.

Different timescales were adopted for the angular coordinate systems defined by B 1950.0 and J2000.0: the Besselian year, adopted by convention, is defined in terms of UT0 and is thus defined by a clock (the Earth) whose imperfections are large and uncorrected; the Julian year, on the other hand, is defined as a tropical year and is therefore a sounder time reference. The fundamental constants describing the Earth's motion, such as the rate of precession, are also known to higher accuracy. For these reasons, adoption of the epoch of J2000.0 will certainly result in better positional measurements of celestial objects.

The details of time keeping are spread through many references, the *Astronomical Almanac* being the starting point for the fundamentals. When one is interested in precision, a careful search of the current literature, including the *Bulletin* of the Bureau International de l'Heure (BIH), should be made, and this should be combined with consultation with authorities in the field.

## A2.3 Further Reading

*Supplementary Material* at www.cambridge.org/ira4.
Johnston, K. J., and de Veght, C. 1999. Reference frames in astronomy. *Ann. Rev. Astron. Astrophys.*, **37**, 97–126.

# Appendix 3  Digitization

The availability of commercial high speed analogue-to-digital converters (ADCs) and field-programmable gate arrays (FPGAs) has led to the ubiquity of digital signal processing (DSP) in radio astronomy. Increasingly, high speed sampling allows the replacement of IF or even RF analogue sub-systems with digital hardware whose characteristics can be altered in software; further down the receiver chain, sophisticated signal manipulations become possible. The stability of digitized signals allows long integrations with no deleterious effect on the achieved noise level, which is often a limitation in analogue systems.

## A3.1  Digitizing the Signal

An analogue signal is continuous in time and amplitude whereas a digital signal is discrete in both parameters. The translation is carried out by an analogue-to-digital converter (ADC) circuit, which quantizes the amplitude and samples the signal in time; this involves making choices about both discretizations. The result of the quantization may be a single bit, representing positive or negative values of the signal waveform with respect to an average value, or the quantization may be into a much greater number of signal levels. In any case the sampling must be sufficiently rapid, so that no information is lost.

### A3.1.1  Amplitude Quantization

Amplitude quantization is achieved with thresholding and/or comparator circuits, the details of which are beyond our present scope. Commercial ADCs typically quantize the amplitude with between 8 and 16 bits, corresponding to $2^8 = 256$ levels or $2^{16} = 65\,536$ levels, although 24-bit ADCs are available for specific applications. There is an obvious trade-off between digital signal fidelity and the number of bits to be dealt with, and the choice depends on the application. If the application requires that the digitized signal be a very close copy of the input analogue signal (as in high fidelity audio or digital imaging) then more bits can be used. Figure A3.1 illustrates an idealized case in which the signal amplitude is sampled at frequent regular time intervals and the amplitude quantization is sufficiently fine to follow the variations reasonably faithfully.

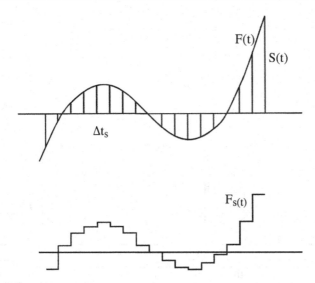

Figure A3.1. A well-sampled waveform produces a discretized wave which contains the original, together with higher-frequency components.

Radio astronomy systems operate at the opposite extreme. There is only limited information available in a noise signal and much cruder quantization is sufficient to capture it with only modest losses in signal-to-noise ratio.

Figure A3.2 shows a time-varying analogue noise signal discretized with single-bit and two-bit amplitude quantization. Sampling with so few bits inevitably means that the digitized signal does not follow the analogue signal perfectly, the differences being termed 'quantization noise'. Equivalently, the discontinous amplitude steps generate spurious frequencies in the digitized spectrum. Such effects may be unacceptable where the amplitude variations contain signal information. For random noise inputs, however, the quantization noise has only the statistical effect of decreasing the signal-to-noise ratio.

When the noise signal obeys Gaussian statistics, the relative power spectrum can be deduced even if only the sign of the signal amplitude, with respect to the average level, is measured, using one-bit sampling. This surprising result, in which the digitized noise signal can be represented by a string of ones and zeros, is known as the van Vleck quantization approximation. Single-bit quantization of the signal degrades the signal-to-noise ratio by only $2/\pi$ (a loss of 36%) compared with that of an ideal correlator but historically this has been a small price to pay for the great simplicity of single-bit data handling. However, since no amplitude information is preserved the total noise power in the band is not available. This makes single-bit systems unsuitable for pure radiometry.

Faster, large-scale integrated circuits allow multi-bit representations and two- to five-bit systems are becoming the norm. As shown in Figure A3.2, a two-bit system represents the signal in terms of four signal levels. The two bits represent the sign of the signal and whether its absolute value exceeds a number that is very nearly the rms level of the noise signal. This means that there are four possible signal levels, and the two-bit system is sometimes called a four-level system. The degradation of the signal-to-noise ratio is less than that of the

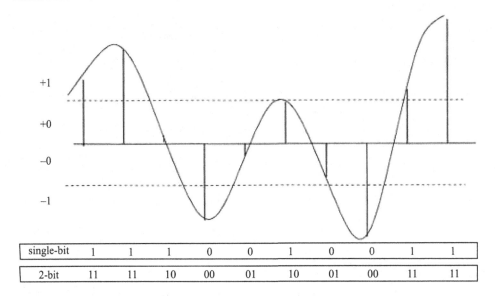

| single-bit | 1 | 1 | 1 | 0 | 0 | 1 | 0 | 0 | 1 | 1 |
| 2-bit | 11 | 11 | 10 | 00 | 01 | 10 | 01 | 00 | 11 | 11 |

Figure A3.2.  Single-bit and two-bit digitization.

single-bit system; the value achieved is 0.88 that of an ideal correlator. A larger number of bits can be used; a three-bit representation would give eight levels of quantization, with a value 0.95 for the signal-to-noise quality factor. A comprehensive discussion of digital sampling is given by TMS, Chapter 8. For current practice in radio astronomy see also van Straten and Bailes (2010).

Digital sampling offers great advantages in the flexibility of signal handling, but the process of quantization is inherently non-linear, which can have important effects that can override purely signal-to-noise ratio considerations. For example, the power in a narrow-band interfering radio signal in any part of the observed spectrum may be spread out over a much larger range of the output spectrum. This effect is reduced by employing more levels of quantization, to represent the input signal more exactly. A digital receiver for pulsar observations in an environment with high levels of RFI might be forced to use 12- or even 16-bit quantization where eight bits or fewer might otherwise be adequate. The low RFI in the SKA sites allows for digitization at the four-bit level, but to allow 'headroom' for long-term RFI mitigation, eight-bit quantization is currently envisaged.

### A3.1.2 Time Quantization and the Nyquist Criterion

We now need to consider the issues associated with sampling the signal in time – as illustrated in Figure A3.1. The sampled signal $F_s(t)$ can be represented by the product of the analogue signal $F(t)$ multiplied by a *sampling function* $S(t)$ consisting of infinitely narrow spikes (mathematical delta functions) at the times when the samples are taken, separated by an interval $\Delta t_s$. How short must $\Delta t_s$ be, alternatively, how often does one have to sample, in order to ensure that no signal information is lost? This question is addressed by the Nyquist–Shannon sampling theorem.

The Nyquist–Shannon sampling theorem states that a band-limited signal, whose spectrum extends up to a maximum frequency $\nu_{max}$, can unambiguously be reconstructed from discrete samples taken at or above the *Nyquist rate* $2\nu_{max}$, twice the maximum frequency in the signal. By sampling the signal at intervals less than $\Delta t_s \leq 1/2\nu_{max}$ (at least two samples per cycle of the highest frequency) the samples will always be sufficient to capture the fastest time variations in the original signal.[1] For reasons which will become clear below the *Nyquist frequency* is defined as half the sampling rate, i.e. $\nu_s/2$.

Although digitization takes place in time-domain Nyquist sampling, what happens if the criterion is not met is best understood in the frequency domain. Figure A3.3 shows a spectrum $F(\nu)$ defined by a low-pass filter in the receiver and idealized as uniform from zero to a maximum frequency $\nu_{max}$; it is therefore at baseband. The original time series $F(t)$ is the Fourier transform of $F(\nu)$. The spectrum $S(\nu)$ of the sampling function $S(t)$ is a series of delta functions with spacing $1/\Delta t_s$ (see Appendix 1). Since the sampled signal $F_s(t) = F(t)S(t)$ its spectrum $F_s(\nu)$ (shown at the bottom of Figure A3.3) is the convolution of $F(\nu)$ and $S(\nu)$ and consists of multiple copies of $F(\nu)$ centred on zero, $\nu_s$, $2\nu_s$, etc. A single copy of $F(\nu)$ is recovered by multiplying $F_s(\nu)$ with an ideal low-pass filter created in software. On inverse Fourier transformation the original signal $F(t)$ could then be recovered perfectly[2] but this is usually not the point of the procedure, and the digital samples are passed on for further manipulation in a digital signal processing (DSP) system.

### A3.1.3 Aliassing

If the time increment $\Delta t_s$ between samples is longer than than $1/2\nu_{max}$, i.e. if the sampling frequency is less than the Nyquist rate $2\nu_{max}$, the signal is said to be undersampled. By definition it contains frequencies greater than half the Nyquist rate and hence above the Nyquist frequency, and these higher signal frequencies appear at incorrect frequencies in the sampled data. This phenomenon is called *aliassing* and is illustrated in Figure A3.4.

In the left-hand picture a sine wave is sampled faster than the Nyquist rate (two samples per shortest cycle in the time domain); thus its frequency is below the Nyquist frequency. No information is lost and the sine wave could be recovered perfectly from the samples. In the right-hand picture the sampling rate is just below the Nyquist criterion and

---

[1] This discussion is idealized. In particular, real-life filters do not cut off infinitely sharply, so their frequency bounds are blurred to an extent depending on the filter design. A significant degree of oversampling (at least ten per cent) compared with the nominal rate is therefore the norm. Alternatively, if the sampling frequency $\nu_s$ is fixed, the signal bandwidth should be restricted before the ADC with an *anti-aliassing filter* so as to ensure significant oversampling.

[2] Another way to look at signal recovery is in the time domain. The Fourier transform of the rectangular filter function extending over $\pm\nu_{max}$ is a sinc function whose first zeros are at $\pm 1/2\nu_{max}$. From the convolution theorem the effect of multiplication in the frequency domain is to convolve in the time domain and hence each delta function can be replaced with a sinc function having the appropriate amplitude for that sample. The sidelobes of the sinc function interpolate between the delta functions, and the original signal is precisely recovered.

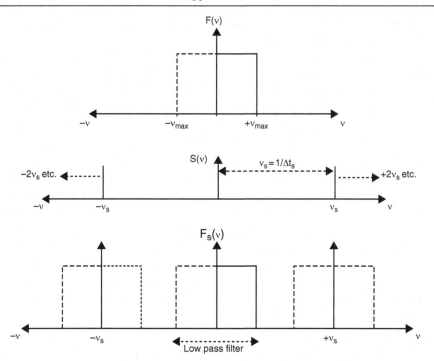

Figure A3.3. Sampling in the frequency domain. (top) A signal whose spectrum extends from zero to $\nu_{max}$ is sampled at intervals $\Delta t_s$. (middle) The spectrum of this sampling process is a series of lines at interval $\nu_s = 1/\Delta t_s$. (bottom) The resultant spectrum is the convolution of these two spectra. It consists of the spectrum of the original signal with multiples centred on zero, $\nu_s$, $2\nu_s$, etc.

Figure A3.4. Aliassing in the time domain. (left) Sampling faster than the Nyquist rate of twice per cycle. (right) Sampling just below the Nyquist rate; the aliassed sine wave (dotted) appears at a lower frequency than the true sine wave.

as a result the frequency of the sampled data is incorrect – it is lower than the input frequency. Information has been lost and the original sine wave cannot be recovered from the samples.

Once again the phenomenon is easier to understand in the frequency domain. In Figure A3.3 notice that the separation of the delta functions in $S(\nu)$ has been chosen to be significantly larger than the width of the signal spectrum $F(\nu)$. This means that the

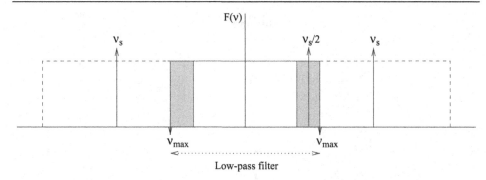

Figure A3.5. Aliassing in the frequency domain with an inadequate sampling rate. The Nyquist frequency is below the maximum signal frequency. The shaded overlap region contains a mixture of aliassed and non-aliassed frequencies.

sampling rate is significantly above $2\nu_{max}$ and the multiple copies in $F_s(\nu)$ are therefore well separated from each other. However, when the sampling rate is too low the multiple copies start to overlap and a high frequency component $\nu_h$ is 'folded back' around the Nyquist frequency, to appear as $\nu_h - \nu_s/2$; because of this effect the Nyquist frequency is sometimes called the folding frequency. Take for example an ADC which samples at a rate of 200 MHz; the Nyquist frequency is 100 MHz and hence a signal at 110 MHz will appear at 90 MHz, a 120 MHz signal at 80 MHz and so on.[3] The aliassed frequencies cannot be filtered out as in Figure A3.3, and information has been lost.

Figure A3.5 illustrates aliassing in the frequency domain. For clarity, only the central region, analogous to that in Figure A3.3, is shown and the sampling rate has been reduced so that the Nyquist frequency is somewhat below the maximum signal frequency. The result is that copies of the signal band start to overlap; this is aliassing caused by undersampling. The top end of the signal band is 'folded back' around the Nyquist frequency $\nu_s/2$ and the shaded region now contains a mixture of aliassed and non-aliassed frequencies. The whole of the original band cannot now be retrieved by filtering as in Figure A3.3.

The reader might note the similarity with analogue mixing described in Section 6.6; the Nyquist frequency acts like the local oscillator, with the aliassed signals appearing at $\nu - \nu_s/2$, i.e. as mirror images of the original frequency around the Nyquist frequency. This is effectively digital mixing, and deliberate undersampling can be used to advantage in system design. To see how this is done we need to extend our thinking.

So far we have assumed that the signal band starts at d.c. (baseband) and so its bandwidth $\Delta\nu \equiv \nu_{max}$. If the same band were centred at a higher frequency then the basic Nyquist criterion would require a sample rate of twice the, now higher, maximum frequency component. In fact this is not necessary. The generalized Nyquist–Shannon sampling theorem says that the sampling rate need only be $2\Delta\nu$ in order to be able to reconstruct the original signal. Heuristically, the reason is that all the information in the band-limited signal is contained

---

[3] Equivalently, one can take the difference between the signal frequency and the sampling rate; thus a sampling rate of 200 MHz and a signal frequency of 110 MHz produces an alias at 90 MHz.

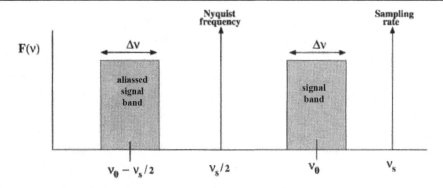

Figure A3.6. Digital mixing. The Nyquist frequency is now below the signal band but is larger than the signal bandwidth. The effect is to create an aliassed copy of the signal band at a lower frequency, as in a heterodyne receiver.

in the lower frequency modulations around the higher centre frequency, as illustrated in Figure 6.3 (Chapter 6).

Digitally mixing a higher frequency signal to a lower frequency signal by undersampling (also called bandpass sampling) can be exploited to avoid the need for an analogue mixing stage. Figure A3.6 illustrates the basic idea. If the signal band is centred at 160 MHz and has a 40 MHz bandwidth and the sampling rate $\nu_s$ is 200 MHz, the aliassed components will appear at $60 \pm 20$ MHz. Note that the sampled output also contains aliases of the original signal at integer multiples of $\nu_s$ (see Figure A3.4), and, using one of these harmonics, the information within a high frequency signal band can be captured with lower sampling rates; however, this concept of 'Nyquist zones' is beyond our present scope.

In conclusion, data which have been insufficiently quantized in time and amplitude are imperfect. The errors are important for high fidelity applications such as audio reproduction and digital imagery but the information in the noise signals encountered in radio astronomy is very limited and the errors only affect the output statistically, except in the face of strong RFI. Aliassing is avoided if the signal bandwidth is less than half the sampling rate but in some cases deliberate undersampling, which has the effect of digital mixing, can be put to good use in system design.

# Appendix 4  Calibrating Polarimeters

### A4.1 Single-Dish Radio Telescopes

The Stokes parameters, the choice of orthogonal polarization bases, and their selection in feed systems were discussed in Sections 7.5 and 7.6. In Sections 7.7 and 7.8 we described a basic polarimeter and noted that the Stokes parameters can be obtained from auto- and cross-correlations of digitized signals. The resulting power outputs then require correction for various cross-coupling effects in the telescope, feed, and receiver systems and for the parallactic angle $\Psi_P$ between the telescope axis and the sky coordinates as defined in Section 11.3. We refer to Heiles (2002) for a highly readable exposition of the issues involved; here we give an illustrative summary.

The required modifications to the Stokes parameters are described by the *Mueller matrix* $M$, which acts as the transfer function between the intrinsic, $s_0$, and measured Stokes vectors $s_m$, i.e. $s_m = Ms_0$; thus

$$
\begin{bmatrix} I \\ Q \\ U \\ V \end{bmatrix}_m = \begin{bmatrix} m_{II} & m_{IQ} & m_{IU} & m_{IV} \\ m_{QI} & m_{QQ} & m_{QU} & m_{QV} \\ m_{UI} & m_{UQ} & m_{UU} & m_{UV} \\ m_{VI} & m_{VQ} & m_{VU} & m_{VV} \end{bmatrix} \begin{bmatrix} I \\ Q \\ U \\ V \end{bmatrix}_0 . \tag{A4.1}
$$

Since there are four Stokes parameters, $M$ has $4 \times 4$ elements (not all of which are independent) to account for the cross-terms between $s_0$ and $s_m$; each element describes the coupling between two Stokes parameters. The elements are real-valued since the Stokes parameters are measurements of power and no phase information is involved. Separate parts of the signal path can be described with individual Mueller matrices, and the overall system matrix is the product of them. Since in general matrices are not commutative the order of multiplication is important (it is the reverse order of occurrence). The main contributions are the polarimeter imperfections and the rotation in parallactic angle; hence $M = M_{pol} \times M_{\Psi_P}$. A further matrix $M_{rot}$ can represent the progress of the signal from the source to the telescope, including Faraday rotation in the ISM and in the ionosphere.

Heiles (2002) gives a typical polarimeter Mueller matrix $M_{pol}$, which we reproduce below to illustrate the range of imperfections which need to be considered; the reader should consult Heiles (2002) for further details.

$$
M_{\mathrm{pol}} =
\begin{bmatrix}
1 & \begin{array}{c} -\,2\epsilon\sin\phi\sin 2\alpha \\ +(\Delta G\cos 2\alpha)/2 \end{array} & 2\epsilon\cos\phi & \begin{array}{c} 2\ \epsilon\sin\phi\cos 2\alpha \\ +(\Delta G\sin 2\alpha)/2 \end{array} \\
\Delta G/2 & \cos 2\alpha & 0 & \sin 2\alpha \\
2\epsilon\cos(\phi+\psi) & \sin 2\alpha\sin\psi & \cos\psi & -\cos 2\alpha\sin\psi \\
2\epsilon\sin(\phi+\psi) & -\sin 2\alpha\cos\psi & \sin\psi & \cos 2\alpha\cos\psi
\end{bmatrix}. \quad (A4.2)
$$

- $\Delta G$ is the intensity calibration error between the orthogonal channels;
- $\psi$ is the phase calibration error between the orthogonal channels;
- $\alpha$ is a measure of the voltage ratio of the polarization ellipse when the feed observes pure linear polarization;
- $\epsilon$ is a measure of imperfection of the feed in producing non-orthogonal polarizations (false correlations) in the two cross-correlated outputs;
- $\phi$ is the phase angle at which the voltage coupling $\epsilon$ occurs; it works with $\epsilon$ to couple $I$ with $Q$, $U$, and $V$.

The Mueller matrix describing the rotation with parallactic angle is

$$
M_{\Psi_P} =
\begin{bmatrix}
1 & 0 & 0 & 0 \\
0 & \cos 2\,\Psi_P & \sin 2\,\Psi_P & 0 \\
0 & -\sin 2\,\Psi_P & \cos 2\,\Psi_P & 0 \\
0 & 0 & 0 & 1
\end{bmatrix}. \quad (A4.3)
$$

The central submatrix is a rotation matrix which only affects $Q$ and $U$; angular rotations have no effect on $I$ or $V$.

The calibration of the polarimeter is achieved by observing a partially polarized source for a period of hours with an alt-azimuth mounted telescope; this automatically covers a range of parallactic angles. Analysis of the actual dependence of $I, Q, U, V$ on $\Psi_P$ yields both $M_{\mathrm{pol}}$ and the intrinsic Stokes parameters of the source.

An example data set obtained by Johnston (2002) is shown in Figure A4.1. Here a pulse peak of a highly polarized pulsar was observed with a polarimeter over a 180° range of parallactic angle. The two large sinusoids are the measured $Q$ and $U$; for an ideal polarimeter they should be in quadrature and have equal amplitudes (see Eqns 7.24 and 7.25). As noted earlier the measured $I$ (top curve) and $V$ (middle curve) should be independent of $\Psi_P$ but clearly they are not; for the analysis of these data see Johnston (2002).

## A4.2  Polarization in Interferometers

In Section 11.2 we outlined the current practical approach to the polarimetric calibration of interferometers. This is adequate for dish-based arrays for limited fields of view and current dynamic ranges. However, when dynamic range requirements exceed $10^6 : 1$ (as for SKA1-mid), polarization leakage effects can be a limiting factor (Section 10.14.4) and

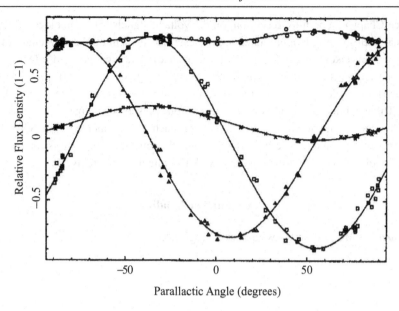

Figure A4.1. Measured Stokes parameters for PSR J1359-6038 as a function of parallactic angle. The sinusoidal lines represent the best fit, which yields the calibration of the polarimeter and the Stokes parameters of the source (Johnston 2002).

for low frequency arrays with wide fields of view (e.g. LOFAR and SKA-low) significant direction-dependent effects become a limitation. In such cases the approximations implicit in the current approach become unacceptable and a new approach is required. This new approach is based on the *measurement equation*, a general formalism introduced by Hamaker *et al.* (1996) explicitly to capture the effects on a signal in its passage from the source via the intervening media and through the receivers and the correlator. The formalism was revisited by Smirnov (2011a,b) and has been dubbed RIME for the radio interferometer measurement equation. The RIME enables the overall interferometric system response to be described systematically in terms of the language of linear algebra and for polarization to be included right from the start. When dealing with power (as in the previous section) the Mueller formalism is appropriate, but when dealing with complex signals, where the phase is important, a more fundamental formalism based on Jones matrices is required.

The relationship between an incoming electric field $\mathbf{e}_{\text{in}}$ and an outgoing field $\mathbf{e}_{\text{out}}$ is captured by a $2 \times 2$ Jones matrix $J$, i.e. $\mathbf{e}_{\text{out}} = J\mathbf{e}_{\text{in}}$:

$$\left[ \begin{array}{c} e_1 \\ e_2 \end{array} \right]_{\text{out}} = \left[ \begin{array}{cc} a & b \\ c & d \end{array} \right] \left[ \begin{array}{c} e_1 \\ e_2 \end{array} \right]_{\text{in}}, \tag{A4.4}$$

where $e_1$ and $e_2$ represent orthogonal bases of the electric field, either linear or circular. The Jones matrix elements may be complex, in order to describe the effects of propagation conditions and the reception chain on the orthogonal components of the electric field (see Chapter 2) or, after reception, on the voltage signals. The combined effects over the signal path to an individual antenna are described by a multiplicative chain of Jones matrices (again taken in reverse order of occurrence). The amalgamated effects of these antenna-based terms on the four cross-correlations for a given baseline (Eqns 11.1) are the polarimetric equivalents of the antenna gains in a non-polarimetric system (Eqns 10.11 and 10.12). The algebraic manipulations are beyond our present scope but are summarized in TMS, Chapter 4. The measurement equation approach is implemented in the CASA[1] data reduction package, and a description of the overall principles and the issues involved in practical calibration can be found in the CASA Documentation Archives.[2]

## A4.3 Further Reading

*Supplementary Material* at www.cambridge.org/ira4.

---

[1] https://casa.nrao.edu/.
[2] https://casa.nrao.edu/casadocs/casa-5.1.0/reference-material/the-measurement-equation-calibration.

# Appendix 5  Spherical Harmonics

It is natural to represent the brightness temperature $T_{B(l,b)}$ as an expansion in spherical harmonics $Y_{l,m}$, the proper orthonormal base functions to use. These are the associated Legendre polynomials (normalized); note that, since $b$ is measured from the celestial equator, whereas the customary polar angle $\theta$ in the usual spherical polar coordinates is a co-latitude, $\cos\theta$ has become $\sin b$ in the expressions below:

$$Y_{l,m}(\theta,\phi) = \left[\frac{(2l+1)(l-|m|)!}{4\pi(l+|m|)!}\right]^{1/2} P_{l,m}(\sin b)e^{im\phi} \times \begin{cases} (-1)^m & m \geq 0, \\ 1 & m < 0. \end{cases} \quad (A5.1)$$

The expansion in spherical harmonics takes the form

$$\frac{T(\theta,\phi)}{T} = \frac{T(\theta,\phi) - T_0}{T_0} = \sum_{l=0}^{l_{max}} \sum_{m=-l}^{l} a_{l,m} Y_{l,m}(\theta,\phi), \quad (A5.2)$$

where the coefficients are derived from the data by calculating

$$a_{l,m} = \int \frac{T(\theta,\phi)}{T} Y^*_{l,m}(\theta,\phi)d\Omega. \quad (A5.3)$$

The integral is taken over the entire celestial sphere; $Y^*_{l,m}$ is the complex conjugate of $Y_{l,m}$. As a result, the spatial map of the brightness fluctuations, $T_F(l,b)$, is represented as a finite sum of spherical harmonics, $l_{max}$ being determined by the angular resolution, which is usually of the order of the half-width half-power effective beamwidth. Although a two-dimensional autocorrelation could be performed on $T_F(l,b)$, followed by a spherical harmonic transform, it is unnecessary since one has already decomposed the brightness distribution into the spherical harmonic $(l,m)$ space that is dual to the angular $(l,b)$ space (these are analogous to the dual time and frequency spaces of signal analysis). The coefficients $a_{l,m}$ represent the wave amplitude for a given spherical harmonic $(l,m)$. Multiplying by the complex conjugate $a^*_{l,m}$ then gives the power in a given mode:

$$C_{l,m} = a_{l,m}a^*_{l,m}. \quad (A5.4)$$

The final step in the analysis relies upon the isotropy of the Universe. All directions are equivalent, so each of the $2l + 1$ coefficients for a given $l$ can be summed to include waves travelling in different directions on the sphere. This set of $l$-coefficients constitutes the *angular power spectrum, $C_l$*:

$$C_l = \frac{1}{2l + 1} \Sigma_m a_{l,m} a_{l,m}^*. \qquad (A5.5)$$

Since it turns out that the resulting spectrum follows a power law over several orders of magnitude, it is convenient to multiply the coefficients by $l(l + 1)/2\pi$ to keep the spectrum within a convenient range when plotting the results. The ordinates are then in units of temperature squared.

# References

Abbott, B. P., and 1103 others. 2017. GW170608: Observation of a 19 solar-mass binary black hole coalescence. *ApJ*, **851**, L35.

Adam, R., and 50 others. 2017. Sub-structure and merger detection in resolved NIKA Sunyaev–Zel'dovich images of distant clusters. ArXiv e-prints.

Ade, P. A. R., and 277 others. 2015. Planck 2013 results. XXXII. The updated Planck catalogue of Sunyaev–Zel'dovich sources. *A&A*, **581**, A14.

Ade, P. A. R., and 235 others. 2016. Planck 2015 results. XXIV. Cosmology from Sunyaev–Zel'dovich cluster counts. *A&A*, **594**, A24.

Akgiray, A., and Weinreb, S. 2012. Ultrawideband square and circular quad-ridge horns with near-constant beamwidth. In: *Proc. IEEE International Conf. on Ultra-Wideband*, p. 518.

Alpar, M. A., Cheng, K. S., and Pines, D. 1989. Vortex creep and the internal temperature of neutron stars – linear and nonlinear response to a glitch. *ApJ*, **346**, 823–832.

Alpher, R. A., Bethe, H., and Gamov, G. 1948. The origin of the chemical elements. *Phys. Rev.*, **73**, 803–804.

Altenhoff, W. J., Downes, D., Pauls, T., and Schraml, J. 1979. Survey of the galactic plane at 4.875 GHz. *A&AS*, **35**, 23–54.

Alves, M. I. R., and seven others. 2015. The HIPASS survey of the Galactic plane in radio recombination lines. *MNRAS*, **450**, 2025–2042.

Allen, C. W. 2000. *Allen's Astrophysical Quantities*, 4th edn. Springer.

Antonucci, R. R. J., and Miller, J. S. 1985. Spectropolarimetry and the nature of NGC 1068. *ApJ*, **297**, 621–632.

Archibald, A. M., and 17 others. 2009. A radio pulsar/X-ray binary link. *Science*, **324**, 1411.

Avision, S., and George, S. J. 2013. A graphical tool for demonstrating the techniques of radio interferometry. *Eur. J. Phys.*, **34**, 7. arXiv:1211.0228.pdf.

Baade, W., and Minkowski, R. 1954. Indentification of the radio sources in Cassiopeia, Cygnus A, and Puppis A. *ApJ*, **119**, 206–214.

Baars, J. W. M. 2007. *The Paraboloidal Reflector Antenna in Radio Astronany and Communications*. Springer.

Baars, J. W. M., and Kärcher, H. J. 2017. *Radio Telescope Reflectors*. Springer.

Baars, J. W. M., and six others. 1973. The synthesis radio telescope at Westerbork. *IEEE Proc.*, **61**, 1258–1266.

Baars, J. W. M., Genzel, R., Pauliny-Toth, I. I. K., and Witzel, A. 1977. The absolute spectrum of CAS A – an accurate flux density scale and a set of secondary calibrators. *A&A*, **61**, 99–106.

Backer, D. C., Kulkarni, S. R., Heiles, C., Davis, M. M., and Goss, W. M. 1982. A millisecond pulsar. *Nature*, **300**, 615–618.

Bahcall, J. N., and seven others. 1995. Hubble Space Telescope and MERLIN Observations of the jet in 3C 273. *ApJ*, **452**, L91.

Baldwin, J. E., and Warner, P. J. 1976. Aperture synthesis without phase measurements. *MNRAS*, **175**, 345–353.

Baldwin, J. E., and Warner, P. J. 1978. Phaseless aperture synthesis. *MNRAS*, **182**, 411–422.

Bambi, C. 2017. Astrophysical black holes: a compact pedagogical review. ArXiv e-prints.

Bartel, N., and 23 others. 1994. The shape, expansion rate and distance of supernova 1993J from VLBI measurements. *Nature*, **369**, 584.

Bassa, C. G., and 12 others. 2014. A state change in the low-mass X-ray binary XSS J12270-4859. *MNRAS*, **441**, 1825–1830.

Bastian, T. S. 1994. Stellar flares. *Space Sci. Rev.*, **68**, 261–274.

Bastian, T. S., Benz, A. O., and Gary, D. E. 1998. Radio emission from solar flares. *Ann. Rev. Astr. Ap.*, **36**, 131–188.

Bates, S. D., Lorimer, D. R., and Verbiest, J. P. W. 2013. The pulsar spectral index distribution. *MNRAS*, **431**, 1352–1358.

Battye, R. A., Browne, I. W. A., Dickinson, C., Heron, G., Maffei, B., and Pourtsidou, A. 2013. H I intensity mapping: a single dish approach. *MNRAS*, **434**, 1239–1256.

Baudry, A., and eight others 2018. Vibrationally excited water emission at 658 GHz from evolved stars. *A&A*, **609**, A25.

Beck, R. 1996. The structure of interstellar magnetic fields as derived from polarization observations in radio continuum. In: Roberge, W. G., and Whittet, D. C. B. (eds.), *Polarimetry of the Interstellar Medium*, p. 475. Astronomical Society of the Pacific Conference Series, vol. 97.

Beck, R. 2016. Magnetic fields in spiral galaxies. *Astron. Astrophys. Rev.*, **24**(Dec.), 4.

Becker, R. H., White, R. L., and Helfand, D. J. 1995. The FIRST survey: faint images of the radio sky at twenty centimeters. *ApJ*, **450**, 559.

Bell, A. R., Gull, S. F., and Kenderdine, S. 1975. New radio map of Cassiopeia A at 5 GHz. *Nature*, **257**, 463–465.

Bennett, A. S. 1962. The revised 3C catalogue of radio sources. *Mem. RAS*, **68**, 163.

Bennett, C. L., and nine others. 1996. Four-year COBE DMR cosmic microwave background observations: maps and basic results. *ApJ*, **464**, L1.

Bennett, C. L., and 20 others. 2013. Nine-year Wilkinson microwave anisotropy probe (WMAP) observations: final maps and results. *ApJS*, **208**, 20.

Bentz, M. C., and Katz, S. 2015. The AGN black hole mass database. *PASP*, **127**, 67.

Benz, A. O., Monstein, C., and Meyer, H. 2005. Callisto, a new concept for solar radio spectrometers. *Sol. Phys.*, **226**, 143–151.

Beswick, R. 2006. Radio supernovae. In: *Proc. 8th European VLBI Network Symp.*, p. 51.

Beuermann, K., Kanbach, G., and Berkhuijsen, E. M. 1985. Radio structure of the Galaxy – thick disk and thin disk at 408 MHz. *A&A*, **153**, 17–34.

Bhatnagar, S., Cornwell, T. J., Golap, K., and Uson, J. M. 2008. Correcting direction-dependent gains in the deconvolution of radio interferometric images. *A&A*, **487**, 419–429.

Bhatnagar, S., Rau, U., and Golap, K. 2013. Wide-field wide-band interferometric imaging: the WB A-projection and hybrid algorithms. *ApJ*, **770**, 91.

Bicknell, G. V., Mukherjee, D., Wagner, A. Y., Sutherland, R. S., and Nesvadba, N. P. H. 2018. Relativistic jet feedback – II. Relationship to gigahertz peak spectrum and compact steep spectrum radio galaxies. *MNRAS*, **475**, 3493–3501.

Biggs, A., and Browne, I. 2017. Gravitational lens time delays using polarization monitoring. *Galaxies*, **5**, 76.

Bignami, G. F., Caraveo, P. A., and Mereghetti, S. 1993. Understanding GEMINGA: past and future observations. In: Friedlander, M., Gehrels, N., and Macomb, D. J. (eds.), *Compton Gamma-Ray Observatory: St. Louis, MO 1992*, pp. 233–237. American Institute of Physics Conference Series, vol. 280.

Bilous, A. V., and 26 others. 2016. A LOFAR census of non-recycled pulsars: average profiles, dispersion measures, flux densities, and spectra. *A&A*, **591**, A134.

Binney, J. 1992. WARPS. *Ann. Rev. Astr. Ap.*, **30**, 51–74.

Binney, J., and Tremaine, S. 1987. *Galactic Dynamics*. Princeton University Press.

Binney, J., Gerhard, O. E., Stark, A. A., Bally, J., and Uchida, K. I. 1991. Understanding the kinematics of Galactic centre gas. *MNRAS*, **252**, 210–218.

Biretta, J. A., Moore, R. L., and Cohen, M. H. 1986. The evolution of the compact radio source in 3C 345. I – VLBI observations. *ApJ*, **308**, 93–109.

Birkinshaw, M. 1999. The Sunyaev–Zel'dovich effect. *Phys. Rep.*, **310**, 97–195.

Blake, G. A., Masson, C. R., Phillips, T. G., and Sutton, E. C. 1986. The rotational emission-line spectrum of Orion A between 247 and 263 GHz. *ApJS*, **60**, 357–374.

Bland-Hanothoin, J., and Gerhard, O. 2016. The Galaxy in context: structural, kinematic and integrated properties. *Ann. Rev. Astron. Astrophys.*, **54**, 529–596.

Blandford, R. D., and Znajek, R. L. 1977. Electromagnetic extraction of energy from Kerr black holes. *MNRAS*, **179**, 433–456.

Blandford, R. D., McKee, C. F., and Rees, M. J. 1977. Super-luminal expansion in extragalactic radio sources. *Nature*, **267**, 211–216.

Bleem, L. E., and seven others. 2015. A new reduction of the Blanco Cosmology Survey: an optically selected galaxy cluster catalog and a public release of optical data products. *ApJS*, **216**, 20.

Blitz, L., Binney, J., Lo, K. Y., Bally, J., and Ho, P. T. P. 1993. The centre of the Milky Way. *Nature*, **361**, 417–424.

Boboltz, D. A., Diamond, P. J., and Kemball, A. J. 1997. R Aquarii: first detection of circumstellar SiO maser proper motions. *ApJ*, **487**, L147–L150.

Boggess, N. W., and 17 others. 1992. The cosmic background explorer (COBE): mission and science overview. *Highlights of Astronomy*, **9**, 273.

Bøifot, A. M. 1991. Classification of ortho-mode transducers. *European Trans. Telecommun.*, **2**, 503–510.

Bonato, M., and 14 others. 2018. ALMACAL IV: a catalogue of ALMA calibrator continuum observations. *MNRAS*, **478**, 1512–1519.

Boorman, J. A., McLean, D. J., Sheridan, K. V., and Wild, J. P. 1961. The spectral components of 150 major solar radio events (1952–1960). *MNRAS*, **123**, 87.

Booth, R. S., Norris, R. P., Porter, N. D., and Kus, A. J. 1981. Observations of a circumstellar shell around the OH/IR star OH127.8-0.0. *Nature*, **290**, 382–384.

Bowman, J. D., Rogers, A. E. E., Monsalve, R. A., Mozdzen, T. J., and Mahesh, N. 2018. An absorption profile centred at 78 megahertz in the sky-averaged spectrum. *Nature*, **555**, 67–70.

Bracewell, R. N. 1961. Interferometry and spectral sensitivity island diagram. *IRE Trans. Ant. Propag.*, **9**, 59.

Bracewell, R. N. 1962. Radio astronomy techniques. *Handbuch der Phys.*, **54**, 42.

Bracewell, R. N., and Roberts, J. A. 1954. Aerial smoothing in radio astronomy. *Australian J. Phys.*, **7**, 615.

Bradley, L. D., Kaiser, M. E., and Baan, W. A. 2004. Physical conditions in the narrow-line region of M51. *ApJ*, **603**, 463.

Braun, R. 2013. Understanding synthesis imaging dynamic range. *A&A*, **551**, A91.

Bridle, A. H., and Schwab, F. R. 1989. Wide field imaging I: bandwidth and time-average smearing. *ASPC*, **6**, 247.

Bridle, A. H., Hough, D. H., Lonsdale, C. J., Burns, J. O., and Laing, R. A. 1994. Deep VLA imaging of twelve extended 3CR quasars. *AJ*, **108**, 766–820.

Brouw, W. N., and Spoelstra, T. A. T. 1976. Linear polarization of the galactic background at frequencies between 408 and 1411 MHz. Reductions. *A&AS*, **26**, 129.

Brown, J. C., and seven others. 2007. Rotation measures of extragalactic sources behind the Southern Galactic Plane: new insights into the large-scale magnetic field of the inner Milky Way. *ApJ*, **663**(July), 258–266.

Browne, I. W. A., and 21 others. 2003. The Cosmic Lens All-Sky Survey – II. Gravitational lens candidate selection and follow-up. *MNRAS*, **341**, 13–32.

Bryerton, E. W., Morgan, M. A., and Pospieszalski, M. W. 2013. Ultra low noise cryogenic amplifiers for radio astronomy. In: *Proc. 2013 IEEE Radio and Wireless Symp. (RWS)*.

Burke, B. F., and Franklin, K. L. 1955. Observations of a variable radio source associated with the planet Jupiter. *J. Geophys. Res.*, **60**, 213–217.

Burn, B. J. 1966. On the depolarization of discrete radio sources by Faraday dispersion. *MNRAS*, **133**, 67.

Burrows, A. 2000. Supernova explosions in the Universe. *Nature*, **403**, 727–733.

Burton, W. B. 1988. The structure of our galaxy derived from observations of neutral hydrogen. In: Verschuur, G. L., and Kellerman, K. I. (eds.), *Galactic and Extragalactic Radio Astronomy*, pp. 295–358. Springer.

Cane, H. V. 1978. A 30 MHz map of the whole sky. *Australian J. Phys.*, **31**, 561.

Cane, H. V. 1979. Spectra of the non-thermal radio radiation from the galactic polar regions. *MNRAS*, **189**, 465–478.

Carilli, C. L., and Walter, F. 2013. Cool gas in high-redshift galaxies. *Ann. Rev. Astr. Ap.*, **51**, 105–161.

Carilli, C. L., and 64 others. 2018. HI 21cm cosmology and the bi-spectrum: closure diagnostics in massively redundant interferometric arrays. ArXiv e-prints.

Carlstrom, J. E., Holder, G. P., and Reese, E. D. 2002. Cosmology with the Sunyaev–Zel'dovich effect. *Ann. Rev. Astr. Ap.*, **40**, 643–680.

Carlstrom, J. E., and 43 others. 2011. The 10 meter South Pole Telescope. *PASP*, **123**, 568.

Chael, A. A., Johnson, M. D., Bouman, K. L., Blackburn, L. L., Akiyama, K., and Narayan, R. 2018. Interferometric imaging directly with closure phases and closure amplitudes. ArXiv e-prints.

Chapman, J. M., and Cohen, R. J. 1986. MERLIN observations of the circumstellar envelope of VX Sagittarius. *MNRAS*, **220**, 513–528.

Chatterjee, S., and 24 others. 2017. A direct localization of a fast radio burst and its host. *Nature*, **541**, 58–61.

Cherepashchuk, A. M., and 20 others. 2005. INTEGRAL observations of SS433: results of a coordinated campaign. *A&A*, **437**, 561–573.

Cheung, A. C., Rank, D. M., Townes, C. H., Knowles, S. H., and Sullivan, III, W. T. 1969. Distribution of ammonia density, velocity, and rotational excitation in the region of Sagittarius B2. *ApJ*, **157**, L13.

Christiansen, W. N., and Hogböm, J. A. 1985. *Radiotelescopes*, 2nd edn. Cambridge University Press.

Clarricoats, P. J. B., and Olver, A. D. 1984. *Corrugated Horns for Microwave Antennas*. IEEE.

Clark, B. G. 2003. A review of the history of VLBI. In: Zensus, J. A., Cohen, M. H., and Ros, E. (eds.), *Radio Astronomy at the Fringe*, p. 1. Astronomical Society of the Pacific Conference Series, vol. 300.

Clemens, D. P. 1985. Massachusetts–Stony Brook galactic plane CO survey – the galactic disk rotation curve. *ApJ*, **295**, 422–428.

Cohen, M. H., and Shaffer, D. B. 1971. Positions of radio sources from long-baseline interferometry. *AJ*, **76**, 91.

Cohen, M. H., and nine others. 1977. Radio sources with superluminal velocities. *Nature*, **268**, 405–409.

Cohen, R. J. 1989. Compact maser sources. *Rep. Progr. Phys.*, **52**, 881–943.

Cohen, R. J., Brebner, G. C., and Potter, M. M. 1990. Magnetic field decay in the bipolar outflow source Cepheus-A. *MNRAS*, **246**, 3P.

Condon, J. J. 1984. Cosmological evolution of radio sources. *ApJ*, **287**, 461–474.

Condon, J. J. 1988. *Radio Sources and Cosmology*, pp. 641–678. Springer.

Condon, J. J. 1992. Radio emission from normal galaxies. *Ann. Rev. Astr. Ap.*, **30**, 575–611.

Condon, J. J. 1997. Errors in elliptical Gaussian fits. *PASP*, **109**, 166–172.

Condon, J. J. 2007 (Dec.). Deep radio surveys. In: Afonso, J., Ferguson, H. C., Mobasher, B., and Norris, R. (eds.), *Deepest Astronomical Surveys*, p. 189. Astronomical Society of the Pacific Conference Series, vol. 380.

Condon, J. J. 2017. In *Proc.* Conf. on *The Many Facets of Extragalactic Radio Surveys: Towards New Scientific Challenges*, 20–23 October 2015, Bologna, Italy. Online at https://pos.sissa.it/267/004/pdf.

Condon, J. J., and Matthews, A M. 2018. ACDM cosmology for astronomers. arXiv:18404.10047v1.

Condon, J. J., and Ransom, S. M. 2016. *Essential Radio Astronomy*. Princeton University Press.

Condon, J. J., Anderson, M. L., and Helou, G. 1991. Correlations between the far-infrared, radio, and blue luminosities of spiral galaxies. *ApJ*, **376**, 95–103.

Condon, J. J., Broderick, J. J., Seielstad, G. A., Douglas, K., and Gregory, P. C. 1994. A 4.85 GHz sky survey. 3: Epoch 1986 and combined (1986 + 1987) maps covering $0°$ to $\leq 75°$. *Astron. J.*, 107, 1829.

Condon, J. J., and six others. 1998. The NRAO VLA Sky Survey. *AJ*, **115**, 1693–1716.

Condon, J. J., and eight others. 2012. Resolving the radio source background: deeper understanding through confusion. *ApJ*, **758**, 23.

Conway, J. E., Cornwell, T. J., and Wilkinson, P. N. 1990. Multi-frequency synthesis – a new technique in radio interferometric imaging. *MNRAS*, **246**, 490.

Conway, R. G., and Kronberg, P. P. 1969. Interferometric measurement of polarization distribution in radio sources. *MNRAS*, **142**, 11.

Cornwell, T. J. 1981. *VLA Scientific Memorandum*, **135**.

Cornwell, T. J. 1986. Synthesis imaging: self-calibration. In: Perley, R. A., Schwab, F. R., and Bridle, A. H. (eds.), *Synthesis Imaging*, pp. 137–147. NRAO.

Cornwell, T. J. 1987. Radio-interferometric imaging of weak objects in conditions of poor phase stability – the relationship between speckle masking and phase closure methods. *A&A*, **180**, 269–274.

Cornwell, T. J. 1988. Radio-interferometric imaging of very large objects. *A&A*, **202**, 316–321.

Cornwell, T. J. 2008. Multiscale CLEAN deconvolution of radio synthesis images. *IEEE J. Sel. Topics Signal Process.*, **2**, 793–801.

Cornwell, T. J., and Perley, R. A. 1992. Radio-interferometric imaging of very large fields – the problem of non-coplanar arrays. *A&A*, **261**, 353–364.

Cornwell, T. J., and Wilkinson, P. N. 1981. A new method for making maps with unstable radio interferometers. *MNRAS*, **196**, 1067–1086.

Cornwell, T. J., Golap, K., and Bhatnagar, S. 2008. The noncoplanar baselines effect in radio interferometry: the W-Projection Algorithm. *IEEE J. Sel. Topics Signal Process.*, **2**, 647–657.

Cotton, W. D. 1995. Fringe fitting. In: Zensus, J. A., Diamond, P. J., and Napier, P. J. (eds.), *Very Long Baseline Interferometry and the VLBA*, p. 189. Astronomical Society of the Pacific Conference Series, vol. 82.

Cotton, W. D., and nine others. 1979. 3C 279 – the case for 'superluminal' expansion. *ApJ*, **229**, L115–L117.

Cox, D. P., and Reynolds, R. J. 1987. The local interstellar medium. *Ann. Rev. Astr. Ap.*, **25**, 303–344.

Crawford, A. B., Hogg, D. C., and Hunt, L. E. 1961. A horn-reflector antenna for space communication. *Bell Syst. Tech. J.*, **40**, 1095–1116.

Crill, B. P., and 36 others. 2003. BOOMERANG: a balloon-borne millimeter-wave telescope and total power receiver for mapping anisotropy in the cosmic microwave background. *ApJS*, **148**, 527–541.

Damashek, M., Backus, P. R., Taylor, J. H., and Burkhardt, R. K. 1982. Northern Hemisphere pulsar survey – a third radio pulsar in a binary system. *ApJ*, **253**, L57–L60.

Dame, T. M., and Thaddeus, P. 1985. A wide-latitude CO survey of molecular clouds in the northern Milky Way. *ApJ*, **297**, 751–765.

Dame, T. M., and eight others. 1987. A composite CO survey of the entire Milky Way. *ApJ*, **322**, 706–720.

Dame, T. M., Hartmann, D., and Thaddeus, P. 2001. The Milky Way in molecular clouds: a new complete CO survey. *ApJ*, **547**, 792–813.

Davenport, W. D., and Rost, W. L. 1958. *An Introduction to the Theory of Random Signals and Noise*. McGraw Hill.

Davies, R. D., Dickinson, C., Banday, A. J., Jaffe, T. R., Górski, K. M., and Davis, R. J. 2006. A determination of the spectra of Galactic components observed by the Wilkinson Microwave Anisotropy Probe. *MNRAS*, **370**, 1125–1139.

Davis, R. J., Muxlow, T. W. B., and Conway, R. G. 1985. Radio emission from the jet and lobe of 3C273. *Nature*, **318**, 343–345.

De Young, D. S. 2002. *The Physics of Extragalatic Radio Sources*. University of Chicago Press.

de Zotti, G., Massardi, M., Negrello, M., and Wall, J. 2010. Radio and Millimeter continuum surveys and their astrophysical implications. *A & AR*, **18**, 1.

Deller, A. T., and ten others. 2011 (Feb.). DiFX2: A more flexible, efficient, robust and powerful software correlator. Astrophysics Source Code Library.

Dermer, C. D., and Giebels, B. 2016. Active galactic nuclei at gamma-ray energies. *Comptes Rendus Phys.*, **17**, 594–616.

Deshpande, A. A., and Rankin, J. M. 2001. The topology and polarization of sub-beams associated with the 'drifting' sub-pulse emission of pulsar B0943+10 – I. Analysis of Arecibo 430- and 111-MHz observations. *MNRAS*, **322**, 438–460.

Diamond, P. J. 1995. VLBI data reduction in practice. In: Zensus, J. A., Diamond, P. J., and Napier, P. J. (eds.), *Very Long Baseline Interferometry and the VLBA*, p. 227. Astronomical Society of the Pacific Conference Series, vol. 82.

Dickel, J. R., and Willis, A. G. 1980. The radio emission of the supernova remnants CTB1 and the Cygnus Loop. *A&A*, **85**, 55–65.

Dickey, J. M., and Lockman, F. J. 1990. H I in the Galaxy. *Ann. Rev. Astr. Ap.*, **28**, 215–261.

Dickinson, C. 2016. CMB foregrounds – a brief review. ArXiv e-prints, June.

Diep, P. N., and nine others. 2016. CO and HI emission from the circumstellar envelopes of some evolved stars. In: Qain, L., and Li, D. (eds.), *Frontiers in Radio Astronomy and FAST Early Sciences Symp. 2015*, p. 61. Astronomical Society of the Pacific Conference Series, vol. 502.

Doeleman, S. S., and 27 others. 2008. Event-horizon-scale structure in the supermassive black hole candidate at the Galactic centre. *Nature*, **455**, 78–80.

Dougherty, S. M., Bode, M. F., Lloyd, H. M., Davis, R. J., and Eyres, S. P. 1995. High-resolution radio images of the symbiotic star R Aquarii. *MNRAS*, **272**, 843–849.

Dowell, J., Taylor, G. B., Schinzel, F. K., Kassim, N. E., and Stovall, K. 2017. The LWA1 Low Frequency Sky Survey. *MNRAS*, **469**, 4537–4550.

Duin, R. M., and Strom, R. G. 1975. A multifrequency study of the radio structure of 3C10, the remnant of Tycho's supernova. *A&A*, **39**, 33–42.

Duncan, R. C., and Thompson, C. 1992. Formation of very strongly magnetized neutron stars – implications for gamma-ray bursts. *ApJ*, **392**, L9–L13.

Ekers, R. D. 1983. In: Kellermann, K. I., and Sheets, B. (eds.), *Serendipitous Discoveries in Radio Astronomy, Proc. NRAO Workshop*, p. 154.

Elitzur, M. 1992. Astronomical masers. *Ann. Rev. Astr. Ap.*, **30**, 75–112.

Ellingson, S. W., Craig, J., Dowell, J., Taylor, G. B., and Helmboldt, J. F. 2013. Design and commissioning of the LWA1 radio telescope. ArXiv e-prints.

Emerson, D. T., Klein, U., and Haslam, C. G. T. 1979. A multiple beam technique for overcoming atmospheric limitations to single-dish observations of extended radio sources. *A&A*, **76**, 92–105.

Endres, C. P., Schlemmer, S., Schilke, P., Stutzki, J., and Müller, H. S. P. 2016. The Cologne database for molecular spectroscopy, CDMS, in the Virtual Atomic and Molecular Data Centre, VAMDC. *J. Molecular Spectroscopy*, **327**, 95–104.

Engels, D., Etoka, S., Gérard, E., and Richards, A. 2015 (Aug.). Phase-lag distances of OH masing AGB stars. In: Kerschbaum, F., Wing, R. F., and Hron, J. (eds.), *Why Galaxies Care about AGB Stars III: A Closer Look in Space and Time*, p. 473. Astronomical Society of the Pacific Conference Series, vol. 497.

Espinoza, C., Lyne, A., Stappers, B., and Kramer, M. 2011. Glitches in the rotation of pulsars. In: Burgay, M., D'Amico, N., Esposito, P., Pellizzoni, A., and Possenti, A. (eds.), pp. 117–120. American Institute of Physics Conference Series, vol. 1357.

Espinoza, C. M., Lyne, A. G., and Stappers, B. W. 2017. New long-term braking index measurements for glitching pulsars using a glitch-template method. *MNRAS*, **466**, 147–162.

Ettori, S., and six others. 2013. Mass profiles of galaxy clusters from X-ray analysis. *Space Sci. Rev.*, **177**, 119–154.

Ewen, H. I., and Purcell, E. M. 1951. Observation of a line in the Galactic radio spectrum: radiation from Galactic hydrogen at 1,420 Mc./sec. *Nature*, **168**, 356.

Eyres, S. P. S., Davis, R. J., and Bode, M. F. 1996. Nova Cygni 1992 (V1974 Cygni): MERLIN observations from 1992 to 1994. *MNRAS*, **279**, 249–256.

Fabian, A. C. 2012. Observational evidence of active galactic nuclei feedback. *Ann. Rev. Astr. Ap.*, **50**, 455–489.

Fabrika, S. 2004. The jets and supercritical accretion disk in SS433. *Astrophys. Space Phys. Rev.*, **12**, 1–152.

Fan, L., Knudsen, K. K., Fogasy, J., and Drouart, G. 2017. ALMA detections of CO emission in the most luminous, heavily dust-obscured quasars at $z > 3$. ArXiv e-prints.

Fanaroff, B. L., and Riley, J. M. 1974. The morphology of extragalactic radio sources of high and low luminosity. *MNRAS*, **167**, 31P–36P.

Fender, R. P., and seven others. 1999. MERLIN observations of relativistic ejections from GRS 1915+105. *MNRAS*, **304**, 865–876.

Fermi, E. 1949. On the origin of the cosmic radiation. *Phys. Rev.*, **75**, 1169–1174.

Fey, A. L., and 30 others. 2015. The second realization of the international celestial reference frame by very long baseline interferometry. *AJ*, **150**, 58.

Fich, M., and Tremaine, S. 1991. The mass of the Galaxy. *Ann. Rev. Astr. Ap.*, **29**, 409–445.

Fixsen, D. J., and ten others. 1996. A balloon-borne millimeter-wave telescope for cosmic microwave background anisotropy measurements. *ApJ*, **470**, 63.

Frail, D. A., Vasisht, G., and Kulkarni, S. R. 1997. The changing structure of the radio nebula around the soft gamma-ray repeater SGR 1806-20. *ApJ*, **480**, L129–L132.

Frater, R. H., Goss, W. M., and Wendt, H. W. 2013. Bernard Yarnton Mills AC FAA. 8 August 1920 – 25 April 2011. *Biographical Memoirs of Fellows of the Royal Society*, **59**, 215–239.

Garrett, M. A., Calder, R. J., Porcas, R. W., King, L. J., Walsh, D., and Wilkinson, P. N. 1994. Global VLBI observations of the gravitational lens system 0957+561A, B. *MNRAS*, **270**, 457.

Garrington, S. T., Leahy, J. P., Conway, R. G., and Laing, R. A. 1988. A systematic asymmetry in the polarization properties of double radio sources with one jet. *Nature*, **331**, 147–149.

Gehrels, N., and Chen, W. 1993. The Geminga supernova as a possible cause of the local interstellar bubble. *Nature*, **361**, 706.

Gentile, G., Salucci, P., Klein, U., and Granato, G. L. 2007. NGC 3741: the dark halo profile from the most extended rotation curve. *MNRAS*, **375**, 199–212.

Georgelin, Y. M., and Georgelin, Y. P. 1976. The spiral structure of our Galaxy determined from H II regions. *A&A*, **49**, 57–79.

Gérard, E., and Le Bertre, T. 2006. Circumstellar atomic hydrogen in evolved stars. *AJ*, **132**, 2566–2583.

Ghez, A. M., and seven others. 2005. Stellar orbits around the Galactic Center black hole. *ApJ*, **620**, 744–757.

Giacconi, R., Gursky, H., Kellogg, E., Schreier, E., and Tananbaum, H. 1971. Discovery of periodic X-ray pulsations in Centaurus X-3 from UHURU. *ApJ*, **167**, L67.

Gillessen, S., and six others. 2009. Monitoring stellar orbits around the massive black hole in the Galactic Center. *ApJ*, **692**, 1075–1109.

Ginzburg, V. L., and Syrovatskii, S. I. 1969. Developments in the theory of synchrotron radiation and its reabsorption. *Ann. Rev. Astr. Ap.*, **7**, 375.

Girard, J. N., and 73 others. 2016. Imaging Jupiter's radiation belts down to 127 MHz with LOFAR. *A&A*, **587**, A3.

Gizani, N. A. B., and Leahy, J. P. 1999. The environment of Hercules A. *New Astron. Rev.*, **43**, 639–642.

Goldreich, P., and Julian, W. H. 1969. Pulsar electrodynamics. *ApJ*, **157**, 869.

Goldsmith, P. F., and 34 others. 2011. Herschel measurements of molecular oxygen in Orion. *ApJ*, **737**, 96.

Golla, G., and Hummel, E. 1994. The intrinsic magnetic field orientation in NGC 4631. *A&A*, **284**, 777–792.

Gordon, M. A., and Sorochenko, R. L. 2007. *Radio Recombination Lines*. Springer.

Gray, M. 2012. *Maser Sources in Astrophysics*. Cambridge University Press.

Gray, R. O. 1998. The absolute flux calibration of Strömgren UVBY photometry. *AJ*, **116**, 482–485.

Gregory, P. C., and Condon, J. J. 1991. The 87GB catalog of radio sources covering delta between O and + 75 deg at 4.85 GHz. *Astrophys. J. Suppl.*, 75, 1011.

Griffith, M. R., and Wright, A. E. 1993. The Parkes–MIT–NRAO (PMN) surveys. I – The 4850 MHz surveys and data reduction, *Astron. J.*, 105, 1666.

Güdel, M. 2002. Stellar radio astronomy: probing stellar atmospheres from protostars to giants. *Ann. Rev. Astr. Ap.*, **40**, 217–261.

Gugliucci, N. E., Taylor, G. B., Peck, A. B., and Giroletti, M. 2005. Dating COINS: kinematic ages for compact symmetric objects. *ApJ*, **622**, 136–148.

Hafez, Y. A., and 22 others. 2008. Radio source calibration for the Very Small Array and other cosmic microwave background instruments at around 30 GHz. *MNRAS*, **388**, 1775–1786.

Hallinan, G., and nine others. 2007. Periodic bursts of coherent radio emission from an ultracool dwarf. *ApJ*, **663**, L25–L28.

Hamaker, J. P., Bregman, J. D., and Sault, R. J. 1996. Understanding radio polarimetry. I. Mathematical foundations. *A&AS*, **117**, 137–147.

Han, J. L. 2007. Magnetic fields in our Galaxy on large and small scales. In: Chapman, J. M., and Baan, W. A. (eds.), *Astrophysical Masers and their Environments, Proc. IAU Symp.*, vol. 242, pp. 55–63.

Han, J. L. 2017. Observing interstellar and intergalactic magnetic fields. *Ann. Rev. Astr. Ap.*, **55**, 111–157.

Han, J. L., Manchester, R. N., and Qiao, G. J. 1999. Pulsar rotation measures and the magnetic structure of our Galaxy. *MNRAS*, **306**, 371–380.

Han, J. L., Manchester, R. N., van Straten, W., and Demorest, P. 2018. Pulsar rotation measures and large-scale magnetic field reversals in the Galactic disk. *ApJS*, **234**, 11.

Handa, T., Sofue, Y., Nakai, N., Hirabayashi, H., and Inoue, M. 1987. A radio continuum survey of the Galactic plane at 10 GHz. *PASJ*, **39**, 709–753.

Hankins, T. H., Jones, G., and Eilek, J. A. 2015. The Crab Pulsar at centimeter wavelengths. I. Ensemble characteristics. *ApJ*, **802**, 130.

Hankins, T. H., Eilek, J. A., and Jones, G. 2016. The Crab Pulsar at centimeter wavelengths. II. Single pulses. *ApJ*, **833**, 47.

Harper, G. M., Brown, A., and Lim, J. 2001. A spatially resolved, semiempirical model for the extended atmosphere of $\alpha$ Orionis (M2 Iab). *ApJ*, **551**, 1073–1098.

Harris, A. I. 2005. Spectroscopy with multichannel correlation radiometers. *Rev. Scientific Instrum.*, **76**, 054503.

Harris, S., and Wynn-Williams, C. G. 1976. Fine radio structure in W3. *MNRAS*, **174**, 649–659.

Harrison, E. R. 1970. Fluctuations at the threshold of classical cosmology. *Phys. Rev. D*, **1**, 2726–2730.

Haslam, C. G. T., Salter, C. J., Stoffel, H., and Wilson, W. E. 1982. A 408 MHz all-sky continuum survey. II – The atlas of contour maps. *A&AS*, **47**, 1.

Hazard, C., and Walsh, D. 1959a. A comparison of an interferometer and total-power survey of discrete sources of radio-frequency radiation. In: Bracewell, R. N. (ed.), *Proc. URSI Symp. 1: Paris Symp. on Radio Astronomy*, p. 477. IAU Symposium, vol. 9.

Hazard, C., and Walsh, D. 1959b. An experimental investigation of the effects of confusion in a survey of localized radio sources. *MNRAS*, **119**, 648.

Healy, F., O'Brien, T. J., Beswick, R., Avison, A., and Argo, M. K. 2017. Multi-epoch radio imaging of $\gamma$-ray Nova V959 Mon. *MNRAS*, **469**, 3976–3983.

Hecht, E. 1970. Note on an operational definition of the Stokes parameters. *Am. J. Phys.*, **38**, 1156.

Heckman, T. M., and Best, P. N. 2014. The coevolution of galaxies and supermassive black holes: insights from surveys of the contemporary universe. *Ann. Rev. Astr. Ap.*, **52**, 589–660.

Heiles, C. 1980. Is the intercloud medium pervasive? *ApJ*, **235**, 833–839.

Heiles, C. 1995. The galactic B-field (GBF). In: Ferrara, A., McKee, C. F., Heiles, C., and Shapiro, P. R. (eds.), *The Physics of the Interstellar Medium and Intergalactic Medium*, p. 507. Astronomical Society of the Pacific Conference Series, vol. 80.

Heiles, C. 2002. A heuristic introduction to radioastronomical polarization. In: Stanimirovic, S., Altschuler, D., Goldsmith, P., and Salter, C. (eds.), *Single-Dish Radio Astronomy: Techniques and Applications*, pp. 131–152. Astronomical Society of the Pacific Conference Series, vol. 278.

Heiles, C., Chu, Y.-H., Reynolds, R. J., Yegingil, I., and Troland, T. H. 1980. A warm magnetoactive plasma in a large volume of space. *ApJ*, **242**, 533.

Heiles, C., Goodman, A. A., McKee, C. F., and Zweibel, E. G. 1993. Magnetic fields in star-forming regions – observations. In: Levy, E. H., and Lunine, J. I. (eds.), *Protostars and Planets III*, pp. 279–326. University of Arizedona Press.

Helou, G., Soifer, B. T., and Rowan-Robinson, M. 1985. Thermal infrared and nonthermal radio – remarkable correlation in disks of galaxies. *ApJ*, **298**, L7–L11.

Henstock, D. R., Browne, I. W. A., Wilkinson, P. N., Taylor, G. B., Vermeulen, R. C., Pearson, T. J., and Readhead, A. C. S. 1995. The second Caltech–Jodrell Bank VLBI survey. II. Observations of 102 of 193 sources. *ApJS*, **100**, 1.

Herbst, E., and van Dishoeck, E. F. 2009. Complex organic interstellar molecules. *Ann. Rev. Astr. Ap.*, **47**, 427–480.

Herrnstein, J. R., and 8 others. 1999. A geometric distance to the galaxy NGC4258 from orbital motions in a nuclear gas disk. *Nature*, **400**, 539–541.

Hewish, A., Bell, S. J., Pilkington, J. D. H., Scott, P. F., and Collins, R. A. 1968. Observation of a rapidly pulsating radio source. *Nature*, **217**, 709–713.

Hey, J. S., Parsons, S. J., and Phillips, J. W. 1946. A new intense source of radio-frequency radiation in the constellation of Cassiopeia. *Proc. Roy. Soc. London*, **A192**, 425.

HI4PI Collaboration, Ben Bekhti, N., and 19 others. 2016. HI4PI: A full-sky H I survey based on EBHIS and GASS. *A&A*, **594**, A116.

Hill, R. J., and Clifford, S. F. 1981. *Radio Sci.*, **16**, 77.

Hinshaw, G., and 20 others. 2013. Nine-year Wilkinson microwave anisotropy probe (WMAP) observations: cosmological parameter results. *ApJS*, **208**, 19.

Hirabayashi, H. 2005 (Dec.). VSOP mission results and space VLBI mission studies. In: Romney, J., and Reid, M. (eds.), *Future Directions in High Resolution Astronomy*, p. 561. Astronomical Society of the Pacific Conference Series, vol. 340.

Hirabayashi, H., and 56 others. 2000. The VLBI Space Observatory Programme and the radio-astronomical satellite HALCA. *PASJ*, **52**, L955–L965.

Hobbs, G. B., and Dai, S. 2017. A review of pulsar timing array gravitational wave research. ArXiv e-prints.

Hobbs, G. B., Edwards, R. T., and Manchester, R. N. 2006. TEMPO2, a new pulsar-timing package – I. An overview. *MNRAS*, **369**, 655–672.

Högbom, J. A. 1974. Aperture synthesis with a non-regular distribution of interferometer baselines. *A&AS*, **15**, 417.

Hoglund, B., and Mezger, P. G. 1965. Hydrogen emission line $n_{110} \rightarrow n_{109}$: detection at 5009 megahertz in Galactic H II regions. *Science*, **150**, 339–340.

Holdaway, M. A. 1999. Mosaicing with interferometric arrays. In: Taylor, G. B., Carilli, C. L., and Perley, R. A. (eds.), *Synthesis Imaging in Radio Astronomy II*, p. 401. Astronomical Society of the Pacific Conference Series, vol. 180.

Honma, M., and 22 others. 2018. In: *Proc. IAU Symp.*, vol. 336, p. 162.

Hovatta, T., Nieppola, E., Tornikoski, M., Valtaoja, E., Aller, M. F., and Aller, H. D. 2008. Long-term radio variability of AGN: flare characteristics. *A&A*, **485**, 51–61.

Hu, W., and Dodelson, S. 2002. Cosmic microwave background anisotropies. *Ann. Rev. Astr. Ap.*, **40**, 171–216.

Hu, W., and White, M. 1997. A CMB polarization primer, *New Astronomy*, **2**, 323.

Hubble, E. 1929. A relation between distance and radial velocity among extra-galactic nebulae. *Proc. Nat. Acad. Sci.*, **15**, 168–173.

Humphreys, E. M. L., Reid, M. J., Moran, J. M., Greenhill, L. J., and Argon, A. L. 2013. Toward a new geometric distance to the active galaxy NGC 4258. III. Final results and the Hubble constant. *ApJ*, **775**, 13.

Imai, M., and nine others. 2016. The beaming structures of Jupiter's decametric common S-bursts observed from the LWA1, NDA, and URAN2 radio telescopes. *ApJ*, **826**, 176.

Isliker, H., and Benz, A. O. 1994. Non-linear properties of the dynamics of bursts and flares in the solar and stellar coronae. *A&A*, **285**, 663.

Jackson, N. 2015. The Hubble constant. *Living Rev. Relativity*, **18**, 2.

Jackson, N., and 78 others. 2016. LBCS: the LOFAR long-baseline calibrator survey. *A&A*, **595**, A86.

Jahoda, K., Lockman, F. J., and McCammon, D. 1990. Galactic H I and the interstellar medium in Ursa Major. *ApJ*, **354**, 184–189.

Jarvis, M., and 12 others. 2015. The star-formation history of the Universe with the SKA. In: *Proc. Conf. on Advancing Astrophysics with the Square Kilometre Array (AASKA'14)*, vol. 68.

Jauncey, D. L. 1967. Re-examination of the source counts for the 3C revised catalogue. *Nature*, **216**, 877–878.

Jeffrey, R. M., Blundell, K. M., Trushkin, S. A., and Mioduszewski, A. J. 2016. Fast launch speeds in radio flares, from a new determination of the intrinsic motions of SS 433's jet bolides. *MNRAS*, **461**, 312–320.

Jennison, R. C. 1958. A phase sensitive interferometer technique for the measurement of the Fourier transforms of spatial brightness distributions of small angular extent. *MNRAS*, **118**, 276.

Johnston, S. 2002. Single dish polarisation calibration. *PASA*, **19**, 277–281.

Johnston, S., and Karastergiou, A. 2017. Pulsar braking and the $P$–$\dot{P}$ diagram. *MNRAS*, **467**, 3493–3499.

Johnston, S., and Kerr, M. 2017. *MNRAS*, **474**, 4029.

Kalberla, P. M. W., and Kerp, J. 2009. The H I distribution of the Milky Way. *Ann. Rev. Astr. Ap.*, **47**, 27–61.

Kantharia, N. G., and six others. 2007. Giant metrewave radio telescope observations of the 2006 outburst of the nova RS Ophiuchi: first detection of emission at radio frequencies. *ApJ*, **667**, L171–L174.

Karako-Argaman, C., and 19 others. 2015. Discovery and follow-up of rotating radio transients with the Green Bank and LOFAR telescopes. *ApJ*, **809**, 67.

Kaspi, V. M., and Beloborodov, A. M. 2017. Magnetars. *Ann. Rev. Astr. Ap.*, **55**, 261–301.

Kassim, N. E., and nine others. 2007. The 74 MHz system on the Very Large Array. *ApJS*, **172**, 686–719.

Kauffmann, J. 2016. Central molecular zone of the Milky Way: star formation in an extreme environment. In: Jablonka, P., André, P., and van der Tak, F. (eds.), *From Interstellar Clouds to Star-Forming Galaxies: Universal Processes?*, pp. 163–166. IAU Symposium, vol. 315.

Keane, E. F., and 41 others. 2016. The host galaxy of a fast radio burst. *Nature*, **530**, 453–456.

Keane, E. F., and Petroff, E. 2015. Fast radio bursts: search sensitivities and completeness. *MNRAS*, **447**, 2852–2856.

Kellermann, K. I., and Owen, F. N. 1988. Radio galaxies and quasars. In: Verschuur, G. L., and Kellerman, K. I. (eds.), *Galactic and Extragalactic Radio Astronomy*, pp. 563–602. Springer.

Kellermann, K. I., and Pauliny-Toth, I. I. K. 1969. The spectra of opaque radio sources. *ApJ*, **155**, L71.

Kellermann, K. I., and Wall, J. V. 1987. Radio source counts and their interpretation. In: Hewitt, A., Burbidge, G., and Fang, L. Z. (eds.), *Observational Cosmology*, pp. 545–562. IAU Symposium, vol. 124.

Kellermann, K. I., Condon, J. J., Kimball, A. E., Perley, R. A., and Ivezić, Ž. 2016. Radio-loud and radio-quiet QSOs. *ApJ*, **831**, 168.

Kim, J.-Y., and eight others. 2018. The limb-brightened jet of M87 down to 7 Schwarzschild radii scale. ArXiv e-prints.

Kitayama, T. 2014. Cosmological and astrophysical implications of the Sunyaev–Zel'dovich effect. *Prog. Theor. Exp. Phys.*, **2014**(6), 06B111.

Komissarov, S. S., and Gubanov, A. G. 1994. Relic radio galaxies: evolution of synchrotron spectrum. *A&A*, **285**, 27–43.

Konovalenko, A., and 71 others. 2016. The modern radio astronomy network in Ukraine: UTR-2, URAN and GURT. *Exp. Astron.*, **42**, 11–48.

Kormendy, J., and Ho, L. C. 2013. Coevolution (or not) of supermassive black holes and host galaxies. *Ann. Rev. Astr. Ap.*, **51**, 511–653.

Kormendy, J., and Norman, C. A. 1979. Observational constraints on driving mechanisms for spiral density waves. *ApJ*, **233**, 539–552.

Korngut, P. M., and eight others. 2011. MUSTANG high angular resolution Sunyaev–Zel'dovich effect imaging of substructure in four galaxy clusters. *ApJ*, **734**, 10.

Kovalev, Y. Y., and 19 others. 2016. RadioAstron observations of the quasar 3C273: a challenge to the brightness temperature limit. *ApJ*, **820**, L9.

Kovetz, E. D., and 47 others. 2017. Line-intensity mapping: 2017 status report. ArXiv e-prints.

Kramer, M. 2017. George Darwin Lecture, 2016. *Astron. & Geophys.*, **58**, 3.31.

Kramer, M., and 14 others. 2006. Tests of general relativity from timing the double pulsar. *Science*, **314**, 97–102.

Kramer, M., and six others. 1999. The characteristics of millisecond pulsar emission. III. From low to high frequencies. *ApJ*, **526**, 957–975.

Kronberg, P. P. 1994. Extragalactic magnetic fields. *Rep. Progr. Phys.*, **57**, 325–382.

Kulkarni, S. R., and Heiles, C. 1988. Neutral hydrogen and the diffuse interstellar medium. In: *Galactic and Extragalactic Radio Astronomy*, pp. 95–153. Springer.

Kuo, C. Y., and eight others. 2015. The Megamaser Cosmology Project. VI. Observations of NGC 6323. *ApJ*, **800**, 26.

Ladd, E. F., Deane, J. R., Sanders, D. B., and Wynn-Williams, C. G. 1993. Luminous radio-quiet sources in W3(main). *ApJ*, **419**, 186.

Lamarre, J.-M., and 94 others. 2010. Planck pre-launch status: the HFI instrument, from specification to actual performance. *A&A*, **520**, A9.

Landecker, T. L., and eight others. 2006 (June). The Canadian Galactic Plane Survey: Arcminute imaging of polarization structure at 1.4 GHz. In: *American Astronomical Society Meeting Abstracts*, no. 208 p. 125. Bulletin of the American Astronomical Society, vol. 38.

Landon, J., and eight others. 2010. Phased array feed calibration, beamforming, and imaging. *AJ*, **139**, 1154–1167.

Lattimer, J. M., and Prakash, M. 2001. Neutron star structure and the equation of state. *ApJ*, **550**, 426–442.

Law, C. J., Yusef-Zadeh, F., and Cotton, W. D. 2008. A wide-area VLA continuum survey near the Galactic Center at 6 and 20 cm wavelengths. *ApJS*, **177**, 515–545.

Leahy, J. P., and Perley, R. A. 1995. The jets and hotspots of 3C 390.3. *MNRAS*, **277**, 1097–1114.

Leipski, C., Falcke, H., Bennert, N., and Hüttemeister, S. 2006. The radio structure of radio-quiet quasars. *A&A*, **455**, 161–172.

Levy, G. S., and nine others. 1986. Very long baseline interferometric observations made with an orbiting radio telescope. *Science*, **234**, 187–189.

Lewin, W., and van der Klis, M. 2010. *Compact Stellar X-ray Sources*. Cambridge University Press.

Liao, S., Qi, Z., Bucciarelli, B., Guo, S., Cao, Z., and Tang, Z. 2018. The properties of the quasars astrometric solution in Gaia DR2. ArXiv e-prints.

Linfield, R. P., and 14 others. 1989. VLBI using a telescope in Earth orbit. II – brightness temperatures exceeding the inverse Compton limit. *ApJ*, **336**, 1105–1112.

Lister, M. L., and seven others. 2018. MOJAVE XV. VLBA 15 GHz total intensity and polarization maps of 437 parsec-scale AGN jets from 1996 to 2017. *ApJS*, **234**, 12.

Lockman, F. J. 1989. A survey of radio H II regions in the northern sky. *ApJS*, **71**, 469–479.

Longair, M. S. 1966. On the interpretation of radio source counts. *MNRAS*, **133**, 421.

Longair, M. S. 1994. *High Energy Astrophysics*, 2nd edn. Cambridge University Press.

Longair, M. S. 2006. *The Cosmic Century*. Cambridge University Press.

Longmore, S. N., Burton, M. G., Barnes, P. J., Wong, T., Purcell, C. R., and Ott, J. 2007. Multiwavelength observations of southern hot molecular cores traced by methanol masers – I. Ammonia and 24-GHz continuum data. *MNRAS*, **379**, 535–572.

Lonsdale, C. J. 2005. Configuration considerations for low frequency arrays. In: Kassim, N., Perez, M., Junor, W., and Henning, P. (eds.), *From Clark Lake to the Long Wavelength Array: Bill Erickson's Radio Science*, p. 399. Astronomical Society of the Pacific Conference Series, vol. 345.

Lonsdale, C. J., Diamond, P. J., Thrall, H., and Smith, H. E. 2006. VLBI images of 49 radio supernovae in Arp 220. *ApJ*, **647**, 185–193.

Lorimer, D. R. 2008. Binary and millisecond pulsars. *Living Rev. Relativity*, **11**.

Lorimer, D. R., and Kramer, M. 2005. *Handbook of Pulsar Astronomy*. Cambridge University Press.

Lorimer, D. R., Bailes, M., McLaughlin, M. A., Narkevic, D. J., and Crawford, F. 2007. A bright millisecond radio burst of extragalactic origin. *Science*, **318**, 777.

Lowe, S. R., and seven others. 2007. 30 GHz flux density measurements of the Caltech–Jodrell flat-spectrum sources with OCRA-p. *A&A*, **474**, 1093–1100.

Lynden-Bell, D. 1969. Galactic nuclei as collapsed old quasars. *Nature*, **223**, 690–694.

Lynden-Bell, D., and Rees, M. J. 1971. On quasars, dust and the galactic centre. *MNRAS*, **152**, 461.

Lyne, A. G., and Manchester, R. N. 1988. The shape of pulsar radio beams. *MNRAS*, **234**, 477–508.

Lyne, A. G., and McKenna, J. 1989. PSR 1820-11 – a binary pulsar in a wide and highly eccentric orbit. *Nature*, **340**, 367–369.

Lyne, A. G., Pritchard, R. S., Graham-Smith, F., and Camilo, F. 1996. Very low braking index for the Vela pulsar. *Nature*, **381**, 497–498.

Lyne, A. G., and 11 others. 2004. A double-pulsar system: a rare laboratory for relativistic gravity and plasma physics. *Science*, **303**, 1153–1157.

Lyne, A. G., Hobbs, G., Kramer, M., Stairs, I., and Stappers, B. 2010. Switched magnetospheric regulation of pulsar spin-down. *Science*, **329**, 408.

Lyne, A. G., Jordan, C. A., Graham-Smith, F., Espinoza, C. M., Stappers, B. W., and Weltevrede, P. 2015a. 45 years of rotation of the Crab pulsar. *MNRAS*, **446**, 857–864.

Lyne, A. G., Stappers, B. W., Keith, M. J., Ray, P. S., Kerr, M., Camilo, F., and Johnson, T. J. 2015b. The binary nature of PSR J2032+4127. *MNRAS*, **451**, 581–587.

Lyne, A. G., and 40 others. 2017. Two long-term intermittent pulsars discovered in the PALFA survey. *ApJ*, **834**, 72.

Lyons, R. G. 2011. *Understanding Digital Signal Processing*, 3rd edn. Prentice Hall.

Madau, P., and Dickinson, M. 2014. Cosmic star-formation history. *Ann. Rev. Astr. Ap.*, **52**, 415–486.

Mäkinen, K., Lehto, H. J., Vainio, R., and Johnson, D. R. H. 2004. Proper motion analysis of the jet of R Aquarii. *A&A*, **424**, 157–164.

Manchester, R. N., and eight others. 1996. The Parkes Southern Pulsar Survey. I. Observing and data analysis systems and initial results. *MNRAS*, **279**, 1235–1250.

Mancuso, C., Lapi, A., Prandoni, I., Obi, I., Gonzalez-Nuevo, J., Perrotta, F., Bressan, A., Celotti, A., and Danese, L. 2017. Galaxy evolution in the radio band: the role of star-forming galaxies and active galactic nuclei. *ApJ*, **842**, 95.

Mann, G., Jansen, F., MacDowall, R. J., Kaiser, M. L., and Stone, R. G. 1999. A heliospheric density model and type III radio bursts. *A&A*, **348**, 614–620.

Marcaide, J. M., and 17 others. 2009. A decade of SN 1993J: discovery of radio wavelength effects in the expansion rate. *A&A*, **505**, 927–945.

Marscher, A. P. 2006. Probing the compact jets of blazars with light curves, images, and polarization. In: Miller, H. R., Marshall, K., Webb, J. R., and Aller, M. F. (eds.), *Blazar Variability Workshop II: Entering the GLAST Era*, p. 155. Astronomical Society of the Pacific Conference Series, vol. 350.

Marti, J., Paredes, J. M., and Estalella, R. 1992. Modelling Cygnus X-3 radio outbursts – particle injection into twin jets. *A&A*, **258**, 309–315.

Marti, J., Rodriguez, L. F., and Reipurth, B. 1995. Large proper motions and ejection of new condensations in the HH 80-81 thermal radio jet. *ApJ*, **449**, 184.

Marven, C., and Ewers, G. 1996 *A Simple Approach to Digital Signal Processing*. Wiley.

Massi, M., and six others. 1997. Baseline errors in European VLBI network measurements. III. The dominant effect of instrumental polarization. *A&A*, **318**, L32–L34.

Mather, J. C., and 22 others. 1994. Measurement of the cosmic microwave background spectrum by the COBE FIRAS instrument. *ApJ*, **420**, 439–444.

Matsushita, S., and 11 others. 2017. ALMA long baseline campaigns: phase characteristics of atmosphere at long baselines in the millimeter and submillimeter wavelengths. *PASP*, **129**, 035004.

Max-Moerbeck, W., Richards, J. L., Hovatta, T., Pavlidou, V., Pearson, T. J., and Readhead, A. C. S. 2014. A method for the estimation of the significance of cross-correlations in unevenly sampled red-noise time series. *MNRAS*, **445**, 437–459.

May, T., Zakosarenko, V., Kraysa, E., Esch, W., Solveig, A., Gremuend, H.-P., and Heinz, E. 2012. *Rev. Sci. Instruments*, **83**, 114502.

Mazets, E. P., and eight others. 1979. Venera 11 and 12 observations of gamma-ray bursts – the Cone experiment. *Soviet Astron. Letters*, **5**, 163–167.

McCready, L. L., Pawsey, J. L., and Payne-Scott, R. 1947. Solar radiation at radio frequencies and its relation to sunspots. *Proc. Roy. Soc. London Series A*, **190**, 357–375.

McKee, J. W., and 19 others. 2016. A glitch in the millisecond pulsar J0613-0200. *MNRAS*, **461**, 2809–2817.

McLaughlin, M., and 12 others. 2013. Transient radio neutron stars. ATNF proposal.

McLean, D. J., and Labrum, N. R. (eds.) 1985. *Solar Radiophysics: Studies of Emission from the Sun at Metre Wavelengths*. Cambridge University Press.

McLean, D. J., and Sheridan, K. V. 1985. The quiet sun at metre wavelengths. In: *Solar Radiophysics: Studies of Emission from the Sun at Metre Wavelengths*, pp. 443–466. Cambridge University Press.

Melia, F., and Falcke, H. 2001. The supermassive black hole at the Galactic Center. *Ann. Rev. Astr. Ap.*, **39**, 309–352.

Mellier, Y. 1999. Probing the universe with weak lensing. *Ann. Rev. Astr. Ap.*, **37**, 127–189.

Melrose, D. B. 2017. Coherent emission mechanisms in astrophysical plasmas. ArXiv e-prints.

Menn, W., and 63 others. 2013. The PAMELA space experiment. *Adv. Space Res.*, **51**, 209–218.

Mennella, A., and 85 others. 2010. Planck pre-launch status: low frequency instrument calibration and expected scientific performance. *A&A*, **520**, A5.

Mennella, A., and 160 others. 2011. Planck early results. III. First assessment of the low frequency instrument in-flight performance. *A&A*, **536**, A3.

Menten, K. M., and Young, K. 1995. Discovery of strong vibrationally excited water masers at 658 GHz toward evolved stars. *ApJ*, **450**, L67.

Mevius, M., and 26 others. 2016. Probing ionospheric structures using the LOFAR radio telescope. *Radio Sci.*, **51**, 927–941.

Mezger, P. G., and Henderson, A. P. 1967. Galactic H II regions. I. Observations of their continuum radiation at the frequency 5 GHz. *ApJ*, **147**, 471.

Michilli, D., and 19 others. 2018. An extreme magneto-ionic environment associated with the fast radio burst source FRB 121102. *Nature*, **553**, 182–185.

Miley, G. 1980. The structure of extended extragalactic radio sources. *Ann. Rev. Astr. Ap.*, **18**, 165–218.

Miller-Jones, J. C. A., Blundell, K. M., Rupen, M. P., Mioduszewski, A. J., Duffy, P., and Beasley, A. J. 2004. Time-sequenced multi-radio frequency observations of Cygnus X-3 in flare. *ApJ*, **600**, 368–389.

Minier, V., Booth, R. S., and Conway, J. E. 1999. Observations of methanol masers in star-forming regions. *New Astron. Rev.*, **43**, 569–573.

Mirabel, I. F., and Rodríguez, L. F. 1994. A superluminal source in the Galaxy. *Nature*, **371**, 46–48.

Miyoshi, M., Moran, J., Herrnstein, J., Greenhill, L., Nakai, N., Diamond, P., and Inoue, M. 1995. Evidence for a black hole from high rotation velocities in a sub-parsec region of NGC4258. *Nature*, **373**, 127–129.

Moffett, D. A., and Hankins, T. H. 1996. Multifrequency radio observations of the Crab Pulsar. *ApJ*, **468**, 779.

Moran, J. M. 1989. Introduction to VLBI. In: Felli, M., and Spencer, R. E. (eds.), pp. 27–45. NATO Advanced Science Institutes (ASI) Series C, vol. 283.

Morganti, R. 2017. The many routes to AGN feedback. ArXiv e-prints.

Morris, M., and Serabyn, E. 1996. The Galactic Center environment. *Ann. Rev. Astr. Ap.*, **34**, 645–702.

Moskalenko, I. V., and Strong, A. W. 1998. Production and propagation of cosmic-ray positrons and electrons. *ApJ*, **493**, 694–707.

Müller, H. S. P., Schlöder, F., Stutzki, J., and Winnewisser, G. 2005. The Cologne Database for Molecular Spectroscopy, CDMS: a useful tool for astronomers and spectroscopists. *J. Molecular Structure*, **742**, 215–227.

Muxlow, T. W. B., Pedlar, A., Wilkinson, P. N., Axon, D. J., Sanders, E. M., and de Bruyn, A. G. 1994. The Structure of Young Supernova Remnants in M82. *MNRAS*, **266**, 455.

Muxlow, T. W. B., and ten others. 2005. High-resolution studies of radio sources in the Hubble Deep and Flanking Fields. *MNRAS*, **358**, 1159–1194.

Muxlow, T. W. B., Beswick, R. J., Richards, A. M. S., and Thrall, H. J. 2006. Starburst galaxies. In: *Proc. 8th European VLBI Network Symp.*, p. 31.

Myers, S. T., and ten others. 2003. A fast gridded method for the estimation of the power spectrum of the cosmic microwave background from interferometer data with application to the cosmic background imager. *ApJ*, **591**, 575–598.

Nagar, N. M., Wilson, A. S., Mulchaey, I. S., and Gallimore, J. F. 1999. Radio structures of Seyfert galaxies. *ApJS*, **120**, 209.

Neininger, N. 1992. The magnetic field structure of M 51. *A&A*, **263**, 30–36.

Netterfield, C. B., Jarosik, N., Page, L., Wilkinson, D., and Wollack, E. 1995. The anisotropy in the cosmic microwave background at degree angular scales. *ApJ*, **445**, L69–L72.

Newburgh, L. B., and 34 others. 2014. Calibrating CHIME: a new radio interferometer to probe dark energy. In: *Ground-Based and Airborne Telescopes V*, p. 91454V. Proceedings of SPIE, vol. 9145.

Nikolic, B., Bolton, R. C., Graves, S. F., Hills, R. E., and Richer, J. S. 2013. Phase correction for ALMA with 183 GHz water vapour radiometers. *A&A*, **552**, A104.

Norris, R. P. 2017a. Discovering the unexpected in astronomical survey data. *PASA*, **34**, e007.

Norris, R. P. 2017b. Extragalactic radio continuum surveys and the transformation of radio astronomy. *Nature Astron.*, **1**, 671–678.

O'Brien, T. J., and eight others. 2006. An asymmetric shock wave in the 2006 outburst of the recurrent nova RS Ophiuchi. *Nature*, **442**, 279–281.

O'Dea, C. P. 1998. The compact steep-spectrum and gigahertz peaked-spectrum radio sources. *PASP*, **110**, 493–532.

Offringa, A. R., and 52 others. 2014. WSCLEAN: an implementation of a fast, generic wide-field imager for radio astronomy. *MNRAS*, **444**, 606–619.

O'Gorman, E., and six others. 2015. Temporal evolution of the size and temperature of Betelgeuse's extended atmosphere. *A&A*, **580**, A101.

O'Gorman, E., and six others. 2017. The inhomogeneous submillimeter atmosphere of Betelgeuse. *A&A*, **602**, L10.

Ohm, E. A. 1961. *Bell Syst. Techn. J.*, **40**, 1065.

Olausen, S. A., and Kaspi, V. M. 2014. The McGill Magnetar Catalog. *ApJS*, **212**, 6.

Oort, J. H., Kerr, F. J., and Westerhout, G. 1958. The galactic system as a spiral nebula (Council Note). *MNRAS*, **118**, 379.

Oppenheimer, A. V., and Lim, J. S. 1981. *Proc. IEEE*, **69**, 529.

Oswianik, I., Conway, J. E., and Polatidis, A. G. 1999. The youngest lobe-dominated radio sources. *New Astron. Rev.*, **43**, 669–673.

Owen, F. N., O'Dea, C. P., Inoue, M., and Eilek, J. A. 1985. VLA observations of the multiple jet galaxy 3C 75. *ApJ*, **294**, L85–L88.

Ozel, F., and Freire, P. 2016. Masses, radii, and the equation of state of neutron stars. *Ann. Rev. Astr. Ap.*, **54**, 401–440.

Padovani, P. 2016. The faint radio sky; radio astronomy becomes mainstream. *Astron. Astrophys. Rev.*, **24**, 13.

Padovani, P., and ten others. 2017. Active galactic nuclei: what's in a name? *Astron. Astrophys. Rev.*, **25**, 2.

Paine, S. 2017. https://doi.org/10.5281/zenodo.438726.

Pardo, J. R., Cernicharo, J., and Serabyn, E. 2001. Atmospheric transmission at microwaves (ATM): an improved model for millimeter/submillimeter applications. *IEE Trans. Antennas and Propagation,* **49**, 1983.

Paresce, F. 1984. On the distribution of interstellar matter around the sun. *AJ*, **89**, 1022–1037.

Partridge, R. B. 1995. 3 K: *The Cosmic Microwave Background Radiation*. Cambridge University Press.

Patruno, A., and nine others. 2014. A new accretion disk around the missing link binary system PSR J1023+0038. *ApJ*, **781**, L3.

Pavelin, P. E., Davis, R. J., Morrison, L. V., Bode, M. F., and Ivison, R. J. 1993. Radio observations of the classical nova Cygni 92 eighty days after outburst. *Nature*, **363**, 424–426.

Pearson, T. J., and Readhead, A. C. S. 1984. Image formation by self-calibration in radio astronomy. *Ann. Rev. Astr. Ap.*, **22**, 97–130.

Pearson, T. J., and seven others. 1981. Superluminal expansion of quasar 3C273. *Nature*, **290**, 365–368.

Pedlar, A., and six others. 1999. VLBI observations of supernova remnants in Messier 82. *MNRAS*, **307**, 761–768.

Peel, M. W., Dickinson, C., Davies, R. D., Clements, D. L., and Beswick, R. J. 2011. Radio to infrared spectra of late-type galaxies with Planck and Wilkinson Microwave Anisotropy Probe data. *MNRAS*, **416**, L99–L103.

Penzias, A. A., and Wilson, R. W. 1965. A measurement of excess antenna temperature at 4080 Mc/s. *ApJ*, **142**, 419–421.

Perley, R. A. 1999. Imaging with non-coplanar arrays. In: Taylor, G. B., Carilli, C. L., and Perley, R. A. (eds.), *Synthesis Imaging in Radio Astronomy II*, p. 383. Astronomical Society of the Pacific Conference Series, vol. 180.

Perley, R. A., and Butler, B. J. 2013. An accurate flux density scale from 1 to 50 GHz. *ApJS*, **204**, 19.

Perley, R. A., and Meisenheimer, K. 2017. High-fidelity VLA imaging of the radio structure of 3C 273. *A&A*, **601**, A35.

Perley, R. A., Willis, A. G., and Scott, J. S. 1979. The structure of the radio jets in 3C449. *Nature*, **281**, 437–442.

Perley, R. A., Dreher, J. W., and Cowan, J. J. 1984. The jet and filaments in Cygnus A. *ApJ*, **285**, L35–L38.

Peterson, W. M., Mutel, R. L., Lestrade, J.-F., Güdel, M., and Goss, W. M. 2011. Radio astrometry of the triple systems Algol and UX Arietis. *ApJ*, **737**, 104.

Phillips, N., Hills, R., Bastian, T., Hudson, H., Marson, R., and Wedemeyer, S. 2015. Fast single-dish scans of the Sun using ALMA. In: Iono, D., Tatematsu, K., Wootten, A., and Testi, L. (eds.), *Revolution in Astronomy with ALMA: The Third Year*, p. 347. Astronomical Society of the Pacific Conference Series, vol. 499.

Phillips, R. B., Straughn, A. H., Doeleman, S. S., and Lonsdale, C. J. 2003. R Cassiopeiae: relative strengths of SiO masers at 43 and 86 GHz. *ApJ*, **588**, L105–L108.

Planck Collaboration, Adam, R., and 369 others. 2016a. Planck 2015 results. I. Overview of products and scientific results. *A&A*, **594**, A1.

Planck Collaboration. 2016b. Planck 2015 results. X. Diffuse component separation: foreground maps. *A&A*, **594**, A10.

Planck Collaboration. 2016c. Cosmological parameters. *A&A*, **594**, A13.

Polatidis, A., and six others. 1999. Compact symmetric objects in a complete flux density limited sample. *New Astron. Rev.*, **43**, 657–661.

Popov, M. V., and 15 others. 2017. PSR B0329+54: substructure in the scatter-broadened image discovered with RadioAstron on baselines up to 330 000 km. *MNRAS*, **465**, 978–985.

Pospieszalski, M. W. 1989. Modeling of noise parameters of MESFETs and MODFETs and their frequency and temperature dependence. *IEEE Trans. Microwave Theory Techniques*, **37**, 1340–1350.

Punsly, B., and Rodriguez, J. 2016. A temporal analysis indicates a mildly relativistic compact jet in GRS 1915+105. *ApJ*, **823**, 54.

Quireza, C., Rood, R. T., Balser, D. S., and Bania, T. M. 2006. Radio recombination lines in Galactic H II regions. *ApJS*, **165**, 338–359.

Radhakrishnan, V., and Cooke, D. J. 1969. Magnetic poles and the polarization structure of pulsar radiation. *ApJ*, **3**, 225.

Rand, R. J., and Kulkarni, S. R. 1990. M51 – molecular spiral arms, giant molecular associations, and superclouds. *ApJ*, **349**, L43–L46.

Rand, R. J., and Lyne, A. G. 1994. New rotation measures of distant pulsars in the inner Galaxy and magnetic field reversals. *MNRAS*, **268**, 497.

Rau, U., and Cornwell, T. J. 2011. A multi-scale multi-frequency deconvolution algorithm for synthesis imaging in radio interferometry. *A&A*, **532**, A71.

Raymond, J. C. 1984. Observations of supernova remnants. *Ann. Rev. Astr. Ap.*, **22**, 75.

Readhead, A. C. S., and Wilkinson, P. N. 1978. The mapping of compact radio sources from VLBI data. *ApJ*, **223**, 25–36.

Readhead, A. C. S., Walker, R. C., Pearson, T. J., and Cohen, M. H. 1980. Mapping radio sources with uncalibrated visibility data. *Nature*, **285**, 137–140.

Readhead, A. C. S., Lawrence, C. R., Myers, S. T., Sargent, W. L. W., Hardebeck, H. E., and Moffet, A. T. 1989. A limit of the anisotropy of the microwave background radiation on arc minute scales. *ApJ*, **346**, 566–587.

Readhead, A. C. S., Taylor, G. B., Pearson, T. J., and Wilkinson, P. N. 1996. Compact symmetric objects and the evolution of powerful extragalactic radio sources. *ApJ*, **460**, 634.

Reber, G. 1944. Cosmic static. *ApJ*, **100**, 279.

Rees, M. J. 1966. Appearance of relativistically expanding radio sources. *Nature*, **211**, 468–470.

Reich, P., and Reich, W. 1988. A map of spectral indices of the Galactic radio continuum emission between 408 MHz and 1420 MHz for the entire northern sky. *A&AS*, **74**, 7–23.

Reid, H. A. S., and Ratcliffe, H. 2014. A review of solar type III radio bursts. *Res. Astron. Astrophys.*, **14**, 773–804.

Reiner, M. J., Jackson, B. V., Webb, D. F., Kaiser, M. L., Cliver, E. W., and Bougeret, J. L. 2004. Wind/WAVES and SMEI observations of ICMEs. In: *AGU Fall Meeting Abstracts*.

Remazeilles, M., Dickinson, C., Banday, A. J., Bigot-Sazy, M.-A., and Ghosh, T. 2015. An improved source-subtracted and destriped 408-MHz all-sky map. *MNRAS*, **451**, 4311–4327.

Reynolds, S. P., and Chevalier, R. A. 1984. A new type of extended nonthermal radio emitter – detection of the old nova GK Persei. *ApJ*, **281**, L33–L35.

Reynolds, S. P., and Gilmore, D. M. 1986. Radio observations of the remnant of the supernova of A.D. 1006. I – Total intensity observations. *AJ*, **92**, 1138–1144.

Richards, E. A., 2000, The nature of radio emission from distant galaxies: the 1.4 GHZ observations. *Astrophys. J.*, 533, 611.

Richards, E. A., Kellermann, K. I., Fomalont, E. B., Windhorst, R.A., and Partridge, R. B, 1998. Radio emission from galaxies in the Hubble Deep Field. *Astron. J.*, 116, 1039.

Richards, A. M. S., and ten others. 2013. e-MERLIN resolves Betelgeuse at $\lambda$ 5 cm: hotspots at 5 $R_*$. *MNRAS*, **432**, L61–L65.

Richards, J. L., Hovatta, T., Max-Moerbeck, W., Pavlidou, V., Pearson, T. J., and Readhead, A. C. S. 2014. Connecting radio variability to the characteristics of gamma-ray blazars. *MNRAS*, **438**, 3058–3069.

Rickett, B. J. 1990. Radio propagation through the turbulent interstellar plasma. *Ann. Rev. Astr. Ap.*, **28**, 561–605.

Roger, R. S., Costain, C. H., Landecker, T. L., and Swerdlyk, C. M. 1999. The radio emission from the Galaxy at 22 MHz. *A&AS*, **137**, 7–19.

Rogers, A. E. E. 1970. Very long baseline interferometry with large effective bandwidth for phase-delay measurements. *Radio Sci.*, **5**, 1239–1247.

Rogers, A. E. E. 1983. *VLB Array Memo.*, **253**.

Rogers, A. E. E., and nine others. 1974. The structure of radio sources 3C 273B and 3C 84 deduced from the 'closure' phases and visibility amplitudes observed with three-element interferometers. *ApJ*, **193**, 293–301.

Rogers, A. E. E., Dudevoir, K. A., and Bania, T. M. 2007. Observations of the 327 MHz deuterium hyperfine transition. *AJ*, **133**, 1625–1632.

Romero, G. E., Boettcher, M., Markoff, S., and Tavecchio, F. 2017. Relativistic jets in active galactic nuclei and microquasars. *Space Sci. Rev.*, **207**, 5–61.

Romney, J. D. 1999. Cross correlators. In: Taylor, G. B., Carilli, C. L., and Perley, R. A. (eds.), *Synthesis Imaging in Radio Astronomy II*, p. 57. Astronomical Society of the Pacific Conference Series, vol. 180.

Rood, R. T., Bania, T. M., and Wilson, T. L. 1984. The 8.7 GHz hyperfine line of He-3(+) in galactic H II regions. *ApJ*, **280**, 629–647.

Rookyard, S. C., Weltevrede, P., Johnston, S., and Kerr, M. 2017. On the difference between γ-ray-detected and non-γ-ray-detected pulsars. *MNRAS*, **464**, 2018–2026.

Rowan-Robinson, M. 1968. The determination of the evolutionary properties of quasars by means of the luminosity–volume test. *MNRAS*, **138**, 445.

Rowson, B. 1963. High resolution observations with a tracking radio interferometer. *MNRAS*, **125**, 177.

Rubin, V. C., Burstein, D., Ford, W. K., Jr., and Thonnard, N. 1985. Rotation velocities of 16 SA galaxies and a comparison of Sa, Sb, and Sc rotation properties. *ApJ*, **289**, 81–98.

Rumsey, V. H. 1966. *Frequency Independent Antennas*. Academic Press.

Ruze, J. 1966. Antenna tolerance theory – a review. *Proc. IEEE*, **54**, 633.

Ryden, B. 2017. *Introduction to Cosmology*, 2nd edn. Cambridge University Press.

Ryle, M. 1962. The new Cambridge radio telescope. *Nature*, **194**, 517–518.

Ryle, M. 1972. The 5-km radio telescope at Cambridge. *Nature*, **239**, 435–438.

Ryle, M., and Neville, A. C. 1962. A radio survey of the North Polar region with a 4.5 minute of arc pencil-beam system. *MNRAS*, **125**, 39.

Ryle, M., Smith, F. G., and Elsmore, B. 1950. A preliminary survey of the radio stars in the Northern Hemisphere. *MNRAS*, **110**, 508.

Sanna, A., Reid, M. J., Dame, T. M., Menten, K. M., and Brunthaler, A. 2017. Mapping spiral structure on the far side of the Milky Way. *Science*, **358**, 227–230.

Santos-Costa, D., Bolton, S. J., and Sault, R. J. 2009. Evidence for short-term variability of Jupiter's decimetric emission from VLA observations. *A&A*, **508**, 1001–1010.

Saripalli, L., Subrahmanyan, R., and Udaya Shankar, N. 2002. A case for renewed activity in the giant radio galaxy J0116-473. *ApJ*, **565**, 256–264.

Saripalli, L., Subrahmanyan, R., and Udaya Shankar, N. 2003. Renewed activity in the radio galaxy PKS B1545-321: twin edge-brightened beams within diffuse radio lobes. *ApJ*, **590**, 181–191.

Sault, R. J., and Wieringa, M. H. 1994. Multi-frequency synthesis techniques in radio interferometric imaging. *A&AS*, **108**, 585–594.

Scheuer, P. A. G. 1957. A statistical method for analysing observations of faint radio stars. *Proc. Cambridge Phil. Soc.*, **53**, 764–773.

Scheuer, P. A. G. 1968. Amplitude variations in pulsed radio sources. *Nature*, **218**, 920–922.

Scheuer, P. A. G. 1974. Models of extragalactic radio sources with a continuous energy supply from a central object. *MNRAS*, **166**, 513–528.

Schilizzi, R. T., Burke, B. F., Jordan, J. F., and Hawkyard, A. 1984 (Sept.). The QUASAT mission: an overview. In: Burke, W. R. (ed.), *QUASAT: A VLBI Observatory in Space*. ESA Special Publications, vol. 213.

Schmidt, M. 1968. Space distribution and luminosity functions of quasi-stellar radio sources. *ApJ*, **151**, 393.

Schneider, P., Ehlers, J., and Falco, E. E. 1992. *Gravitational Lenses*. Springer.

Schödel, R., and 22 others. 2002. A star in a 15.2-year orbit around the supermassive black hole at the centre of the Milky Way. *Nature*, **419**, 694–696.

Scholz, P., and 33 others. 2015. Timing of five millisecond pulsars discovered in the PALFA survey. *ApJ*, **800**, 123.

Schuk, H., and Behrend, D. 2012. VLBI: a fascinating technique for geodesy and astronomy. *J. Geophys.*, **61**, 68.

Schuk, H., and Bohm, J. 2013. Very long baseline interferometry for geodesy and astronomy. In: Xu, G. (ed.), *Sciences of Geodesy, volume II*, p. 339. Springer.

Schwab, F. R. 1980 (Jan.). Adaptive calibration of radio interferometer data. In: Rhodes, W. T. (ed.), *Proc. 1980 Int. Optical Computing Conf. I*, pp. 18–25. *Proc. SPIE*, vol. 231.

Schwab, F. R. 1984. Relaxing the isoplanatism assumption in self-calibration; applications to low-frequency radio interferometry. *AJ*, **89**, 1076–1081.

Schwab, F. R., and Cotton, W. D. 1983. Global fringe search techniques for VLBI. *AJ*, **88**, 688–694.

Schwarz, U. J. 1978. Mathematical–statistical description of the iterative beam removing technique (method CLEAN). *A&A*, **65**, 345.

Seaquist, E. R. 1989. Radio emission from novae. In: Bode, M. F., and Evans, A. (eds.), *Classical Novae*, pp. 143–161. Cambridge University Press.

Seaquist, E. R., Bode, M. F., Frail, D. A., Roberts, J. A., Evans, A., and Albinson, J. S. 1989. A detailed study of the remnant of nova GK Persei and its environs. *ApJ*, **344**, 805–825.

Shemar, S. L., and Lyne, A. G. 1996. Observations of pulsar glitches. *MNRAS*, **282**, 677–690.

Shklovskii, I. S. 1960. *Cosmic Radio Waves*. Harvard University Press.

Simpson, C. 2017. Extragalactic radio surveys in the pre-Square Kilometre Array era. *Royal Soc. Open Sci.*, **4**, 170522.

Sironi, G., and six others. 1990. The absolute temperature of the sky and the temperature of the cosmic background radiation at 600 MHz. *ApJ*, **357**, 301–308.

Smirnov, O. M. 2011a. Revisiting the radio interferometer measurement equation. I. A full-sky Jones formalism. *A&A*, **527**, A106.

Smirnov, O. M. 2011b. Revisiting the radio interferometer measurement equation. II. Calibration and direction-dependent effects. *A&A*, **527**, A107.

Smith, D. A., Guillemot, L., Kerr, M., Ng, C., and Barr, E. 2017. Gamma-ray pulsars with Fermi. ArXiv e-prints.

Smith, E. K. 1982. *Radio Sci.*, **17**, 455.

Smith, E. K., and Weintraub, S. 1953. The constants in the equation of atmospheric refractive index at radio frequencies. *Proc. IRE*, **41**, 1035.

Smith, F. G. 1952. The determination of the position of a radio star. *MNRAS*, **112**, 497.

Smoot, G. F., Gorenstein, M. V., and Muller, R. A. 1977. Detection of anisotropy in the cosmic blackbody radiation. *Phys. Rev. Lett.*, **39**, 898–901.

Snellen, I., and Schilizzi, R. 2002. On the lives of extra-galactic radio sources: the first 100 000 years. *New Astron. Rev.*, **46**, 61–65.

Sokoloff, D. D., Bykov, A. A., Shukurov, A., Berkhuijsen, E. M., Beck, R., and Poezd, A. D. 1998. Depolarization and Faraday effects in galaxies. *MNRAS*, **299**, 189–206.

Sovers, O. J., Fanselow, J. L., and Jacobs, C. S. 1998. Astrometry and geodesy with radio interferometry: experiments, models, results. *Rev. Mod. Phys.*, **70**, 1393–1454.

Sparke, L. S., and Gallagher, J. S., III. 2007. *Galaxies in the Universe: An Introduction*, 2nd edn., Chapter 9. Cambridge University Press.

Spencer, R. E. 1996. Energetics of radio emitting X-ray binary stars. In: Taylor, A. R., and Paredes, J. M. (eds.), *Radio Emission from the Stars and the Sun*, p. 252. Astronomical Society of the Pacific Conference Series, vol. 93.

Spencer, R. E., Vermeulen, R. C., and Schilizzi, R. T. 1993. VLBI and MERLIN Observations of the moving knots in SS 433. In: Errico, L., and Vittone, A. A. (eds.), *Stellar Jets and Bipolar Outflows*, p. 203. Astrophysics and Space Science Library, vol. 186.

Spitler, L. G., and 23 others. 2016. A repeating fast radio burst. *Nature*, **531**, 202–205.

Staveley-Smith, L., and ten others. 1996. The Parkes 21 cm multibeam receiver. *PASA*, **13**, 243–248.

Sturrock, P. A. 1971. A model of pulsars. *ApJ*, **164**, 529.

Sullivan, W. T. 2009. *Cosmic Noise: A History of Early Radio Astronomy*. Cambridge University Press.

Sunyaev, R. A., and Zel'dovich, I. B. 1980. Microwave background radiation as a probe of the contemporary structure and history of the universe. *Ann. Rev. Astr. Ap.*, **18**, 537–560.

Tadhunter, C. 2016. Radio AGN in the local universe: unification, triggering and evolution. *Astron. Astrophys. Rev.*, **24**, 10.

Taylor, G. B., and six others. 1994. The second Caltech–Jodrell Bank VLBI survey. 1: Observations of 91 of 193 sources. *ApJS*, **95**, 345–369.

Taylor, J. H., and Cordes, J. M. 1993. Pulsar distances and the galactic distribution of free electrons. *ApJ*, **411**, 674–684.

Taylor, J. H., and Huguenin, G. R. 1971. Observations of rapid fluctuations of intensity and phase in pulsar emissions. *ApJ*, **167**, 273.

Taylor, J. H., and Weisberg, J. M. 1989. Further experimental tests of relativistic gravity using the binary pulsar PSR 1913+16. *ApJ*, **345**, 434–450.

Terzian, Y., and Parrish, A. 1970. Observations of the Orion Nebula at low radio frequencies. *Astrophys. Lett.*, **5**, 261.

Thompson, A. R., and Bracewell, R. N. 1974. Interpolation and Fourier transformation of fringe visibilities. *AJ*, **79**, 11–24.

(TMS) Thompson, A. R., Moran, J. M., and Swenson, G. W. Jr. 2017. *Interferometry and Synthesis in Radio Astronomy*, 3rd edn. Wiley.

Thorne, K, S., and Blandford, D. 2018. *Modern Classical Physics*, Chapter 8. Princeton University Press.

Tillman, R. H., Ellingson, S. W., and Brendler, J. 2016. Practical limits in the sensitivity–linearity trade-off for radio telescope front ends in the HF and VHF-low bands. *J. Astronom. Instrum.*, **5**, 1650004.

Tingay, S. J., and 60 others. 2013. The Murchison Widefield Array: the Square Kilometre Array precursor at low radio frequencies. *PASA*, **30**, e007.

Townes, C. H. 1957. Microwave and radio-frequency resonance lines of interest to radio astronomy. In: van de Hulst, H. C. (ed.), *Radio Astronomy*, p. 92. IAU Symposium, vol. 4.

Townes, C. H., and Schawlow, A. L. 1955. *Microwave Spectroscopy*. Dover.

Tudose, V., and ten others. 2010. Probing the behaviour of the X-ray binary Cygnus X-3 with very long baseline radio interferometry. *MNRAS*, **401**, 890–900.

Turner, E. L., Ostriker, J. P., and Gott, III, J. R. 1984. The statistics of gravitational lenses – the distributions of image angular separations and lens redshifts. *ApJ*, **284**, 1–22.

Twiss, R. Q., Carter, A. W. L., and Little, A. G. 1960. Brightness distribution over some strong radio sources at 1427 Mc/s. *The Observatory*, **80**, 153.

Uchida, K. I., Morris, M. R., Serabyn, E., and Bally, J. 1994. AFGL 5376: a strong, large-scale shock near the Galactic center. *ApJ*, **421**, 505–516.

Urry, C. M., and Padovani, P. 1995. Unified schemes for radio-loud active galactic nuclei. *PASP*, **107**, 803.

Uyaniker, B., Fürst, E., Reich, W., Reich, P., and Wielebinski, R. 1999. A 1.4 GHz radio continuum and polarization survey at medium Galactic latitudes. II. First section. *A&AS*, **138**, 31–45.

van der Hulst, J. M., Punzo, D., and Roerdink, J. B. T. M. 2017 (June). 3-D interactive visualisation tools for H I spectral line imaging. In: *Proc. IAU Symposium*, pp. 305–310. IAU Symposium Series, vol. 325.

van der Tak, F., de Pater, I., Silva, A., and Millan, R. 1999. Time variability in the radio brightness distribution of Saturn. *Icarus*, **142**, 125–147.

van Dishoeck, E. F., Jansen, D. J., and Phillips, T. G. 1993. Submillimeter observations of the shocked molecular gas associated with the supernova remnant IC 443. *A&A*, **279**, 541–566.

van Dyk, S. D., Weiler, K. W., Sramek, R. A., Rupen, M. P., and Panagia, N. 1994. SN 1993J: the early radio emission and evidence for a changing presupernova mass-loss rate. *ApJ*, **432**, L115–L118.

van Haarlem, M. P., and 200 others. 2013. LOFAR: The LOw-Frequency ARray. *A&A*, **556**, A2.

van Straten, W., and Bailes, M. 2010. DSPSR: digital signal processing for pulsar astronomy. arXiv: 1008.393.

van Straten, W., Manchester, R. N., Johnston, S., and Reynolds, J. E. 2010. PSRCHIVE and PSRFITS: definition of the Stokes parameters and instrumental basis conventions. *PASA*, **27**, 104–119.

Vernstrom, T., Scott, D., Wall, J. V., Condon, J. J., Cotton, W. D., and Perley, R. A. 2016a. Deep 3-GHz observations of the Lockman Hole North with the Very Large Array – I. Source extraction and uncertainty analysis. *MNRAS*, **461**, 2879–2895.

Vernstrom, T., Scott, D., Wall, J. V., Condon, J. J., Cotton, W. D., Kellermann, K. I., and Perley, R. A. 2016b. Deep 3-GHz observations of the Lockman Hole North with the Very Large Array – II. Catalogue and μJy source properties. *MNRAS*, **462**, 2934–2949.

Verschuur, G. L. 1989. Measurements of the 21 centimeter Zeeman effect in high-latitude directions. *ApJ*, **339**, 163–170.

Vlemmings, W. H. T., Harvey-Smith, L., and Cohen, R. J. 2006. Methanol maser polarization in W3(OH). *MNRAS*, **371**, L26–L30.

von Hoerner, H. 1967. Design of large steerable antennas. *Astron. J.*, **72**, 35.

Walker, R. C. 1989. Calibration methods. In: Felli, M., and Spencer, R. E. (eds.), *NATO Advanced Science Institutes (ASI) Series C*, pp. 141–162. NATO Advanced Science Institutes (ASI) Series C, vol. 283.

Walker, R. C., Hardee, P. E., Davies, F. B., Ly, C., and Junor, W. 2018. The structure and dynamics of the subparsec jet in M87 based on 50 VLBA observations over 17 years at 43 GHz. *ApJ*, **855**, 128.

Wall, J. V. 1994. Populations of extragalactic radio sources. *Australian J. Phys.*, **47**, 625–655.

Wall, J. V., and Jenkins, C. R. 2012. *Practical Statistics for Astronomers*. Cambridge University Press.

Walsh, C. 2017. Organic molecules in protoplanetary disks: new insights and directions with ALMA. In: *American Astronomical Society Meeting Abstracts*, p. 208.02. American Astronomical Society Meeting Abstracts Series, vol. 230.

Walsh, D., Carswell, R. F., and Weymann, R. J. 1979. 0957+561 A, B – twin quasistellar objects or gravitational lens. *Nature*, **279**, 381–384.

Wang, Y., and Mohanty, S. D. 2017. Pulsar timing array based search for supermassive black hole binaries in the Square Kilometer Array era. *Phys. Rev. Lett.*, **118**, 151104.

Weinberg, D. H. 2017. On the deuterium-to-hydrogen ratio of the interstellar medium. *Ap. J.*, **851**, 25.

Weinreb, S., Barrett, A. H., Meeks, M. L., and Henry, J. C. 1963. Radio observations of OH in the interstellar medium. *Nature*, **200**, 829–831.

Welch, W. J., and six others. 2017. New cooled feeds for the Allen Telescope Array. *PASP*, **129**, 045002.

Wevers, B. M. H. R., van der Kruit, P. C., and Allen, R. J. 1986. The Palomar–Westerbork survey of northern spiral galaxies. *A&AS*, **66**, 505–662.

White, R. L., Becker, R. H., and Gregg, M. D. *et al.* 2000, The FIRST bright quasar survey. II. 60 nights and 1200 spectra later. *Astrophys. J. Suppl.*, 126, 133.

White, S. D. M., Efstathiou, G., and Frenk, C. S. 1993. The amplitude of mass fluctuations in the universe. *MNRAS*, **262**, 1023–1028.

Wielebinski, R., and Beck, R. (eds.). 2005. *Cosmic Magnetic Fields*. Lecture Notes in Physics, vol. 664. Springer.

Wielebinski, R., and Krause, F. 1993. Magnetic fields in galaxies. *Astron. Astrophys. Rev.*, **4**, 449–485.

Wild, J. P., Smerd, S. F., and Weiss, A. A. 1963. Solar bursts. *Ann. Rev. Astr. Ap.*, **1**, 291.

Wilkins, D. R., and Gallo, L. C. 2015. The Comptonization of accretion disk X-ray emission: consequences for X-ray reflection and the geometry of AGN coronae. *MNRAS*, **448**, 703–712.

Wilkinson, P. N. 1989a. An introduction to closure phase and self-calibration. In: Felli, M., and Spencer, R. E. (eds.), *Very Long Baseline Interferometry, Techniques and Applications*, pp. 69–93. NATO Advanced Science Institutes (ASI) Series C, vol. 283.

Wilkinson, P. N. 1989b. An introduction to deconvolution in VLBI. In: Felli, M., and Spencer, R. E. (eds.), *Very Long Baseline Interferometry, Techniques and Applications*, pp. 183–197. NATO Advanced Science Institutes (ASI) Series C, vol. 283.

Wilkinson, P. N. 1991. The hydrogen array. In: Cornwell, T. J., and Perley, R. A. (eds.), *Proc. IAU Colloq. 131: Radio Interferometry. Theory, Techniques, and Applications*, pp. 428–432. Astronomical Society of the Pacific Conference Series, vol. 19.

Wilkinson, P. N., and Woodall, P. 1991. Numerical experiments with low SNR data in radio interferometry. In: Cornwell, T. J., and Perley, R. A. (eds.), *IAU Colloq. 131: Radio Interferometry. Theory, Techniques, and Applications*, pp. 272–275. Astronomical Society of the Pacific Conference Series, vol. 19.

Wilkinson, P. N., Polatidis, A. G., Readhead, A. C. S., Xu, W., and Pearson, T. J. 1994. Two-sided ejection in powerful radio sources: The compact symmetric objects. *ApJ*, **432**, L87–L90.

Wilkinson, P. N., Kellermann, K. I., Ekers, R. D., Cordes, J. M., and W. Lazio, T. J. 2004. The exploration of the unknown. *New Astron. Rev.*, **48**, 1551–1563.

Wilman, R. J., and ten others. 2008. A semi-empirical simulation of the extragalactic radio continuum sky for next generation radio telescopes. *MNRAS*, **388**, 1335–1348.

Wilner, D. J., and Welch, W. J. 1994. The S140 core: aperture synthesis HCO(+) and SO observations. *ApJ*, **427**, 898–913.

Wilson, T. L., Mezger, P. G., Gardner, F. F., and Milne, D. K. 1970. A survey of H 109 α recombination line emission in Galactic H II regions of the southern sky. *A&A*, **6**, 364–384.

Wilson, T. L., Rohlfs, K., and Hüttemeister, S. 2013. *Tools of Radio Astronomy*, 6th edn. Springer.

Wolleben, M., Landecker, T. L., Reich, W., and Wielebinski, R. 2006. An absolutely calibrated survey of polarized emission from the northern sky at 1.4 GHz. Observations and data reduction. *A&A*, **448**, 411–424.

Wright, A. E., and Barlow, M. J. 1975. The radio and infrared spectrum of early-type stars undergoing mass loss. *MNRAS*, **170**, 41–51.

Wynn-Williams, C. G., Becklin, E. E., and Neugebauer, G. 1972. Infra-red sources in the H II region W3. *MNRAS*, **160**, 1–14.

Yao, J. M., Manchester, R. N., and Wang, N. 2017. A new electron-density model for estimation of pulsar and FRB distances. *ApJ*, **835**, 29.

Yin, J., Yang, J., and Pantaleev, M. 2013. The circular eleven antenna: a new decade bandwidth feed for reflector antennas with high aperture efficiency. *IEEE Trans. Ant. Propag.*, **61**, 3976.

Young, M. D., Manchester, R. N., and Johnston, S. 1999. A radio pulsar with an 8.5-second period that challenges emission models. *Nature*, **400**, 848–849.

Zel'dovich, Y. B., Rakhmatulina, A. K., and Sunyaev, R. A. 1972. The observation of fluctuations of relict radio emission as a method of distinguishing adiabatic perturbations from other forms of perturbations of material density in the universe which lead to galaxy formation. *Radiophys. Quantum Electron.*, **15**, 121–128.

Zensus, J. A. 1997. Parsec-scale jets in extragalactic radio sources. *Ann. Rev. Astr. Ap.*, **35**, 607–636.

Zirin, H., Baumert, B. M., and Hurford, G. J. 1991. The microwave brightness temperature spectrum of the quiet sun. *ApJ*, **370**, 779–783.

Zucca, P., Carley, E. P., Bloomfield, D. S., and Gallagher, P. T. 2014. The formation heights of coronal shocks from 2D density and Alfvén speed maps. *A&A*, **564**, A47.

# Index

absorption coefficient, 26
    negative, 55
active galactic nuclei (AGN), 401
    accretion disk, 421
    brightness temperature, 415
    FR classification, 402
    host galaxies, 419
    kinetic-mode, 426
    optical properties, 416
    radiative model, 420
    radio-loud, 402
    radio-quiet, 429
    torus, 423
    variability, 417
Airy function, 150, 227
Algol ($\beta$ Persei), 324
aliassing, 483
ALMA, 276, 302
ammonia, $NH_3$, 48
angular power spectrum, 492
angular resolution, 166
    Rayleigh's criterion, 166
antenna, radio, 69
    as transmitter, 72
    beam pattern, 146
    beamwidth, 72
    bow-tie dipole, 138
    effective area, 23, 72, 77, 131
    efficiency, 160
    frequency independent, 138
    gain, 72
    Hertzian dipole, 132
    impedance, 134
    log-periodic, 139, 306
    polar diagram, 72
    power gain, 72, 77, 131, 133
    sidelobes, 72, 77, 81
    temperature, 76, 78, 79
aperture
    blockage, 162
    illumination, 160

aperture synthesis, 177, 220
    millimetre wavelengths, 273
apodization, 160
ascending node, 475
astrometry, 283, 285
atmosphere
    absorption, 5, 66
    optical depth, 67
    refraction, 65
    refractivity, 65
    scale height, 67
atomic clocks, 477
Australia Telescope (ATCA), 222
autocorrelation, 70, 110
    spectrometer, 99
autocorrelation function, 472
autocorrelation spectroscopy, 473

balun, 134
barred spirals, 349
baryon acoustic oscillations, 458
baseband, 99
beam smoothing, 166, 168
beam switching, 216
Bell, Jocelyn, 8
binary orbits, 391
binary stars, 324
    Algol, 324
    X-ray, 324
BINGO, 458
birefringence, 60
black body, 76
    blackbody radiation, 21
    blackbody temperature, 76
black hole, 352, 420
bolometer, 82
Bolton, John, 7
Boltzmann constant, 21
Boltzmann distribution, 9
BOOMERANG, 450
bremsstrahlung, 8, 29

brightness, 24, 25
    temperature, 75, 78

calibration, noise sources, 96
CALLISTO, 306, 308
Cambridge One-Mile Telescope, 221
carbon dioxide, 45
Cassegrain telescope feed, 155
Cassiopeia A, 178, 335
celestial coordinates, 475
Chandler wobble, 476
CHIME, 458
CLEAN, 246
closure amplitude, 251
closure density, 445
closure phase, 249
CO spectral line, 351
COBE, 448, 450
coherence, 207
coherence time, 188
compact sources, 411
compact steep spectrum (CSS), 411
compact symmetrical object (CSO), 411
complex visibility, 202
conformal deformation, 164
confusion, 172
convolution theorem, 186, 470
core–jet sources, 407
coronal mass emission, 306
corrugated horn feed, 157
cosmic microwave background (CMB), 3, 8, 9, 21, 442, 445
    angular spectrum, 451
    dipole, 447
    discovery, 446
    polarization, 455
    spectrum, 448
    structure, 448
    temperature, 447
cosmic rays, 338
    energy spectrum, 37, 356
cosmological constant, 443
Coudé telescope feed, 156
Crab Nebula, 336, 367
Crab Pulsar, 336, 373, 374, 387
critical density, 444
cross-correlation, 182, 471
cross-power product, 183, 225
cross-product power, 188
cross-spectrum power density, 186, 189, 472
cyclotron frequency, 32
Cygnus A, 178
Cygnus Loop, 334
Cygnus X-3, 324, 325

dark energy, 444
dark matter, 338, 345
decibel, 75

delta function, 71
Dicke radiometer, 103
differential radiometer, 105
digital autocorrelation, 115
dipole
    folded, 133
    half-wave, 133
    Hertzian, 132
Dirac impulse function, 467
DIRBE infrared radiometer, 448
direction cosines, 170, 204
directional coupler, 96
dirty beam, 227, 228
dispersion constant, 390
dispersion measure, 59, 390
DMR radiometer, 448, 450
Doppler boosting, 414
double quasar
    B0957+561, 461, 464
dynamic range, 263

Earth-rotation synthesis, 177, 221, 225, 229
ecliptic, 476
Eddington limit, 423
effective area, 77
eikonal representation, 17
Einstein ring, 463
Einstein–de Sitter, 444
electromagnetic spectrum, 5
    infrared bands, 74
electromagnetic waves, 15
eMERLIN array, 222, 233
emission measure, 32
emissivity of hydrogen line, 52
equipartition, 355, 410
European VLBI Network (EVN), 288
evolution
    pulsars, 386
    radio sources, 437

Faraday depth, 62
Faraday rotation, 60, 61, 359
Faraday screen, 62
fast Fourier transform (FFT), 116, 470
fast radio bursts, 381
feed systems
    circular horn, 157
    focal plane arrays, 159
field effect transistor (FET), 92
field of view, 131
FIRAS spectrometer, 448
flagging data, 259
flare stars, 324
flux density, 72
Fourier synthesis, 196
Fourier transform, 12, 202

in antenna theory, 149
in smoothing, 170
Fraunhofer diffraction, 146, 147
free–free emission, 8
free–free radiation, 29, 330
 absorption coefficient, 31
 circumstellar, 312
 optical depth, 31
frequency conversion, 217
frequency transfer function, 227
Friedmann, Alexander, 442, 445
fringe fitting, 280
fringe visibility, 194, 225
fringes, 179

galactic centre, 350, 352, 383
galactic coordinates, 340
galactic warp, 349
galaxies
 Local Group, 339
 Local Supercluster, 339
 M51, 345
 M82, 333, 397
 M87, 405
gamma-ray pulsar, 374
Gaunt factor, 30
Gaussian function, 198
Gaussian noise, 69, 70
general relativity, 390
geodesy, 283, 286
geometric optics, 17
geometric time delay, 218, 225
gigahertz peaked spectrum (GPS), 411
globular clusters, 339
GMRT synthesis telescope, 223
gravitational lensing, 461
 time delay, 465
 weak lensing, 465
gravitational waves, 395, 478
gridding data, 242
ground plane, 135
group velocity, 59
gyro-synchrotron radiation, 324
gyrofrequency, 32, 60
 radiation, 32, 307

H II regions, 353, 354
$H_2O$ maser, 423
Hanning function, 114
Harrison–Zel'dovich spectrum, 451
Her A, 428
Herbig–Haro objects, 323
Hertzian dipole, power gain, 133
heterodyne, 98, 217, 278
Hewish, Antony, 8, 12
Hey, James, 7

holography, 164
homologous design, 164, 174
horn antenna, 136
hot spots, 404
Hubble parameter, 443
Hubble's law, 442
Hulse, Russell, 12
hydrogen
 H II regions, 329
 molecular, 45
hydrogen intensity mapping, 457
hydrogen line
 optical depth, 53
 Zeeman splitting, 60
hydroxyl, 315

image quality, 262
impedance matching, 83
impedance of free space, 132
inflation, 445, 455
infrared
 astronomy, 4
 bands, 74
 flux densities, 74
 radio correlation, 399
 spectrum, 398
instrumental time delay, 225
integration, 86
interferometer
 adding, 181
 arrays, 11
 calibration, 215, 216
 correlation, 182
 delay beam, 188
 finite bandwidth, 187
 Michelson, 192
interstellar medium (ISM), 4, 65, 338
inverse Compton scattering, 39, 415, 426
inversion doubling, 48
ionosphere, 58
isoplanatic patch, 211

Jansky, Karl, 3, 4, 7, 352
jansky, unit of flux density, 73
Jansky Very Large Array (JVLA), 222
jets, relativistic, 323–325
Johnson noise, 24
Julian date (JD), 479
 modified (MJD), 479
Jupiter
 radio bursts, 307
 synchrotron radiation, 308

Kirchhoff's law, 28, 31

Laing–Garrington effect, 425
lambda doubling, 47

laser, 10
Lemaître, Georges, 442
local standard of rest, 267, 342
lock-in amplifier, 103
LOFAR, 297
Long Wavelength Array (LWA), 297
loops I, II, and III, 363
low frequency arrays, 297
low frequency imaging, 294
luminosity–volume test, 438

macrolensing, 462
Magellanic Clouds, 339
magnetars, 381, 387
magnetic field units, 32
magnetobremsstrahlung, 32
magnitude scale, 74, 75
mapping, dynamic range, 263
maser, 10, 53
    circumstellar, 313, 316
    hydroxyl, 315
    molecular species, 313
    polarization, 318
    pumping, 53
    saturation, 55
    silicon oxide, 314
    water, 315
Mather, John, 12
mesolensing, 462
methane, 45
methanol, 48, 313, 314
Michelson interferometer, 180
microlensing, 462
microquasars, 324
Milky Way Galaxy
    bar structure, 349
    central molecular zone, 350
    centre, 350
    magnetic field, 359
    radio spectrum, 353
    rotation, 342, 344
    spiral arms, 346
millimetre-wavelength arrays, 276
Moon, 23
mosaicing, 292
multi-bit sampling, 482
Murchison Widefield Array (MWA),
    297

Nasmyth telescope feed, 156
neutron star
    magnetosphere, 369
    structure, 368
    superfluid, 373
    thermal X-rays, 368
NGC 4258, 423
nitrogen, 45

Nobeyama radioheliograph, 302
NOEMA, 276
noise, 71
    bandwidth, 85
    from atmosphere, 80
    Gaussian, 70
    power spectrum, 70, 71
non-thermal radiation, 9
North Galactic Spur, 363
novae, examples
    Cygni 92, 319
    Persei 1901, 322
    R Aquarii, 323
    RS Ophiucus, 320
    V723 Cassiopeia, 319
    V959, 319
novae, properties
    non-thermal radiation, 320
    radio jets, 323
    recurrent, 320
    thermal radiation, 318
Nyquist frequency, 485
Nyquist sampling rate, 483
Nyquist–Shannon sampling theorem, 483

offset telescope feed, 162
Oort, Jan, 7
optical depth, 28
    ionized hydrogen, 31
Orion Nebula, 31, 354
orthomode transducers, 124
OVRA, 306
oxygen, molecular, 45, 47

Palmer, Henry, 8
parallactic angle, 291
partition function, 52
Pawsey, Joseph, 177
peeling, 292
Penzias, Arno, 8, 446
phase-sensitive detector, 103
phase structure function, 212
phase tracking centre, 186
phased array, 131, 139, 141, 182
Planck
    constant, 21
    distribution, 21
    spacecraft, 452
planets, solar system, 307
plasma
    refractive index, 58
    resonant frequency, 58
plerion, 336
point spread function, 72, 160
polarimetry, 118
polarization, 17
    circular, 18

definitions, 19
ellipse, 18
elliptical, 19
linear, 16
synchrotron radiation, 37
polarization imaging, 269
population inversion, 53
position angle, 476
position measurement, 189
positional reference frame, 477
power density, 24
power spectral density, 72
Poynting vector, 132
precipitable water vapour (PWV), 67
probability density function, 70
probability distribution function, 70
propagation
ionosphere, 212
troposphere, 210
pulsar
as clock, 478
binary, 383, 384
binary orbits, 391
braking index, 370
characteristic age, 373
discovery, 367
distance, 390
double, 392
drifting, 380
evolution, 371, 386
gamma rays, 384
glitches, 373
magnetic field, 371
magnetosphere, 374
millisecond, 383, 384
moding, 380
nulling, 380
polar cap, 376
polarization, 377
proper motion, 390
pulse timing, 388
slowdown rate, 370
spectra, 387
surveys, 394
X-ray, 383

quantization amplitude, 480
quasar, 7
brightness temperature, 415
RQQ, radio quiet, 429
3C175, 404
3C273, 407

radiation pattern
far-field, 132
near-field, 132
related to aperture distribution, 146

radiation resistance, 133
radiative transfer, 27
radio frequency interference (RFI), 13
radio galaxy
2352+495, 412
3C 348, 428
3C 75, 407
3C390.3, 405
J0116-473, 408
jets, 405, 407
NGC 1265, 407
spectrum, 409
types FR I and FR II, 404
radio jets, 402
radio recombination lines, 42
radio source counts, 432
radio telescopes, 69
aperture blockage, 162
APEX, 175
Arecibo, 153, 174
Cassegrain, 155
cylindrical paraboloid, 153
Effelsberg, 164
FAST, 153, 174
feed systems, 156
GBT, 162, 174
IRAM 30-metre, 175
JCMT, 175
Kharkov, 144
LOFAR, 144
mountings, 154
MWA, 144
Parkes, 159
phased array, 141, 144
pointing accuracy, 171
SRT, 175
surface accuracy, 163
Tian Ma, 175
UTR-2, 139
RadioAstron, 65, 287
radiometer, 78, 80, 82
Dicke, 103
Rayleigh distance, 146, 158
Rayleigh's theorem, 473
Rayleigh–Jeans spectrum, 22, 31
Reber, Grote, 7
red giant Betelgeuse, 310
reflector telescopes
surface deformations, 164
surface reflectivity, 163
refractive index
air, 65
group, 58
plasma, 58
relativistic beaming, 414
rotation measure, 61, 359
rotational lines, 45

RRATs, 381
Ruze formula, 165
Ryle, Martin, 7, 12, 177, 221

Sag A*, 383
sampling function, 469
Schwarzschild radius, 421
scintillation, 62
  atmospheric, 17
  bandwidth, 64
  refractive, 64
  scattering angle, 63
self-absorption, synchrotron, 39
self-calibration, 251
self-Compton effect, 39
septum plate, 125
Seyfert galaxies, 429
Sgr A*, 430
Shapiro delay, 465
shot noise, 69
sidelobes, 72, 81, 145, 170
sinc function, 468
single-bit sampling, 481
SIS mixer diode, 93
SKA, 439
Slipher, Vesto, 442
Smoot, George, 12
smoothing function, 168
snapshot mode, 222
soft gamma-ray repeater (SGR), 367, 382
solar
  chromosphere, 301
  corona, 59
  flare, 305
source counts, 431
  modern, 435
source size, 192
source surveys, 433, 438
South Pole Telescope, 453
space VLBI, 287
spatial frequency transfer function, 227
spatial sensitivity diagram, 228
specific
  brightness, 24, 73
  emissivity, 26, 51
  flux, 72
  power, 72
spectral distribution function, 467
spectral index, 353
  radio galaxy, 409
  sign convention, 37
spectral line, 50
  emissivity, 53
  imaging, 266
  line profile, 50
  width parameter, 50

spectral power density, 473
spectrometer, 82
  autocorrelation, 110
  multichannel, 109
spectrum, 72
spherical harmonics, 491
spillover, 155, 158
spiral arms, 360
square-law detector, 86
SS433, 325
star-formation region (SFR), 312, 315
star-forming galaxies, 397
stars
  binary, 313
  variable, 324
state temperature, 52
Stefan constant, 21
stellar wind, 312
stimulated emission, 10
Stokes parameters, 118
Strömgren sphere, 329
strong lensing, 462
Sub-Millimetre Array (SMA), 276
Sun
  brightness, 301
  chromosphere, 301
  corona, 303, 306
  photosphere, 302
  radio bursts, 303
Sunyaev–Zel'dovich effect, 458
superbubble, 364
superluminal GRS 1915+105, 325
superluminal velocity, 325, 328, 412, 414
supermassive black hole(SMBH), 397
supernova, 332, 367
supernova remnants, 332, 353, 363, 398
  SNR 1006, 334
  SNR 1054 Crab, 334
  SNR 1574 Tycho, 334
  SNR 1604 Kepler, 334
  brightness, 354
  IC443, 334
superresolution, 166, 248
surface brightness, minimum, 276
synchro-Compton radiation, 40
synchrotron radiation, 4, 32, 409
  characteristic frequency, 33
  critical frequency, 34
  electron energy loss rate, 35
  electron lifetime, 36
  Milky Way Galaxy, 355
  polarization, 34, 37, 357
  power-law spectrum, 36
  self-absorption, 39
  spectral index, 37
  spectrum, 34

synthesis beam, 227
synthesis imaging
    calibration, 238
    deconvolution, 245
system equivalent flux density (SEFD), 215
system noise temperature, 88

Taylor, Joe, 12
temperature, 8
    brightness, 10
    kinetic, 9
    state, 9
    system noise, 11
temporal coherence, 207
thermal radiation, 8, 9
    Moon, 23
    planets, 307
time, 477
    UTC and TAI, 479
timing noise, 381
troposphere, 158
twin-beam radiometer, 158

$u, v$ plane, 202
    filling, 229
    incomplete coverage, 237
    support, 235
    support domain, 228

van Cittert–Zernike theorem, 210, 224
van der Hulst, H., 7
van Vleck digital approximation, 481
variance, 70
Vela Nebula, 334

Vela Pulsar, 373, 387
Venus, 307
vernal equinox, 475
Very Large Array (VLA), 235
Very Long Baseline Array (VLBA), 288
very long baseline interferometry (VBLI), 235, 277
    imaging, 282
    space-based, 416
    techniques, 277, 278
visibility
    amplitude, 197
    phase, 194
visibility function, 206
Vivaldi dipole, 139
VX Sgr, 318

water vapour, 80
wave equation, 15
waveguide modes, 138
weak lensing, 462, 465
Westerbork Synthesis Radio Telescope (WSRT), 221
white dwarf, 318
wide-field imaging, 288
Wiener–Khinchin theorem, 111, 207, 473
Wilson, Robert, 8, 446
window function, 113
WMAP, 451

X-ray
    astronomy, 4
    binaries, 324
    bursts (XRBs), 382
    from pulsar, 374

Zeeman splitting, 55, 60, 314, 315, 318

Printed in the United States
by Baker & Taylor Publisher Services